Georg Lenz, Laura Pasqualucci (Eds.)

Malignant Lymphomas

Georg Lenz, Laura Pasqualucci (Eds.)

Malignant Lymphomas

Biology and Molecular Pathogenesis

DE GRUYTER

Editors
Georg Lenz, M. D.
University Professor
Head Translational Oncology
University Hospital Münster
Albert-Schweitzer-Campus 1
48149 Münster, Germany
e-mail: georg.lenz@ukmuenster.de

Laura Pasqualucci, M. D.
Associate Professor of Pathology and Cell Biology
Institute for Cancer Genetics
Herbert Irving Comprehensive Cancer Center
Columbia University
1130 St. Nicholas Ave
New York, NY 10032, United States of America
e-mail: lp171@columbia.edu

The book has 70 figures and 11 tables.

ISBN 978-3-11-037119-2
e-ISBN (PDF) 978-3-11-036691-4
e-ISBN (EPUB) 978-3-11-047880-8

Library of Congress Cataloging-in-Publication Data
A CIP catalog record for this book has been applied for at the Library of Congress.

Bibliographic information published by the Deutsche Nationalbibliothek
The Deutsche Nationalbibliothek lists this publication in the Deutsche Nationalbibliografie; detailed bibliographic data are available on the Internet at http://dnb.dnb.de.

Cover image: Ugreen/iStock/Thinkstock
Typesetting: PTP Protago-TEX-Production, Berlin
Printing and binding: CPI books GmbH, Leck
♾ Printed on acid-free paper
Printed in Germany

www.degruyter.com

Preface

Over the past decade, we have witnessed a tremendous improvement in our understanding of the biology and pathogenesis of lymphoid malignancies, spurred by advances in genomics, immunology, and molecular biology.

As a result, a number of diagnostic categories have been refined or added to the updated World Health Organization classification of Lymphoid Malignancies, which currently lists over 40 different types of lymphomas, each defined by distinctive phenotypes. Additionally, the characterization of the genomic landscape of an increasing number of lymphoid cancers has provided the foundation for a new era of clinical trials, aiming to design and test more effective, and possibly less toxic strategies tailored to the genetic aberrations and dysregulated pathways that are present in individual patients.

This textbook focuses on the molecular pathogenesis of the most common neoplasms derived from various stages of lymphocyte development in the mature B and T cell compartment. Our goal was not to be as inclusive as possible, but rather to provide a concise review of the patho-biological mechanisms associated with these diseases, which may serve as a reference text for students, experimental researchers and practicing clinicians.

The book opens with an introductory chapter that provides a historical view of the various lymphoma classifications, whereas two separate chapters focus on the normal B and T cell development. Indeed, the relationship between the unique genetic and transcriptional events that take place in these cells as they progress through differentiation and the various subtypes of lymphoma represents a fundamental concept for our current understanding of lymphomagenesis. Fourteen chapters cover individual B cell non-Hodgkin lymphomas, T cell and NK-cell non-Hodgkin lymphomas, Hodgkin lymphoma and also multiple myeloma, a malignancy recapitulating the phenotype of terminally differentiated plasma cells. Finally, the last chapter discusses these themes in a clinical context and presents the implications of more recent findings for the development of improved therapeutic strategies.

Although the road to cure will still face several challenges, the impressive wealth of new information accumulated to date offers unprecedented opportunities and is expected to have major impacts on the way we approach these diseases clinically.

Acknowledgements

We are indebted to all of the authors whose expertise and dedication have made the preparation of this textbook possible. We would also like to thank our colleagues and collaborators, as well as all post-doctoral fellows and students who, through their work and passion, have provided instrumental contributions to the current knowledge in lymphoma research. Finally, we are grateful to the De Gruyter publisher and especially to Dr. Britta Nagl, who enthusiastically supported the development of this textbook from the start.

Contents

List of contributing authors

Chapter 1
Wolfram Klapper, MD
Professor of Pathology
Head of Hematopathology Section
Department of Pathology
Hematopathology Section and
Lymph Node Registry
Universitätsklinikum Schleswig-Holstein
Kiel, Germany
e-mail: wklapper@path.uni-kiel.de

Chapter 2
Ulf Klein, PhD
Assistant Professor of Microbiology and Immunology
Departments of Pathology & Cell Biology and Microbiology & Immunology
Herbert Irving Comprehensive Cancer Center
Columbia University
New York, NY, United States of America
e-mail: uk30@cumc.columbia.eduflnstitute

Nilushi S. De Silva, PhD
Postdoctoral Research Scientist
Institute for Cancer Genetics
Department of Microbiology and Immunology
Columbia University
New York, NY, United States of America

Chapter 3
Shan Zha, MD, PhD
Assistant Professor of Pathology & Cell Biology and Pediatrics
Institute for Cancer Genetics
Herbert Irving Comprehensive Cancer Center
Departments of Pathology & Cell Biology and Pediatrics
Columbia University
New York, NY, United States of America
e-mail: sz2296@cumc.columbia.edu

Jennifer L. Crowe, M. Ph.
Ph.D. Candidate
Graduate Program of Pathobiology and Molecular Medicine
Columbia University
New York, NY, United States of America

Chapter 4
Beiying Dai, M. Sc.
Ph.D. Candidate
Translational Oncology
University Hospital Münster
Münster, Germany
e-mail: beying.dai@ukmuenster.de

André Weilemann, Dipl. Biol.
Ph.D. Candidate
Translational Oncology
University Hospital Münster
Münster, Germany
e-mail: andre.weilemann@ukmuenster.de

Georg Lenz, MD
University Professor
Head Translational Oncology
University Hospital Münster
Münster, Germany
e-mail: georg.lenz@ukmuenster.de

Chapter 5
Jessica Okosun, MD, PhD
Clinical Research Fellow
Centre for Haemato-Oncology
Barts Cancer Institute
Queen Mary University of London
London, UK
e-mail: j.e.okosun@qmul.ac.uk

Jude Fitzgibbon, PhD
Professor of Personalized Cancer Medicine
Centre for Haemato-Oncology
Barts Cancer Institute
Queen Mary University of London
London, UK
e-mail: j.fitzgibbon@qmul.ac.uk

Chapter 6
Ming-Qing Du, MD, PhD
Professor of Oncological Pathology
Honorary Consultant Molecular Pathologist
Division of Molecular Histopathology
Department of Pathology
University of Cambridge
Addenbrooke's Hospital
Cambridge, UK
e-mail: mqd20@cam.ac.uk

Chapter 7
David Dominguez-Sola, MD, PhD
Assistant Professor of Oncological Sciences and Pathology
Department of Pathology
Mount Sinai School of Medicine
New York, NY, United States of America
e-mail: david.dominguez-sola@mssm.edu

Riccardo Dalla-Favera, MD
URIS Professor of Pathology & Cell Biology
Director, Institute for Cancer Genetics
Herbert Irving Comprehensive Cancer Center
Department of Pathology & Cell Biology
Department of Genetics & Development
Department of Microbiology & Immunology
Columbia University
New York, NY, United States of America
e-mail: rd10@cumc.columbia.edu

Chapter 8
Laura Pasqualucci, MD
Associate Professor of Pathology & Cell Biology
Institute for Cancer Genetics
Herbert Irving Comprehensive Cancer Center
Department of Pathology & Cell Biology
Columbia University
New York, NY, United States of America
e-mail: lp171@columbia.edu

Chapter 9
Christian Steidl, MD, PhD
Assistant Professor of Pathology and Laboratory Medicine
Department of Pathology and Laboratory Medicine
Centre for Lymphoid Cancer
British Columbia Cancer Agency
BC Cancer Research Centre
University of British Columbia
Vancouver, Canada
e-mail: CSteidl@bccancer.bc.ca

Randy D. Gascoyne, MD
Clinical Professor of Pathology and Laboratory Medicine
Department of Pathology and Laboratory Medicine
Centre for Lymphoid Cancer
British Columbia Cancer Agency
BC Cancer Research Centre
University of British Columbia
Vancouver, Canada
e-mail: rgascoyn@bccancer.bc.ca

Chapter 10
James L. Rubenstein, MD, PhD
Professor of Medicine
Division of Hematology/Oncology
University of California San Francisco
San Francisco, CA, United States of America
e-mail: jamesr@medicine.ucsf.edu

Chapter 11
Davide Rossi, MD, PhD
Assistant Professor of Hematology
Division of Hematology
Department of Translational Medicine
Amedeo Avogadro University of Eastern Piedmont
Novara, Italy
e-mail: davide.rossi@med.uniupo.it

Gianluca Gaidano, MD, PhD
Professor of Hematology
Chair, Division of Hematology
Department of Translational Medicine
Amedeo Avogadro University of
Eastern Piedmont
Novara, Italy
e-mail: gaidano@med.uniupo.it

Chapter 12

Zachary R. Hunter, PhD
Instructor of Medicine
Harvard Medical School
Boston, MA, United States of America

Jorge J. Castillo, MD
Assistant Professor of Medicine
Harvard Medical School
Boston, MA, United States of America

Steven P. Treon, MD, PhD
Associate Professor of Medicine
Director, Bing Center for Waldenstrom's
Macroglobulinemia
Dana Farber Cancer Institute
Harvard Medical School
Boston, MA, United States of America
e-mail: Steven_Treon@dfci.harvard.edu

Chapter 13

Marta Chesi, PhD
Assistant Professor of Biochemistry and
Molecular Biology
Department of Biochemistry and
Molecular Biology
Mayo Clinic Cancer Center
Scottsdale, AZ, United States of America

Leif Bergsagel, MD
David F. and Margaret T. Grohne Professor of
Novel Therapeutics for Cancer Research
Consultant, Division of Hematology/Oncology
Professor of Medicine
Mayo Clinic College of Medicine
Scottsdale, AZ, United States of America
email: bergsagel.leif@mayo.edu

Chapter 14

Ralf Küppers, PhD
Professor of Molecular Genetics
Chair, Department of Molecular Genetics
Institute of Cell Biology (Cancer Research)
University of Duisburg-Essen
Essen, Germany
e-mail: ralf.kueppers@uk-essen.de

Chapter 15

Brunangelo Falini, MD
Professor of Hematology
Director, Institute of Hematology and Bone
Marrow Transplantation
University of Perugia
Ospedale S. Maria della Misericordia
Perugia, Italy
e-mail: brunangelo.falini@unipg.it

Stefano Pileri, MD
Professor of Pathology
Director of Hematopathology
European Institute of Oncology
Milan, Italy
Bologna University School of Medicine
Bologna, Italy
e-mail: stefano.pileri@ieo.it;
stefano.pileri@unibo.it

Chapter 16

José Rodríguez Cortés, PhD
Postdoctoral Research Scientist
Institute for Cancer Genetics
Columbia University
New York, NY, United States of America

Adolfo Ferrando, MD, PhD
Assistant Professor of Pediatrics and
Pathology & Cell Biology
Institute for Cancer Genetics
Herbert Irving Comprehensive Cancer Center
Departments of Pediatrics and Pathology &
Cell Biology
Columbia University
New York, NY, United States of America
email: af2196@cumc.columbia.edu

Teresa Palomero, PhD
Associate Professor of Pathology & Cell Biology
Institute for Cancer Genetics
Herbert Irving Comprehensive Cancer Center
Department of Pathology & Cell Biology
Columbia University
New York, NY, United States of America
e-mail: tp2151@columbia.edu

Chapter 17
Wing C. Chan, MD
Professor of Pathology
Department of Pathology
City of Hope Medical Center
Duarte, United States of America
e-mail: jochan@coh.org

Chapter 18
Ethel Cesarman, MD, PhD
Professor of Pathology and Laboratory Medicine
Department of Pathology and
Laboratory Medicine
Weill Cornell Medical College
New York, NY, United States of America
e-mail: ecesarm@med.cornell.edu

Chapter 19
Loic Ysebaert, MD, PhD
Assistant Professor of Hematology
Institut Universitaire du Cancer de
Toulouse-Oncopole
Toulouse, France

Roch Houot, MD, PhD
Professor of Hematology
University Hospital of Rennes,
Rennes, France

Franck Morschhauser, MD, PhD
Professor of Hematology
Centre Hospitalier Regional Universitaire
de Lille,
Université de Lille 2
Lille, France
e-mail: franck.morschhauser@chru-lille.fr

Wolfram Klapper

1 Classification of lymphomas – past, present and future

1.1 Classification – what is it good for?

"Classification is the language of medicine." This is the first sentence in the World Health Organization (WHO) book on hematopoietic and lymphoid tumors. Classification is used as a technical term in many fields of science to describe the diversity of objects. Generally, classification describes how elements are recognized and differentiated. To translate this into medicine, classification describes the diagnostic criteria and the differential diagnosis – or simply the disease. Similar to the words in our language, the implicit meaning of a diagnosis as defined by a classification is a prerequisite for medicine, since it is the fixed point for treatment and the common ground for research.

Furthermore, the term classification implies that it helps us to understand the elements, as their arrangement reflects a scientific concept. Therefore, one should assume that the features used for medical classification should not only be based on clinical presentation and histopathology but should also include pathomechanisms. However, so far in the field of medicine, including the lymphoid tumors, molecular processes/pathomechanisms as an organizing principle have only been selectively applied in classifications. One example is the category of "myeloid and lymphoid neoplasms with eosinophilia and abnormalities of PDGFRA, PDGFRB or FGFR1" in which primarily the genetic aberrations and the pathomechanisms induced by them define the disease, but not the morphology or clinical presentation.

In biology, classification belongs to the field of taxonomy and its aim is to establish categories of species. This concept is based on the initial work by Carl Linnaeus (Carl von Linné), who established a classification of species based on shared physiological features. It was further developed by a number of scientists, including Charles Darwin, so as to also reflect the principle of common descent. An important feature of most biological classifications is that they are hierarchically organized. The hierarchy is based on morphological, biochemical or evolutionary relatedness. This hierarchy is reflected in the binomial nomenclature that, for example, is applied to bacterial species. Medical classifications of diseases, such as the classification of lymphoma, differ in several ways from biological classifications. The most obvious difference is the lack of a systematic nomenclature, since most disease names are not systematic but historical in nature. A second obvious difference between medical and biological classifications is the lack of a hierarchical regulatory principle. As described below, historical lymphoma classifications were searching for such a hierarchal principle of classification (e.g. by defining high- and low-grade lymphomas). Nevertheless, in the current WHO classification it is stated that there is currently no evidence for a hierarchical classification principle for hematopoietic and lymphoid neoplasms, and it regards it as a list of diseases that is arranged arbitrarily, e.g. by predominant clinical presentation. Later we will discuss the possibility that future developments in molecular-genetic analysis could in fact reveal such hierarchical principles based on molecular pathomechanisms that might profoundly change our current ideas as to how to classify lymphoid tumors.

1.2 The history of lymphoma classification

The history of lymphoma classification is not a straightforward story. Nevertheless, it has been described both compellingly and in detail [1–3]. This current chapter provides only a brief overview of the developments in lymphoma classification from the 19th century to the present.

Why look at the history at all? First, understanding the history of lymphoma classification also helps to understand current classifications. The origins of disease definitions and the nomenclature used for the majority of lymphomas are historical in nature and are primarily based on morphological features. Since then, entities have continuously been either confirmed, redefined or subdivided on the basis of phenotypic features, especially molecular-genetic ones. Second, knowing the history of lymphoma classification will help us to appreciate a worldwide consensus. Many scientists and pathologists who developed our current lymphoma classification started their career at a time before the current WHO classification. This experience made them aware of the importance of worldwide consensus in classification. However, it should not be taken for granted that future classifications will also represent a worldwide consensus. An increasing number of molecular-genetic tests in diagnostics and disease definition may not be available in many areas of the world. This is a challenge that future WHO classifications will have to face.

1.2.1 19th century: Tumors of lymphoid cells and their relation to leukemia

Most of the literature dates the beginning of lymphoma classification to the work of Thomas Hodgkin in 1832 [4]. Hodgkin described a progressive lymphadenopathy in patients that he distinguished from tuberculosis, the most frequent cause of lymph node swelling at that time, and also from other known infectious diseases and metastasis. Although Hodgkin did not draw definite conclusions about the neoplastic nature of the disease, he did certainly describe a lymphoma entity, which we now refer to as Hodgkin lymphoma. In the 19th century and until the first decades of the twentieth century, the origin and the functional properties of lymphocytes and the lymphoid tissues were largely unknown. After the first description of a leukemia in 1845, Rudolf

Virchow reported tumors of lymphoid tissue that seemed to be composed of cells similar to the leukemias he had been observing and used the term "lymphosarcoma" to describe these "aleukemic leukemias." In the second half of the 19th century, Theodor Billroth used the term "malignant lymphoma", and from then on the term began to be used for most neoplasms of lymphoid cells (reviewed in [1]). However, at that time a large group of lymphomas were not recognized as lymphoid in nature and were referred to as "reticulum cell sarcoma". This name was derived from the generally accepted hypothesis that lymphocytes are small differentiated cells that develop from large (undifferentiated) cells, the latter being part of the reticulohistiocytic system, a system of cells that are phagocytically active like histiocytes and that form the fundamental structure of lymphoid tissue. The possibility of a reverse differentiation from small lymphocytes to large cells was denied. As a consequence of this incorrect assumption about the origin of lymphocytes, most lymphomas in the group of large cell lymphomas, which are composed of large blasts (e.g. diffuse large B-cell lymphoma), were called "reticulum cell sarcoma" at that time. It took until the 1960s–1970s before it was proved that "reticulum cell sarcomas" are in fact tumors derived from lymphoid cells (reviewed in [1] and [3]).

1.2.2 20th century: Description of individual diseases and historical lymphoma classifications

The first half of the 20th century was characterized by the description of individual entities. The entities that were characterized at this time included the follicular lymphoma described by Gall in 1940[5] (previously described by Brill and Symers [6, 7] as "giant lymph follicle hyperplasia"). Similarly, extranodal neoplasms and lymphoid neoplasms involving the bone marrow, multiple myeloma, Waldenström's macroglobulinemia, mycosis fungoides, and Sézary syndrome were characterized as entities in the mid-1940s [1]. Probably one of the most fascinating discoveries was the identification of Burkitt lymphoma in African children by the surgeon Dennis Burkitt, a prime example of how precise clinical observation provides deep insights into the pathogenesis of a disease [8]. Despite the fact that the definitions of the diseases were mostly the individual achievements of one or two researchers and were only based on clinical or morphological observations, they all turned out to be very valid and all of these entities, with modified diagnostic criteria, are still part of the current WHO classification. Nevertheless, in addition to the descriptions of individual lymphoma entities, researchers in the 20th century attempted to establish a systematic classification of lymphoid tumors for two reasons. First, the individually described entities did not cover the whole spectrum of neoplasms observed in lymphoid tissue, and second, they wanted to apply the principles of scientific classification from biology (see above) to human cancer. These historical classifications were published between 1934 and 1942, with the most relevant one being that by Robb-Smith and Gall and Malloy (reviewed in

[1] and [3]). In contrast to the general opinion, these classifications were complex and consisted of a large number of entities [3]. The major deficiencies of these historical classifications were (i) ambiguous terminology (that did not always clearly distinguish between reactive and neoplastic conditions), (ii) incompleteness (not all tumors of the lymphatic system were covered), (iii) poor reproducibility (due to their nature as single-center concepts, poor illustration and complex criteria) and (iv) neglect of the biology of the lymphatic system (even becoming contradictory with growing knowledge of the function of the lymphatic system).

1.2.3 The second half of the 20th century: Modern lymphoma classifications, controversy and consensus

A milestone in the development of lymphoma classification was that published by Henry Rappaport in 1956. This classification system overcame most of the deficits of the historical classification in that it used clear terminology (separating reactive from neoplastic conditions), was comprehensive (most of the neoplasms of the lymphoid tissue could be assigned to one of the categories) and was finally reproducible by other pathologists (due to the excellent illustrations in the publication). The initial version of the Rappaport classification distinguished two growth pattern of lymphomas, diffuse and nodular ("follicular") and five major entities defined by the cytology occurring in any of the two growth patterns: "lymphocytic type, well differentiated," "lymphocytic type, poorly differentiated", "mixed type (lymphocytic and reticulum cell)," "reticulum cell type" and "Hodgkin's type." The classification was developed in a large cohort of clinically well-characterized lymphoma patient samples at the Armed Forces Institute of Pathology. Understandably, this classification became widely accepted among pathologists in the United States. The success of this classification scheme was enhanced by the prognostic value of the categories, with diffuse lymphomas being more aggressive in their clinical course than nodular lymphomas. However, in the late 1960s and 1970s one feature of the Rappaport classification became more and more controversial among hematopathologists: the lack of accordance with the biology of the lymphatic system. It were two milestone findings about lymphocyte biology in the sixties that challenged the historical and the Rappaport classifications. First, phytohemagglutinin (PHA) was shown to transform small nonproliferating lymphocytes into large proliferating blasts in cell culture, demonstrating that lymphoid cells possess the capability to differentiate from small to large cells and vice versa [9]. This cast doubt on the origin of lymphocytes from large reticulum cells. Instead, it appeared more likely that lymphoid cells also differentiate *in vivo* from small cells to blasts and vice versa, and that the large cells observed under the microscope in lymphoid tissue might, like the small lymphocytes, be lymphoid in nature. Consequently, tumors of lymphoid tissues composed of large cells might in fact be lymphomas rather than "reticulum cell sarcomas." The second milestone publication that raised questions about the Rappaport classification was the discovery of two types of lymphoid cells: B cells and T cells. Experiments in chickens revealed that antibody producing plasma cells and cells of the follicles were derived from the *bursa fabricii* (B cells), whereas the periarteriolar lymphoid cells in the spleen seemed to be derived from the thymus (T cells) [10]. This concept of two types of lymphoid cells was soon transferred to human tissues and human lymphoma specimens, as immunoglobulins were detected in extracts of tumors that were at that time classified on the basis of their morphology as "reticulum cell sarcomas" [11]. With these experiments, it became evident that "reticulum cell sarcomas" are lymphomas, and that the concept of B cells and T cells can be transferred to neoplasms of lymphoid cells as well. As a consequence, new classifications were developed, unfortunately most of them independently from each other and some even in competition with each other. In 1974 four suggestions for classifications were published, three of which even appeared in the same journal (reviewed in [3]), provoking ironic comments [12].

Two of the classifications published in the seventies stood out since they also incorporated the concept of B cells and T cells: the classification by Lukes and Collins [14] and the Kiel classification by a group of European pathologists led by Lennert [14]. Nevertheless, techniques like immunohistochemistry, with which to determine the B or T-cell nature of lymphoid cells in daily practice, were not available at that time. Moreover, the classifications distinguished B- and T-cell lymphomas but were based primarily on morphology assessed on conventional hematoxilin and eosin or Giemsa stained slides. A long-standing controversy started that split the world of lymphoma researchers and hemato-oncologists. In the United States, the Rappaport classification remained the most frequently used classification scheme and the Kiel or Lukes and Collins classification never achieved popularity, whereas in Europe the Kiel classification became the most frequently used method to diagnose and name lymphomas. The reasons why the Rappaport classification remained popular were its wide acceptance among pathologists in the USA, based on its practicability and its proven clinical relevance, the latter being of prime importance for hemato-oncologists. The new classifications, such as the Kiel classification, were biologically more sound, but the entities did not match existing entities in the Rappaport classification; the terminology used was completely different and, most importantly, evidence of the clinical/prognostic relevance was lacking or was less solid than for the Rappaport classification. However, the application of different classifications created considerable problems, since scientific data from clinical trials, histopathological and molecular studies were simply not comparable. As a consequence, the National Institute of Health in the United States initiated a large study involving all major groups that had developed classification schemes to compare the feasibility and the clinical relevance of these approaches in a large number of cases. The result was disappointing, since the reproducibility of most classifications was poor and none of them proved to be superior at predicting outcome [15]. There were multiple reasons why the biologically sound classifications lacked superiority and were interpreted differently by the authors of the various classifications [16]. As a result of the study, a preliminary classification was suggested that was basically composed of the categories of the Rappaport classification using terminology adopted from Lukes and Collins. This classification scheme was named "working formulation" to emphasize its preliminary nature [15], but it was not a consensus of all groups participating in the preceding comparative study, and it never gained widespread acceptance. Instead, the Kiel classification continued to be the classification predominantly used in Europe.

Scientific stagnation lasted until the late nineties, when a new generation of pathologists from the United States, Europe and Asia came together as the International Lymphoma Study Group. The members of the group felt that a worldwide consensus on classification was needed to release the field from scientific stagnation, and that such a consensus could not be developed by a single person or center but only by a multinational group of experts. The classification suggested by this group was not only a compromise between the competing concepts but also incorporated newly recognized lymphoma entities, such as lymphomas of the mucosa associated tissue, and was named revised European-American classification of lymphomas (REAL) [17]. The REAL classification was the starting point that allowed members of the American Society for Hematopathology and the European Association for Haematopathology, who largely overlapped with the authors of the REAL classification, to develop the first WHO classification of lymphomas in 2000. This WHO classification was very rapidly accepted worldwide as the new standard [18]. Since then this classification has been continuously updated [19].

1.3 The principles of the current WHO classification

The WHO "blue books" (the name is based on the color of the book cover) are published by the WHO and revised every 6–10 years. These books represent the current worldwide consensus on how to define and diagnose neoplasms. Therefore, the blue books also define the principles of classification. Our current WHO classification aims to describe disease entities which are clearly defined, clinically distinctive, nonoverlapping (mutually exclusive), covering all known entities (collectively exhaustive) [19]. In contrast to historical classifications, the disease is not solely defined by pathological assessment. Instead, all available information that is useful for defining the disease should be used. The main features currently used are morphology (cytology or histology), immunophenotype (immunohistochemistry or flow cytometry), genetics (cytogenetics, fluorescence in situ hybridization and molecular genetics) and clinical features (the most important ones being localization, age and blood cell counts). The relative importance of each feature for the definition of a disease differs according to the disease and our current knowledge. The differential weighting of features is most evident in the definition of myeloproliferative neoplasms and plasma cell myeloma, which contain a listing of minor and major criteria and rules for how to sum them up for a definite diagnosis.

There are two prime classification principles implemented in the current WHO classification. First, neoplasms are classified according to the cell lineage. The lineage describes the assumed cell of origin: myeloid, histiocytic/dendritic or lymphoid. Myeloid refers to all cells of the granulocytic (neutrophil, eosinophil and basophil), monocytic/macrophage, erythroid, megakaryocytic and mast cell lineage. Within each lineage, immature (precursor cell) and mature neoplasms are distinguished. This principle of classification is most relevant for lymphoid neoplasms, in which immature neoplasms show features of developing B and T cells in the bone marrow and thymus, such as an incomplete B-cell and T-cell receptor rearrangement and expression of proteins involved in this process. The one most frequently detected for diagnosis and classification is terminal deoxynucleotidyl transferase (TdT). Mature lymphoid neoplasms have completed their receptor rearrangement at the genetic level and lack expression of TdT. Both immature and mature lymphoid tumors are further subdivided into B-cell lymphomas and T-cell lymphomas. A specific group of mature lymphomas is the group of Hodgkin lymphomas. Within the large group of mature B-cell lymphomas there is no further classification principle. The categories "high-grade" and "low-grade" lymphomas that are frequently used clinically are based on the historical Kiel classification and describe morphological features, with "high-grade" lymphomas being composed of blasts, and "low-grade" lymphomas resembling small cells. In contrast, the categories of "aggressive" and "indolent" describe clinical behavior and fast and slow tumor growth, respectively. However, these categories are not a classifying principle of the WHO. Instead, the diseases are listed in the "blue book" according to their predominant (although not exclusive) clinical presentation as leukemic/disseminated, extranodal, nodal indolent and nodal aggressive. The authors of the WHO classification stress that the order is arbitrary.

The principles of the WHO classification are not strictly applied to all diseases. Some elements in the current version of the classification are remarkable since they might indicate trends for future developments of the classification. First, a group of entities is defined by genetic features only ("myeloid and lymphoid neoplasms associated with eosinophilia and abnormalities of PDGFRA, PDGFRB, or FGFR1"), breaking the rule that the prime principle of classification of hematopoietic tumors is the lineage (cell of origin). Second, some disease categories of lymphomas are defined that show features intermediate between two well-defined entities. One of the two currently existing "grey zone" lymphomas cannot be classified with currently available methods. There is a "diagnostic grey zone" which exists between Burkitt lymphoma and diffuse large B-cell lymphoma. However, the two lymphomas have different pathogenetic mechanisms and do not occur simultaneously or subsequently as clonally related diseases. The "grey zone" nature of some lymphomas is simply based on insufficient diagnostic technology. The second "grey zone" group is a true "biological grey zone" between primary mediastinal large B-cell lymphoma and classical

Hodgkin lymphoma. Lymphomas in this category share features of two entities and occur simultaneously or subsequently as clonally related tumors. It is a matter of debate whether using the "grey zone" lymphomas as a bridging category artificially maintains the principle of nonoverlapping (mutually exclusive) entities, although overlap in fact exists. Such a biological overlap might also be envisioned between nodular lymphocyte predominant Hodgkin lymphoma and T-cell and histiocyte rich B-cell lymphoma according to current disease definitions [20]. In such cases, one might discuss whether lumping the entities into one category and stressing its morphological heterogeneity instead of introducing more "grey zones" might in fact be biologically more adequate.

The classification of lymphomas is developed by scientists including molecular biologists, chemists and physicians. Among physicians, pathologists and geneticists are very prominent groups that influence the research on classification, simply because of their diagnostic work in patient care. The distinction between disease categories is thus based on an assumed biology of the disease reflected by morphology, phenotype or molecular-genetic features. Distinguishing two diseases should ideally correlate with differences in clinical presentation; the distinguished diseases are not necessarily differentially treated at the current time. However, if classification is implemented as a "language of medicine", it needs to fulfill more criteria than simply being biologically sound and clinically distinct. Classifications of diseases are applied in patient care and influence individual decisions related to the choice of therapy and thus also the individual fate of patients every day. From the perspective of a physician treating patients, and probably also from the perspective of the patient him/herself, the classification must be developed for the benefit of the patients. With the growing insights into pathomechanisms and the development of pathomechanism-specific ("targeted") therapy, in some fields of oncology the presence of genetic lesions/biomarkers influences the management and fate of the patient more than the pathological classification. In the WHO classification the clinical advisory board ensures that the classification is clinically useful and will be accepted in daily practice.

1.4 Future lymphoma classification – towards new hierarchical classification principles?

Classification needs to develop continuously on the basis of ongoing scientific progress. At the same time, stringency is required to allow continuity in research and the adaptation of patient management strategies to new definitions of diseases. The latter is achieved by the life cycle of classifications, which is determined by the frequency of updates of the WHO blue books, and is about eight years in the field of lymphomas. Certainly, information technology would allow a faster and more flexible system of updates, e.g. by introducing changes in the definition of individual diseases or subtypes when sufficient scientific evidence is available and not only after eight years. However, the WHO classifications are not intended to represent a synopsis of the most recent scientific development, but rather a worldwide standard for longer periods of time that will allow for continuity in research. In fact, the history of lymphoma classification impressively demonstrates the difficulties that arise from too rapid changes in classification schemes, the most important being the potential loss of comparability of clinical study data.

A frequently quoted example illustrating a deficit of our current classification is the category of diffuse large B-cell lymphoma, not otherwise specified. This group of lymphomas is large (about 30 % of all lymphomas in adults in the United States and Europe), but on the other hand it is extremely heterogeneous. The heterogeneity is evident in histopathology, immunophenotype, genetics and molecular features such as the gene expression defining the activated B cell-like (ABC) and germinal-center-like (GCB) subtypes. Most importantly, the clinical course under current therapy is also heterogeneous. On the one hand there is a subgroup of chemotherapy-refractory lymphomas, whereas in other patients the disease is curable. Therefore, many researchers have contemptuously called diffuse large B-cell lymphoma a "wastebasket" category. In fact, molecular analysis has revealed many clinically highly relevant subgroups. The ABC and GCB types of lymphomas are not only prognostically relevant groups under current therapy, they also differ in their pathogenetic mechanisms and thus can potentially be treated differently in the future [21]. As a consequence, it is being discussed whether in future classifications diffuse large B-cell lymphomas should be split into two diseases. However, basic pathological techniques such as morphology and immunophenotyping are not able to identify these subtypes among the diffuse large B-cell lymphomas with sufficient accuracy. Thus, gene expression analysis remains the gold standard at the current time [22]. This technology will be available only in countries with highly developed medical systems, and much of the world will be unable to apply a classification that calls for distinguishing the two subtypes of diffuse large B-cell lymphoma. The same scenario might be envisioned with progressing molecular diversification of other disease entities: existing entities are split into subgroups on the basis of molecular and genetic features. To enable an application of future classifications in all parts of the world, "wastebasket" entities such as diffuse large B-cell lymphoma, not otherwise specified, and peripheral T-cell lymphoma, not otherwise specified, which can be identified by widely available techniques such as histopathology will still be needed. Molecular diversification within these "wastebasket" entities will allow us to establish a diagnostic hierarchy of classification. This hierarchy is not a biological principle, such as common origin in biological classifications, but simply defines levels of diagnostic depth that diversify the large histopathologically defined entities.

Hierarchical features of classifications might develop from the progressing insights into pathogenetic mechanisms that are discussed in the other chapters of this book. In addition to a potential "diagnostic hierarchy," two biological aspects might introduce a hierarchy into future classifications: the description of early lymphoid proliferations [23] and the insights into the order of pathogenetic steps occurring during lymphomagenesis detected by genome analysis [23, 24]. This progressive understanding of molecular and genetic pathomechanisms is accompanied by the development of drugs that specifically target these mechanisms. In this scenario, the pathomechanism defined, e.g. by a mutation, might become a more relevant principle of classification than the lineage (cell or origin), when it comes to patient management. One first example for such future classes of diseases defined primarily by genetic lesions that define pathomechanisms is the category of "myeloid and lymphoid neoplasms associated with eosinophilia and abnormalities of PDGFRA, PDGFRB, or FGFR1" [19]. A similar development can be observed in the field of solid tumors, such as lung tumors, and even organ-overlapping classifications of diseases based on shared pathomechanisms might be envisioned in the future of lymphoma classification [25, 26].

References

[1] Trumper L, Brittinger G, Diehl V, Harris N. Non-Hodgkin's lymphoma: A history of classification and clinical observations. In: Mauch PM, Armitage JO, Coiffer B, Dalla-Favera R, Harris N, editors. Non-Hodgkin's Lymphoma. Philadelphia, PA: Lippincott, Williams & Wilkins; 2004:3–19.

[2] Jaffe ES, Harris NL, Stein H, Isaacson PG. Classification of lymphoid neoplasms: The microscope as a tool for disease discovery. Blood 2008;112(12):4384–4399.

[3] Harris N, Ferry J. Classification of Non-Hodgkin's Lymphomas. In: Knowles DM, editor. Neoplastic Hematopathology. 2 ed. Lippicott Williams & Wilkins; 2001:691–754.

[4] Hodgkin. On some morbid appearances of the absorbent glands and spleen. Med Chir Trans 1832;17:68–114.

[5] Gall EA, Morrison HR, Scott AT. The follicular type of malignant lymphoma: A survey of 63 cases. Arch Int Med 1941;14:2073–2090.

[6] Brill NE, Baehr G, Rosenthal N. Generalized giant lymph follicle hyperplasia of lymph nodes and spleen; a hitherto undescribed type. Am J Med 1952;13(5):570–574.

[7] Symmers D. Follicular lymphadenopathy with splenomegaly, a new recognized disease of the lymphatic system. Arch Path 1927;3:816–820.

[8] Burkitt D. A sarcoma involving the jaws in African children. Br J Surg 1958;46(197):218–223.

[9] Nowell PC. Phytohemagglutinin: An initiator of mitosis in cultures of normal human leukocytes. Cancer Res 1960;20:462–466.

[10] Cooper MD, Peterson RD, Good RA. Delineation of the thymic and bursal lymphoid systems in the chicken. Nature 1965;205:143–146.

[11] Stein H, Lennert K, Parwaresch MR. Malignant lymphomas of B-cell type. Lancet 1972;2(7782):855–857.

[12] Kay HE. Letter: Classification of non-Hodgkin's lymphomas. Lancet 1974;2(7880):586.

[13] Lukes RJ, Collins RD. Immunologic characterization of human malignant lymphomas. Cancer 1974;34(4 Suppl):suppl-503.

[14] Gérard-Marchant R, Hamlin I, Lennert K, Rilke F, Stansfeld AG, van Unnik JA. Letter: Classification of non-Hodgkin's lymphomas. Lancet 1974;2(7877):405–408.

[15] National Cancer Institute sponsored study of classifications of non-Hodgkin's lymphomas: Summary and description of a working formulation for clinical usage. The Non-Hodgkin's Lymphoma Pathologic Classification Project. Cancer 1982;49(10):2112–2135.

[16] Lennert K. History of the Europaen Association for Haematopathology. Berlin, Heidelberg, New York: Springer; 2006.

[17] Harris NL, Jaffe ES, Stein H et al. A revised European-American classification of lymphoid neoplasms: A proposal from the International Lymphoma Study Group. Blood 1994;84(5):1361–1392.

[18] Jaffe E, Harris N, Stein H, Vardiman JW. Pathology and genetics of tumours of haematopoetic and lymphoid tissues. Lyon, France: IARC Press; 2001.

[19] Swerdlow SH, Campo E, Harris N et al. WHO Classification of Tumors of the Haematopoietic and Lymphoid Tissues. Lyon: IARC; 2008.

[20] Hartmann S, Doring C, Jakobus C et al. Nodular lymphocyte predominant Hodgkin lymphoma and T cell/histiocyte rich large B cell lymphoma – endpoints of a spectrum of one disease? PLoS One 2013;8(11):e78812.

[21] Shaffer AL, III, Young RM, Staudt LM. Pathogenesis of human B cell lymphomas. Annu Rev Immunol 2012;30:565–610.

[22] Masque-Soler N, Szczepanowski M, Kohler CW, Spang R, Klapper W. Molecular classification of mature aggressive B-cell lymphoma using digital multiplexed gene expression on formalin-fixed paraffin-embedded biopsy specimens. Blood 2013;122(11):1985–1986.

[23] Ganapathi KA, Pittaluga S, Odejide OO, Freedman AS, Jaffe ES. Early lymphoid lesions: conceptual, diagnostic and clinical challenges. Haematologica 2014;99(9):1421–1432.

[24] Salaverria I, Siebert R. The gray zone between Burkitt's lymphoma and diffuse large B-cell lymphoma from a genetics perspective. J Clin Oncol 2011;29(14):1835–1843.

[25] Heim D, Budczies J, Stenzinger A et al. Cancer beyond organ and tissue specificity: Next-generation-sequencing gene mutation data reveal complex genetic similarities across major cancers. Int J Cancer 2014;135(10):2362–2369.

[26] Alexandrov LB, Nik-Zainal S, Wedge DC et al. Signatures of mutational processes in human cancer. Nature 2013;500(7463):415–421.

Ulf Klein and Nilushi S. De Silva

2 Mature B-cell development and the germinal center reaction

2.1 Introduction

B lymphocytes produce antibodies that are crucial for protection against microorganisms. The affinity of the antibodies for the pathogen increases during the course of an immune response, thus allowing efficient binding and elimination of the antigen together with the generation of immunological memory against the pathogen. The importance of B lymphocytes (or B cells) for immunity is exemplified by the immunodeficiency syndromes that occur if B cells are absent or function abnormally. On the other hand, the processes that enable maturation of the antibody response – somatic hypermutation and class switch recombination – are inherently dangerous as they can lead to erroneous alterations in the genomic DNA that may activate oncogenes or inactivate tumor suppressors.

Each B cell expresses one specific antigen receptor on its surface that recognizes one particular antigenic structure. When a B cell encounters antigen, binding of this antigen to the surface receptor, commonly known as the B-cell receptor (BCR), in conjunction with other external, costimulatory signals leads to the activation of the B cell. During this *humoral* immune response, the gene encoding the BCR is modified to achieve optimal antigen binding, and eventually the B cell differentiates into a plasma cell that produces antibodies (or immunoglobulins), the secreted form of the B-cell receptor, that recognize and eliminate the pathogen. In addition, a specialized B lymphocyte is generated that is termed a memory B cell which can recognize the same pathogen many years later and respond quickly by secreting highly specific antibodies against the pathogen.

The antibodies produced by B cells consist of two identical heavy and two identical light chains. Each chain consists of a variable (V) region that mediates binding to antigen and a constant (C) region that determines the effector functions of the antibody. The V region of an antibody is assembled via a process of somatic DNA recombination that juxtaposes separate gene segments. In the case of the light chain, variable (V) and joining (J) gene segments are combined and in the case of the heavy chain, V, diversity (D) and J gene segments are combined. The process by which this occurs is also called VDJ recombination and is explained in more detail within the context of the T-cell receptor in a different chapter of this book. As there are many different V, D and J gene segments encoded in the germline DNA, this recombination process generates a huge repertoire of highly diverse antibodies that can potentially recognize any foreign antigen.

When a particular pathogen invades the body, only a few B cells out of this huge repertoire of B cells with diverse antibody specificities are activated following binding of their BCR to their cognate antigen. In an adaptive (or acquired) immune response, these antigen-activated B cells interact with antigen-specific T cells in secondary lymphoid tissues (lymph nodes or spleen) and migrate to specific microenvironments within those tissues where they rapidly proliferate. During this proliferation, the specificity of the BCR for the antigen is altered by somatically mutating the DNA encoding the V region of the immunoglobulin (Ig) gene in a process called somatic hypermutation. This process creates several B cell clones expressing BCRs with different affinities towards the immunizing antigen, and those clones with high-affinity BCRs are selected and propagated. Many of those B cells also switch from the Ig class they initially express, IgM, to other Ig classes (i.e. IgG, IgA, IgE) via class switch recombination. This process results in the acquisition of specialized effector functions that are tailored to optimally eliminate the specific pathogen in cooperation with other immune cell types. The DNA-modifying processes of somatic hypermutation and class switch recombination take place in a particular microenvironment within the lymph nodes called the germinal center (GC). The GC is a highly organized, transient structure in which various immune cell types interact with each other to coordinate the generation of B cells expressing high-affinity antibodies against their respective antigen by somatic hypermutation and affinity-based selection [1, 2]. Eventually, B cells with a modified BCR are instructed to differentiate into plasma cells or memory B cells.

The generation of humoral memory in the GC reaction is the hallmark of T-cell dependent immune responses. That is, T-cell interactions and signals are required for B cells to undergo the GC reaction. B cells can also be instructed to differentiate into antibody-secreting cells via a T-cell independent (TI) fashion. Two TI immune responses can be defined based on the particular immunizing antigens. TI-1 responses are characterized by the binding of molecules to receptors on B cells (other than the BCR) that result in the activation and proliferation of those B cells. For example, bacterial lipopolysaccharide (LPS) activates B cells via signaling through toll-like receptor 4 (TLR4). TLR4 is expressed on most B cells, and TLR4 stimulation leads to polyclonal B cell activation and differentiation into antibody-secreting cells. TI-2 responses involve BCR-crosslinking compounds such as specific polysaccharides that are found as repetitive epitopes in the cell wall of encapsulated bacteria. B cells activated in TI-2 responses secrete antibodies that facilitate the opsonization of such bacteria to enable uptake by phagocytic cells that normally are unable to eliminate encapsulated pathogens. Finally, there are distinct subsets of B cells with specialized immunological functions, including B1 cells and marginal zone B cells (see below), that respond to particular types of antigens or pathogens. As most mature B-cell lymphomas derive from the malignant transformation of B cells that have passed through the GC, during which the IgV genes were diversified by somatic hypermutation [3–5], this chapter predominantly focuses on the GC B-cell reaction.

2.2 The peripheral B-cell repertoire

Newly generated B cells expressing BCRs assembled via VDJ recombination that are not self-reactive exit the bone marrow, and the majority of these B cells recirculate through the secondary lymphoid system as naïve $IgM^{+}IgD^{+}$ B cells. Since these B cells have not 'seen' exogenous antigen, they are also referred to as antigen-inexperienced B cells or as follicular B cells, since they predominantly localize in the primary follicles of lymph nodes and spleen (see below). Naïve B cells are the precursors of GC B cells that participate in T-cell dependent immune responses to exogenous antigens. The primary function of a naïve B cell is to recognize foreign antigen and during the course of the ensuing GC response to modify their IgV region to generate antibodies that bind the same antigen with greatly increased affinity. As will be discussed in the following paragraphs, naïve B cells that are activated by antigen differentiate first into GC B cells whose Ig genes become somatically hypermutated and often switch to other Ig classes. These GC B cells then differentiate into plasma cells and memory B cells that express these modified Ig genes. Plasma cells migrate to the bone marrow where they secrete large amounts of high-affinity antibodies into the peripheral blood, thereby clearing the body of the infectious agent as well as providing long-term protection by secreting antibodies for extended periods of time [6, 7]. Memory B cells are long-lived, localized at the entry points of pathogens, and in antibody recall responses quickly differentiate into plasma cells [8, 9], thereby effectively protecting the body against repeated infection with the same pathogen. Thus, the GC reaction is the basis for the affinity maturation of the humoral immune response.

A subset of bone-marrow derived, newly generated B cells home to a particular site in the spleen, the marginal zone [10]. The primary function of these marginal zone B cells is to respond quickly to common blood-borne pathogens in order to prevent a potentially lethal sepsis [11]. The B cells recognize conserved structures on the microorganisms and are localized at the sites where the blood enters the spleen. Activated marginal zone B cells quickly differentiate into plasma cells that secrete antibodies that bind to the pathogen and, with the help of cells of the innate immune system, eliminate it. It is know from studies in mice that marginal zone B cells are phenotypically and functionally distinct from both naïve B cells and memory B cells [12, 13]; a commonality between marginal zone and memory B cells is their ability to rapidly respond to pathogens by differentiating into antibody-secreting cells. In mice, marginal zone B cells do not express somatically mutated IgV genes, in line with the notion that they represent a lineage distinct from naïve B cells that enter the GC response upon exposure to antigen [12, 13]. However, in humans, the majority of marginal zone B cells express somatically mutated IgV genes [14], indicating that they are GC-derived, although the precise cellular derivation of these cells is debated [15–17]. The transcription factor NOTCH2 has been identified as an important regulator of marginal zone B-cell development [18, 19]. Moreover, continued NOTCH2 activation seems to be crucial for the retention of B cells in the marginal zone microenvironment [20].

B-1 cells, characterized by the cell surface expression of the CD5 antigen, represent a distinct subset of B cells that secrete natural antibodies mainly against bacterial antigens, such as those with repetitive carbohydrate structures [21]. The B-1 cell subset is the predominant B-cell type in cord blood (although only a fraction of the $CD5^+$ B cells represent mature B1 cells) and in adulthood comprises a small fraction of B cells that localize in the peritoneal and pleural cavities, but also in the lymphoid system. Whereas in mice B-1 cells generally express germline-encoded antibody-specificities [21], in humans, a subset of B-1 cells expresses somatically mutated IgV regions [22]. Human B-1 cells display phenotypic, genetic and functional differences compared to the other B cell subsets that strongly indicate a unique immunological function for these cells that remains to be characterized in detail [22].

2.3 The germinal center reaction

2.3.1 Initiation of the GC reaction

The initiation of the GC reaction involves the coordinated interplay of different immune cell types that, in addition to antigen-specific naïve B cells, include antigen-specific T cells and dendritic cells. B cells activated by exogenous antigen [23] migrate to a specific area in the lymphoid tissue – the interfollicular area or the border of the T-cell/B-cell zone (Fig. 2.1). Here they undergo prolonged interactions with antigen-specific T cells that have been activated through priming by dendritic cells [24, 25]. During the course of this interaction, the B cells proliferate and eventually follow two major developmental routes. The first route involves the migration of a subset of these B cells to the medullary chords of the lymph nodes where they differentiate into plasmablasts that are characterized by the secretion of IgM or IgG antibodies that bind to the pathogen with low affinity [26]. Via this route, the immune system has some capacity to eliminate pathogens with specific antibodies as soon as a few days after infection. The second developmental pathway activated B cells may follow is the GC pathway of differentiation. Importantly, only B cells with the highest relative affinity towards the immunizing antigen among the pool of antigen-specific B cells are driven into the GC reaction [27, 28]. The basis for this selectivity is the successful intraclonal competition of GC precursor B cells for signals provided by antigen-specific T cells [29]. B cells committed to differentiate into a GC B cell upregulate the transcriptional repressor B-cell lymphoma-6 (BCL6) [30, 31], the master regulator of the GC reaction [32]. BCL6, in part mediated by a particular repression domain (RD2) within the protein, establishes a transcriptional program in the GC precursor B cell that allows the B cell to migrate to the center of the follicle [33], where the GC begins to form within and around a network of follicular dendritic cells (FDCs) [1] (Fig. 2.1). Specifically, BCL6 facilitates the movement of the GC precursor B cells towards the inner follicle by downregulating the expression of the chemokine receptor Epstein–Barr virus-induced G-protein

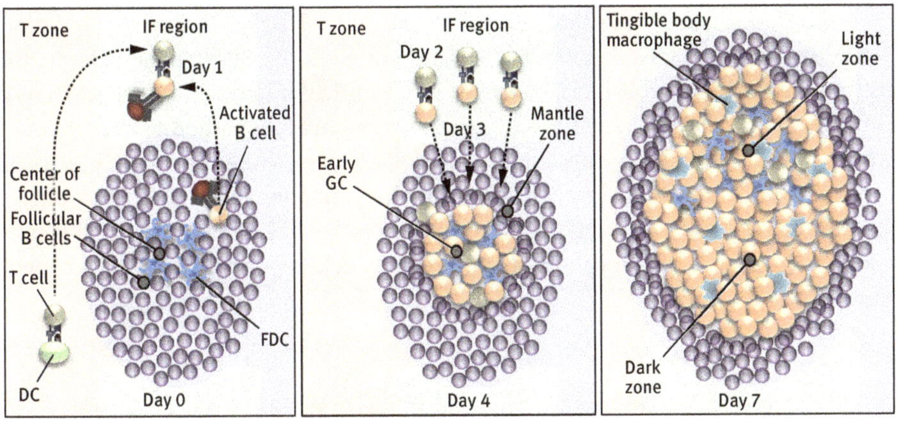

Fig. 2.1: **Initiation of the GC B-cell reaction and formation of the mature GC.** Day 0: B cells and T cells are activated in the lymph node by recognizing their cognate antigen in the follicle and T-cell zone (T zone), respectively. DC, dendritic cells; FDC, follicular dendritic cells; IF region, interfollicular region. Day 1: The activated T cells and B cells migrate to the IF region where they begin to interact. Day 2: T and B cells form prolonged interactions that lead to the full activation of B cells. T cells acquire the phenotype of Tfh cells. Day 3 (not shown): Tfh cells migrate into the follicle. Day 4: B cells migrate from the IF region into the center of the follicle where they fill the network of FDCs, start to proliferate and push the follicular B cells aside to form the early GC. The early GC is comprised of B-cell blasts that are surrounded by a mantle zone consisting of follicular B cells. Day 5–6 (not shown): The GC expands due to the rapid proliferation of the B-cell blasts. Day 7: Dark zones (DZ) and light zones (LZ) form, thereby establishing the mature GC. The DZ is predominantly composed of densely packed B-cell blasts; the LZ contains Tfh cells and FDCs in addition to the B cells. Tingible body macrophages localize throughout the GC and help to remove GC B cells that undergo apoptosis as they were not positively selected.

coupled receptor-2 (EBI2) and sphingosine-1-phosphate receptor type-1 (S1PR1) [31, 33], both of which are associated with mediating B-cell migration out of the follicle [34, 35]. BCL6-positive GC precursor B cells also upregulate CXCR4 [31], a chemokine receptor specifically found on GC B cells that localize in the dark zone of the GC (see below) [36, 37]. In addition to establishing this migratory program, upregulation of BCL6 expression has been associated with increased integrin-dependent conjugation of B and T cells [31].

Evidence suggests that the induction of BCL6 expression in B cells following immunization is mediated by the concerted action of several transcription factors. These include myocyte enhancer-binding factor-2B (MEF2B) which activates BCL6 transcription and whose upregulation in antigen-activated B cells precedes BCL6 expression [38], as well as interferon regulator factor-4 and 8 (IRF4 and IRF8) that have both been associated with activating BCL6 transcription [39, 40]. IRF4 is required for the initiation of the GC response within both B cells and T cells [40–42]; it is expressed in GC precursor B cells and then quickly downregulated in the proliferating GC B cells. In contrast, IRF8 and MEF2B continue to be expressed in proliferating GC B cells.

Antigen-specific T cells that control the entry of B cells into the GC initiation phase, and later the selection of somatically mutated GC B cells with high affinity for antigen within the established GC, are termed T follicular helper (Tfh) cells [43, 44]. These cells which are characterized by BCL6 expression and a particular immunophenotype ($CXCR5^+PD\text{-}1^+GL7^+$) migrate into the follicle one day before the GC precursor B cells [30, 31]. Tfh cells are involved in guiding the formation of the early GC, which is the earliest GC structure that arises from the coalescence of antigen-activated B cells that have migrated to the center of the B-cell follicle around day 4 after antigen-encounter [1].

2.3.2 GC formation and establishment of the mature GC

Upon their entry into the follicle at day 4 after immunization, antigen-activated B cells grow, begin to proliferate and differentiate into blasts that fill the network of FDCs in the center of the follicle [1] (Fig. 2.1). As a consequence, the follicular IgM^+IgD^+ B cells are pushed aside, resulting in the formation of the mantle zone surrounding the developing GC. In the following two days, the rapid proliferation of the B-cell blasts leads to a dramatic increase in the size of the GC. At day 7, the GC is considered mature and is characteristically polarized into two distinct microenvironments, known as the dark zone (DZ) and the light zone (LZ) [1, 2] (Fig. 2.1). The DZ consists of densely packed B-cell blasts that rapidly proliferate. The LZ contains B cells but also several other immune cell types, including Tfh cells, FDCs and macrophages [2]. Historically, the cells populating the DZ and the LZ were respectively defined as centroblasts and centrocytes. These definitions were based on morphological criteria in human secondary lymphoid tissue. Since studies in mice failed to reveal a difference in the morphology of GC B cells within the two separate zones, it seems more appropriate to define GC B-cell subsets by their microenvironmental location and immunophenotypic or functional characteristics [2, 45].

A defining feature of early GC B cells at day 4 after immunization is the expression of the c-MYC proto-oncogene [46]. c-MYC is generally known to have an important role in the proliferation of various cell types. Acting through transcriptional as well as nontranscriptional mechanisms, c-MYC regulates cell-cycle progression, cellular metabolism and telomere maintenance, thereby facilitating cellular proliferation [47]. The finding that c-MYC is expressed along with BCL6 in early GC B cells was surprising, as c-MYC mRNA and protein could not be detected in the vast majority of GC B cells [48, 49] due to its repression by BCL6 [50]. It was therefore believed that c-MYC and BCL6 are expressed in a mutually exclusive fashion. However, using sophisticated experimental tools, it was demonstrated that c-MYC is expressed in GC precursor B cells and early GC B cells at day 4 past immunization; it is then quickly downregulated in the proliferating blasts until it is re-expressed in a small subset of GC B cells in the LZ at day 7 (see below) [46, 50]. In agreement with this expression pattern, transgenic mouse studies in which *MYC* was specifically deleted in GC precursor B cells demonstrated that c-MYC is required for GC formation *in vivo* [46, 50]. It appears therefore that c-MYC regulates a cyclin D2 (*Ccnd2*)-dependent proliferation program that is replaced by the characteristic *Ccnd3*-dependent proliferation program of the DZ B cells [51, 52] after day 4 of the GC reaction. As for how c-MYC is required in conjunction with BCL6 during GC formation while these transcriptional regulators have mutually exclusive roles in later stages during the GC response remains a point of confusion. It is possible that in GC precursor B cells, BCL6-mediated repression of target genes occurs primarily via the RD2-domain (see above), whereas in DZ cells, BCL6 exerts its repressive functions via a different domain located at the N-terminal portion of the protein named BTB-domain [53]. Thus, BCL6 could control different sets of target genes during the different cellular stages of GC development through the predominant use of distinct protein domains. As summarized in Tab. 2.1, several other transcription factors besides BCL6, c-MYC, IRF4, IRF8 and MEF2B exert critical roles during GC formation. These include OCA-B [54, 55], MEF2C [56, 57], E2A (encoded by *TCF3*) [58], the Rho-Rac GTP-exchange factor family member DOCK8 [59], EBF1 [60], the Ets-protein SPI-B [61], BACH2 [62], a member of the BTB-basic region leucine zipper (bZip) family of transcription factors, and nuclear factor-κB (NF-κB) [63].

Proliferating DZ B cells are characterized by a unique cell proliferation and cell survival program. Similar to the use of a specific D-type cyclin-associated proliferation program mediated by CCND3 (see above) [51, 52], GC precursor B cells and the proliferating B-cell blasts depend on one particular anti-apoptotic protein, MCL1, for their survival [64]. In contrast, BCL2 has no role in the GC reaction since the gene is repressed by BCL6 and thus is not expressed in GC B cells [65], and BCL-X_L has been shown to be dispensable for the survival of GC B cells [64].

In summary, one week after pathogen encounter, the concerted activity of several transcriptional regulators culminates in the establishment of a mature GC comprised of a specialized DZ/LZ microenvironment that supports close interactions of the various cell types of the GC and the rapid trafficking of cells within this structure [1, 2]. Importantly, day 7 marks the timepoint when somatically mutated memory B cells and plasma cells selected for increased affinity against the antigen start to be released into the circulation [66].

2.3.3 Somatic hypermutation

Somatic hypermutation is a DNA-modifying process that introduces mostly single nucleotide exchanges at a mutation rate of 10^{-3} per base/generation into the Ig variable (IgV) region genes of the B cell [67]. Somatic hypermutation occurs in the rapidly proliferating DZ B cells and is the basis for the affinity maturation of T-cell dependent antibody responses [68]. The mutations are nonrandom in the sense that transitions occur more frequently than transversions, and mutation hotspots exist, as most prominently

Tab. 2.1: Functional requirement of molecules during the initiation and formation of the mature GC.

Molecule	Early initiation* d0–d1	d2	Late initiation* d3–4	Proliferation/Establishment* d5–d7	>d7
Canonical NF-κB	+	–	–	–	+ (subset)
IRF4	+	–	–	–	+ (subset)
OCA-B	+	–	–	–	–
MEF2C	+	–	–	–	–
c-MYC	+	+	+	–	+ (subset)
MEF2B	–	+	+	+	+
IRF8	–	+	+	+	+
BCL6	–	+	+	+	+
MCL1	–	+	+	+	+
E2A	–	–	+	+	+
EBF1	–	–	–	+	+
SPI-B	–	–	–	+	+
DOCK8	–	–	–	+	+
BACH2	–	–	–	+	+

* The formation of the GC can be subdivided into three separate phases: (1) An early initiation phase (d0–2) that includes the antigen-activation of the T and B cells and the long-lived interactions of T and B cells in the interfollicular (IF) region (see Fig. 2.1). (2) A late initiation phase (d3–4) that is characterized by the migration of the T and B cells from the IF region to the follicle and the appearance of the early GC at day 4. (3) A proliferation/establishment phase (d5–7) in which the GC grows until it forms a dark and a light zone characteristic of the mature GC. Functional requirements of molecules involved in the initiation of the GC reaction (d0–7) in relation to time, as suggested by the phenotypes resulting from the deletion of the corresponding molecules in antigen-activated B cells or GC B cells (see text); or, for MEF2B, its expression pattern in antigen-activated B-cells *in vivo*; for IRF8, its high expression in GC B cells and its role in the induction of BCL6 expression; and for BACH2, its expression in established GC B cells. A group of molecules comprising (canonical) NF-κB subunits, IRF4, OCA-B and MEF2C seem to exert their critical functions in GC formation shortly after B-cell activation. In contrast, c-MYC is required throughout the early and late initiation phases. A group of molecules including MEF2B, IRF8, BCL6 and MCL1 exert their functions in GC formation and later also in GC maintenance from day 2 onwards. E2A was found to be dispensable during the antigen-activation phase, but the inability to form early GCs suggests a role in the late initiation phase. Finally, the deficiencies of EBF1, SPI-B and DOCK8, and presumably also BACH2, do not affect the formation of early GCs; however, the size of the GCs is severely reduced, suggesting roles in the proliferation/establishment phase. NF-κB subunits, IRF4, and c-MYC are re-expressed in subsets of light zone B cells in the mature GC.

exemplified by the RGYW motif. However, most of the mutations will not increase the affinity against the immunizing antigen as they may encode replacement mutations resulting in the loss of antigen binding or proper folding of the antibody chains, or the mutations may introduce stop codons that terminate translation of the antibody chain. GC B cells that have acquired such unfavorable mutations are efficiently eliminated. As will be discussed in the next section, GC B cells with favorable mutations that increase the affinity of the BCR to antigen are positively selected to either undergo additional rounds of hypermutation and selection to further increase affinity to antigen, or to differentiate into post-GC memory B cells or plasma cells.

Somatic hypermutation requires the activity of activation-induced cytidine deaminase (AID) [69], an enzyme that deaminates cytidines on DNA and thereby initiates error-prone DNA repair that ultimately results in the introduction of hypermutations in the IgV region [67]. While somatic mutations are normally introduced into a region around 2 kilobases downstream of the transcriptional start site of the IgV region, the process occasionally affects non-Ig genes and when this occurs specifically in lymphoma it is referred to as aberrant somatic hypermutation [70]. The activity of AID in non-Ig loci has been associated with the introduction of aberrant somatic hypermutations in several proto-oncogenes; this may lead to the deregulated expression of these proto-oncogenes, promoting lymphomagenesis. For example, the regulatory region of the *BCL6* proto-oncogene is the most frequently targeted non-Ig region (at a mutation frequency of about 1/10th of that of rearranged IgV genes) and is often found mutated in GC-derived lymphomas. Besides *BCL6*, various other genes with diverse functions in GC biology were found to be the target of aberrant somatic hypermutation [71], highlighting the importance of this oncogenic mechanism in GC lymphomagenesis. Finally, genomic studies on mouse models of B-cell lymphomas were able to formally demonstrate a link between AID activity and the occurrence of chromosomal translocations [72, 73]. Thus, the benefit of somatic hypermutation in the establishment of immunity to pathogens has a downside, as occasional errors in this process may lead to the oncogenic transformation of B cells.

2.3.4 B cell movements within the zonal structure of the GC

It has long been established that somatic hypermutation generates a variety of B-cell clones that bind to the immunizing antigen with different affinities. Antibody mutants arising in the DZ move to the LZ where these mutants are positively selected for enhanced binding to antigen and instructed to recycle to the DZ (Fig. 2.2). Iterative rounds of mutation and selection enabled by the recirculation of GC B cells between DZ and LZ ultimately result in the appearance of memory B cells and plasma cells with very high-affinity antibodies against the invading microorganism. The chemokine receptors CXCR4 and CXCR5 are critically involved in the organization of the GC into a DZ and a LZ [36]. CXCR4 is specifically expressed on DZ B cells. The analysis of mice whose GC B cells are deficient for CXCR4 and thus restricted to the LZ revealed that the DZ to LZ transition occurred normally without DZ cues [74]. This finding suggests that the developmental states of DZ and LZ B cells are determined by a cell intrinsic program rather than through microenvironment-mediated signals. However, CXCR4-deficient GC B cells that were restricted to the LZ accumulated less somatic hypermutations in their rearranged IgV genes than their CXCR4-proficient counterparts [74], demonstrating that the polarization of the GC into DZ and LZ microenvironments to spatially separate hypermutation from selection ensures the optimal generation of high-affinity antibody mutants.

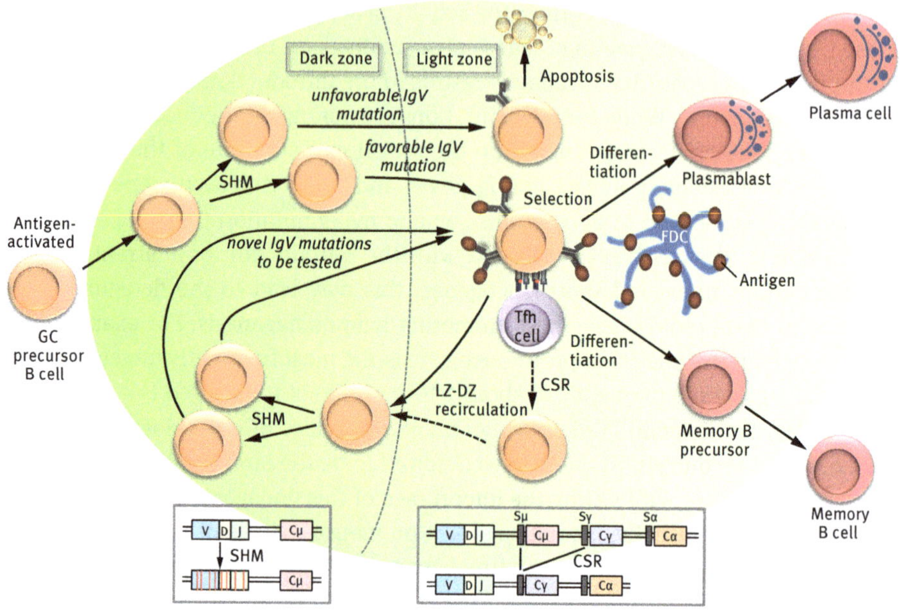

Fig. 2.2: **Dynamics of the GC B-cell reaction and selection of high-affinity antibody mutants.** Antigen-activated GC precursor B cells form the early GC at day 4 after antigen encounter. The B cells differentiate into blasts that undergo rapid clonal expansion over the next few days until the formation, at day 7, of the mature GC that is characterized by the dark and the light zone. The proliferation is coupled with the introduction of base-pair changes into the V(D)J region of the rearranged immunoglobulin variable (IgV) region genes by the somatic hypermutation (SHM) process, and some of these mutations result in a change in the amino acid sequence of the V(D)J region. Dark zone (DZ) B cells subsequently traffic to the light zone (LZ), where the modified antigen-receptor of the LZ B cell, under the guidance of immune cells including T-follicular helper (Tfh) cells and follicular dendritic cells (FDCs), is selected for improved antigen binding. Newly generated LZ B cells that express an unfavorable antibody are negatively selected and undergo apoptosis. LZ B cells that produce a favorable antibody are positively selected, and a subset of those cells is instructed to recirculate to the DZ. LZ B cells may undergo Ig class switch recombination (CSR) before they reenter the DZ. Upon return to the DZ, the cells undergo further proliferation and SHM which, potentially, may result in the generation of antibody mutants with even higher affinity towards the antigen. Recycling between DZ and LZ allows for the occurrence of iterative rounds of IgV gene mutation and selection. This process is the basis for the generation of memory B cells and plasma cells that produce high-affinity antibodies within a short time. A subset of antigen-selected LZ B cells differentiates into memory B-cell precursors or plasmablasts, the precursor cells of the long-lived memory B cells and plasma cells, respectively.

2.3.5 Selection of high-affinity GC B cells

The selection of high-affinity GC B cells is driven by Tfh cells within the LZ [2] (Fig. 2.2). GC B cells bind and internalize antigen acquired from FDCs via the BCR and present the processed antigen on major histocompatibility complexes-II (MHCII) to Tfh cells.

BCRs with higher affinity capture more antigen compared to those with lower affinity, which in turn results in a higher cell surface density of MHCII-peptide presentation [75]. As a result, those GC B cells receive the greatest share of Tfh-cell help, which drives the selection of GC B cells with high-affinity BCRs. Tfh cells continuously 'patrol' the LZ B cells and establish the longest cell-cell contacts with those B cells that present the highest density of antigen on MHCII molecules to Tfh cells [76]. Mechanistically, through these long cellular interactions, Tfh cells instruct the B cells to divide more frequently upon reentry into the DZ compared to their competitors that express BCRs with lower affinity [77]. Since somatic hypermutation and cell division are linked, the GC B cells that proliferate most harbor more affinity-enhancing IgV gene mutations than their less dividing competitors. Following several rounds of DZ-LZ recirculation, this process ensures that the B cells with the highest affinity towards the antigen are selected out of the large pool of B cells with heterogeneous affinities within only a few days of the mature GC having been established.

The signals provided by Tfh cells to LZ B cells during the process of selection are only beginning to be understood. It is known that the gene expression program of LZ B cells is enriched for 'signatures' typically observed upon B-cell activation *in vitro*, including CD40, NF-κB, BCR and c-MYC signatures [75]. Moreover, the interaction between Tfh and GC B cells was found to induce calcium signaling as well as the expression of certain cytokines in Tfh cells [76]. Finally, specific dynamic interactions between Tfh cells and GC B cells have been identified that ultimately result in the acquisition of a greater share of T-cell help by high-affinity GC B cells [78].

The realization that Tfh cells are the major factors guiding the selection of LZ B cells was somewhat surprising since it had long been assumed that the selection of high-affinity B cells was primarily driven by direct competition for antigen. In this model, high-affinity BCRs would compete most efficiently for direct acquisition of antigen from FDCs, and this binding would deliver survival signals to the B cell [1]. However, it was recently found that antigen-mediated BCR-signaling is largely inhibited in GC B cells (this signaling is distinct from the tonic BCR-signaling that is required for the maintenance of GC B cells) [79]. Thus, it now appears that within the GC microenvironment, the principal role of the BCR is not to direct B cell activation via BCR signaling, but rather the BCR is required to bind and internalize antigen for subsequent presentation on MHCII complexes [2].

The GC reaction is a highly complex process and several additional intricacies feed into the selection mechanism described above. One of these processes of fine-tuning GC selection involves limiting antigen access via antibody masking. Low affinity antibodies secreted from plasma cells generated early in the immune response are able to enter GCs and limit access to antigen deposited on FDCs during the ongoing GC response. This masking enables only those GC B cells expressing BCRs with higher affinity to compete for antigen acquisition [80]. Over time, as antibody mutants with increasing affinity arise, the stringency of the selection increases. Another important issue regards the maintenance of B-cell tolerance. While the possibility exists that somatic hypermutation may lead to the generation of self-reactive BCRs, the tolerance mechanisms acting in GCs are only incompletely understood. However, there is experimental evidence from transgenic mouse models that auto-reactive GC B cells are cleared when self-antigen was ubiquitous or expressed in close proximity to the GC [81]. Conversely, newly generated antibody mutants that recognize rare antigens or those expressed selectively in certain tissues were not eliminated within the GC but rather positively selected [81].

2.3.6 Class switch recombination

The process of class switch recombination allows a B cell to switch from IgM and IgD to IgG, IgA or IgE. Via this process, antibodies with specialized effector functions are generated. For example, some switched isotypes are more efficient at opsonizing antigen. Class switch recombination requires AID and involves nonhomologous end-joining processes [82, 83] and can occur in T-cell dependent and T-independent immune responses. Evidence suggests that during the GC reaction, this process occurs in the LZ where GC B cells receive signals from Tfh cells, such as CD40–CD154 stimulation, and other receptor-mediated costimulatory interactions as well as signaling through cytokine receptors that determine the specific Ig class. CD40 activation, presumably in an indirect fashion, may also induce the IRF4 transcription factor that is required for optimal upregulation of AID expression [84, 85]. Following class switching, the LZ cells can either recycle to the DZ and undergo additional rounds of somatic hypermutation, or exit the GC as memory B cells or plasma cells.

2.3.7 GC maintenance

GC B cells are characterized by a specialized physiology that is vastly different not only from other B-cell subpopulations, but also in comparison to any other cell type in the organism [86]. DZ B cells proliferate extremely fast, at an estimated cell cycle rate of ~ 8 hours. Since proliferation and somatic hypermutation are linked, this enables the rapid generation of a large repertoire of antibody mutants from which high-affinity variants are selected. In addition, as the somatic hypermutation and class switch recombination processes deliberately introduce DNA breaks into Ig loci, the machinery that senses and responds to DNA damage must be suppressed in GC B cells. Moreover, a defining feature of GC B cells is that they are exceptionally prone to undergo apoptosis, which allows the rapid elimination of newly generated antibody mutants with unfavorable somatic mutations. Lastly, the GC reaction is extremely dynamic as it involves coordinated interzonal migration and a rapid change of cellular states (i.e. DZ to LZ cell, LZ to DZ cell or LZ into memory B-cell precursor or plasmablast). Therefore, it is evident that the events occurring in B cells within the GC require tight molecular control that involves the coordination of transcriptional, nontranscriptional, epigenetic and microRNA-mediated regulation.

2.3.7.1 BCL6, a master regulator of the GC reaction

Besides the critical role of the transcriptional repressor BCL6 in GC initiation that is discussed in an earlier section, BCL6 is required for the maintenance of the GC reaction and is the major regulator of the specialized physiology of the DZ B cells [32] (Fig. 2.3). BCL6 was originally cloned from the breakpoint of a chromosomal translocation in B-cell lymphomas and has subsequently been shown in knockout mouse models to be necessary for GC B-cell development [87, 88]. On the other hand, the deregulated expression of BCL6 in transgenic mice causes increased GC formation [89]. BCL6 protein is expressed in all DZ B cells and in the majority of LZ B cells. Target gene repression by BCL6 is mediated by binding to specific DNA motifs along with the recruitment of class I and II histone deacetylase complexes, either directly or via interactions with co-repressors [33, 53, 90]. In the DZ cells, BCL6 controls a complex transcriptional network that modulates various different signaling pathways [91, 92]. The rapid proliferation of DZ cells is at least in part mediated by the repression of genes by BCL6 that are involved in negative cell cycle regulation. BCL6 also suppresses the anti-apoptotic gene *BCL2* in GC B cells [65], thereby establishing a program that allows efficient elimination of B cells with unfavorable IgV gene mutations. Moreover, the suppression of the p53, ATM and ATR-mediated DNA damage response by BCL6 allows the GC B cell to cope with the physiological genotoxic stress that it is exposed to during the DNA-modifying processes of somatic hypermutation and class switch recombination [32]. Under normal conditions, activation of these stress sensors would lead to growth arrest, as observed in other cell types. BCL6 also inhibits the expression of B-cell activation associated genes such as CD274 (PD-L1) and CD80, which interacts with CD28-expressing T cells in the LZ. In addition, components of the BCR and CD40 signaling pathways are downmodulated by BCL6. Through the repression of B-cell activation genes, BCL6 prevents premature activation by Tfh cells before DZ cells have completed proliferative expansion and somatic hypermutation. A further critical function of BCL6 is the prevention of premature plasma cell differentiation by directly repressing the gene encoding the master plasma cell regulator B lymphocyte-induced protein-1 (BLIMP1; gene name *PRDM1*) [93], and of memory B-cell differentiation [94]. As will be further discussed below, the inhibition of terminal differentiation is ensured by several different mechanisms that act in parallel. Lastly, BCL6 represses microRNAs mir-155 and mir-361 which downregulate AID expression and thereby indirectly promotes the AID-driven processes of somatic hypermutation and class switch recombination [95].

2.3.7.2 Other transcriptional regulators

The transcription factor BACH2, expressed in GC B cells and required for GC formation [62], was shown to cooperate with BCL6 in the repression of many genes in DZ cells [96]. Another transcription factor that is highly expressed in GC B cells is IRF8 [39], implicating a role for this factor in GC physiology in addition to its presumed function in the initiation of the GC reaction (see above). However, IRF8-deficiency alone did not impair GC B-cell development [97]. Rather, the combined ablation of IRF8 and the transcription factor PU.1 resulted in increased plasma cell differentiation, indicating that these transcriptional regulators cooperate in blocking premature terminal differentiation of GC B cells [98].

2.3.7.3 Nontranscriptional regulation of the DZ phenotype

It is becoming increasingly clear that epigenetic mechanisms also contribute to the maintenance of the phenotype of DZ B cells [99, 100] (Fig. 2.3). Polycomb repressive complexes 1 and 2 (PRC1 and PRC2) are epigenetic regulators that modulate gene repression. Enhancer of zeste-homolog-2 (EZH2), a histone methyltransferase that methylates lysine 27 of histone 3, is the catalytic component of PRC2. This H3K27me3 histone mark is associated with silenced gene expression. EZH2 is highly expressed in GC B cells [101, 102], and its ablation in those cells results in impaired GC development and the upregulation of IRF4 and BLIMP1 [99, 100], important regulators of plasma cell differentiation. Besides the repressive H3K27me3 mark, the promoters of several genes that have roles in terminal differentiation (including *IRF4* and *PRDM1*) also exhibit the activating mark H3K4me3. These genes are therefore characterized by the possession of bivalent chromatin domains, similar to what has previously been described for embryonic and tissue-specific stem cells [103]. This finding suggests that in GC B cells, critical developmental factors are poised for activation or repression. In the GC microenvironment, within which dynamic changes in cellular phenotype occur, this chromatin configuration would allow a transient or reversible suppression of transcription factors that are directly involved in terminal differentiation. Signals that instruct the B cell to exit the GC would lead to the removal of H3K27me3, thus committing the B cell towards terminal differentiation.

Regulation via microRNA-mediated mechanisms are also involved in the GC reaction and the maintenance of the DZ phenotype (Fig. 2.3). Ablation of the RNaseIII enzyme dicer, which is a central component of the microRNA-processing machinery, in antigen-activated B cells strongly impaired the formation of GCs [104]. The same phenotype was observed upon deletion of miR-155 in B cells [105, 106]. Since it was observed that miR-155 is only expressed in few GC B cells, and based on the *in vitro* observation that miR-155 expression is upregulated by CD40 stimulation [107], one may hypothesize that miR-155 expression in LZ cells (that are subjected to CD40 stimulation) is involved in the termination of the DZ phenotype. This scenario would be consistent with the observation that BCL6 represses *miR-155* in GC B cells [95]. An-

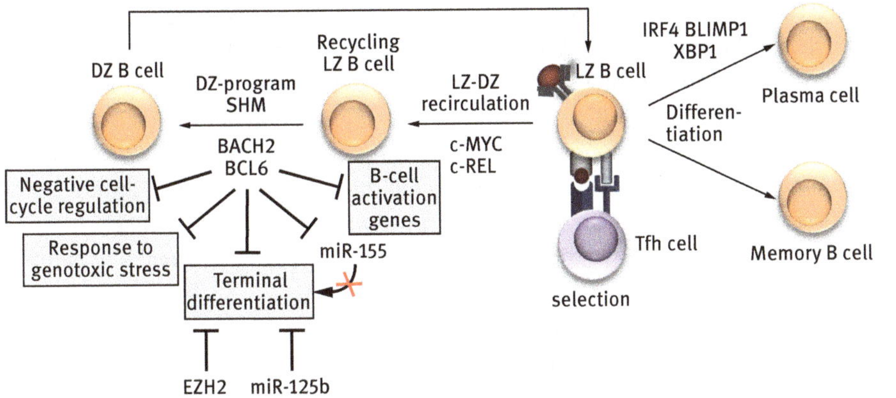

Fig. 2.3: **Molecular control of GC B-cell development.** Roles of transcriptional regulators during GC B-cell development. BCL6 establishes the specialized dark zone (DZ) program that allows somatic hypermutation (SHM) of the rearranged IgV genes. BACH2 cooperates with BCL6 in the repression of many genes in DZ cells. BCL6 is involved in the suppression of several cellular processes in the DZ cell. It represses genes involved in negative cell cycle regulation, inhibits the response to genotoxic stress and blocks B-cell activation and terminal differentiation. BCL6 also indirectly represses plasma cell-associated genes via downregulation of miR-155 expression. In addition, terminal differentiation is inhibited by nontranscriptional mechanisms which involve the activities of EZH2, the core catalytic component of polycomb repressive complex 2, and miR-125b. Evidence suggests that c-MYC and c-REL are required for the recirculation of antigen-selected light zone (LZ) B cells back into the DZ and thus for the maintenance of the GC reaction. Plasma cell differentiation requires the transcription factors IRF4, BLIMP1 and XBP1.

other microRNA, miR-125b, is highly expressed in DZ cells and seems to directly block the differentiation of GC B cells into plasma cells by downregulating BLIMP1 and IRF4 expression in the DZ cells [108]. Several other microRNAs have been identified that play roles in certain aspects of the GC reaction [95, 109, 110], further highlighting the importance of microRNAs in the regulation of GC B-cell function and development. In summary, both transcriptional and nontranscriptional mechanisms are involved in the establishment and maintenance of the complex DZ phenotype.

2.3.7.4 LZ-DZ recirculation

The recirculation of antigen-selected LZ cells back to the DZ where they undergo additional cell divisions continuously fuels the GC reaction. This iterative process is the basis for the maintenance of the GC reaction. Several transcription factors with a possible regulatory role in this process have been identified, including c-MYC and a specific NF-κB subunit, i.e. c-REL [46, 50, 111] (Fig. 2.3). c-MYC, which is absent in proliferating DZ cells within mature GCs, is however expressed in a small subset of LZ cells [46, 50]. Experimental evidence from the analysis of transgenic mouse models suggests that c-MYC-expressing LZ cells are undergoing positive selection for high-affinity antigen-binding. Given that positive selection of a B cell within the LZ instructs reentry into the DZ which is followed by cell division, the finding that the inhibition of c-MYC function in mature GCs led to the collapse of established GCs indicates that c-MYC is critically involved in sustaining the GC reaction. Similarly, the deletion of the NF-κB transcription factor c-REL in GC B cells resulted in the dissolution of established GCs [111]. Thus, at the point of selection of antigen-specific B cells, both c-MYC and c-REL induce a program that permits LZ-DZ recirculation and consequently GC maintenance, although the extent to which these two transcription factors interact in this process remains to be determined.

Other factors have been implicated in the maintenance of the GC reaction. Similar to what has been observed for c-REL, B cell-specific deletion of the transmembrane protein tyrosine phosphatase CD45 in B cells caused the disintegration of established, mature GCs, whereas early GC development remained normal [112]. Also, the transcription factors EBF1 and SPI-B that are involved in the formation of the GC (discussed above) appear to have an additional role in GC maintenance [60, 61].

2.3.8 Termination of the GC reaction and post-GC B-cell differentiation

Before a positively selected B cell can exit the GC, it must terminate BCL6 expression. By switching off BCL6, the repression of BCL6 target genes that are incompatible with the specialized physiology of the DZ cell is alleviated [91, 92]. De-repression of *PRDM1* (encoding the plasma cell regulator BLIMP1) [93] is required for plasma cell differentiation, and memory B-cell differentiation also requires cessation of BCL6 expression [94]. Several mechanisms cooperate in the downregulation of BCL6. CD40 stimulation, via NF-κB, induces upregulation of IRF4 that represses *BCL6* transcription [113]. In addition, BCR signaling causes MAPK-mediated BCL6 phosphorylation, resulting in degradation of the protein [114]. In addition, an F-box only protein 11 (FBXO11)-mediated process tags BCL6 by ubiquitination, targeting the protein to the proteasome [115]. Of note, BCL6 activity can be modulated by acetylation [116]. This chemical modification interferes with the interaction of BCL6 with co-repressors and thus impairs the repressive function of this complex. Indeed, acetylation is a fast and reversible protein-modifying process that may be ideally suited for the modulation of BCL6 activity during the rapid phenotypic changes that occur during the GC reaction, before transcriptional and posttranscriptional mechanisms ultimately converge to terminate BCL6 expression.

Several transcription factors have been identified that exert important functions during plasma cell development. The functional inactivation of the transcriptional repressor paired box-5 (PAX5), a major determinant of the mature B-cell phenotype (naïve, GC and memory B cells), is required for the differentiation of a B cell into a plasma cell [117]. Indeed, downregulation of PAX5 expression in plasma cells was found to be critical for the establishment of the plasma cell differentiation program [118, 119]. The differentiation of a GC B cell into a plasma cell that secretes huge amounts of antibody is a complex process that is coordinated by the activity of the transcriptional regulators BLIMP1, IRF4 and XBP1, all of which are required for the establishment of the plasma cell transcriptional program [84, 85, 120–124] (Fig. 2.3). IRF4, which is expressed at much higher levels in plasma cells than in other B-cell differentiation stages, directly activates *PRDM1* (the gene encoding BLIMP1) transcription [85] in addition to repressing *BCL6* [113], as discussed above. BLIMP1 represses the B-cell specific transcriptional program, in part via repressing the *BCL6* and *PAX5* genes as well as cell proliferation [121]. BLIMP1 acts upstream of X-box binding protein-1 (XBP1) which encodes a transcription factor that establishes the characteristic secretory phenotype of plasma cells [124]. Plasma cells generated during the GC reaction migrate to the bone marrow where they home to specialized microenvironments that provide these cells with survival signals and allow them to secrete high-affinity antibodies for extended periods of time [6].

While we know increasingly more about the control of plasma cell development, the transcriptional regulation of memory B-cell development remains undefined. Gene expression profiling analyses have uncovered only relatively few differences between the transcriptomes of memory and naïve B cells [49]. However, memory B cells, in contrast to the naïve B cells, are rapidly activated upon renewed antigen encounter, thus demonstrating the existence of functional differences that remain to be defined and may be underpinned by nontranscriptional and/or epigenetic regulatory mechanisms. Memory B cells can have two fates. They can rapidly differentiate into plasma cells that secrete high-affinity antibodies that enable pathogen elimination before an infection is established. Alternatively, memory B cells can reenter the GC reaction and undergo additional rounds of somatic hypermutation to generate new antibody mutants with refined affinities to the antigen. The importance of the latter scenario is exemplified by the extremely high mutation rates found in memory B cells of HIV-infected individuals that secrete neutralizing antibodies with broad diversity [125].

2.4 Concluding remarks

As will be discussed in several chapters of this textbook, the genes encoding key regulators of B-cell development are frequently targeted by genetic aberrations, leading to dysregulated activity which promotes lymphomagenesis. Therefore, deciphering the molecular control of B-cell differentiation is expected to directly translate into a better understanding of the oncogenic mechanisms that promote B-cell lymphomagenesis. Of particular interest are the developmental processes occurring during the GC reaction, since errors in the mechanisms of somatic hypermutation and class switch recombination frequently cause oncogenic mutations in proto-oncogenes or tumor suppressors with important roles during GC differentiation. Significant progress has been made in unraveling the complexity of the GC reaction and in identifying the functions of key transcriptional and epigenetic regulators during GC development. This knowledge helps us to better understand how the dysregulation of those factors contributes to cellular transformation and in turn may provide leverage points for the development of novel, more effective therapies. Nevertheless, much remains to be learned about the precise mechanisms that control the selection of high-affinity B cells and the cues that determine whether a GC B cell differentiates into a memory B cell or a plasma cell. In addition, it is now clear that there are many different subsets of memory B cells and plasma cells that have specific roles in humoral immunity. Insights into the generation and physiology of these subsets may help us to identify the normal counterparts of the heterogeneous B-cell malignancies and to understand the mechanisms that lead to their transformation.

References

[1] MacLennan IC. Germinal centers. Annu Rev Immunol 1994;12:117–139.
[2] Victora GD, Nussenzweig MC. Germinal centers. Annu Rev Immunol 2012;30:429–457.
[3] Klein U, Goossens T, Fischer M, et al. Somatic hypermutation in normal and transformed human B cells. Immunol Rev 1998;162:261–280.
[4] Stevenson F, Sahota S, Zhu D, et al. Insight into the origin and clonal history of B-cell tumors as revealed by analysis of immunoglobulin variable region genes. Immunol Rev 1998;162:247–259.
[5] Kuppers R, Klein U, Hansmann ML, Rajewsky K. Cellular origin of human B-cell lymphomas. N Engl J Med 1999;341:1520–1529.
[6] Tokoyoda K, Hauser AE, Nakayama T, Radbruch A. Organization of immunological memory by bone marrow stroma. Nat Rev Immunol 2010;10:193–200.
[7] Nutt SL, Hodgkin PD, Tarlinton DM, Corcoran LM. The generation of antibody-secreting plasma cells. Nat Rev Immunol 2015;15:160–171.
[8] Kurosaki T, Kometani K, Ise W. Memory B cells. Nat Rev Immunol 2015;15:149–159.
[9] Ahmed R, Gray D. Immunological memory and protective immunity: understanding their relation. Science 1996;272:54–60.
[10] Pillai S, Cariappa A, Moran ST. Marginal zone B cells. Annu Rev Immunol 2005;23:161–196.
[11] Martin F, Kearney JF. B-cell subsets and the mature preimmune repertoire. Marginal zone and B1 B cells as part of a "natural immune memory". Immunol Rev 2000;175:70–79.
[12] Martin F, Kearney JF. Marginal-zone B cells. Nat Rev Immunol 2002;2:323–335.
[13] Pillai S, Cariappa A. The follicular versus marginal zone B lymphocyte cell fate decision. Nat Rev Immunol 2009;9:767–777.
[14] Spencer J, Perry ME, Dunn-Walters DK. Human marginal-zone B cells. Immunol Today 1998;19:421–426.
[15] Weller S, Braun MC, Tan BK, et al. Human blood IgM "memory" B cells are circulating splenic marginal zone B cells harboring a prediversified immunoglobulin repertoire. Blood 2004;104:3647–654.
[16] Seifert M, Kuppers R. Molecular footprints of a germinal center derivation of human IgM+(IgD+)CD27+B cells and the dynamics of memory B cell generation. J Exp Med 2009;206:2659–2669.

[17] Descatoire M, Weller S, Irtan S, et al. Identification of a human splenic marginal zone B cell precursor with NOTCH2-dependent differentiation properties. J Exp Med 2014;211:987–1000.
[18] Saito T, Chiba S, Ichikawa M, et al. Notch2 is preferentially expressed in mature B cells and indispensable for marginal zone B lineage development. Immunity 2003;18:675–685.
[19] Hampel F, Ehrenberg S, Hojer C, et al. CD19-independent instruction of murine marginal zone B-cell development by constitutive Notch2 signaling. Blood 2011;118:6321–6331.
[20] Simonetti G, Carette A, Silva K, et al. IRF4 controls the positioning of mature B cells in the lymphoid microenvironments by regulating NOTCH2 expression and activity. J Exp Med 2013;210:2887–2902.
[21] Hardy RR. B-1 B cell development. J Immunol 2006;177:2749–2754.
[22] Seifert M, Sellmann L, Bloehdorn J, et al. Cellular origin and pathophysiology of chronic lymphocytic leukemia. J Exp Med 2012;209:2183–2198.
[23] Batista FD, Harwood NE. The who, how and where of antigen presentation to B cells. Nat Rev Immunol 2009;9:15–27.
[24] Okada T, Miller MJ, Parker I, et al. Antigen-engaged B cells undergo chemotaxis toward the T zone and form motile conjugates with helper T cells. PLoS Biol 2005;3:e150.
[25] Qi H, Cannons JL, Klauschen F, Schwartzberg PL, Germain RN. SAP-controlled T-B cell interactions underlie germinal centre formation. Nature 2008;455:764–769.
[26] Jacob J, Kelsoe G. In situ studies of the primary immune response to (4-hydroxy-3-nitrophenyl)acetyl. II. A common clonal origin for periarteriolar lymphoid sheath-associated foci and germinal centers. J Exp Med 1992;176:679–687.
[27] Dal Porto JM, Haberman AM, Kelsoe G, Shlomchik MJ. Very low affinity B cells form germinal centers, become memory B cells, and participate in secondary immune responses when higher affinity competition is reduced. J Exp Med 2002;195:1215–1221.
[28] Shih TA, Meffre E, Roederer M, Nussenzweig MC. Role of BCR affinity in T cell dependent antibody responses in vivo. Nat Immunol 2002;3:570–575.
[29] Schwickert TA, Victora GD, Fooksman DR, et al. A dynamic T cell-limited checkpoint regulates affinity-dependent B cell entry into the germinal center. J Exp Med 2011;208:1243–1252.
[30] Kerfoot SM, Yaari G, Patel JR, et al. Germinal center B cell and T follicular helper cell development initiates in the interfollicular zone. Immunity 2011;34:947–960.
[31] Kitano M, Moriyama S, Ando Y, et al. Bcl6 protein expression shapes pre-germinal center B cell dynamics and follicular helper T cell heterogeneity. Immunity 2011;34:961–972.
[32] Basso K, Dalla-Favera R. Roles of BCL6 in normal and transformed germinal center B cells. Immunol Rev 2012;247:172–183.
[33] Huang C, Gonzalez DG, Cote CM, et al. The BCL6 RD2 domain governs commitment of activated B cells to form germinal centers. Cell Rep 2014.
[34] Gatto D, Paus D, Basten A, Mackay CR, Brink R. Guidance of B cells by the orphan G protein-coupled receptor EBI2 shapes humoral immune responses. Immunity 2009;31:259–269.
[35] Pereira JP, Kelly LM, Xu Y, Cyster JG. EBI2 mediates B cell segregation between the outer and centre follicle. Nature 2009;460:1122–1126.
[36] Allen CD, Ansel KM, Low C, et al. Germinal center dark and light zone organization is mediated by CXCR4 and CXCR5. Nat Immunol 2004;5:943–952.
[37] Caron G, Le Gallou S, Lamy T, Tarte K, Fest T. CXCR4 expression functionally discriminates centroblasts versus centrocytes within human germinal center B cells. J Immunol 2009;182:7595–7602.
[38] Ying CY, Dominguez-Sola D, Fabi M, et al. MEF2B mutations lead to deregulated expression of the oncogene BCL6 in diffuse large B cell lymphoma. Nat Immunol 2013;14:1084–1092.

[39] Lee CH, Melchers M, Wang H, et al. Regulation of the germinal center gene program by interferon (IFN) regulatory factor 8/IFN consensus sequence-binding protein. J Exp Med 2006;203:63–72.

[40] Ochiai K, Maienschein-Cline M, Simonetti G, et al. Transcriptional regulation of germinal center B and plasma cell fates by dynamical control of IRF4. Immunity 2013;38:918–929.

[41] Willis SN, Good-Jacobson KL, Curtis J, et al. Transcription factor IRF4 regulates germinal center cell formation through a B cell-intrinsic mechanism. J Immunol 2014;192:3200–3206.

[42] Bollig N, Brustle A, Kellner K, et al. Transcription factor IRF4 determines germinal center formation through follicular T-helper cell differentiation. Proc Natl Acad Sci USA 2012;109:8664–8669.

[43] Crotty S. Follicular helper CD4 T cells (TFH). Annu Rev Immunol 2011;29:621–663.

[44] Ramiscal RR, Vinuesa CG. T-cell subsets in the germinal center. Immunol Rev 2013;252:146–155.

[45] Allen CD, Okada T, Tang HL, Cyster JG. Imaging of germinal center selection events during affinity maturation. Science 2007;315:528–531.

[46] Calado DP, Sasaki Y, Godinho SA, et al. The cell-cycle regulator c-Myc is essential for the formation and maintenance of germinal centers. Nat Immunol 2012;13:1092–1100.

[47] Dang CV. MYC on the path to cancer. Cell 2012;149:22–35.

[48] Shaffer AL, Rosenwald A, Hurt EM, et al. Signatures of the immune response. Immunity 2001;15:375–385.

[49] Klein U, Tu Y, Stolovitzky GA, et al. Transcriptional analysis of the B cell germinal center reaction. Proc Natl Acad Sci USA 2003;100:2639–2644.

[50] Dominguez-Sola D, Victora GD, Ying CY, et al. The proto-oncogene MYC is required for selection in the germinal center and cyclic reentry. Nat Immunol 2012;13:1083–1091.

[51] Peled JU, Yu JJ, Venkatesh J, et al. Requirement for cyclin D3 in germinal center formation and function. Cell Res 2010;20:631–646.

[52] Cato MH, Chintalapati SK, Yau IW, Omori SA, Rickert RC. Cyclin D3 is selectively required for proliferative expansion of germinal center B cells. Mol Cell Biol 2011;31:127–137.

[53] Hatzi K, Melnick A. Breaking bad in the germinal center: how deregulation of BCL6 contributes to lymphomagenesis. Trends Mol Med 2014;20:343–352.

[54] Kim U, Qin XF, Gong S, et al. The B-cell-specific transcription coactivator OCA-B/OBF-1/Bob-1 is essential for normal production of immunoglobulin isotypes. Nature 1996;383:542–547.

[55] Schubart DB, Rolink A, Kosco-Vilbois MH, Botteri F, Matthias P. B-cell-specific coactivator OBF-1/OCA-B/Bob1 required for immune response and germinal centre formation. Nature 1996;383:538–542.

[56] Khiem D, Cyster JG, Schwarz JJ, Black BL. A p38 MAPK-MEF2C pathway regulates B-cell proliferation. Proc Natl Acad Sci USA 2008;105:17067–17072.

[57] Wilker PR, Kohyama M, Sandau MM, et al. Transcription factor Mef2c is required for B cell proliferation and survival after antigen receptor stimulation. Nat Immunol 2008;9:603–612.

[58] Kwon K, Hutter C, Sun Q, et al. Instructive role of the transcription factor E2A in early B lymphopoiesis and germinal center B cell development. Immunity 2008;28:751–762.

[59] Randall KL, Lambe T, Johnson AL, et al. Dock8 mutations cripple B cell immunological synapses, germinal centers and long-lived antibody production. Nat Immunol 2009;10:1283–1291.

[60] Vilagos B, Hoffmann M, Souabni A, et al. Essential role of EBF1 in the generation and function of distinct mature B cell types. J Exp Med 2012;209:775–792.

[61] Su GH, Chen HM, Muthusamy N, et al. Defective B cell receptor-mediated responses in mice lacking the Ets protein, Spi-B. EMBO J 1997;16:7118–7129.

[62] Muto A, Tashiro S, Nakajima O, et al. The transcriptional programme of antibody class switching involves the repressor Bach2. Nature 2004;429:566–571.

[63] Jacque E, Schweighoffer E, Visekruna A, et al. IKK-induced NF-kappaB1 p105 proteolysis is critical for B cell antibody responses to T cell-dependent antigen. J Exp Med 2014;211:2085–2101.

[64] Vikstrom I, Carotta S, Luthje K, et al. Mcl-1 is essential for germinal center formation and B cell memory. Science 2010;330:1095–1099.

[65] Saito M, Novak U, Piovan E, et al. BCL6 suppression of BCL2 via Miz1 and its disruption in diffuse large B cell lymphoma. Proc Natl Acad Sci USA 2009;106:11294–11299.

[66] Blink EJ, Light A, Kallies A, Nutt SL, Hodgkin PD, Tarlinton DM. Early appearance of germinal center-derived memory B cells and plasma cells in blood after primary immunization. J Exp Med 2005;201:545–554.

[67] Di Noia JM, Neuberger MS. Molecular mechanisms of antibody somatic hypermutation. Annu Rev Biochem 2007;76:1–22.

[68] Rajewsky K. Clonal selection and learning in the antibody system. Nature 1996;381:751–758.

[69] Muramatsu M, Kinoshita K, Fagarasan S, Yamada S, Shinkai Y, Honjo T. Class switch recombination and hypermutation require activation-induced cytidine deaminase (AID), a potential RNA editing enzyme. Cell 2000;102:553–763.

[70] Pasqualucci L, Neumeister P, Goossens T, et al. Hypermutation of multiple proto-oncogenes in B-cell diffuse large-cell lymphomas. Nature 2001;412:341–346.

[71] Pasqualucci L, Dalla-Favera R. The genetic landscape of diffuse large B-cell lymphoma. Semin Hematol 2015;52:67–76.

[72] Klein IA, Resch W, Jankovic M, et al. Translocation-capture sequencing reveals the extent and nature of chromosomal rearrangements in B lymphocytes. Cell 2011;147:95–106.

[73] Chiarle R, Zhang Y, Frock RL, et al. Genome-wide translocation sequencing reveals mechanisms of chromosome breaks and rearrangements in B cells. Cell 2011;147:107–119.

[74] Bannard O, Horton RM, Allen CD, An J, Nagasawa T, Cyster JG. Germinal center centroblasts transition to a centrocyte phenotype according to a timed program and depend on the dark zone for effective selection. Immunity 2013;39:912–924.

[75] Victora GD, Schwickert TA, Fooksman DR, et al. Germinal center dynamics revealed by multiphoton microscopy with a photoactivatable fluorescent reporter. Cell 2010;143:592–605.

[76] Shulman Z, Gitlin AD, Weinstein JS, et al. Dynamic signaling by T follicular helper cells during germinal center B cell selection. Science 2014;345:1058–1062.

[77] Gitlin AD, Shulman Z, Nussenzweig MC. Clonal selection in the germinal centre by regulated proliferation and hypermutation. Nature 2014;509:637–640.

[78] Liu D, Xu H, Shih C, et al. T-B-cell entanglement and ICOSL-driven feed-forward regulation of germinal centre reaction. Nature 2014.

[79] Khalil AM, Cambier JC, Shlomchik MJ. B cell receptor signal transduction in the GC is short-circuited by high phosphatase activity. Science 2012;336:1178–1181.

[80] Zhang Y, Meyer-Hermann M, George LA, et al. Germinal center B cells govern their own fate via antibody feedback. J Exp Med 2013;210:457–464.

[81] Chan TD, Wood K, Hermes JR, et al. Elimination of germinal-center-derived self-reactive B cells is governed by the location and concentration of self-antigen. Immunity 2012;37:893–904.

[82] Honjo T, Kinoshita K, Muramatsu M. Molecular mechanism of class switch recombination: linkage with somatic hypermutation. Annu Rev Immunol 2002;20:165–196.

[83] Stavnezer J, Guikema JE, Schrader CE. Mechanism and regulation of class switch recombination. Annu Rev Immunol 2008;26:261–292.

[84] Klein U, Casola S, Cattoretti G, et al. Transcription factor IRF4 controls plasma cell differentiation and class-switch recombination. Nat Immunol 2006;7:773–782.
[85] Sciammas R, Shaffer AL, Schatz JH, Zhao H, Staudt LM, Singh H. Graded expression of interferon regulatory factor-4 coordinates isotype switching with plasma cell differentiation. Immunity 2006;25:225–236.
[86] Klein U, Dalla-Favera R. Germinal centres: role in B-cell physiology and malignancy. Nat Rev Immunol 2008;8:22–33.
[87] Ye BH, Cattoretti G, Shen Q, et al. The BCL-6 proto-oncogene controls germinal-centre formation and Th2-type inflammation. Nat Genet 1997;16:161–170.
[88] Dent AL, Shaffer AL, Yu X, Allman D, Staudt LM. Control of inflammation, cytokine expression, and germinal center formation by BCL-6. Science 1997;276:589–592.
[89] Cattoretti G, Pasqualucci L, Ballon G, et al. Deregulated BCL6 expression recapitulates the pathogenesis of human diffuse large B cell lymphomas in mice. Cancer Cell 2005;7:445–455.
[90] Huang C, Hatzi K, Melnick A. Lineage-specific functions of Bcl-6 in immunity and inflammation are mediated by distinct biochemical mechanisms. Nat Immunol 2013;14:380–388.
[91] Ci W, Polo JM, Cerchietti L, et al. The BCL6 transcriptional program features repression of multiple oncogenes in primary B cells and is deregulated in DLBCL. Blood 2009;113:5536–5548.
[92] Basso K, Saito M, Sumazin P, et al. Integrated biochemical and computational approach identifies BCL6 direct target genes controlling multiple pathways in normal germinal center B cells. Blood 2010;115:975–984.
[93] Shaffer AL, Yu X, He Y, Boldrick J, Chan EP, Staudt LM. BCL-6 represses genes that function in lymphocyte differentiation, inflammation, and cell cycle control. Immunity 2000;13:199–212.
[94] Kuo TC, Shaffer AL, Haddad J, Jr., Choi YS, Staudt LM, Calame K. Repression of BCL-6 is required for the formation of human memory B cells in vitro. J Exp Med 2007;204:819–830.
[95] Basso K, Schneider C, Shen Q, et al. BCL6 positively regulates AID and germinal center gene expression via repression of miR-155. J Exp Med 2012;209:2455–2465.
[96] Huang C, Geng H, Boss I, Wang L, Melnick A. Cooperative transcriptional repression by BCL6 and BACH2 in germinal center B-cell differentiation. Blood 2014;123:1012–1020.
[97] Feng J, Wang H, Shin DM, Masiuk M, Qi CF, Morse HC, 3rd. IFN regulatory factor 8 restricts the size of the marginal zone and follicular B cell pools. J Immunol 2011;186:1458–1466.
[98] Carotta S, Willis SN, Hasbold J, et al. The transcription factors IRF8 and PU.1 negatively regulate plasma cell differentiation. J Exp Med 2014;211:2169–2181.
[99] Beguelin W, Popovic R, Teater M, et al. EZH2 is required for germinal center formation and somatic EZH2 mutations promote lymphoid transformation. Cancer Cell 2013;23:677–6792.
[100] Caganova M, Carrisi C, Varano G, et al. Germinal center dysregulation by histone methyltransferase EZH2 promotes lymphomagenesis. J Clin Invest 2013;123:5009–5022.
[101] Raaphorst FM, van Kemenade FJ, Fieret E, et al. Cutting edge: polycomb gene expression patterns reflect distinct B cell differentiation stages in human germinal centers. J Immunol 2000;164:1–4.
[102] Velichutina I, Shaknovich R, Geng H, et al. EZH2-mediated epigenetic silencing in germinal center B cells contributes to proliferation and lymphomagenesis. Blood 2010;116:5247–5255.
[103] Bernstein BE, Mikkelsen TS, Xie X, et al. A bivalent chromatin structure marks key developmental genes in embryonic stem cells. Cell 2006;125:315–326.
[104] Xu S, Guo K, Zeng Q, Huo J, Lam KP. The RNase III enzyme Dicer is essential for germinal center B-cell formation. Blood 2012;119:767–776.
[105] Thai TH, Calado DP, Casola S, et al. Regulation of the germinal center response by microRNA-155. Science 2007;316:604–608.

[106] Vigorito E, Perks KL, Abreu-Goodger C, et al. microRNA-155 regulates the generation of immunoglobulin class-switched plasma cells. Immunity 2007;27:847–859.

[107] van den Berg A, Kroesen BJ, Kooistra K, et al. High expression of B-cell receptor inducible gene BIC in all subtypes of Hodgkin lymphoma. Genes Chromosomes Cancer 2003;37:20–28.

[108] Gururajan M, Haga CL, Das S, et al. MicroRNA 125b inhibition of B cell differentiation in germinal centers. Int Immunol 2010;22:583–592.

[109] de Yebenes VG, Bartolome-Izquierdo N, Nogales-Cadenas R, et al. miR-217 is an oncogene that enhances the germinal center reaction. Blood 2014;124:229–239.

[110] Schneider C, Setty M, Holmes AB, et al. MicroRNA 28 controls cell proliferation and is downregulated in B-cell lymphomas. Proc Natl Acad Sci USA 2014;111:8185–8190.

[111] Heise N, De Silva NS, Silva K, et al. Germinal center B cell maintenance and differentiation are controlled by distinct NF-kappaB transcription factor subunits. J Exp Med 2014;211:2103–2118.

[112] Huntington ND, Xu Y, Puthalakath H, et al. CD45 links the B cell receptor with cell survival and is required for the persistence of germinal centers. Nat Immunol 2006;7:190–1908.

[113] Saito M, Gao J, Basso K, et al. A signaling pathway mediating downregulation of BCL6 in germinal center B cells is blocked by BCL6 gene alterations in B cell lymphoma. Cancer Cell 2007;12:280–2892.

[114] Niu H, Ye BH, Dalla-Favera R. Antigen receptor signaling induces MAP kinase-mediated phosphorylation and degradation of the BCL-6 transcription factor. Genes Dev 1998;12:1953–1961.

[115] Duan S, Cermak L, Pagan JK, et al. FBXO11 targets BCL6 for degradation and is inactivated in diffuse large B-cell lymphomas. Nature 2012;481:90–93.

[116] Bereshchenko OR, Gu W, Dalla-Favera R. Acetylation inactivates the transcriptional repressor BCL6. Nat Genet 2002;32:606–613.

[117] Kallies A, Hasbold J, Fairfax K, et al. Initiation of plasma-cell differentiation is independent of the transcription factor Blimp-1. Immunity 2007;26:555–566.

[118] Delogu A, Schebesta A, Sun Q, Aschenbrenner K, Perlot T, Busslinger M. Gene repression by Pax5 in B cells is essential for blood cell homeostasis and is reversed in plasma cells. Immunity 2006;24:269–281.

[119] Nera KP, Kohonen P, Narvi E, et al. Loss of Pax5 promotes plasma cell differentiation. Immunity 2006;24:283–293.

[120] Shapiro-Shelef M, Lin KI, McHeyzer-Williams LJ, Liao J, McHeyzer-Williams MG, Calame K. Blimp-1 is required for the formation of immunoglobulin secreting plasma cells and pre-plasma memory B cells. Immunity 2003;19:607–620.

[121] Shaffer AL, Lin KI, Kuo TC, et al. Blimp-1 orchestrates plasma cell differentiation by extinguishing the mature B cell gene expression program. Immunity 2002;17:51–62.

[122] Mittrucker HW, Matsuyama T, Grossman A, et al. Requirement for the transcription factor LSIRF/IRF4 for mature B and T lymphocyte function. Science 1997;275:540–543.

[123] Reimold AM, Iwakoshi NN, Manis J, et al. Plasma cell differentiation requires the transcription factor XBP-1. Nature 2001;412:300–307.

[124] Shaffer AL, Shapiro-Shelef M, Iwakoshi NN, et al. XBP1, downstream of Blimp-1, expands the secretory apparatus and other organelles, and increases protein synthesis in plasma cell differentiation. Immunity 2004;21:81–93.

[125] Scheid JF, Mouquet H, Feldhahn N, et al. Broad diversity of neutralizing antibodies isolated from memory B cells in HIV-infected individuals. Nature 2009;458:636–640.

Shan Zha and Jennifer L. Crowe

3 T-cell development

3.1 Introduction

Vertebrate immune responses include both innate and adaptive immunities. Innate immunity (e.g. skin barrier, eosinophils, natural killer cells) is conserved throughout evolution, presents prior to an encounter with pathogens, generally lacks specificity, and serves to provide a rapid and usually short-lived defense against microbes. Adaptive immunity, on the other hand, is unique to vertebrates, and has the ability to launch highly specific, long-lasting protection against diverse infectious agents for the host, known as "memory". T lymphocytes (T cells) and B lymphocytes (B cells) together form the basic units of adaptive immunity. Both T cells and B cells originate from hematopoietic cells in the bone marrow, with B-cell development occurring in the bone marrow and T-cell development in the thymus. In addition to the major component of cell-mediated immunity, T cells also regulate the antibody-mediated immunity carried out primarily by B cells. A smaller subset of T cells also participates in innate immunity. In this chapter, we will discuss the development, maturation, and activation of T cells with an emphasis on the mechanism underling the development of T-cell lymphomas, which, together with natural killer (NK) cell neoplasms, accounts for ~ 12 % of all non-Hodgkin's lymphomas (NHLs) worldwide [1].

3.2 Overview

The defining feature of T cells, which distinguishes them from other lymphocytes, such as B or NK cells, is the expression of the T-cell receptor (TCR). Each T cell possesses a uniquely rearranged TCR composed of a disulfide-bound heterodimer of one α and one β chain or one δ and one γ chain. In human and mouse, approximately 95 % of peripheral T cells are αβ T cells and ~ 2–5 % are γδ T cells [2]. Based on the expression of co-receptor molecules, mature αβ T cells are further classified into $CD4^+$ and $CD8^+$ T cells, which exert different effector functions, as discussed in Section 3.3. All four TCR chains (α, β, γ and δ) are assembled from collections of germline gene segments during early T-cell development. This process ensures the diversity of the TCRs for antigen recognition. However, the gene assembly processes and the proliferation between each round of gene rearrangement are also a major source of genomic instability in developing T cells, and will be discussed in Section 3.4. While this chapter primarily focuses on the development of αβ T cells, the rearrangement of TCRδ is recognized as a major source of genomic instability and will also be discussed in Section 3.4.

In contrast to 'direct' binding of the B-cell receptor (BCR) to the corresponding antigen, the peripheral αβTCR recognizes peptides bound to specific major histocompatibility complexes (MHC), referred to as "MHC restricted", which is the basis for "self-restriction". In addition, classically adaptive αβ T cells are also self-tolerant – i.e. they do not have high affinity to respond to self-antigens. Both self-restriction and self-tolerance are established through a selection process in the thymus, which has the unique structure to support the development of T cells. In this context, while both αβ and γδ T cells originate from common progenitors in the bone marrow, T-cell development occurs primarily in the thymus [2]. The special features of the thymus and the intra-thymic development of T cells will be discussed in Section 3.5. Notably γδ TCRs are not MHC restricted [3] and have the ability to recognize antigens with unique pathogen-associated molecular patterns (PAMPs), like glycoproteins or lipids [1]. Thus, γδ T cells can be considered part of innate immunity, quickly recognizing antigens and secreting cytokines to propagate the immune response [4, 5]. The development of most innate γδ T cells also occurs in the thymus, but can also occur extra-thymically [6].

The TCR interacts degeneratively with the MHC peptide complex with low affinity. As such, each TCR can recognize more than one MHC peptide complex, and each MHC peptide complex can be recognized by more than one TCR. The interaction between αβTCR and MHC peptide complex is also highly dynamic and is modulated by a large number of co-stimulatory molecules. The cytoplasmic portion of the TCR is also extremely short and insufficient for signal transduction. Therefore, the variable αβTCR is always associated with the invariable cluster of differentiation 3 (CD3) dimeric signaling modules. Upon engagement with the specific MHC peptide complex, the signals received by the TCR are transmitted through the co-expressed CD3s, leading to the activation of an intracellular signaling cascade [7]. Activating mutations or loss of the negative regulatory factors along the TCR signaling cascade leads to antigen independent proliferation of T cells, underlying a subset of T-cell lymphomas. Therefore, in the last section (Section 3.6) we will briefly discuss the CD3-TCR complex, the co-stimulatory signals and signal transduction downstream of the TCR activation.

3.3 αβT-cell repertoire

3.3.1 Antigen presentation and MHC molecules

Peripheral TCRs only recognize peptides in complex with the MHC. MHC contributes the majority of the amino acids on the MHC peptide complex surface that are recognized by TCRs. In both human and mouse, there are two kinds of MHC molecules – class I and class II [6]. The MHC class I molecule is expressed in almost all nucleated cells, which present proteasome-generated, cytosolic peptides from within the cell on the cell surface. The MHC class I molecule is a noncovalent heterodimer consisting of one polymorphic α chain encoded by the HLA locus and one invariable β chain encoded by the nonpolymorphic β2-microglobulin (*B2M*) gene. The peptide binding groove of the MHC class I molecule is closed on both ends and can only accommodate peptides that are 7–9 amino acids long. Cytosolic peptides are first imported into the endoplasmic reticulum (ER), where they are loaded onto the MHC class I molecule, stabilizing the MHC complex. The ER uptake of cytosolic peptides is mediated by a complex consisting of TAP1/2, tapasin, calreticulin, calnexin, and Erp57 [17]. Defects in any of these proteins compromises MHC class I peptide complex loading and have been utilized by cancer cells or virus-infected cells to escape T-cell mediated surveillance [17, 18].

In contrast to the ubiquitous expression of MHC class I molecules, the MHC class I molecule is only expressed in antigen presenting cells (APCs), including dendritic cells (DCs), mononuclear phagocytes, thymic epithelial cells, a subset of endothelial cells and B cells. MHC class II is a heterodimer of two polymorphic chains, α and β, which are both encoded by the HLA loci [17]. The peptide binding groove of the MHC class II molecule is open at one end and can accommodate peptides as long as 15–25 amino acids. The antigens presented by class II molecules are derived from extracellular proteins taken up by either phagocytosis or endocytosis (in contrast to the cytosolic peptides presented by MHC class I). The class II MHC is transported from the ER to the Golgi, where it is fused with late endosomes containing endocytosed, degraded proteins and peptide fragments.

3.3.2 Peripheral αβTCR repertoires

Based on the expression of co-receptors, peripheral αβ T cells can be divided into two large categories with different effector functions. $CD8^+$ T cells are cytotoxic T cells (CTLc) and $CD4^+$ T cells are helper T cells (T_H) (Fig. 3.1). The TCR of CTLs interacts directly with peptides bound to MHC class I molecules and the CD8 co-receptor interacts with the MHC class I molecule in a region independent from antigen-binding, stabilizing the interaction between the TCR and the MHC class I peptide complex. High affinity TCR-MHC class I peptide complex and CD8-MHC class I interaction, together with proper co-stimulatory signals, such as CD28-B7 (also called CD80, or CD86 for B7-2), activate CTLs. Activated CTLs release cytotoxins, including perforin, granzymes and granulysin, directly into the cytoplasm of targeted cells, or turn on the expression of the FAS ligand (FASL or CD95L), which in turn binds the FAS (CD95) molecule expressed on tumor cells, virus-infected cells or transplanted cells and eventually induces apoptosis in the targeted cells [8]. TCRs on the surface of $CD4^+$ T_H cells, on the other hand, recognize peptides of antigens that are associated with class II MHC molecules on the surface of antigen presenting cells (APCs). Activated T_H cells express cell surface molecules, which engage with other immune cells directly to modulate their functions, such as CD40 ligand and CD40 on B cells, and/or release cytokines, which modulate the function and maturation of both B cells and CTLs. Depending on their specific transcriptional programs and the type of cytokines secreted, T_H cells are further classified as T_H1, T_H2, T_H17 and T_{Reg}. For example, T_H1 differentiation is

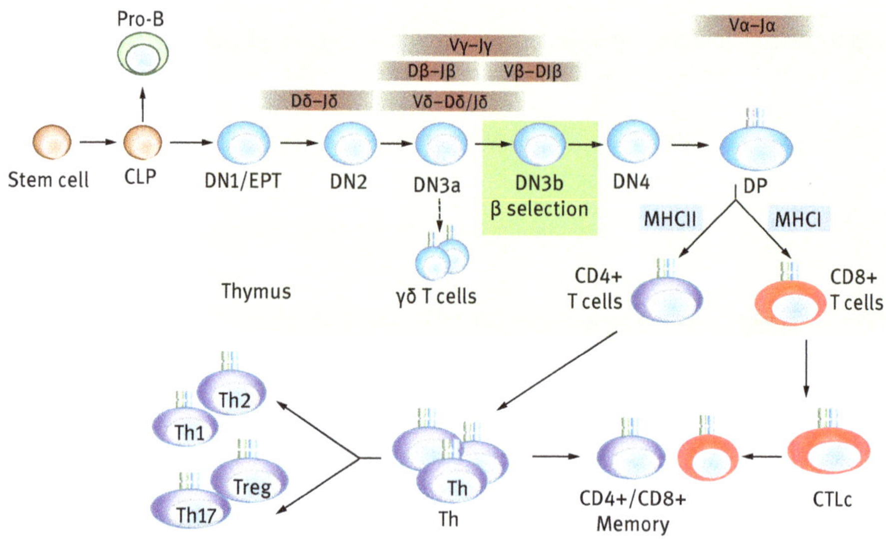

Fig. 3.1: **Classification and development of T cells.** T cells, like other hematopoietic cells, are originally derived from hematopoietic stem cells, which give rise to common lymphoid progenitors (CLPs). CD4-CD8-double negative (DN) thymocytes enter the thymus and retain the potential to differentiate to myeloid and natural killer (NK) cells. Notch expression starts in the DN1 cells and increases with the engagement of delta ligand on the thymic epithelial cells. During the late DN2 stage, BCL11b expression occurs and correlates with commitment to the T-cell lineage. Among the four TCR genes, TCRδ rearrangement starts first in the late DN1 and early DN2 stage with Dδ-Jδ first, followed by Vδ-DJδ. TCRβ rearrangement starts in DN3 cells and continues until β-selection in the DN3b stage. TCRγ rearrangement occurs at the same time as TCRβ rearrangement. After passing the β-selection checkpoint, T-cells enter rapid proliferation in the DN4 cell stage. At the transition to the $CD4^+CD8^+$ double positive (DP) T-cell stage, DN T-cells turn on either CD8 or CD4 expression first, resulting in immature single positive (ISP) cells. TCRα rearrangement occurs in the DP T-cells. Engagement with either MHC I or MHC II molecules determines the differentiation to $CD8^+$ or $CD4^+$ T cells, respectively, and the departure from the thymus to the peripheral blood. In the periphery, $CD4^+$ cells further differentiate to different kinds of T helper (T_H) cells, classified by the type of cytokines they produce. $CD8^+$ T-cells further develop into cytotoxic T-cells. A subset of T_H or cytotoxic T-cells becomes memory T-cells. NK cells can be generated from the CLP in the bone marrow and a subset of NK-like cytotoxic T-cells can also be induced from TCR rearranged T-cells.

driven by IL-12, which drives expression of STAT4 and T-bet transcriptional factors [9]. Functional T_H1 cells secrete IFN-gamma to recruit and activate macrophages, $CD8^+$ CTLs and other cells that protect against ‘intracellular’ pathogens [10]. T_H2 differentiation, on the other hand, is driven by IL-4 through STAT6 and GATA3, which leads to the secretion of IL-4, IL-5, IL-9, IL-10 and IL-13 [11]. These cytokines recruit eosinophils, basophils, neutrophils, mast cells and B cells to fight against ‘extracellular’ parasites including helminths [12]. IL-10 also suppresses T_H1 differentiation and reinforces T_H2 differentiation together with the IL-4 autocrine loop [13]. Furthermore, a subset of

$CD4^+$ T cells turn on expression of FoxP3 and becomes regulatory T cells (T_{Reg}, also known as suppressive T cells), which are responsible for turning off the immune response [14]. Defects in T_{Reg} cells underlie several autoimmune diseases [15]. Finally, a small fraction of both $CD4^+$ T_H cells and $CD8^+$ CTL cells become antigen-specific $CD45RO^+$ memory T cells that persist after the infection has been cleared [16].

3.4 The T-cell receptor

3.4.1 The variable chains of the T-cell receptor

The ability of T cells to specifically interact with various MHC bound peptides requires great diversity in TCR gene products. Unlike conventional proteins, which are encoded by a single germline gene, the variable regions of the TCR gene and the related immunoglobulin gene (Ig, also called B-cell receptor, BCR) are encoded by a large set of noncontinuous gene segments. In developing lymphocytes, somatic assembly of these germline gene segments through a process known as V(D)J recombination [19] forms the variable region exon, which is then spliced to the constant region exons to form the TCR or Ig gene products that are unique to each T or B cell, respectively.

The human genome encodes four TCR gene loci: α(14q11), β(7q34), γ(7p1), or δ(14q11); and three immunoglobulin loci: Ig heavy (IgH, 14q32), Igκ(2p11) and Igλ(22q11). The TCRβ, TCRδ and IgH genes contain clusters of variable (V), diversity (D) and joining (J) segments, while the TCRα, TCRγ, Igκ and Igλ only contain clusters of V and J gene segments (Fig. 3.2). The variable region exons assembled through V(D)J recombination in the TCRβ and α loci each have three hypervariable complementarity determining regions (CDRs), conferring antigen specificity. Structural analyses suggest that the CDR3 of the TCRα/β chain, which is formed by the DJ junctions in the TCRβ chain and by the J region in the TCRα chain, directly contacts the peptide, while the CDR1 and CDR2 mainly contact MHC molecules. In addition, the variable region exon, TCRβ, TCRγ as well as Igκ and Igλ loci also contain more than one set of constant region exons (Fig. 3.2).

3.4.2 The order and regulation of V(D)J recombination

V(D)J recombination is a highly regulated process with ordered rearrangement coupled with feedback regulation and selection at each step. In each individual T cell, the four TCR genes are rearranged and expressed in the order of δ, γ, β and α. As such, TCRβ genes always rearrange before TCRα genes. Within the TCRβ gene, Dβ-Jβ joining always occurs before Vβ to DβJβ joining. This ordered rearrangement is controlled by gene specific enhancer and insulator elements that regulate stage-specific transcription of germline V, D and J gene segments (Fig. 3.2). Active transcription is required for

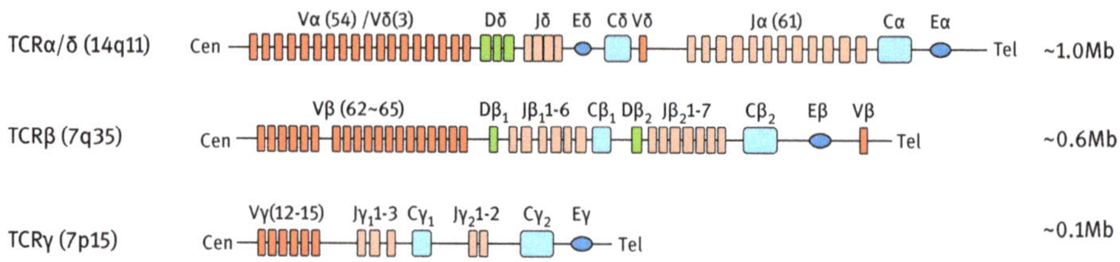

Fig. 3.2: **Diagram of the human TCR locus.** The chromosomal location and the centromere (cen) and telomere (tel) orientation of the human TCRα/δ, β, γ loci are shown in this figure. The size of the locus is marked at the end. The number in brackets (for example, 54 in Vα) indicates the number of different V or J segments in a cluster. We note that not all segments are functional: some contain internal stop codons or nonfunctional RSS elements rendering them pseudo-segments. While there are four types of T-cell receptor chains (α, β, δ, γ), there are only three loci, because the TCRδ locus is interspersed in the TCRα locus. The human TCRα/δ locus is complex and includes 54 Vα gene segments, of which 45 are functional. This locus also contains 61 Jα segments, the majority of which are functional. The human TCRβ was the first TCR locus to be fully mapped and sequenced. It spans ~ 0.6 MB and contains 62–65 V genes, of which 39–41 are functional. The TCRγ locus is the simplest one: it spans ~ 0.1 MB and contains 12–15 V genes, of which only 4–6 are functional. While the majority of V-D-J rearrangements are deletional events within the TCR loci, there is one Vβ and one Vδ segment located at the 3′ end of the corresponding constant (C) region, which require inversional V(D)J recombination for TCR assembly. The developmental stage specific and allele specific rearrangement of the TCR loci is in part regulated by the enhancer (E) located at the 3′ end of each loci.

the initiation of V(D)J recombination and is a major component of "accessibility" control. Similar to TCRβ, D_H-J_H rearrangement also precedes the V_H-DJ_H rearrangement at the IgH locus in developing B cells [20]. The TCRδ locus is a notable exception for this "ordered" rule, in that Vδ-Jδ, Dδ-Jδ and Vδ-Dδ rearrangements at the TCRδ locus occur without apparent order [20, 21].

The accessibility of the TCR locus is also subject to feedback regulation. Successful rearrangement and expression of the in-frame TCRβ gene, paired with a pre-TCRα chain that is constantly expressed in immature T cells, suppresses the rearrangement of the second TCRβ allele in a mechanism known as allelic exclusion. This mechanism ensures that V to DJ or V-J recombination occurs only one allele at a time, making certain that each T cell expresses only one TCRβ protein. The same rule also governs the rearrangement of the IgH, Igκ and Igλ loci in developing B cells. There are exceptions, however. The D-J rearrangement is not allelically excluded, and V-J rearrangement in the TCRα locus is also not allelically excluded. This lack of regulation via allelic exclusion in the TCRα locus allows for concurrent rearrangement at both alleles, resulting in dual expression of both TCRα alleles in some cells [20].

Finally, nonproductive recombination can shut down further recombination in that allele. For instance, nonproductive Vβ to DβJβ recombination destroys the rearranged DβJβ cassette and permanently disables the rearranging TCRβ allele. Similarly, a nonproductive IgH allele cannot be further rearranged. In this context, the TCRα locus contains a large number of Jα segments (50), which facilitates multiple successive

Vα to Jα rearrangements on each allele. The physical organization of the TCRα/δ locus also regulates the commitment of the two T-cell lineages. In both human and mouse, the TCRδ gene resides between the Vα and Jα gene segments of the TCRα locus. As such, successful rearrangement of Vα to Jα would delete the embedded TCRδ gene, resulting in permanent commitment to the αβTCR lineage.

3.4.3 Molecular mechanism of V(D)J recombination

At the molecular level, V(D)J recombination is initiated by the lymphocyte-specific recombination activating genes (RAG) RAG-1 and RAG-2; these molecules form a complex (RAG) that recognizes conserved recombination signal sequences (RSSs) flanking the germline V, D, and J segments, brings them together (forming a synapse) and then introduces double strand breaks (DSBs) between the participating gene segments and their flanking RSSs [17]. RSSs are composed of conserved heptamer and nanomer sequences with a spacer of 12 or 23 nucleotides between them. RSSs with a 23 nucleotide spacer can only be rearranged next to an RSS with a 12 nucleotide spacer, referred to as the 12–23 rule, which is governed by the structure of the RAG1/2 complex [22]. RAG contains a C-terminal motif that interacts with histone markers associated with actively transcribed regions, ensuring that cleavage only occurs in actively transcribed V, D and J gene segments [23]. At the biochemistry level, the RAG cleavage reaction generates two kinds of ends: two blunt 5′-phosphorylated signal ends (SEs) and two covalently-sealed hairpin coding ends (CE) [24, 25]. In a second reaction phase, the two SEs are precisely joined to form a signal joint (SJ); while the two CEs are modified and joined via a process that involves hairpin opening, potential loss of nucleotides, and potential *de novo* addition of palindromic (“P”) elements or nontemplated nucleotides (N regions) to form the coding joint (CJ) [22, 25, 26]. Other than N region addition, which is carried out by terminal deoxynucleotidyl transferase (TdT), the joining phase of V(D)J recombination is carried out by the ubiquitously expressed nonhomologous end-joining (NHEJ) pathway of DNA double stand break repair [27]. Specifically, Ku70 and Ku80 recognize and form a dimer on SEs or CEs. The Ku dimer facilitates repair by protein recruitment: it recruits DNA-PKcs and Artemis to open the hairpin of the CEs as well as the ligation complex of Ligase4, Xrcc4 and likely XLF to the break, culminating in break resolution and the formation of SJs or CJs [26]. When the hairpin is not opened at the apex, polymerases μ or λ polish the ends by adding nucleotides complementary to the open hairpin, resulting in palindromic (P) elements at the junction. TdT can also add N-nucleotides randomly to the ends during TCRβ and IgH rearrangement [25]. Defects in the initiation (e.g. RAG) or the completion of V(D)J recombination (e.g. NHEJ) abrogate lymphocyte development and lead to severe combined immunodeficiency [28]. Together, the combination of V, D and J segments within the α and β TCRs and the end processing events during the repair phase of V(D)J recombination result in an estimated $> 10^{15}$ combinatorial diversity of the two TCR chains [29].

3.4.4 Tumorigenic potential of TCR recombination

Genomic rearrangements involving DNA double strand break intermediates could be dangerous in quickly proliferating progenitor cells. Translocations between the TCR locus and proto-oncogenes, which put the strong promoter/enhancer of the TCR loci in proximity of proto-oncogenes and lead to deregulation of proto-oncogene expression, are characteristic of immature T-cell malignancies [30]. The existence of cryptic RSSs, which are similar in sequence to bona fide RSSs but are outside of the TCR locus, contributes to a significant number of characteristic chromosomal translocations in T-cell acute lymphoblastic leukemias (T-ALLs) [31], such as the *TCR-LMO2* translocations. Defects in DNA repair, as in the case of genomic instability syndromes (e.g. ATM, NBS) would also increase the number of DSBs and the frequency of misrepair, eventually leading to a greatly increased risk for lymphomas and leukemia [32, 33].

Among the four TCR loci, the TCRα/δ locus is most frequently involved in oncogenic chromosomal translocations in immature T cells. This is in part due to the special features of the TCRα/δ locus: (i) the TCRα/δ locus is the only TCR locus that actively rearranges in both double negative (TCRδ) and double positive (TCRα) T cells; (ii) both alleles of the TCRα/δ can rearrange at the same time (this locus is not allelically excluded); (iii) TCRδ rearrangement is not ordered: V-J can directly join and V-J joining can occur before DJ joining; (iv) TCRα permits successful rearrangements at the same locus. As a result, chromosomal translocations involving the TCRα/δ loci contribute to over 70 % of all oncogenic translocations in T-ALL. Among them, the first rearranged TCRδ locus is involved in over 90 % of all TAL1 translocations in T-ALL [31].

3.5 Thymus and early T-cell development

3.5.1 The thymus

The thymus (Fig. 3.3) is a primary lymphoid organ found in all vertebrates, which emerged evolutionarily in parallel with V(D)J recombination [34]. During early embryonic development, endodermal cells at the anterior foregut migrate into neural crest-derived mesenchymal cells, where expression of transcription factor Foxn1 within epithelial cells dictates the thymus anlage [35]. While the differentiation of cortical and medullary epithelium occurs independently of the influx of hematopoietic components, including thymocyte progenitors [36], the arrival of the hematopoietic components (macrophages, dendritic cells and T-cell progenitors) dramatically shapes the organization of the primary thymus anlage through direct ligand and receptor interactions. For example, tumor necrosis factor receptor (TNFR) family members, such as the lymphotoxin-β receptor (LTβR) on thymocytes, interact with ligands on medullary thymic epithelial cells (mTECs), dictating mTEC cellularity and architecture. Further,

Notch-1, expressed on pro- and pre-T cells, interacts with Notch ligand on mTECs, inducing profound changes in the microenvironment [37].

Since the thymus cannot support hematopoietic stem cell self-renewal, it requires a continuous migration of early T-cell precursors (ETP) from the bone marrow for thymic lymphopoiesis. ETPs enter the thymus at the post-capillary venules near the cortical-medullary junction driven by a combination of thymic-specific adhesion molecules and chemoattractants (Fig. 3.3). Chemoattractants act as guides, instructing the ETP to move towards high concentrations of various chemokines (CXCL12, CCL21 and CCL25) via respective ETP ligands (CXCR4, CCR7 and CCR9). Once near the thymus, adhesion molecules work to retain the ETPs at the thymic entrance. Namely, P-selectin expressed on TECs interacts with CD44 and P-selectin receptors on the T-cell precursors [38–40], bringing them in closer proximity to TECs, which they then engage via the ICAM-1 and VCAM-1 adhesion molecules, tethering the precursors securely to the entrance of the thymus, at the cortical-medullary junction. As the ETPs progress and develop through the thymus, they will be further guided by strength of TCR signaling and direct interactions between thymocytes and TECs. Together, these signals guide the T cells through a well-organized journey within the thymus, tightly coupled to their developmental stages [41, 42] (Fig. 3.3).

3.5.2 Early T-cell development in the thymus

Thymocytes can be divided into four groups based on the surface expression of the CD4 and CD8 co-receptors (Fig. 3.1): (i) $CD4^-CD8^-$ double negative (DN), (ii) $CD4^+CD8^+$ double positive (DP) cells, (iii) and (iv) $CD4^+CD8^-$ or $CD4^-CD8^+$ single positive (SP) cells. DN cells are the most immature population and SP cells are the mature T cells that are ready to leave the thymus. In addition, DN cells can be further divided into four populations (Fig. 3.1), DN 1–4, based on surface expression of CD44 and CD25 (also called IL-2Rα). ETPs from the bone marrow enter the thymus as $CD25^-CD44^+$ DN I cells (Fig. 3.3), still expressing the stem cell marker c-Kit [41, 43]. DN I (ETPs) are multipotent, possessing T, myeloid, and NK cell potential. Engagement of NOTCH 1 on DN I cells via Delta ligand 4 (DL4) represses the differentiation to B, myeloid and NK cells *in vivo* [44–47]. In addition to NOTCH, IL-7 is secreted by the thymic microenvironment to engage the IL-7 receptors (IL-7R) on DN1 cells, propagating signaling that results in the rapid expansion and proliferation of the DN1 population [48]. Neither IL-7 signaling nor NOTCH signaling are sufficient for the proliferation of DN1 cells; both are required. In fact, it has been shown that Notch signaling helps maintain IL-7 signaling, by inducing expression of IL-7R [49]. Additionally, IL-7 signaling prevents apoptosis in the developing and rapidly expanding T cell by inducing BCL2 expression [50].

DN1 cells migrate to the subcapsular zone of the cortex [41] via an interaction with the chemokine CXCL12 [51], where they encounter thymic epithelial and fibroblast T cells (Fig. 3.3). These stromal cells induce the downregulation of CD44 expression

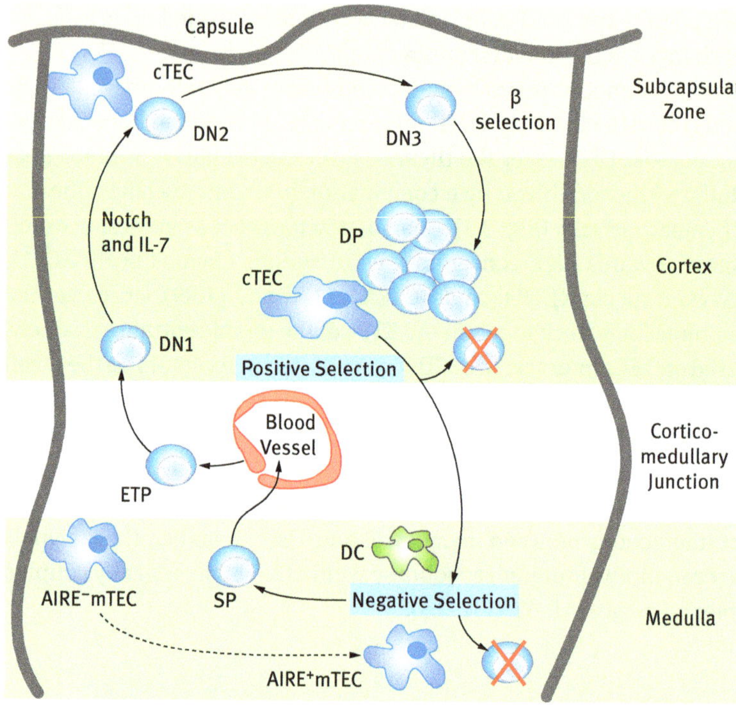

Fig. 3.3: **Architecture of the thymus.** The thymus provides an important environment to shape T-cell development. Early T-cell progenitors (ETPs) arrive initially in the cortico-medullary junction via a blood vessel, after exiting the bone marrow. These early double negative (DN) cells migrate to the cortex, where they encounter Notch signaling, which represses the potential for differentiation to non-T-cell lineages, and IL-7, which is secreted by cortical thymic epithelial cells (cTECs) and induces DN1 cell proliferation. Next, DN1 cells migrate to the subcapsular zone of the cortex, where they further engage with cTECS and the thymic microenvironment to form DN2 cells. At this stage, TCR recombination begins at the TCRδ, γ and β loci. DN2 cells transition into DN3 cells, where they undergo β-selection to eliminate nonfunctional TCRβ chains. After β-selection, DN3 cells mature to DN4 cells, which proliferate in the cortex and induce expression of both CD4 and CD8 co-receptors, to become double positive (DP) cells. DP cells undergo positive selection in the cortico-medullary junction, followed by negative selection in the medulla, fueled by dendritic cells (DCs) and mTECs expressing the AIRE transcription factor. Finally, selected T cells downregulate CD4 or CD8 to become single positive for either co-receptor, and exit the thymus.

and induce NOTCH-mediated CD25 (IL2Rα) expression, forming $cKit^+CD44^+CD25^+$ DN2 cells [43]. Notch signaling also induces Bcl11b expression, which works with NOTCH to restrict DN cells to the T-cell lineage [45, 52, 53, 55]. NOTCH and IL-7 signaling continues to be important in DN2 cells, during the initiation of V(D)J recombination at the TCRδ, γ and β loci [56, 57]. IL-7 signaling activates STAT5, which binds to the TCRγ locus, promoting its accessibility for recombination [58–60]. The type of TCR expressed by the T cell (αβ versus γδ) is solidified at the transition from the DN2 to the $cKit^-CD44^-CD25^+$ DN3 stage, making an irreversible commitment to the T-cell lineage. In-frame rearrangement and expression of the δ and γ chain is essential, but not sufficient, for γ/δ T-cell lineage commitment. TCR signal strength plays an important role in lineage commitment, with weaker or stronger TCR signals favoring αβ or γδ lineage fates, respectively [61].

γδ T cells mature in a NOTCH-independent manner, and undergo thymic selection. αβ T cells, however, undergo further recombination events and NOTCH-dependent maturation before they can undergo thymic selection. Prior to committing to the rearrangement of the TCRα chain, the TCRβ chain is first tested in a process called β-selection. The rearranged TCRβ chain is paired with the pre-TCRα, which is expressed from a nonrearranged locus, in DN3 cells to form the pre-TCR [62, 63], which mediates β-selection. During this process, T cells with successfully rearranged, functional TCRβ chains will express the pre-TCR on their surface and engage in signal transduction via the CD3 complex, resulting in cell proliferation. Any T cells without successful TCRβ rearrangement or surface expression of the pre-TCR will be eliminated by apoptosis [64]. Engagement of NOTCH1 via the Delta ligand is critical for DN3 cells to maintain cellular metabolism at a level that supports survival after pre-TCR mediated induction of proliferation [65] and to protect against apoptosis [66]. Successful β selection downregulates NOTCH expression [67] and, together with pre-TCR signaling, leads to maturation to the $CD44^-CD25^-$ DN4 stage. Proliferation is important in DN4 cells to create a large, clonal pool of T cells with the same β chain and the potential to form a diverse pool of αβ TCRs [64]. After this clonal expansion event, pre-TCR signaling induces expression of both CD4 and CD8 co-receptors and the RAG endonuclease [66], forming $CD4^+CD8^+$ double positive (DP) αβ T cells, and initiating TCRα rearrangement, respectively. At this stage, the αβTCR pool is diverse and has the potential to interact with a range of different MHC and MHC peptide complexes, including self-antigens.

NOTCH plays a very important role in DN T-cell development and differentiation, and its expression must be tightly regulated. Over 60 % of T-cell lymphoblastic leukemia has mutations [68], translocations [69], amplifications or blocks in NOTCH-1 turnover leading to constitutively active NOTCH signaling [70]. Expression of the intracellular portion of NOTCH itself is sufficient to drive T-ALL in mouse models [71, 72]. NOTCH is normally downregulated immediately after β-selection [67]. When it remains active, T cells become blocked at the DP stage [73], further promoting oncogenic transformation of immature T cells.

3.5.3 Positive selection

Thymocytes with successful *TCRα* rearrangement and surface expression of TCRα/β undergo positive and negative selection to ensure the development of only T cells with self-restricted and self-tolerant TCRs, i.e. a TCR that can recognize self-MHC bound to foreign peptides. Positive selection ensures that only T cells whose TCR binds “self”

MHCs can survive, while negative selection ensures the T cells do not react too strongly against self-antigen. Positive selection occurs first in the cortex of the thymus. Only T cells with TCRs recognizing self-MHC loaded with peptides on cortical thymic epithelial cells (cTECs) can survive (Fig. 3.3). T cells with TCRs that fail to bind self-MHC will be allowed to rearrange the TCRα locus again or to die. Antigen presentation by cTECs is the first step of positive selection. cTECs support a higher rate of macro-autophagy to ensure the diversity of peptides for antigen presentation [74, 75]. Both the proteasome, which generates cytosolic peptides that are loaded onto class I MHC molecules, and the lysosome, which generates proteolytic fragments from extracellular peptides that are loaded onto class II MHC molecules, play important roles in positive selection. A specialized thymo-proteasome, with a catalytic subunit specific to the thymus [76, 77], and the thymic lysosome contain two proteases only found within cTECs. Cathepsin L and thymus specific serine protease (TSSP) are responsible for preparing peptides for presentation on class I and II MHC molecules, respectively [77–79]. Self-peptides derived from auto-antigens are also present on the MHC molecules of the cTECs. Any DP T cell that can bind the peptides loaded on self-MHC complexes of the cTEC will receive pro-survival signals. Any T cell that cannot recognize the self-MHC complex, whether because of a lack of rearranged TCR or because of an unreactive TCR, will not receive survival signals and will therefore die of neglect [77, 80].

3.5.4 CD4 vs CD8 lineage commitment

At the time of positive selection, DP T cells express both CD4 and CD8 co-receptors that are critical for positive selection. Eventually, thymocytes need to downregulate the expression of either CD4 or CD8, to ultimately become a mature SP T cell. In the end, αβ thymocytes with TCRs recognizing a class I MHC molecule express CD8 only, while those with a TCR recognizing a class II MHC molecule become $CD4^+$αβ T cells. Several models have been proposed and disproved for the $CD4^+$ vs $CD8^+$ cell fate determination. The stochastic model is unlikely, since the efficiency of SP cell selection *in vivo* is far greater than what would be predicted from complete stochastic action [81, 82]. The "instructive model", which dictates that "correct" TCR-MHC and co-receptor-MHC pairing would suppress the expression of the "wrong" co-receptor [83, 84], is also unlikely based on studies on ITAM deletion cells [85]. Later, the strength [86] and the duration [87] of the TCR signal were also proposed. Stronger CD4-MHCII signaling promotes positive selection without affecting the CD4/CD8 lineage commitment, providing a simple explanation for a higher number of $CD4^+$ over $CD8^+$ T cells (~ 2:1). The most recent experiments suggest that DP T cells transition through an uncommitted $CD4^+CD8^{low}$ stage after positive selection, which is the progenitor of both $CD4^+$ and $CD8^+$ SP T cells [88]. This finding gave rise to the "kinetic model", which suggests that the purpose of downregulating CD8 is to assess the functionality of TCR signaling without the help of that co-receptor. Persistence of signaling commits the cells to the

$CD4^+$ lineage, while transient loss of signaling commits the cell to the $CD8^+$ lineage. TCR signaling suppresses IL-7 signaling, so a cessation of TCR signaling upon CD8 downregulation induces IL-7 signaling, which then mediates the upregulation of CD8 and the downregulation of CD4, to commit the T cell to the $CD8^+$ fate [88–90]. At this stage, the positively selected thymocyte is singly positive for either CD4 or CD8, and has a unique αβ TCR that is self-restricted, but not yet self-tolerant.

3.5.5 Negative selection and exit from the thymus

Chemokine signaling by CCR7 induces the migration of SP T cells cells from the thymic cortex to the medulla, where they undergo negative selection to establish central tolerance [91, 92]. Medullary thymic epithelial cells (mTECs) and antigen presenting dendritic cells express MHC molecules loaded with self-peptide on their surface, along with costimulatory molecules: these surface presentations are required to eliminate T cells with self-reactive TCRs [93–96]. Additionally, both mTECs and DCs present tissue-specific antigens (TSA), like insulin [97, 98], which is regulated by the transcription factor AIRE [97, 99] (Fig. 3.3). The presentation of antigens from the periphery exposes T cells to as diverse a pool of "self" antigen as possible, so that mature T cells will not react to self-antigens in the periphery [100]. Strong binding of the TCR to either the MHC complex alone or the TSA-peptide-MHC complex on the mTECs or DCs triggers apoptosis in T cells. These signals are mediated by the FAS ligand, via the TNF apoptotic pathway [101]. T cells expressing TCRs that weakly bind self-MHC molecules will not receive apoptotic signals. Negative selection prevents autoimmunity, i.e. the initiation of an immune response against self-tissue. Given the stochastic nature of TCR rearrangements, positive and negative selection are essential to ensure a robust, MHC restricted T-cell response and to eliminate self-reactive T cells. Indeed, in the absence of AIRE, the transcription factor required for TSA presentation in mTECs, severe autoimmunity occurs [99]. Finally, T cells that survive both positive and negative selection begin to express sphingosine-1-phosphate receptor 1 (S1P1), to allow the mature cell to leave the thymus. The SIP1 ligand is highly concentrated in the blood, and attracts the SIP1 receptor, promoting the mature T cell to leave the thymus and enter the circulation [102, 103].

3.6 TCR signaling – T-cell activation

3.6.1 TCR signaling complex

Tonic TCR signaling provides essential survival cues that guide the development of immature thymocytes and that are also important for the survival and the homeostatic proliferation of naive T cells in the periphery. Tonic TCR signals are generated by low-affinity TCR/self-peptide/MHC complex interactions. When T cells encounter high-affinity MHC peptide complexes on APCs in the periphery, T cells will stop, bind the high-affinity MHC peptide complex, and induce a signaling cascade resulting in its proliferation and differentiation. This series of actions originates from TCR engagement at the immunological synapse. In addition, the TCR complex and its co-factors are also important for sub-lineage differentiation, i.e. T_H cells, memory and regulatory T-cell differentiation, and is responsible for terminating the immune response.

The TCR complex (Fig. 3.4) is formed by the paired αβTCR together with the co-expressed CD3 molecule and a co-receptor (CD4 or CD8), and is modulated by co-stimulatory factors, e.g. B7. CD3 is an Ig-like protein that directly interacts with the TCR through the transmembrane domain [104–106]. The extracellular domains of CD3 are dimers formed by combinations of five possible polypeptide chains, resulting in ζζ (predominant), ζη, γε, or δε dimers [107–110, 110]. The cytoplasmic domain of CD3 contains the immunoreceptor tyrosine-based activation motifs (ITAMS). There is one ITAM in each of the γ, δ, and ε chains, and 3 ITAMS in each ζ chain [110]. In addition to CD3, the two most well-known co-receptors on T cells are CD4, which binds to MHC class II molecules, and CD8, which binds to MHC class I molecules. Like CD3 and the TCR itself, CD4 and CD8 have Ig-like structures. The long cytoplasmic tails of CD4 or CD8 facilitate signal transduction and propagation [111] by their association with the lymphocyte specific SRC family tyrosine kinase, p56-LCK [112, 113].

3.6.2 Activation and survival of naive T cells

After SP T cells exit the thymus, they circulate through the blood and lymph nodes until they encounter an antigen. These circulating T cells are known as naive T cells, existing in the resting, G0 cell cycle stage. They are characterized by condensed chromatin, little cytoplasm, and low levels of transcription. Contact of the TCR with the MHC peptide complex results in the activation of these naive cells into either effector or memory T cells.

Upon TCR engagement, the first activated molecules are the SRC family kinase member LCK (also known as p56-LCK), which is associated with the CD4/CD8 co-receptors, and FYN (also known as p59-FYN), which is associated with CD3. MHC-TCR and MHC-CD4/8 interactions bring LCK and FYN in close proximity to CD3, leading to phosphorylation of the ITAM on the CD3ζ chain, and enabling the recruitment of ζ chain associated protein kinase of 70 kDa (ZAP70) [114, 115]. ZAP70 is then activated by LCK phosphorylation. Activated Zap70 phosphorylates four tyrosines on the linker for activation of T cells (LAT), which then recruits other signaling molecules to achieve three major signaling outcomes – increased intracellular Ca^{2+} to regulate adhesion and mobility, activation of mitogen-activated protein kinase (MAPK) and the nuclear factor-κB (NF-κB) pathway for proliferation [116, 117], and activation of the mTOR pathway for differentiation and metabolic changes (Fig. 3.4) [117–120].

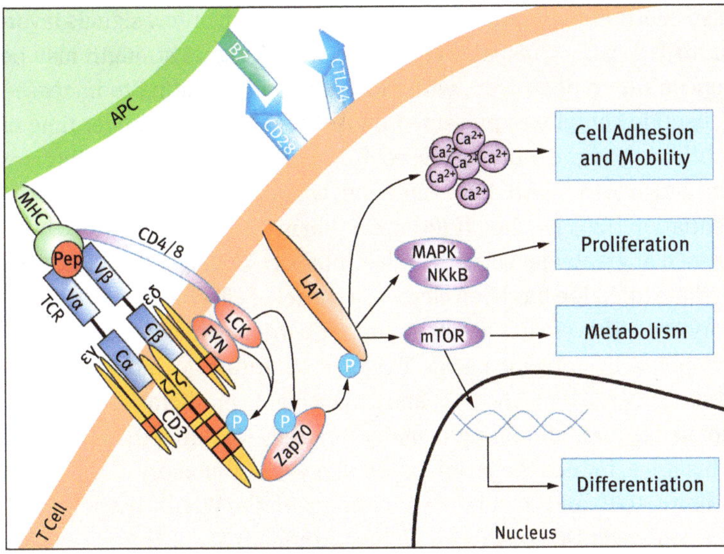

Fig. 3.4: **Interaction between peptide-MHC/TCR complexes and TCR signaling.** Antigen presenting cells (APCs) present antigen peptides (Pep) via the major histocompatibility complex (MHC), upon interaction with foreign bodies. The T cell recognizes a presenting APC via its T-cell receptor (TCR), which has specific affinity for the particular antigen peptide and MHC molecule. TCR binding to the MHC peptide complex initiates a signaling cascade mediated by several proteins. Co-receptors CD4 or CD8, for instance, bind the MHC and activate the Src family kinase LCK. Furthermore, TCR signaling is propagated through the CD3 co-receptor (yellow), resulting in the phosphorylation of the immunoreceptor tyrosine-based activation motifs (ITAMS, red boxes) of CD3 by FYN. These phosphorylation events recruit ζ chain associated protein kinase of 70 kDa (ZAP70), which is phosphorylated by LCK, resulting in its activation. Activated Zap70 phosphorylates four tyrosines on the linker for activation of T cells (LAT), which in turn has effects on several cellular changes, including: (i) an increase in intracellular calcium to regulate cell adhesion and mobility; (ii) activation of the MAPK and NF-κB pathway to induce proliferation; (iii) activation of the mTOR pathway to regulate cell metabolism and induce further T-cell differentiation. This interaction is further mediated by B7, a ligand expressed by APCs that interacts with the CD28 receptor on T cells, resulting in IL-2 expression. Conversely, TCR signaling can be shut down via B7/CTLA4 interaction: upon B7 stimulation, CTLA4 will dephosphorylate CD3, shutting down TCR signaling.

3.6.3 Activated T-cell differentiation

Activated T cells can differentiate into a variety of effector cells. This differentiation decision is orchestrated by both the signaling strength of the TCR and cytokine signaling. Cytokines are produced by other immune cells at the time of antigen presentation, to control the response of the T cells. Specific cytokine combinations induce differentiation to specific effector cell types, such as IL-12 and IFNγ for T_H1, and IL-4, IL-2, IL-7 and TSLP (thymic stromal lymphopoietin) for T_H2. Cytokines control signal transducer and activator (STAT) protein activation via post-transcriptional modification, which in turn regulates expression of transcription factors key to T-cell differentiation [121]. TCR signal intensity also contributes to differentiation fate, with high or low signals favoring differentiation to T_H1 or T_H2, respectively [122, 123]. TCR signal strength may be affected by the concentration of antigen, with more antigen translating to increased numbers of MHC peptide complexes presented on APCs, as well as a longer time of presentation, contributing to generally increased TCR signaling. Strong signaling, for instance, has been shown to induce IL-12R expression, which pushes the cell down the path of T_H1 differentiation [124]. Further, IRF4 expression in $CD8^+$ T cells has also been shown to be expressed at greater levels with higher intensity TCR signaling [125, 126].

After an immediate infection has been cleared, a subset of effector cells will differentiate into memory cells. This creates a very important barrier against reinfection, by ensuring a swift response by antigen specific T cells in the future. Memory cells also have self-renewal capacities, a long lifespan, and can be highly proliferative [127, 128]. Cytokines are responsible for maintaining memory T cells by signaling their survival. IL-7 and IL-15, for instance, induce the proliferation of precursor memory $CD8^+$ T cells [129, 130]. Furthermore, IL-10 and IL-21 in combination with the WNT signaling pathway induce STAT3-mediated $CD8^+$ memory T-cell maturation [130–132].

3.6.4 Shutting down activated T cells

Regulation of T-cell activation is crucial to preventing overactive or underactive T cells. Loss of T-cell signaling at a too early stage will prevent a complete clearance of antigen, while prolonged T-cell activation after the antigen is gone can cause inflammatory and autoimmunity problems. Therefore, a regulatory network of co-stimulatory molecules cooperates to ensure the proper degree of T-cell activation. Furthermore, crosstalk between the T cells and APCs is required to coordinate the timing of B-cell activation. This crosstalk is regulated by a trio of co-stimulatory receptors and ligands: B7, CD28, and CTLA4. APCs express B7, which is a ligand of both CD28 and CTLA (Fig. 3.4). CD28/B7 engagement has actually been shown to be required for TCR activation [133, 134], because CD28 is required for IL-2 expression, which drives the initial proliferation of T cells [135]. On the other end of the spectrum, cytotoxic T-lymphocyte antigen 4 (CTLA4) regulates the inhibition of T-cell activation. CTLA4 can also bind B7, and will bind the CD3ζ chain resulting in its dephosphorylation, effectively cutting off signaling from the TCR [136]. Blocking CTLA4-B7 interaction enhances the immune response and has been used for anti-cancer immune therapy. Similarly, engagement of programmed death-ligand 1 (PD-L1) with is receptor PD-1 on the surface of activated $CD8^+$ T cells transmits inhibitory signals to reduce the proliferation of $CD8^+$ T cells, suppress the immune response during pregnancy and control autoimmune diseases, and is also utilized by cancer cells to escape immune surveillance. Recently, blocking PD-1 and PD-1L interaction has yielded promising results in anti-tumor immune therapy [137].

3.7 Concluding remarks

Great strides have been made in recent years to understand at a molecular level the complexity of events in a T cell's lifespan, and the role it can play in human diseases. This chapter has discussed the development of a T cell, and the complex role the thymus plays in its regulation. Furthermore, this chapter has detailed the complexity of TCR activation and signaling, as well as their dependence on co-receptors and co-stimulatory events to maintain or turn off the T-cell response. Finally, this chapter has provided an understanding of the molecular contexts at every step of the way, highlighting the complexity of a seemingly simple cell.

T-cell lymphomas often arise from the malignant transformation of T cells, especially mature T cells, in which oncogenic mutations (e.g. ALK or FYN mutations) shortcut the pathway for T-cell activation, leading to antigen independent proliferation. In addition to being the progenitor of T-cell lymphomas, T helper cells play an important role in the survival and manifestation of common types of mature B-cell lymphomas, especially of germinal center derived follicular lymphoma, which will be detailed in the B-cell development section. Finally, escaping T-cell immunity has been an important survival strategy for both B-cell lymphomas (e.g. downregulation of β2M – a part of the MHC I – in DLBCL) and solid tumors. Together, the intricate interactions between T and B cells, and within T-cell subtypes, create the homeostatic environment essential for effective immune defense. Many details of T-cell development and T-cell subtype differentiation are yet to be determined. The oncogenic driving events for many T-cell lymphomas are yet to be discovered. A better understanding of the normal regulation of T-cell function and development would provide the strategy for future immunotherapy for lymphomas and other cancer types.

References

[1] Inghirami G, et al. Peripheral T-cell and NK cell lymphoproliferative disorders: cell of origin, clinical and pathological implications. Immunol Rev 2015;263(1):124–159.
[2] Miller JF. Immunological function of the thymus. Lancet 1961;2(7205):748–749.
[3] Girardi M. Immunosurveillance and immunoregulation by gammadelta T-cells. J Invest Dermatol 2006;126(1):25–31.
[4] Brandes M, Willimann K, Moser B. Professional antigen-presentation function by human gammadelta T-cells. Science 2005;309(5732):264–268.
[5] Smith AL, Hayday AC. An alphabeta T-cell-independent immunoprotective response towards gut coccidia is supported by gammadelta cells. Immunology 2000;101(3):325–332.
[6] Heinonen KM, Perreault C. Development and functional properties of thymic and extrathymic T-lymphocytes. Crit Rev Immunol 2008;28(5):441–466.
[7] Brownlie RJ, Zamoyska R. T-cell receptor signalling networks: branched, diversified and bounded. Nat Rev Immunol 2013;13(4):257–269.
[8] Huber M, Lohoff M. IRF4 at the crossroads of effector T-cell fate decision. Eur J Immunol 2014;44(7):1886–1895.

[9] Zhu J, et al. The transcription factor T-bet is induced by multiple pathways and prevents an endogenous Th2 cell program during T_H1 cell responses. Immunity 2012;37(4):660–673.

[10] Stout RD, Bottomly K. Antigen-specific activation of effector macrophages by IFN-gamma producing (TH1) T-cell clones. Failure of IL-4-producing (TH2) T-cell clones to activate effector function in macrophages. J Immunol 1989;142(3):760–765.

[11] Serre K, et al. IL-4 directs both CD4 and CD8 T-cells to produce Th2 cytokines in vitro, but only CD4 T-cells produce these cytokines in response to alum-precipitated protein in vivo. Mol Immunol 2010;47(10):1914–1922.

[12] Zambrano-Zaragoza JF, et al. T_H17 cells in autoimmune and infectious diseases. Int J Inflam 2014;2014:651503.

[13] Cope A, et al. The T_H1 life cycle: molecular control of IFN-gamma to IL-10 switching. Trends Immunol 2011;32(6):278–286.

[14] Josefowicz SZ, Lu LF, Rudensky AY. Regulatory T-cells: mechanisms of differentiation and function. Annu Rev Immunol 2012;30:531–564.

[15] Taams LS, et al. Regulatory T-cells in human disease and their potential for therapeutic manipulation. Immunology 2006;118(1):1–9.

[16] Farber DL, Yudanin NA, Restifo NP. Human memory T-cells: generation, compartmentalization and homeostasis. Nat Rev Immunol 2014;14(1):24–35.

[17] Neefjes J, et al. Towards a systems understanding of MHC class I and MHC class II antigen presentation. Nat Rev Immunol 2011;11(12):823–836.

[18] Challa-Malladi M, et al. Combined genetic inactivation of beta2-Microglobulin and CD58 reveals frequent escape from immune recognition in diffuse large B-cell lymphoma. Cancer Cell 2011;20(6):728–740.

[19] Bassing CH, Swat W, Alt FW. The mechanism and regulation of chromosomal V(D)J recombination. Cell 2002;109 Suppl:S45–55.

[20] Jung D, Alt FW. Unraveling V(D)J recombination; insights into gene regulation. Cell 2004;116(2):299–311.

[21] Wilson A, de Villartay JP, MacDonald HR. T-cell receptor delta gene rearrangement and T early alpha (TEA) expression in immature alpha beta lineage thymocytes: implications for alpha beta/gamma delta lineage commitment. Immunity 1996;4(1):37–45.

[22] Kim MS, et al. Crystal structure of the V(D)J recombinase RAG1-RAG2. Nature 2015;518(7540):507–511.

[23] Liu Y, et al. A plant homeodomain in RAG-2 that binds Hypermethylated lysine 4 of histone H3 is necessary for efficient antigen-receptor-gene rearrangement. Immunity 2007;27(4):561–571.

[24] Rooney S, Chaudhuri J, Alt FW. The role of the non-homologous end-joining pathway in lymphocyte development. Immunol Rev 2004;200:115–131.

[25] Dudley DD, et al. Mechanism and control of V(D)J recombination versus class switch recombination: similarities and differences. Adv Immunol 2005;86:43–112.

[26] Rooney S, Chaudhuri J, Alt FW. The role of the non-homologous end-joining pathway in lymphocyte development. Immunol.Rev. 2004;200:115–131.

[27] Lieber MR. The mechanism of double-strand DNA break repair by the nonhomologous DNA end-joining pathway. Annu Rev Biochem 2010;79:181–211.

[28] Schwarz K, et al. RAG mutations in human B-cell-negative SCID. Science 1996;274(5284):97–99.

[29] Nikolich-Zugich J, Slifka MK, Messaoudi I. The many important facets of T-cell repertoire diversity. Nat Rev Immunol 2004;4(2):123–132.

[30] Kuppers R, Dalla-Favera R. Mechanisms of chromosomal translocations in B-cell lymphomas. Oncogene 2001;20(40):5580–5594.

[31] Larmonie NS, et al. Breakpoint sites disclose the role of the V(D)J recombination machinery in the formation of T-cell receptor (TCR) and non-TCR associated aberrations in T-cell acute lymphoblastic leukemia. Haematologica 2013;98(8):1173–1184.
[32] Marculescu R, et al. Recombinase, chromosomal translocations and lymphoid neoplasia: targeting mistakes and repair failures. DNA Repair (Amst) 2006;5(9-10):1246–1258.
[33] Marculescu R, et al. V(D)J-mediated translocations in lymphoid neoplasms: a functional assessment of genomic instability by cryptic sites. J Exp Med 2002;195(1):85–98.
[34] Boehm T, Bleul CC. The evolutionary history of lymphoid organs. Nat Immunol 2007;8(2):131–135.
[35] Boehm T. Thymus development and function. Curr Opin Immunol 2008;20(2):178–184.
[36] Jenkinson WE, et al. Development of functional thymic epithelial cells occurs independently of lymphostromal interactions. Mech Dev 2005;122(12):1294–1299.
[37] Rodewald HR. Thymus organogenesis. Annu Rev Immunol 2008;26:355–388.
[38] Rossi FM, et al. Recruitment of adult thymic progenitors is regulated by P-selectin and its ligand PSGL-1. Nat Immunol 2005;6(6):626–634.
[39] Wu L, Kincade PW, Shortman K. The CD44 expressed on the earliest intrathymic precursor population functions as a thymus homing molecule but does not bind to hyaluronate. Immunol Lett 1993;38(1):69–75.
[40] Scimone ML, et al. A multistep adhesion cascade for lymphoid progenitor cell homing to the thymus. Proc Natl Acad Sci USA 2006;103(18):7006–7011.
[41] Lind EF, et al. Mapping precursor movement through the postnatal thymus reveals specific microenvironments supporting defined stages of early lymphoid development. J Exp Med 2001;194(2):127–134.
[42] Petrie HT, Zuniga-Pflucker JC. Zoned out: functional mapping of stromal signaling microenvironments in the thymus. Annu Rev Immunol 2007;25:649–679.
[43] Godfrey DI, et al. A developmental pathway involving four phenotypically and functionally distinct subsets of CD3-CD4-CD8- triple-negative adult mouse thymocytes defined by CD44 and CD25 expression. J Immunol 1993;150(10):4244–4252.
[44] Bell JJ, Bhandoola A. The earliest thymic progenitors for T-cells possess myeloid lineage potential. Nature 2008;452(7188):764–767.
[45] Feyerabend TB, et al. Deletion of Notch1 converts pro-T-cells to dendritic cells and promotes thymic B-cells by cell-extrinsic and cell-intrinsic mechanisms. Immunity 2009;30(1):67–79.
[46] Wada H, et al. Adult T-cell progenitors retain myeloid potential. Nature 2008;452(7188):768–772.
[47] Hozumi K, et al. Delta-like 4 is indispensable in thymic environment specific for T-cell development. J Exp Med 2008;205(11):2507–1213.
[48] Kondo M, et al. Cell-fate conversion of lymphoid-committed progenitors by instructive actions of cytokines. Nature 2000;407(6802):383–386.
[49] Garcia-Peydro M, de Yebenes VG, Toribio ML. Notch1 and IL-7 receptor interplay maintains proliferation of human thymic progenitors while suppressing non-T-cell fates. J Immunol 2006;177(6):3711–3720.
[50] von Freeden-Jeffry U, et al. The earliest T lineage-committed cells depend on IL-7 for Bcl-2 expression and normal cell cycle progression. Immunity 1997;7(1):147–154.
[51] Plotkin J, et al. Critical role for CXCR4 signaling in progenitor localization and T-cell differentiation in the postnatal thymus. J Immunol 2003;171(9):4521–4527.
[52] Li L, Leid M, Rothenberg EV. An early T-cell lineage commitment checkpoint dependent on the transcription factor Bcl11b. Science 2010;329(5987):89–93.
[53] Tydell CC, et al. Molecular dissection of prethymic progenitor entry into the T-lymphocyte developmental pathway. J Immunol 2007;179(1):421–438.

[54] Li L, et al. A far downstream enhancer for murine Bcl11b controls its T-cell specific expression. Blood 2013;122(6):902–911.
[55] Avram D, Califano D. The multifaceted roles of Bcl11b in thymic and peripheral T-cells: impact on immune diseases. J Immunol 2014;193(5):2059–2065.
[56] Balciunaite G, et al. The role of Notch and IL-7 signaling in early thymocyte proliferation and differentiation. Eur J Immunol 2005;35(4):1292–1300.
[57] Capone M, Hockett RD Jr, Zlotnik A. Kinetics of T-cell receptor beta, gamma, and delta rearrangements during adult thymic development: T-cell receptor rearrangements are present in CD44(+)CD25(+) Pro-T thymocytes. Proc Natl Acad Sci USA 1998;95(21):12522–12527.
[58] Corcoran AE, et al. Impaired immunoglobulin gene rearrangement in mice lacking the IL-7 receptor. Nature 1998;391(6670):904–907.
[59] Ye SK, et al. The IL-7 receptor controls the accessibility of the TCRgamma locus by Stat5 and histone acetylation. Immunity 2001;15(5):813–823.
[60] Lee HC, et al. Induction of germline transcription in the human TCR gamma locus by STAT5. J Immunol 2001;167(1):320–326.
[61] Kreslavsky T, et al. alphabeta versus gammadelta fate choice: counting the T-cell lineages at the branch point. Immunol Rev 2010;238(1):169–181.
[62] Groettrup M, et al. A novel disulfide-linked heterodimer on pre-T-cells consists of the T-cell receptor beta chain and a 33 kd glycoprotein. Cell 1993;75(2):283–294.
[63] Fehling HJ, et al. Crucial role of the pre-T-cell receptor alpha gene in development of alpha beta but not gamma delta T-cells. Nature 1995;375(6534):795–798.
[64] Kreslavsky T, et al. beta-Selection-induced proliferation is required for alphabeta T-cell differentiation. Immunity 2012;37(5):840–853.
[65] Ciofani M, Zuniga-Pflucker JC. Notch promotes survival of pre-T-cells at the beta-selection checkpoint by regulating cellular metabolism. Nat Immunol 2005;6(9):881–888.
[66] Ciofani M, et al. Obligatory role for cooperative signaling by pre-TCR and Notch during thymocyte differentiation. J Immunol 2004;172(9):5230–5239.
[67] Yashiro-Ohtani Y, et al. Pre-TCR signaling inactivates Notch1 transcription by antagonizing E2A. Genes Dev 2009;23(14):1665–1676.
[68] Weng AP, et al. Activating mutations of NOTCH1 in human T-cell acute lymphoblastic leukemia. Science 2004;306(5694):269–271.
[69] Ellisen LW, et al. TAN-1, the human homolog of the Drosophila notch gene, is broken by chromosomal translocations in T lymphoblastic neoplasms. Cell 1991;66(4):649–661.
[70] Tosello V, Ferrando AA. The NOTCH signaling pathway: role in the pathogenesis of T-cell acute lymphoblastic leukemia and implication for therapy. Ther Adv Hematol 2013;4(3):199–210.
[71] Pear WS, et al. Exclusive development of T-cell neoplasms in mice transplanted with bone marrow expressing activated Notch alleles. J Exp Med 1996;183(5):2283–2291.
[72] Zweidler-McKay PA, Pear WS. Notch and T-cell malignancy. Semin Cancer Biol 2004;14(5):329–340.
[73] Izon DJ, et al. Notch1 regulates maturation of CD4+ and CD8+ thymocytes by modulating TCR signal strength. Immunity 2001;14(3):253–264.
[74] Nedjic J, et al. Macroautophagy, endogenous MHC II loading and T-cell selection: the benefits of breaking the rules. Curr Opin Immunol 2009;21(1):92–97.
[75] Nedjic J, et al. Autophagy in thymic epithelium shapes the T-cell repertoire and is essential for tolerance. Nature 2008;455(7211):396–400.
[76] Murata S, et al. Regulation of CD8+ T-cell development by thymus-specific proteasomes. Science 2007;316(5829):1349–1353.
[77] Klein L, et al. Positive and negative selection of the T-cell repertoire: what thymocytes see (and don't see). Nat Rev Immunol 2014;14(6):377–391.

[78] Nakagawa T, et al. Cathepsin L: critical role in Ii degradation and CD4 T-cell selection in the thymus. Science 1998;280(5362):450–453.
[79] Gommeaux J, et al. Thymus-specific serine protease regulates positive selection of a subset of CD4+ thymocytes. Eur J Immunol 2009;39(4):956–964.
[80] Szondy Z, et al. Thymocyte death by neglect: contribution of engulfing macrophages. Eur J Immunol 2012;42(7):1662–1667.
[81] Leung RK, et al. Deletion of the CD4 silencer element supports a stochastic mechanism of thymocyte lineage commitment. Nat Immunol 2001;2(12):1167–1173.
[82] Itano A, Robey E. Highly efficient selection of CD4 and CD8 lineage thymocytes supports an instructive model of lineage commitment. Immunity 2000;12(4):383-389.
[83] Itano A, et al. The cytoplasmic domain of CD4 promotes the development of CD4 lineage T-cells. J Exp Med 1996;183(3):731–741.
[84] Seong RH, Chamberlain JW, Parnes JR. Signal for T-cell differentiation to a CD4 cell lineage is delivered by CD4 transmembrane region and/or cytoplasmic tail. Nature 1992;356(6371):718–720.
[85] Holst J, et al. Scalable signaling mediated by T-cell antigen receptor-CD3 ITAMs ensures effective negative selection and prevents autoimmunity. Nat Immunol 2008;9(6):658-66.
[86] Erman B, et al. Coreceptor signal strength regulates positive selection but does not determine CD4/CD8 lineage choice in a physiologic in vivo model. J Immunol 2006;177(10):6613–6625.
[87] Yasutomo K, et al. The duration of antigen receptor signalling determines CD4+ versus CD8+ T-cell lineage fate. Nature 2000;404(6777):506-10.
[88] Brugnera E, et al. Coreceptor reversal in the thymus: signaled CD4+8+ thymocytes initially terminate CD8 transcription even when differentiating into CD8+ T-cells. Immunity 2000;13(1):59–71.
[89] Yu Q, et al. In vitro evidence that cytokine receptor signals are required for differentiation of double positive thymocytes into functionally mature CD8+ T-cells. J Exp Med 2003;197(4):475–487.
[90] Singer A, Adoro S, Park JH. Lineage fate and intense debate: myths, models and mechanisms of CD4- versus CD8-lineage choice. Nat Rev Immunol 2008;8(10):788–801.
[91] Lederberg J. Genes and antibodies. Science 1959;129(3364):1649–1653.
[92] Kurobe H, et al. CCR7-dependent cortex-to-medulla migration of positively selected thymocytes is essential for establishing central tolerance. Immunity 2006;24(2):165–177.
[93] Foy TM, et al. An essential role for gp39, the ligand for CD40, in thymic selection. J Exp Med 1995;182(5):1377–1388.
[94] Heath WR, Carbone FR. Cross-presentation in viral immunity and self-tolerance. Nat Rev Immunol 2001;1(2):126–134.
[95] Gallegos AM, Bevan MJ. Central tolerance to tissue-specific antigens mediated by direct and indirect antigen presentation. J Exp Med 2004;200(8):1039–1049.
[96] Hogquist KA, Baldwin TA, Jameson SC. Central tolerance: learning self-control in the thymus. Nat Rev Immunol 2005;5(10):772–782.
[97] Derbinski J, et al. Promiscuous gene expression in medullary thymic epithelial cells mirrors the peripheral self. Nature Immunology 2001;2(11):1032–1039.
[98] Smith KM, et al. Pancreatic gene expression in rare cells of thymic medulla: evidence for functional contribution to T-cell tolerance. International Immunology 1997;9(9):1355–1365.
[99] Anderson MS, et al. Projection of an immunological self shadow within the thymus by the aire protein. Science 2002;298(5597):1395–1401.
[100] Anderson G, Takahama Y. Thymic epithelial cells: working class heroes for T-cell development and repertoire selection. Trends Immunol 2012;33(6):256–263.

[101] Castro JE, et al. Fas modulation of apoptosis during negative selection of thymocytes. Immunity 1996;5(6):617–627.
[102] Matloubian M, et al. Lymphocyte egress from thymus and peripheral lymphoid organs is dependent on S1P receptor 1. Nature 2004;427(6972):355–360.
[103] Allende ML, et al. Expression of the sphingosine 1-phosphate receptor, S1P1, on T-cells controls thymic emigration. J Biol Chem 2004;279(15):15396–15401.
[104] Blumberg RS, et al. Assembly and function of the T-cell antigen receptor. Requirement of either the lysine or arginine residues in the transmembrane region of the alpha chain. J Biol Chem 1990;265(23):14036–14043.
[105] Cosson P, et al. Membrane protein association by potential intramembrane charge pairs. Nature 1991;351(6325):414–416.
[106] Wucherpfennig KW, et al. Structural biology of the T-cell receptor: insights into receptor assembly, ligand recognition, and initiation of signaling. Cold Spring Harb Perspect Biol 2010;2(4):a005140.
[107] Blumberg RS, et al. Structure of the T-cell antigen receptor: evidence for two CD3 epsilon subunits in the T-cell receptor-CD3 complex. Proc Natl Acad Sci USA 1990;87(18):7220–7224.
[108] Sussman JJ, et al. Failure to synthesize the T-cell CD3-zeta chain: structure and function of a partial T-cell receptor complex. Cell 1988;52(1):85–95.
[109] Kuhns MS, Davis MM, Garcia KC. Deconstructing the form and function of the TCR/CD3 complex. Immunity 2006;24(2):133–139.
[110] Love PE, Hayes SM. ITAM-mediated signaling by the T-cell antigen receptor. Cold Spring Harb Perspect Biol 2010;2(6):a002485.
[111] Wittlich M, et al. Structural characterization of the transmembrane and cytoplasmic domains of human CD4. Biochim Biophys Acta 2007;1768(12):2949–2960.
[112] Veillette A, et al. The CD4 and CD8 T-cell surface antigens are associated with the internal membrane tyrosine-protein kinase p56lck. Cell 1988;55(2):301–308.
[113] Barber EK, et al. The CD4 and CD8 antigens are coupled to a protein-tyrosine kinase (p56lck) that phosphorylates the CD3 complex. Proc Natl Acad Sci USA 1989;86(9):3277–3281.
[114] Au-Yeung BB, et al. The structure, regulation, and function of ZAP-70. Immunol Rev 2009;228(1):41–57.
[115] Rhee I, Veillette A. Protein tyrosine phosphatases in lymphocyte activation and autoimmunity. Nat Immunol 2012;13(5):439–447.
[116] Oeckinghaus A, Ghosh S. The NF-kappaB family of transcription factors and its regulation. Cold Spring Harb Perspect Biol 2009;1(4):a000034.
[117] Kannan A, et al. Signal transduction via the T-cell antigen receptor in naive and effector/memory T-cells. Int J Biochem Cell Biol 2012;44(12):2129–2134.
[118] Waickman AT, Powell JD. Mammalian target of rapamycin integrates diverse inputs to guide the outcome of antigen recognition in T-cells. J Immunol 2012;188(10):4721–4729.
[119] Thomson AW, Turnquist HR, Raimondi G. Immunoregulatory functions of mTOR inhibition. Nat Rev Immunol 2009;9(5):324–337.
[120] Bi Y, Liu G, Yang R. mTOR regulates T-cell differentiation and activation in immunity and autoimmunity. Crit Rev Eukaryot Gene Expr 2011;21(4):313–322.
[121] Zhu J, Paul WE. Peripheral CD4+ T-cell differentiation regulated by networks of cytokines and transcription factors. Immunol Rev 2010;238(1):247–262.
[122] Tubo NJ, Jenkins MK. TCR signal quantity and quality in CD4 T-cell differentiation. Trends Immunol 2014;35(12):591–596.
[123] Yamane H, Paul WE. Early signaling events that underlie fate decisions of naive CD4(+) T-cells toward distinct T-helper cell subsets. Immunol Rev 2013;252(1):12–23.

[124] van Panhuys N, Klauschen F, Germain RN. T-cell-receptor-dependent signal intensity dominantly controls CD4(+) T-cell polarization In Vivo. Immunity 2014;41(1):63–74.

[125] Nayar R, et al. Graded levels of IRF4 regulate CD8+ T-cell differentiation and expansion, but not attrition, in response to acute virus infection. J Immunol 2014;192(12):5881–5893.

[126] Man K, et al. The transcription factor IRF4 is essential for TCR affinity-mediated metabolic programming and clonal expansion of T-cells. Nat Immunol 2013;14(11):1155–1165.

[127] Kaech SM, Wherry EJ. Heterogeneity and cell-fate decisions in effector and memory CD8+ T-cell differentiation during viral infection. Immunity 2007;27(3):393–405.

[128] Joshi NS, Kaech SM. Effector CD8 T-cell development: a balancing act between memory cell potential and terminal differentiation. J Immunol 2008;180(3):1309–1315.

[129] Surh CD, Sprent J. Homeostasis of naive and memory T-cells. Immunity 2008;29(6):848–862.

[130] Kaech SM, Cui W. Transcriptional control of effector and memory CD8+ T-cell differentiation. Nat Rev Immunol 2012;12(11):749–761.

[131] Cui W, et al. An interleukin-21-interleukin-10-STAT3 pathway is critical for functional maturation of memory CD8+ T-cells. Immunity 2011;35(5):792–805.

[132] Siegel AM, et al. A critical role for STAT3 transcription factor signaling in the development and maintenance of human T-cell memory. Immunity 2011;35(5):806–818.

[133] Schweitzer AN, Sharpe AH. Studies using antigen-presenting cells lacking expression of both B7-1 (CD80) and B7-2 (CD86) show distinct requirements for B7 molecules during priming versus restimulation of Th2 but not T_H1 cytokine production. J Immunol 1998;161(6):2762–2771.

[134] Salomon B, Bluestone JA. Complexities of CD28/B7: CTLA-4 costimulatory pathways in autoimmunity and transplantation. Annu Rev Immunol 2001;19:225–252.

[135] Jenkins MK, et al. CD28 delivers a costimulatory signal involved in antigen-specific IL-2 production by human T-cells. J Immunol 1991;147(8):2461–2466.

[136] Lee KM, et al. Molecular basis of T-cell inactivation by CTLA-4. Science 1998;282(5397):2263–2266.

[137] Kim JW, Eder JP. Prospects for targeting PD-1 and PD-L1 in various tumor types. Oncology (Williston Park) 2014;28 Suppl 3:15–28.

Beiying Dai, Andre Weilemann, and Georg Lenz

4 Mantle cell lymphoma

4.1 Introduction

Mantle cell lymphoma (MCL) is a B-cell malignancy that represents roughly 5–10 % of all lymphoma cases in adults with a yearly incidence of approximately 0.4 per 100,000 persons [1, 2]. It occurs predominantly in elderly males at a median age of 65 years at diagnosis [1]. Although MCL is rare, it contributes disproportionately to lymphoma mortality due to its frequent aggressive clinical course and lack of curative therapeutic strategies [3]. However, a minority of MCL patients displays an indolent clinical behavior [3].

MCL can be divided into three histological subtypes based on either a mantle zone, nodular or diffuse growth pattern. Using cytological criteria two variants can be distinguished, the classic and the blastoid variant [4–6]. The latter constitutes approximately 10 % of all cases and is associated with poor survival [6]. Immunophenotypically, MCLs are characterized by expression of various B-cell specific markers including CD19, CD20, CD22, and CD79A as well as the T-cell antigen CD5. CD10 and BCL-6 are usually not expressed, whereas CD23 is occasionally expressed at low levels [7]. MCL is characterized by the virtue of a specific gene expression profile that distinguishes it from other lymphoma subtypes [8]. From a prognostic point of view, the proliferative activity of the lymphoma cells seems to be the most important determinant of patient survival, as measurement of cell proliferation has been shown to identify high-risk patients with poor overall survival [8–10].

Recently, significant progress has been achieved in the understanding of the molecular events that drive MCL pathogenesis. The genetic hallmark of this lymphoma subtype is the translocation t(11;14)(q13;32) that leads to overexpression of the cell cycle regulator cyclin D1. This aberration is detectable in the vast majority of primary MCL cases [2, 11, 12]. In addition, MCLs are characterized by dysregulation of the DNA damage response machinery as well as by constitutive activation of various oncogenic signaling pathways such as the nuclear factor-kappa B (NF-κB) cascade [2, 11–13]. These insights did not only help to improve the understanding of the biology of this entity, but have additionally led to the development of novel therapeutic approaches utilizing the detected oncogenic addictions. Nevertheless, the molecular pathogenesis of MCL is still not fully elucidated and further efforts are warranted to fully appreciate the molecular events that drive MCL lymphomagenesis. This chapter will summarize our current understanding of MCL biology and a special emphasis is put on biologic findings that might impact therapeutic strategies of affected patients.

4.2 Molecular pathogenesis of MCL

4.2.1 Overexpression of cyclin D1 and cell cycle dysregulation

The characteristic genetic hallmark of MCL is the chromosomal translocation t(11;14)(q13;q32) that juxtaposes the *CCND1* gene to the immunoglobulin heavy chain (IgH) locus, resulting in constitutive overexpression of the cell cycle regulator cyclin D1 (Fig. 4.1). As a member of the highly conserved cyclin family, cyclin D1 is normally expressed at low levels in normal B cells and interacts with the cyclin dependent kinases CDK4 and CDK6 by forming active kinase complexes [11, 12]. The dimerization of cyclin D1 with CDK4 and CDK6 phosphorylates the tumor suppressor protein retinoblastoma 1 (RB1), thereby inhibiting its activity as a molecular brake on G1 to S phase progression. Cyclin D1/CDK4/CDK6 complexes can further enhance cell cycle progression by promoting cyclin E/CDK2 activation [2, 11, 12]. Cyclin D1 might have additional functions involved in the molecular pathogenesis of MCL. Different studies have revealed cell cycle independent functions of cyclin D1 such as contributing to chromosomal instability or promoting cell survival by improving DNA repair [14–16]. However, functional analyses are required to investigate whether these functions are involved and relevant for MCL biology.

Apart from the described chromosomal translocations affecting *CCND1*, additional molecular mechanisms have been identified that deregulate cyclin D1 expression in MCL (Fig. 4.1). Gains and/or amplifications of the translocated 11q13.3 region have been detected in both MCL cell lines and primary patient samples [17]. In addition, other cyclin D1 aberrations have been identified. While some primary MCL samples express a full length cyclin D1 mRNA, several strongly proliferative MCLs have high *CCND1* mRNA levels and preferentially express short *CCND1* mRNA isoforms. These shorter isoforms are caused either by *CCND1* 3'untranslated region (UTR) genomic deletions or by point mutations. Whereas the longer isoform includes an extended 3'UTR that harbors an mRNA destabilizing element, the shorter isoforms lack the mRNA destabilizing element. Thus, the short *cyclin D1* mRNA isoforms are more stable and accumulate to higher levels [18, 19].

Although the vast majority of MCLs are characterized by overexpression of cyclin D1, this is not present in all MCL cases. Intriguingly, cyclin D1-negative MCL cases are histologically indistinguishable from cyclin D1-positive samples and have the same gene expression profile [2, 20]. Chromosomal translocations juxtaposing the *CCND2* gene frequently to an Ig gene as partner have been detected in roughly half of these cyclin D1-negative MCLs, suggesting that they may be an alternative oncogenic event to *CCND1* translocations (Fig. 4.1) [21, 22].

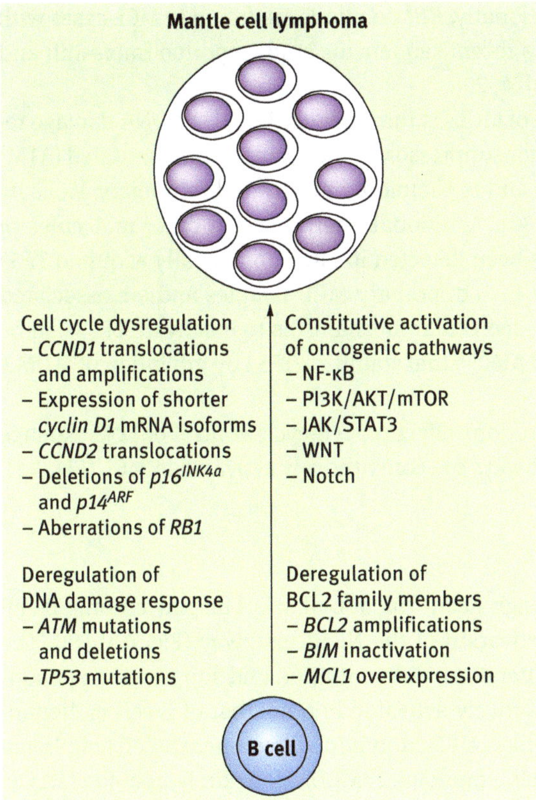

Fig. 4.1: **MCL pathogenesis is driven by different molecular events.** Deregulation of cell cycle, DNA damage response, and BCL2 family members as well as constitutive activation of various oncogenic signaling pathways play a role in MCL lymphomagenesis.

4.2.2 Additional genetic aberrations and deregulated signaling pathways

Various genetic aberrations have been identified in the last couple of years that affect different critical biologic processes such as cell cycle control, apoptosis, DNA damage response, and cell survival. Especially, recent studies utilizing next-generation sequencing approaches significantly contributed to our improved understanding of the driving events in MCL lymphomagenesis (Fig. 4.1) [23, 24].

Besides the described cyclin D1 aberrations, abnormalities that deregulate cell cycle control are homozygous deletions affecting p16^{INK4a} and p14^{ARF}, which have been found in a proportion of MCL cases characterized by an aggressive clinical course [25]. p16$^{\mathrm{INK4a}}$ inhibits the interaction between CDK4 and CDK6 with cyclin D1 and thus suppresses the inactivation of RB1 [26]. Additionally, amplifications and/or overexpression of *BMI1*, that acts as a transcriptional repressor of the p16^{INK4A} locus have been detected and *CDK4* has been found to be amplified and overexpressed in a fraction of highly aggressive MCLs [27, 28]. Finally, *RB1* deletions are found in MCL cases with an aggressive clinical course, and a recent sequencing study reported frameshift and nonsense mutations affecting *RB1* [24, 29].

Another characteristic feature of MCLs is the deregulation of the DNA damage response pathway (Fig. 4.1). The tumor suppressor ataxia telangiectasia mutated (ATM) plays a critical role in DNA repair and is commonly inactivated in primary MCLs by mutations and deletions [23, 24, 30]. Additional aberrations in other molecules of the DNA response machinery have been detected in MCL. Somatically acquired *TP53* mutations occur in approximately 15 % of primary MCL samples and are associated with poor survival [31]. Alternative molecular mechanisms to inactivate TP53 can be high expression of MDM2 or loss of p14^{ARF} that stabilizes TP53 by inhibition of MDM2-mediated degradation [28].

Besides recurrent genetic alterations affecting cell cycle control or DNA damage response, various cell survival pathways are constitutively activated in MCL (Fig. 4.1).

4.2.2.1 NF-κB pathway

Recent work demonstrated that a significant fraction of MCLs display chronic B-cell receptor (BCR) signaling driven activation of the NF-κB pathway (Fig. 4.1) [13]. The NF-κB pathway that promotes proliferation, differentiation and suppression of apoptosis has been shown to be constitutively activated in a number of B-cell malignancies, including the activated B-cell-like (ABC) subtype of diffuse large B-cell lymphoma (DLBCL), primary mediastinal B-cell lymphoma (PMBL), Hodgkin lymphoma (HL) as well as multiple myeloma (MM) [2, 32–39]. The NF-κB family consists of five structurally related transcription factors termed RelA (p65), RelB, c-Rel, NF-κB1 (p50 and its precursor p105), and NF-κB2 (p52 and its precursor p100) that regulate their target genes by formation of homo- and heterodimers [40–43]. Due to their critical roles in survival and proliferation, the NF-κB complexes are tightly regulated. Physiologically, the NF-κB members are kept inactive by cytoplasmic association with inhibitory proteins that consist of IκBα, IκBβ, IκBε, as well as the p105 and p100 precursors of p50 and p52, respectively. Activation of NF-κB occurs either through the canonical or the noncanonical (alternative) pathway. Among others, stimulation of BCR, Toll-like receptor (TLR), or tumor necrosis factor (TNF) receptor activates the canonical NF-κB pathway, whereas stimulation of the BAFF receptor or CD40 engages the noncanonical pathway [40–44]. Inhibition of BCR-driven activation of NF-κB using either the Bruton's tyrosine kinase (BTK) inhibitor ibrutinib or the PKCβ inhibitor sotrastaurin induced toxicity in a substantial fraction of MCL models indicating an addiction to this pathway [13]. In contrast, MCL cell lines that were characterized by resistance to these two compounds harbored constitutive activation of the alternative NF-κB pathway. This activity is caused by genetic lesions in components of the alternative NF-κB cascade such as *TRAF2*, *TRAF3*, and *BIRC3* that are detectable in roughly 15 % of all

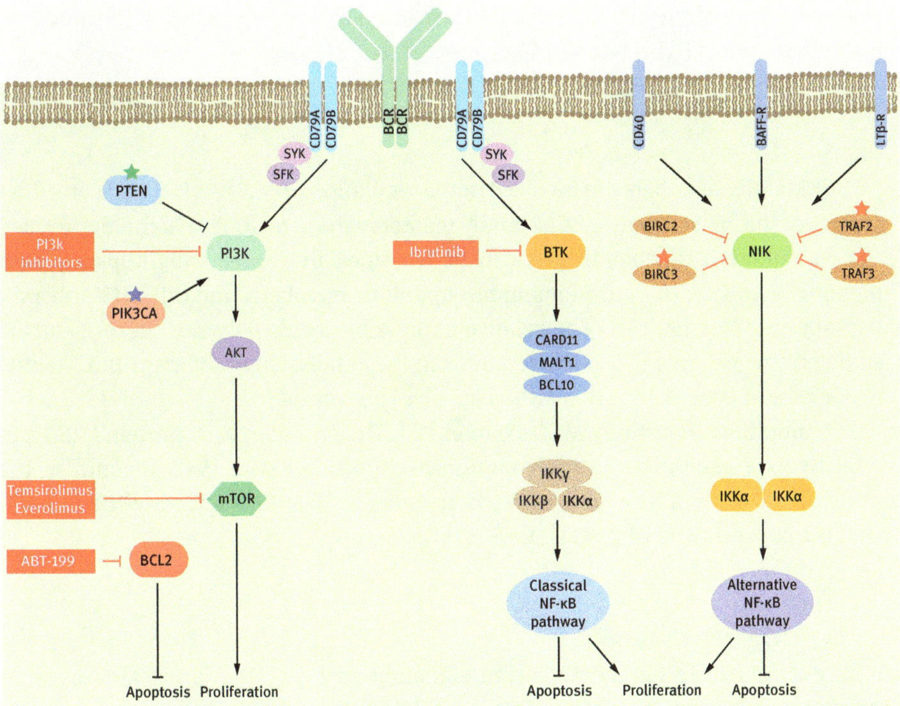

Fig. 4.2: **MCLs are characterized by constitutive activation of various oncogenic signaling pathways that can be targeted therapeutically.** The PI3K/AKT/mTOR and the NF-κB signaling pathway, as well as BCL2 family members, might represent molecular structures that can be targeted by small molecule inhibitors. Asterisks in red indicate genetic aberrations of *TRAF2*, *TRAF3*, and *BIRC3*. Asterisk in green indicates loss of PTEN expression and asterisk in blue indicates amplification of *PIK3CA* locus.

MCLs (Fig. 4.2). In contrast, the molecular mechanisms leading to BCR-driven activation of the canonical NF-κB pathway remain unclear [13].

4.2.2.2 PI3K/AKT/mTOR pathway

Another pro-survival pathway that is constitutively activated in MCL is the phosphoinositide 3-kinase (PI3K)/AKT/mTOR pathway (Fig. 4.1) [45, 46]. PI3K/AKT/mTOR is involved in the regulation of cell proliferation, differentiation, and cell survival. Constitutive activation was found predominantly in blastoid but also in classic MCLs [45]. The molecular mechanisms leading to constitutive activation are not fully elucidated. Chronic BCR signaling might contribute. Alternatively, genomic amplifications of *PIK3CA* that encodes the subunit p110α have been detected in MCL [47]. Loss of PTEN, the major negative regulator of PI3K/AKT signaling, has also been detected in primary MCL cases and cell lines and can be caused by overexpression of the miR17–92 cluster that targets PTEN [45, 48, 49].

4.2.2.3 BCL2 family members

The BCL2 family members are involved in the regulation of cell survival and can either promote or inhibit cell death. BCL2 family members share regions of homology termed BCL2 homology (BH) domain and can be classified into three subgroups: (i) anti-apoptotic members, (ii) multidomain pro-apoptotic members, and (iii) BH3-only pro-apoptotic proteins [50, 51]. The relative expression levels between these members can determine cell fate [52, 53]. Alterations in the expression of different BCL2 family members have been detected in MCL (Fig. 4.1). The chromosomal locus of *BCL2* is frequently amplified in primary MCL samples [17]. The anti-apoptotic protein MCL1 can be highly expressed in MCLs with aggressive clinical behavior [54]. In contrast, the pro-apoptotic protein BIM is inactivated by genomic deletions affecting the *BCL2L11* gene that encodes BIM (Fig. 4.1) [11, 55–57].

4.2.2.4 JAK/STAT3 pathway

As an evolutionarily highly conserved signaling pathway, the Janus kinase-signal transducer and activator of transcription (JAK-STAT) pathway is involved in mediating cell survival and proliferation. Activation of this cascade is initiated with the binding of extracellular ligands such as interleukin, growth factors and interferon that induce the autophosphorylation of JAKs. Phosphorylated JAKs then activate their downstream components including STAT3 via phosphorylation. Phosphorylated STAT3 regulates the transcription of its target genes after translocating into the nucleus [58]. Different studies suggested that STAT3 plays a role in the pathogenesis of MCL (Fig. 4.1). Phosphorylated STAT3 has been detected in primary MCL tumors and seems to be induced by BCR engagement and/or an autocrine interleukin-6 and/or interleukin-10 secretion loop [59, 60]. Along the same lines, suppressor of cytokine signaling 1 (SOCS1), an inhibitor of the JAK/STAT pathway, has been shown to be rarely inactivated by genomic mutations in MCL [61].

4.2.2.5 WNT pathway

WNT proteins are a family of secreted glycoproteins that are involved in the regulation of various critical biologic processes such as cell survival and cell proliferation through both β-catenin-dependent and -independent mechanisms [62]. Activation of the WNT pathway has been described in various forms of solid and hematologic malignancies [56, 62]. Gene expression profiling studies using microarrays suggested a role of WNT signaling in MCL biology (Fig. 4.1) [46]. This was supported by the detection of nuclear accumulation of β-catenin in primary MCL samples [63]. A role of WNT signaling in the pathogenesis of MCL is further supported by data that cell proliferation was reduced and apoptosis induced following either small interfering RNA mediated knockdown of WNT signaling components or by treatment with quercetin, a β-catenin drug inhibitor [64, 65].

4.2.2.6 Notch signaling

A recent study using whole transcriptome sequencing reported somatic *NOTCH1* mutations in 12 % of MCL patients and 20 % of MCL cell lines [66]. These aberrations clustered in the PEST domain and predominantly represented truncating mutations or small frame-shifting indels. Inhibition of NOTCH signaling inhibited cell proliferation and induced apoptosis, suggesting an oncogenic addiction to this pathway. This is further supported as *NOTCH1* mutations were found to be an independent negative prognostic maker that potentially defines a novel MCL subgroup [66].

A recent next-generation sequencing study identified recurrent *NOTCH2* mutations in 9/172 (5.2 %) of primary MCL samples [23]. These mutations were confined to the PEST domain of *NOTCH2* and led to premature stop codons. Interestingly, these *NOTCH2* mutations occurred more frequently in blastoid/pleomorphic MCLs and were associated with adverse survival [23].

4.2.2.7 The role of the microenvironment

Besides acquired tumor-specific genetic aberrations, extracellular signals from the microenvironment, consisting of B cells, stromal cells, antigen-presenting cells (APC) and T cells, might also contribute to lymphomagenesis [67, 68]. MCLs frequently present with extranodal manifestations including involvement of bone marrow or the gastrointestinal tract. This finding indicates a high propensity for dissemination and homing to different tissue compartments. A recent study reported a potential correlation between location of the lymphoma's primary site and overall survival of affected MCL patients, suggesting that differences in tissue tropism might influence MCL progression [69]. In addition, recent data suggest that the interaction with microenvironmental cells might impact lymphoma cell survival and therapy. MCL cells, co-cultured with marrow stromal cells, are characterized by an increased resistance to conventional cytotoxic agents [70]. Residual disease following conventional treatments might be developed in this manner, thereby potentially contributing to relapses. Moreover, expression levels of factors involved in cell-cell crosstalk such as *CCL4* and *TNFSF9* were shown to be elevated in gene expression profiling studies [71].

4.3 Clinical implications

Despite recent improvements in therapy, MCL is still incurable with conventional chemotherapeutic strategies and is characterized by poor prognosis [3, 72]. The only curative approach is allogeneic transplantation. However, median survival of affected patients is in the range of 3–6 years and MCL usually affects elderly patients that often cannot tolerate intensive chemotherapy regimens. Although many MCL patients respond well to first-line immuno-chemotherapy, the vast majority of patients will relapse and eventually succumb to their disease [73, 74]. The lack of more effective regimens especially following relapse emphasizes the urgent need of novel therapeutic approaches. To this end, a better translation of our knowledge on the molecular pathogenesis of MCL into the clinic is indispensable. In the last several years, various new compounds have been detected and successfully introduced into the clinic, while other candidates are currently being tested in clinical trials. In the following section, we will present and discuss clinical translational data on the efficacy of different novel small molecule inhibitors.

4.4 BCR signaling inhibitors

The components of the signaling cascade downstream of the BCR have been identified as potentially attractive targets for the treatment of MCL patients (Fig. 4.2). Whereas fostamatinib and enzastaurin, inhibitors of SYK and PKC-β respectively, showed limited efficacy in MCL patients, the covalent BTK inhibitor ibrutinib proved to be highly effective in relapsed and refractory MCL patients [75–77]. Ibrutinib irreversibly inactivates BTK by selectively binding to the Cys-481 residue in the allosteric inhibitory domain of BTK (Fig. 4.2). Moreover, ibrutinib showed to be rather well tolerated, indicating that it represents a compound that is also suitable for elderly MCL patients. Approximately 70 % of relapsed and refractory MCL patients responded to an ibrutinib monotherapy [76]. These promising results led to the approval of ibrutinib in both the USA and Europe. However, despite these encouraging data still about 30 % of patients are refractory to ibrutinib [76]. In addition, the vast majority of patients will relapse under ibrutinib treatment. This is highly clinically relevant, as MCL patients failing ibrutinib are characterized by poor survival [78]. Different studies investigated the molecular mechanisms of either primary or secondary acquired ibrutinib resistance [13, 79]. In recent work from Rahal and colleagues, ibrutinib-insensitive MCL models were shown to be addicted to the alternative NF-κB pathway and independent of BCR signaling. The addiction to the alternative NF-κB cascade was caused by recurrent mutations affecting negative regulators of alternative NF-κB signaling (Fig. 4.2) [13]. Acquired secondary resistance can be caused by specific mutations affecting the ibrutinib binding site of BTK. The specific C481S *BTK* mutation was detected in MCL cells of an affected patient at progression following a durable response [79].

4.5 PI3K/mTOR inhibitors

The oral PI3Kδ inhibitor idelalisib was recently tested in a phase I study of 40 MCL patients with relapsed or refractory disease. The overall response rate was 40 %, whereas the median duration of response was only 2.7 months. These data suggest that idelalisib has only limited efficacy in MCL [80]. Potentially this is caused by the redundant roles of other PI3K isoforms, as a high *PIK3CA/PIK3CD* ratio identified a subset of primary MCLs resistant to idelalisib [81]. Thus, potentially all PI3K isoforms need to be suppressed to induce higher response rates in MCL (Fig. 4.2).

Inhibition of mTOR using analogs of rapamycin (so called rapalogs), such as temsirolimus or everolimus, shows clinical efficacy in MCL patients [82–84]. In a phase III trial, relapsed or refractory MCL patients were randomized to receive either one of two temsirolimus regimens or investigator's choice therapy [83]. As temsirolimus was superior to investigator's choice treatment, it was approved by the European Medicines Agency (EMA). However, overall rapalogs only induce responses in approximately 20–40 % of MCL patients and these responses are frequently only partial remissions [83]. Thus, rapalogs are only moderately active in MCL.

4.6 BCL2 inhibitors

Encouraging results in different hematologic malignancies have recently been reported for selectively targeting BCL2 with venetoclax (ABT-199), a specific and highly potent BCL2 inhibitor [85]. Interestingly, venetoclax is only moderately active in MCL cell lines, as only two cell lines were affected by BCL2 inhibition, whereas four cell lines were resistant to this compound [86]. In contrast, the clinical activity is potentially higher, as all eight patients responded to venetoclax in a recent phase I study in relapsed or refractory MCL [87].

4.7 Proteasome inhibitors

Bortezomib was the first proteasome inhibitor to enter the clinic and is approved for the treatment of MCL. It is still not fully elucidated how bortezomib acts specifically in MCL [56]. Most likely different molecular mechanisms are involved. Among others, it has been shown that bortezomib induces the upregulation of the BH3-only protein NOXA [88]. Using gene expression profiling, it was recently detected that bortezomib resistance in MCL cell lines is associated with plasmacytic differentiation, including upregulation of the transcription factor IRF4 and the cell surface markers CD38 and CD138. However, in contrast to plasma cells, plasmacytic MCL cells do not increase immunoglobulin secretion [89].

Currently, next-generation proteasome inhibitors such as carfilzomib that have been developed to increase efficacy and to improve the toxicity profile are being tested in clinical trials in MCL.

4.8 Concluding remarks

Although a curative therapy still remains elusive for the vast majority of MCL patients, a promising new era is dawning. Due to our improved understanding of the molecular pathogenesis, various signaling inhibitors targeting different oncogenic pathways have been either approved for clinical use or are currently being tested in different trials. However, a significantly better understanding of the molecular biology of MCL is still required to identify molecular targets in all affected patients. These efforts will eventually lead to a molecular stratification of patients and tailored treatment regimens that are more effective and less toxic compared to our current standard of care.

References

[1] Cortelazzo S, Ponzoni M, Ferreri AJ, Dreyling M. Mantle cell lymphoma. Critical Reviews in Oncology/Hematology 2012;82:78–101.

[2] Nogai H, Dorken B, Lenz G. Pathogenesis of non-Hodgkin's lymphoma. J Clin Oncol 2011;29:1803–1811.

[3] Ghielmini M, Zucca E. How I treat mantle cell lymphoma. Blood 2009;114:1469-76.

[4] Banks PM, Chan J, Cleary ML, et al. Mantle cell lymphoma. A proposal for unification of morphologic, immunologic, and molecular data. The American Journal of Surgical Pathology 1992;16:637–640.

[5] Campo E, Raffeld M, Jaffe ES. Mantle-cell lymphoma. Semin Hematol 1999;36:115–127.

[6] Bernard M, Gressin R, Lefrere F, et al. Blastic variant of mantle cell lymphoma: a rare but highly aggressive subtype. Leukemia 2001;15:1785–1791.

[7] Pileri SA, Falini B. Mantle cell lymphoma. Haematologica 2009;94:1488–1492.

[8] Rosenwald A, Wright G, Wiestner A, et al. The proliferation gene expression signature is a quantitative integrator of oncogenic events that predicts survival in mantle cell lymphoma. Cancer Cell 2003;3:185–197.

[9] Tiemann M, Schrader C, Klapper W, et al. Histopathology, cell proliferation indices and clinical outcome in 304 patients with mantle cell lymphoma (MCL): a clinicopathological study from the European MCL Network. British Journal of Haematology 2005;131:29–38.

[10] Determann O, Hoster E, Ott G, et al. Ki-67 predicts outcome in advanced-stage mantle cell lymphoma patients treated with anti-CD20 immunochemotherapy: results from randomized trials of the European MCL Network and the German Low Grade Lymphoma Study Group. Blood 2008;111:2385–2387.

[11] Jares P, Colomer D, Campo E. Genetic and molecular pathogenesis of mantle cell lymphoma: perspectives for new targeted therapeutics. Nature Reviews Cancer 2007;7:750–762.

[12] Jares P, Colomer D, Campo E. Molecular pathogenesis of mantle cell lymphoma. The Journal of Clinical Investigation 2012;122:3416–3423.

[13] Rahal R, Frick M, Romero R, et al. Pharmacological and genomic profiling identifies NF-kappaB-targeted treatment strategies for mantle cell lymphoma. Nat Med 2014;20:87–92.

[14] Casimiro MC, Crosariol M, Loro E, et al. ChIP sequencing of cyclin D1 reveals a transcriptional role in chromosomal instability in mice. J Clin Invest 2012;122:833–843.

[15] Jirawatnotai S, Hu Y, Michowski W, et al. A function for cyclin D1 in DNA repair uncovered by protein interactome analyses in human cancers. Nature 2011;474:230–234.

[16] Bartek J, Lukas J. DNA repair: Cyclin D1 multitasks. Nature 2011;474:171–172.

[17] Bea S, Salaverria I, Armengol L, et al. Uniparental disomies, homozygous deletions, amplifications, and target genes in mantle cell lymphoma revealed by integrative high-resolution whole-genome profiling. Blood 2009;113:3059–3069.

[18] Wiestner A, Tehrani M, Chiorazzi M, et al. Point mutations and genomic deletions in CCND1 create stable truncated cyclin D1 mRNAs that are associated with increased proliferation rate and shorter survival. Blood 2007;109:4599–4606.

[19] Chen RW, Bemis LT, Amato CM, et al. Truncation in CCND1 mRNA alters miR-16-1 regulation in mantle cell lymphoma. Blood 2008;112:822–829.

[20] Fu K, Weisenburger DD, Greiner TC, et al. Cyclin D1-negative mantle cell lymphoma: a clinicopathological study based on gene expression profiling. Blood 2005;106:4315–4321.

[21] Gesk S, Klapper W, Martin-Subero JI, et al. A chromosomal translocation in cyclin D1-negative/cyclin D2-positive mantle cell lymphoma fuses the CCND2 gene to the IGK locus. Blood 2006;108:1109–1110.

[22] Salaverria I, Royo C, Carvajal-Cuenca A, et al. CCND2 rearrangements are the most frequent genetic events in cyclin D1(-) mantle cell lymphoma. Blood 2013;121:1394–1402.

[23] Bea S, Valdes-Mas R, Navarro A, et al. Landscape of somatic mutations and clonal evolution in mantle cell lymphoma. Proc Natl Acad Sci USA 2013;110:18250–18255.

[24] Zhang J, Jima D, Moffitt AB, et al. The genomic landscape of mantle cell lymphoma is related to the epigenetically determined chromatin state of normal B cells. Blood 2014;123:2988–2996.

[25] Pinyol M, Hernandez L, Cazorla M, et al. Deletions and loss of expression of p16INK4a and p21Waf1 genes are associated with aggressive variants of mantle cell lymphomas. Blood 1997;89:272–280.

[26] Sherr CJ, McCormick F. The RB and p53 pathways in cancer. Cancer Cell 2002;2:103–112.

[27] Bea S, Tort F, Pinyol M, et al. BMI-1 gene amplification and overexpression in hematological malignancies occur mainly in mantle cell lymphomas. Cancer Res 2001;61:2409–24012.

[28] Hernandez L, Bea S, Pinyol M, et al. CDK4 and MDM2 gene alterations mainly occur in highly proliferative and aggressive mantle cell lymphomas with wild-type INK4a/ARF locus. Cancer Res 2005;65:2199–2206.

[29] Pinyol M, Bea S, Pla L, et al. Inactivation of RB1 in mantle-cell lymphoma detected by nonsense-mediated mRNA decay pathway inhibition and microarray analysis. Blood 2007;109:5422–5429.

[30] Schaffner C, Idler I, Stilgenbauer S, Dohner H, Lichter P. Mantle cell lymphoma is characterized by inactivation of the ATM gene. Proc Natl Acad Sci USA 2000;97:2773–2778.

[31] Greiner TC, Moynihan MJ, Chan WC, et al. p53 mutations in mantle cell lymphoma are associated with variant cytology and predict a poor prognosis. Blood 1996;87:4302–4310.

[32] Davis RE, Brown KD, Siebenlist U, Staudt LM. Constitutive Nuclear Factor kappaB activity Is required for survival of Activated B Cell-like Diffuse Large B Cell Lymphoma Cells. J Exp Med 2001;194:1861–1874.

[33] Rosenwald A, Wright G, Leroy K, et al. Molecular diagnosis of primary mediastinal B cell lymphoma identifies a clinically favorable subgroup of diffuse large B cell lymphoma related to Hodgkin lymphoma. J Exp Med 2003;198:851–862.

[34] Savage KJ, Monti S, Kutok JL, et al. The molecular signature of mediastinal large B-cell lymphoma differs from that of other diffuse large B-cell lymphomas and shares features with classical Hodgkin lymphoma. Blood 2003;102:3871–3879.
[35] Schmitz R, Hansmann ML, Bohle V, et al. TNFAIP3 (A20) is a tumor suppressor gene in Hodgkin lymphoma and primary mediastinal B cell lymphoma. J Exp Med 2009;206:981–989.
[36] Bargou RC, Leng C, Krappmann D, et al. High-level nuclear NF-kappa B and Oct-2 is a common feature of cultured Hodgkin/Reed-Sternberg cells. Blood 1996;87:4340–4347.
[37] Bargou RC, Emmerich F, Krappmann D, et al. Constitutive nuclear factor-kappaB-RelA activation is required for proliferation and survival of Hodgkin's disease tumor cells. J Clin Invest 1997;100:2961–2969.
[38] Annunziata CM, Davis RE, Demchenko Y, et al. Frequent engagement of the classical and alternative NF-kappaB pathways by diverse genetic abnormalities in multiple myeloma. Cancer Cell 2007;12:115–130.
[39] Keats JJ, Fonseca R, Chesi M, et al. Promiscuous mutations activate the noncanonical NF-kappaB pathway in multiple myeloma. Cancer Cell 2007;12:131–144.
[40] Jost PJ, Ruland J. Aberrant NF-kappaB signaling in lymphoma: mechanisms, consequences, and therapeutic implications. Blood 2007;109:2700–2707.
[41] Hinz M, Arslan SC, Scheidereit C. It takes two to tango: IkappaBs, the multifunctional partners of NF-kappaB. Immunol Rev 2012;246:59–76.
[42] Ghosh S, Hayden MS. Celebrating 25 years of NF-kappaB research. Immunol Rev 2012;246:5–13.
[43] Nogai H, Wenzel SS, Hailfinger S, et al. IkappaB-zeta controls the constitutive NF-kappaB target gene network and survival of ABC DLBCL. Blood 2013;122:2242–2450.
[44] Bonizzi G, Karin M. The two NF-kappaB activation pathways and their role in innate and adaptive immunity. Trends Immunol 2004;25:280–288.
[45] Rudelius M, Pittaluga S, Nishizuka S, et al. Constitutive activation of Akt contributes to the pathogenesis and survival of mantle cell lymphoma. Blood 2006;108:1668–1676.
[46] Rizzatti EG, Falcao RP, Panepucci RA, et al. Gene expression profiling of mantle cell lymphoma cells reveals aberrant expression of genes from the PI3K-AKT, WNT and TGFbeta signalling pathways. Br J Haematol 2005;130:516–526.
[47] Psyrri A, Papageorgiou S, Liakata E, et al. Phosphatidylinositol 3'-kinase catalytic subunit alpha gene amplification contributes to the pathogenesis of mantle cell lymphoma. Clin Cancer Res 2009;15:5724–5732.
[48] Navarro A, Bea S, Fernandez V, et al. MicroRNA expression, chromosomal alterations, and immunoglobulin variable heavy chain hypermutations in Mantle cell lymphomas. Cancer Res 2009;69:7071–7078.
[49] Rao E, Jiang C, Ji M, et al. The miRNA-17 approximately 92 cluster mediates chemoresistance and enhances tumor growth in mantle cell lymphoma via PI3K/AKT pathway activation. Leukemia 2012;26:1064–1072.
[50] Youle RJ, Strasser A. The BCL-2 protein family: opposing activities that mediate cell death. Nat Rev Mol Cell Biol 2008;9:47–59.
[51] Wenzel SS, Grau M, Mavis C, et al. MCL1 is deregulated in subgroups of diffuse large B-cell lymphoma. Leukemia 2013;27:1381–1390.
[52] Adams JM, Cory S. The Bcl-2 apoptotic switch in cancer development and therapy. Oncogene 2007;26:1324–1337.
[53] Adams JM, Cory S. The Bcl-2 protein family: arbiters of cell survival. Science 1998;281:1322–1326.

[54] Khoury JD, Medeiros LJ, Rassidakis GZ, McDonnell TJ, Abruzzo LV, Lai R. Expression of Mcl-1 in mantle cell lymphoma is associated with high-grade morphology, a high proliferative state, and p53 overexpression. J Pathol 2003;199:90–97.

[55] Mestre-Escorihuela C, Rubio-Moscardo F, Richter JA, et al. Homozygous deletions localize novel tumor suppressor genes in B-cell lymphomas. Blood 2007;109:271–280.

[56] Perez-Galan P, Dreyling M, Wiestner A. Mantle cell lymphoma: biology, pathogenesis, and the molecular basis of treatment in the genomic era. Blood 2011;117:26–38.

[57] Tagawa H, Karnan S, Suzuki R, et al. Genome-wide array-based CGH for mantle cell lymphoma: identification of homozygous deletions of the proapoptotic gene BIM. Oncogene 2005;24:1348–1358.

[58] Yu H, Lee H, Herrmann A, Buettner R, Jove R. Revisiting STAT3 signalling in cancer: new and unexpected biological functions. Nature Reviews Cancer 2014;14:736–746.

[59] Baran-Marszak F, Boukhiar M, Harel S, et al. Constitutive and B-cell receptor-induced activation of STAT3 are important signaling pathways targeted by bortezomib in leukemic mantle cell lymphoma. Haematologica 2010;95:1865–1872.

[60] Lai R, Rassidakis GZ, Medeiros LJ, Leventaki V, Keating M, McDonnell TJ. Expression of STAT3 and its phosphorylated forms in mantle cell lymphoma cell lines and tumours. The Journal of Pathology 2003;199:84–89.

[61] Mottok A, Renne C, Seifert M, et al. Inactivating SOCS1 mutations are caused by aberrant somatic hypermutation and restricted to a subset of B-cell lymphoma entities. Blood 2009;114:4503–4506.

[62] Anastas JN, Moon RT. WNT signalling pathways as therapeutic targets in cancer. Nature Reviews Cancer 2013;13:11–26.

[63] Chung R, Peters AC, Armanious H, Anand M, Gelebart P, Lai R. Biological and clinical significance of GSK-3beta in mantle cell lymphoma – an immunohistochemical study. International Journal of Clinical and Experimental Pathology 2010;3:244–253.

[64] Ortega-Paino E, Fransson J, Ek S, Borrebaeck CA. Functionally associated targets in mantle cell lymphoma as defined by DNA microarrays and RNA interference. Blood 2008;111:1617–1624.

[65] Gelebart P, Anand M, Armanious H, et al. Constitutive activation of the Wnt canonical pathway in mantle cell lymphoma. Blood 2008;112:5171–5179.

[66] Kridel R, Meissner B, Rogic S, et al. Whole transcriptome sequencing reveals recurrent NOTCH1 mutations in mantle cell lymphoma. Blood 2012;119:1963–1971.

[67] Burger JA, Ghia P, Rosenwald A, Caligaris-Cappio F. The microenvironment in mature B-cell malignancies: a target for new treatment strategies. Blood 2009;114:3367–3375.

[68] Burger JA, Ford RJ. The microenvironment in mantle cell lymphoma: cellular and molecular pathways and emerging targeted therapies. Semin Cancer Biol 2011;21:308–3012.

[69] Ambinder AJ, Shenoy PJ, Nastoupil LJ, Flowers CR. Using primary site as a predictor of survival in mantle cell lymphoma. Cancer 2013;119:1570–1577.

[70] Kurtova AV, Tamayo AT, Ford RJ, Burger JA. Mantle cell lymphoma cells express high levels of CXCR4, CXCR5, and VLA-4 (CD49d): importance for interactions with the stromal microenvironment and specific targeting. Blood 2009;113:4604–4613.

[71] Ek S, Bjorck E, Hogerkorp CM, Nordenskjold M, Porwit-MacDonald A, Borrebaeck CA. Mantle cell lymphomas acquire increased expression of CCL4, CCL5 and 4-1BB-L implicated in cell survival. Int J Cancer 2006;118:2092–2097.

[72] Dreyling M, Hiddemann W. Current treatment standards and emerging strategies in mantle cell lymphoma. Hematology Am Soc Hematol Educ Program 2009:542–551.

[73] Vose JM. Mantle cell lymphoma: 2013 Update on diagnosis, risk-stratification, and clinical management. Am J Hematol 2013;88:1082–1088.

[74] Rivera-Rodriguez N, Cabanillas F. Recent advances in the management of mantle cell lymphoma. Curr Opin Oncol 2013;25:716–721.

[75] Morschhauser F, Seymour JF, Kluin-Nelemans HC, et al. A phase II study of enzastaurin, a protein kinase C beta inhibitor, in patients with relapsed or refractory mantle cell lymphoma. Annals of Oncology: Official Journal of the European Society for Medical Oncology/ESMO 2008;19:247–253.

[76] Wang ML, Rule S, Martin P, et al. Targeting BTK with ibrutinib in relapsed or refractory mantle-cell lymphoma. N Engl J Med 2013;369:507–516.

[77] Friedberg JW, Sharman J, Sweetenham J, et al. Inhibition of Syk with fostamatinib disodium has significant clinical activity in non-Hodgkin lymphoma and chronic lymphocytic leukemia. Blood 2010;115:2578–2585.

[78] Cheah CY, Chihara D, Romaguera JE, et al. Patients with mantle cell lymphoma failing ibrutinib are unlikely to respond to salvage chemotherapy and have poor outcomes. Annals of Oncology: Official Journal of the European Society for Medical Oncology / ESMO 2015;26:1175–1179.

[79] Chiron D, Di Liberto M, Martin P, et al. Cell-cycle reprogramming for PI3K inhibition overrides a relapse-specific C481S BTK mutation revealed by longitudinal functional genomics in mantle cell lymphoma. Cancer Discov 2014;4:1022–1035.

[80] Kahl BS, Spurgeon SE, Furman RR, et al. A phase 1 study of the PI3Kdelta inhibitor idelalisib in patients with relapsed/refractory mantle cell lymphoma (MCL). Blood 2014;123:3398–3405.

[81] Iyengar S, Clear A, Bodor C, et al. P110alpha-mediated constitutive PI3K signaling limits the efficacy of p110delta-selective inhibition in mantle cell lymphoma, particularly with multiple relapse. Blood 2013;121:2274–2284.

[82] Hess G. Temsirolimus for the treatment of mantle cell lymphoma. Expert Rev Hematol 2009;2:631–640.

[83] Hess G, Herbrecht R, Romaguera J, et al. Phase III study to evaluate temsirolimus compared with investigator's choice therapy for the treatment of relapsed or refractory mantle cell lymphoma. J Clin Oncol 2009;27:3822–3829.

[84] Renner C, Zinzani PL, Gressin R, et al. A multicenter phase II trial (SAKK 36/06) of single-agent everolimus (RAD001) in patients with relapsed or refractory mantle cell lymphoma. Haematologica 2012;97:1085–1091.

[85] Souers AJ, Leverson JD, Boghaert ER, et al. ABT-199, a potent and selective BCL-2 inhibitor, achieves antitumor activity while sparing platelets. Nature Medicine 2013;19:202–208.

[86] Touzeau C, Brosseau C, Dousset C, Pellat-Deceunynck C, Le Gouill S, Amiot M. Mantle-cell lymphoma (MCL) cells are highly sensitive to ABT-199 but their sensitivity may be altered by the microenvironment via the up-regulation of Bcl-Xl and Bcl2A1. Blood 2013;122.

[87] Davids MS, Seymour JF, Gerecitano JF, et al. The single-agent Bcl-2 Inhibitor ABT-199 (GDC-0199) in patients with relapsed/refractory (R/R) Non-Hodgkin Lymphoma (NHL): Responses observed in all mantle cell lymphoma (MCL) Patients. Blood 2013;122.

[88] Perez-Galan P, Roue G, Villamor N, Montserrat E, Campo E, Colomer D. The proteasome inhibitor bortezomib induces apoptosis in mantle-cell lymphoma through generation of ROS and Noxa activation independent of p53 status. Blood 2006;107:257–264.

[89] Perez-Galan P, Mora-Jensen H, Weniger MA, et al. Bortezomib resistance in mantle cell lymphoma is associated with plasmacytic differentiation. Blood 2011;117:542–552.

Jessica Okosun and Jude Fitzgibbon

5 Follicular lymphoma

5.1 Introduction

Follicular lymphoma (FL) represents one of the commonest lymphomas, second only to diffuse large B-cell lymphoma, with an annual incidence of 4 per 100,000 in the Western world, and accounts for 20–25 % of all cases of NHL. The clinical behavior of FL shows considerable variability. For the majority of patients, the disease follows a prototypical indolent course, relapsing and remitting with eventual refractoriness or resistance to standard therapies. Whilst outcomes have improved in most patient cohorts, in around 30–50 % of individuals, the tumor transforms into a more aggressive lymphoma (transformed FL), usually diffuse large B-cell lymphoma (DLBCL), with this phenotype associated with poorer clinical outcomes even with the use of combination chemo-immunotherapy. Despite the overall survival now exceeding beyond a decade, the malignancy remains incurable. It is anticipated that the elucidation of the genetic and molecular blueprint of these lymphomas will allow the development of molecularly targeted therapies.

This chapter will describe the biology and pathogenesis of FL and transformed FL with a particular focus on recent improvements in understanding the genomics of the malignancy.

5.2 Histology and grading

The tumor is composed of a clonal proliferation of centrocytes (small to medium-sized cells with cleaved inconspicuous nuclei) and centroblasts (large cells with prominent nucleoli). The infiltrated follicles within the lymph nodes lose the normal nodal architecture, have absent or ill-defined mantle zones with the centrocytes and centroblasts randomly distributed within the follicle rather than being polarized to distinct zones as seen in reactive germinal centers. The varying proportion of centroblasts within the FL tumor forms the basis of the grading system: low grade FL (Grade 1 and 2) possess less than 15 centroblasts per high power field whilst high grade FLs (Grade 3) contains greater than 15 centroblasts. Grade 3 FLs can be further subdivided in grade 3A and grade 3B with cumulative evidence suggesting this grade represents a distinct entity that is more molecularly aligned to DLBCL with most clinicians treating this entity equivalently to *de novo* DLBCL [1–5].

5.3 Transformed follicular lymphoma

Transformation from indolent FL to an aggressive lymphoma is a well-recognized complication during the clinical evolution of the malignancy [6]. It is defined by demonstrating histological evidence of progression from a neoplasm of centrocytes and centroblasts to a pattern of predominantly diffuse large cell infiltration with loss of the characteristic follicular architecture. Confirmation of histological transformation (HT) with a repeat biopsy is considered the gold standard approach. However, for those for whom a biopsy is not possible, a clinical diagnosis of HT can be made based on a combination of high risk clinical features including a sudden elevation in serum lactate dehydrogenase (LDH), rapid nodal or new extranodal growth, changes in performance status, new B symptoms or hypercalcemia [7–9]. Recently, ^{18}[F] fluorodeoxyglucose (FDG) positron emission tomography (PET) scans have demonstrated utility in diagnosing transformation [10–12], although this approach still requires prospective validation.

The majority of cases of FL transformation give rise to DLBCL, usually retaining the germinal center phenotype. Less frequently, transformation to B-lymphoblastic lymphoma/leukemia (B-LBL) or *MYC*-driven lymphomas including Burkitt lymphoma (BL), B-cell lymphoma unclassifiable (BCLU) with features intermediate between DLBCL and BL or 'double hit' lymphomas do occur [13–17]. There have also been sporadic cases of transformation to non-GC and even myeloid malignancies such as histiocytic/dendritic cell sarcomas [18, 19]. Interestingly, in a majority of these cases, molecular studies confirmed identical *IGH-BCL2* rearrangements in both the preceding FL and transformed lymphoma suggesting the tumors were clonally related, and thus represented *bona fide* transformation events rather than *de novo* second malignancies. The fact that sequential transformed FL can assume such unusual histological subtypes suggests potential lineage de-differentiation or trans-differentiation from the antecedent FL and raises questions on the exact evolutionary pathways responsible.

5.4 Early steps of FL development: Initial genetic hit t(14;18)

The t(14;18)(q32.3;q21.3) reciprocal translocation is considered the primary genetic hallmark of FL (Fig. 5.1), occurring in 85–90 % of cases, and results in the juxtaposition of the proto-oncogene *BCL2* at 18q21.3 in close vicinity to the active immunoglobulin heavy chain loci, IGH, at chromosome 14q32 [20, 21]. Rare variant translocations t(2;18) and t(18;22) involving the immunoglobulin light chain genes can occur. In most cases, the breaks of the *BCL2* gene are clustered within a 150-bp region termed the major breakpoint region (MBR; 70 %) whilst the remainder occur within the minor breakpoint cluster region (mcr), a region 29 kb downstream of MBR or the intermediate cluster region (icr) [22–25]. This erroneous translocation event is attributed to either defective RAG-mediated double-strand breaks in both partners (IGH and BCL2) or only

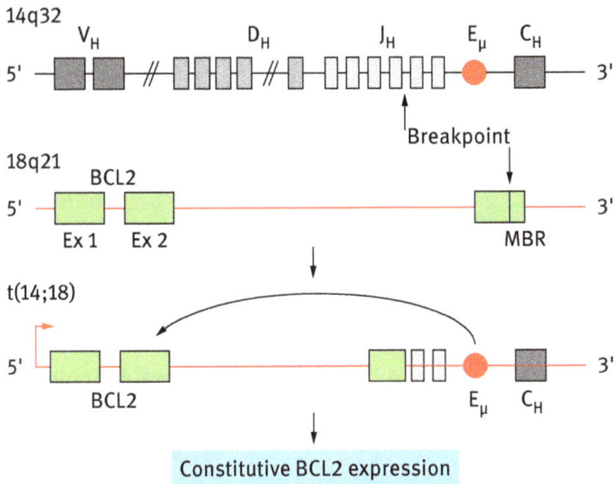

Fig. 5.1: **BCL2-IGH translocation in follicular lymphoma.** Illustrated is the translocation process between the IGH and BCL2 loci. Double-strand breaks (DSBs) are introduced by the RAG enzymes (RAG-1/2) into both the *IGH* and *BCL2* locus followed by aberrant repair of the breaks leading to the translocation. The juxtaposition of the BCL2 loci next to the IgH enhancer (Eμ) promotes constitutive BCL2 over-expression.

at the IGH locus on chromosome 14 and a break at the BCL2 locus possibly linked to inherent fragility at CpG or common fragility sites [26, 27]. The initial t(14;18) arises at the precursor B-cell stage of development when the process of V(D)J recombination occurs in the bone marrow [27, 28]. This chromosomal alteration promotes aberrant constitutive over-expression of the anti-apoptotic protein, BCL2, not normally expressed by GC B cells. As normal GC B cells characteristically undergo rapid apoptosis, the ectopic expression of BCL2 rescues B cells thereby generating a pool of long-lived B cells that exit the bone marrow and migrate to secondary lymphoid organs where they then undergo the germinal center reaction.

5.5 Cell of origin

The tumor cells are highly reminiscent of their normal B-cell counterparts from which they are derived, the germinal center (GC). Morphologically, they resemble normal centrocytes and centroblasts. They characteristically express B-cell surface markers such as CD19, CD20, CD22 and CD79a, as well as typical GC B-cell markers, including BCL6 and CD10. Furthermore, the GC derivation is supported by the presence of key GC-associated features such as ongoing somatic hypermutation (SHM) and class switch recombination (CSR) of their immunoglobulin heavy chain variable (IgV_H) genes. The enzyme activation-induced cytidine deaminase (AID) plays a critical role in mediating both SHM and CSR processes [29] and is expressed by normal GCs and

GC-derived tumors including FL [30, 31]. A direct role for AID in lymphomagenesis has been suggested using AID-deficient mouse models, where the transgenic mice clearly demonstrated absence of CSR- and SHM-mediated structural alterations and an inability to develop BCL6-driven GC-derived lymphomas [32]. As the AID gene locus is not known to be targeted by somatic mutations or copy number variations (CNVs), it is likely AID exerts its effect in promoting lymphomagenesis via indirect mechanisms.

During the course of GC lymphoma development, the tight control of AID targeting becomes defective with the ability to initiate off-target mutations aberrantly (ASHM) at non-Ig loci. In FL, a number of non-Ig genes, particularly proto-oncogenes, are subject to ASHM such as *BCL6, MYC, PAX5, PIM1, RHOH* [33, 34], *SOCS1* [35] and *BCL2* [36]. It remains unclear mechanistically how AID is recruited to these off-target genomic regions to cause ASHM, although there is a suggestion from recent studies that unique epigenetic profiles [37] or B-cell super-enhancers [38, 39] may play a role in AID recruitment to its target genes.

5.6 t(14;18) insufficient for lymphoma development

Several lines of evidence support the t(14;18) translocation as being important but requiring complementary genetic hits for complete malignant transformation of normal B cells to FL. Firstly, the t(14;18) is detectable at low levels in the peripheral blood of normal, healthy individuals at a prevalence of 23–50 % [40–42], without increasing their risk of FL development after long follow-up. Increasing age and pesticide exposure also increase the prevalence of t(14;18) in nonaffected individuals [43, 44]. Interestingly, the population of circulating B cells that carry the t(14;18) in the peripheral blood are GC–experienced memory B cells rather than the naïve B cell subset as originally thought [45]. Secondly, mouse models have further highlighted that additional genetic alterations are required for malignant transformation. Transgenic mouse models expressing Bcl2 under the control of an Igh intronic enhancer (Eμ) [46, 47] only develop lymphoma, usually aggressive lymphomas, after the addition of the oncogene, myc [48]. Finally, about 10–15 % of individuals with FL do not harbor the t(14;18) yet develop tumors that are morphologically similar. The t(14;18) translocation occurs at lower frequencies in other lymphoproliferative disorders indicating that it is not unique to FL.

5.7 Intermediate steps in FL development

Tumor development is recognized as a multistep oncogenic process requiring several genetic events before the manifestation of overt malignancy [49]. In FL, it appears this malignant course can be protracted over periods of several years or indeed over a decade, with emergent data suggestive of potential pre-malignant lesions believed to exist in either the peripheral blood or lymph node compartments (Fig. 5.2).

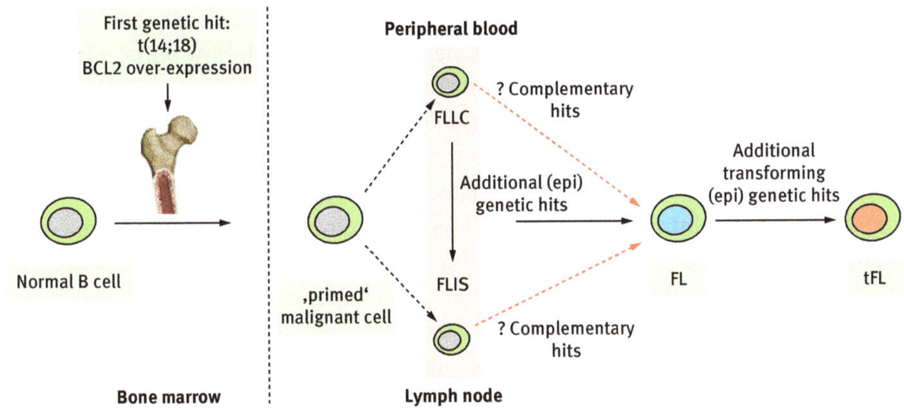

Fig. 5.2: **Current model of follicular lymphoma initiation and progression.** The first genetic hit, t(14;18) occurs in the bone marrow, thought to prime the B cell. Subsequent intermediary steps are believed to occur within the peripheral blood, FLLC, and/or within the lymph node, FLIS, with the acquisition of additional complementary genetic hits [52–54]. Progression to overt FL might occur via these intermediate steps or directly from the 'primed' malignant B cell, with further hits leading to transformed FL.

Precursor lesions identified within the lymph node, termed FL in situ (FLIS), were described over a decade ago [50]. FLIS represents hyperplastic GCs colonized with a variable population of B cells that stain double positive for BCL2 and CD10 with the remainder of the unaffected follicles within the lymph node BCL2 negative. The true incidence of FLIS is unclear as they are typically detected inadvertently following biopsies for suspicious lymphadenopathy, although it occurs in approximately 2 % of all individuals in which lymph nodes are removed for other reasons than lymphoma diagnostics [51]. The risk of progression to overt FL is low (< 5 %) and suggests that the full complement of genetic changes required for tumor development have yet to be acquired. This was verified in two independent studies that compared copy number alterations (CNAs) between FLIS lesions and overt FL [52, 53]. Both studies showed that, although a degree of genomic instability was present in these presumed precursor lesions, this was comparatively lower than overt FL by approximately 6-fold.

Whilst circulating t(14;18) positive B cells referred to as FL-like cells (FLLC) occur in healthy individuals, the level of the t(14;18)-containing FLLCs could predict one's risk of developing FL. Individuals carrying more than one t(14;18)-containing cell per 10,000 cells appear to be at a higher risk of developing FL even after an interval of up to 15 years [54]. In cases that developed FL, molecular backtracking confirmed identical *BCL2-IGH* rearrangements in the pre-diagnostic blood samples, alluding to the presence of these pre-malignant precursors several years prior to the manifestation of overt FL. Moreover, these FLLCs can preferentially accumulate within the GCs of lymph nodes where they exhibit features and reside as nonproliferating cells [55], where it is believed that they undergo multiple rounds of GC re-entries providing an environment where acquisition of genomic instability and genetic alterations can be fostered, leading to the eventual development of overt FL [56]. How purposeful these precursor blood-based FLLCs might be remains unclear. Whilst these findings have the theoretical potential for early detection of pre-malignant lesions, it would seem premature to embark on routine peripheral blood screening in healthy individuals, as the positive predictive value for cancer detection is very low.

These studies intimate varying degrees of heterogeneity in these intermediary states, with some subsequently progressing to overt FL and others not, however the process that drives this selection and commitment remains unknown. A more detailed molecular characterization of these precursors and close longitudinal follow up of individuals will be required to define the natural history of these putative pre-malignant events.

5.8 Existence of a progenitor clone

While we commonly think of the origin of FL as that of a germinal center derivation, the exact definition of the cell of origin covers two notions, one being the final representation of the mature tumor which mimics the normal GC counterparts, centrocytes and centroblasts; however, the other is where the disease begins, the molecular ontogeny. Unlike in leukemias, where the concept of tumor-initiating cells is well established [57], the idea of lymphoma-initiating cells is largely unexplored.

Early evidence for the existence of a tumor-initiating clonogenic population in lymphoma came about from molecular analyses of composite lymphomas, defined as the association of two lymphomas with differing cell of origin at the same anatomical site. The identification of common clone-specific IgH gene rearrangement or translocations between the two lymphomas suggested that each of the lymphomas could be derived from early lymphoid progenitors or precursor cells that subsequently clonally diverge to evolve as distinct lymphoid tumors. In FL, studying SHM, CSR or CNA patterns in sequential tumors revealed that most biopsies within the same individual were clonally related [58–62]. Furthermore, two case reports of donor-derived FL following stem cell transplantation illustrates the existence of a progenitor or lymphoma-initiating population that must have been transferred from the donor to the recipient at the time of the transplantation [59, 63]. In both of these studies, the donor and recipient biopsies shared identical *BCL2-IGH* rearrangements and further genetic alterations including IGH V(D)J rearrangements and somatic mutations, indicating that this progenitor population is likely more genetically complex and evolved than one would expect.

Collectively, these data support the pre-existence of a putative founder or lymphoma-initiating population that has been referred to as the common progenitor cell, CPC [6, 64, 65]. From the studies, one surmises that the population has the following features: (i) based on the bone marrow transplant cases, they can exist and lay dormant for long periods of time; (ii) they retain self-renewal capacity and; (iii) they continue to diversify as the disease evolves likely influenced by the host's tumoral environment. The exact phenotype, plasticity and site of residence of this lymphoma-initiating population remain unclear.

5.9 Molecular genetics of FL

Historically, efforts to decipher the genomic basis of FL have mainly implemented experimental techniques to identify broad chromosomal changes. In recent years, advances in sequencing approaches have improved the resolution, sensitivity and subsequently our ability to characterize both focal and genome-wide aberrations.

5.9.1 Copy number alterations and copy neutral loss of heterozygosity

In addition to t(14;18), almost all cases of FL have additional chromosomal imbalances. Through a combination of conventional cytogenetics, array comparative genomic hybridization (aCGH) and DNA single nucleotide polymorphism (SNP) microarray approaches, deletions in 1p36, 6q, 10q, 13p and 17p and gains in 1q, 2p, 7, 8, 12q, 18q and X are frequent features of FL [66–73]. The variable size of copy number aberrations means several gene loci can be affected simultaneously and indeed, the ability to pinpoint the exact gene(s) targeted within these regions can be difficult. Moreover, the utility of chromosomal alterations as potential prognostic biomarkers is unclear, as studies have demonstrated considerable variability and lack of reproducibility.

Copy neutral loss of heterozygosity (cnLOH) or acquired uniparental disomy (aUPD) arise from mitotic recombination or nondisjunction and provide a mechanism that allows a cell to be rendered homozygous for a pre-existing abnormality without the gene dosage being altered, and therefore cannot usually be detected by conventional cytogenetics, FISH or aCGH techniques. This mechanism was first reported in acute myeloid leukemia (AML) [74, 75]. The implementation of high resolution SNP profiling has allowed the identification of recurrent regions of aUPD in FL including 1p36, 6p, 9p, 12q and 16p [76–78]. aUPD at 1p36 and 16p regions correlate with poorer overall survival and histological transformation [77].

5.9.2 Genetic aberrations affecting key biological and signaling pathways

As the majority of lymphomas are now better characterized genetically, one emergent finding is that a number of genetic aberrations overlap with various other lymphoma subtypes suggesting a degree of relatedness but also some that are uniquely distinct for individual lymphoma subtypes. The prevalence of these aberrations provides a measure of how reliant each of the subtypes is on the distinct oncogenic pathways. A summary of the main biological pathways and genes affecting in FL are illustrated in Fig. 5.3.

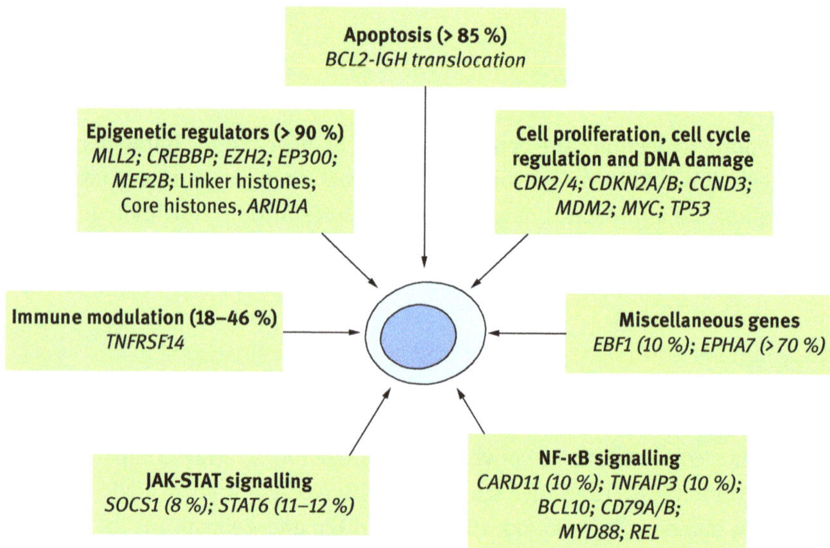

Fig. 5.3: **Summary of the main genetic events and biological pathways leading to the pathogenesis of FL.** Seven functional categories are shown with most commonly identified gene mutations or alterations listed. Each FL tumor can harbor several gene mutations/alterations from within the same or several functional categories.

5.9.3 Alterations in epigenetic regulators

In normal cells, regulation of gene expression is implemented by specific epigenetic mechanisms that switch chromatin, a complex of DNA and histone proteins, between two states: condensed transcriptionally repressed heterochromatin and transcriptionally active euchromatin. This interchange is mediated by a number of processes including DNA methylation, covalent histone post-translational modifications (PTMs), chromatin remodeling and noncoding RNA regulation, of which the former two mechanisms are the best characterized. These epigenetic modifications are very dynamic processes effectuated by families of enzymes and proteins. There is mounting evidence that epigenetic deregulation is becoming an emerging hallmark in all cancers.

Among the most prevalent alterations identified in FL are somatic mutations affecting genes encoding enzymes involved in histone PTMs [79–81]. Almost 90 % of FL tumors harbor mutations in the histone methyltransferase, *MLL2* (also known as *KMT2D*), a frequency that is on par with the t(14;18). *MLL2* is responsible for methylation at lysine 4 on histone 3 (H3K4), leading to an evolutionary conserved activatory histone mark. The majority of the mutations are truncating or frameshift mutations expected to lead to loss of the protein, with more than half of the tumors possessing multiple *MLL2* mutations. Its role in lymphomagenesis has yet to be elucidated although *MLL2* knockouts are embryonically lethal in mice [82]. Moreover, Kabuki syndrome, a rare pediatric disorder associated with multiple congenital anomalies and varying degree of intellectual disability, is caused by pathogenic mutations in *MLL2* with a mutation repertoire that mirrors those seen in FL; however, there is no demonstrable increased risk in lymphoma development [83].

Somatic mutations and or deletions in histone acetyltransferase (HAT) enzymes, *CREBBP* and *EP300*, are also frequent in FL. In *CREBBP*, the mutations are strongly clustered within the catalytic HAT domain, whilst its locus is frequently affected by aUPD, therefore the mutations are typically rendered homozygous with high mutation loads. Functionally, the mutants exhibit reduced ability to acetylate both histone and nonhistone substrates [81], supporting a tumor suppressor role in lymphoma.

Another histone methyltransferase, *EZH2*, is mutated in approximately 25 % of FL [84–87]. EZH2 is a critical enzymatic component of the polycomb repressor complex 2 (PRC2) and is responsible for methylation at histone 3 lysine 27 (H3K27) resulting in a repressive mark and subsequent gene silencing [88, 89]. Notably, the mutations are heterozygous and centered within the catalytic SET domain, the most common being mutations affecting three hotspot amino acid residues, Y646, A682 and A692. Functional studies confirmed the *EZH2* mutants were gain-of-function leading to a global increase in trimethylation of the H3K27 mark. These studies highlighted an important synergy between the wild-type and mutant forms of EZH2. Whilst wild-type EZH2 is efficient at catalyzing the monomethylation of H3K27, mutant EZH2 has an enhanced ability for the subsequent di- and tri-methylation steps [90–93]. Its role in lymphomagenesis is slowly being elucidated. During normal lymphopoiesis, EZH2 expression is undetectable in resting naïve B cells but peaks as it transitions to proliferating GC B cells. Silencing of EZH2 in DLBCL cell lines resulted in acute cell cycle arrest at the G1-S transition stage and upregulation of several cell cycle tumor suppressor target genes, providing clear evidence of its role in facilitating neoplastic B-cell proliferation [94]. More recently, functional analyses proved EZH2 to be essential for GC formation and suggested mutant EZH2 functions as a master regulator of the GC phenotype through repression of specific cell cycle checkpoint genes (CDKN1A) and transcription factors (BLIMP1 and IRF4) via H3K27 trimethylation [95, 96]. Notably, the gain of function nature of the mutations prompted the search for and development of novel targeted inhibitors that have shown promise pre-clinically with preferential activity against mutant lymphoma cell lines and *in vivo* models [97, 98].

Mutations in genes encoding components of chromatin remodeling complexes are also a feature of FL. Genes coding linker histones (H1) are mutated in over a quarter of FL tumors, predominantly missense mutations clustered within the highly conserved globular domain, with the majority affecting the *HIST1H1C* and *HIST1H1E* subtypes. In normal cellular processes, these histones are involved in facilitating the folding of higher-order chromatin structures and regulating access of histone-modifying enzymes and chromatin remodeling complexes to their target genes. The tumorigenic mutation impairs its association with chromatin and is hypothesized to consequently affect regulation of specific genes, although a clear picture of this process is still lacking. Recently, mutations affecting *ARID1A*, encoding a member of the ATP-dependent chromatin remodeling complex SNF/SWI, has been identified in approximately 10–15 % of cases [99]. Early evidence suggests that the mutations, which are mainly truncating, may impair normal DNA damage response mechanisms and render the cells more prone to genomic instability [99].

5.9.4 Alterations in immune modulators

Immune modulation is a mechanism frequently employed by tumors to evade the host's natural immune responses. *TNFRSF14*, a bidirectional signaling molecule involved in B and T cell activation or inhibition, is one of the best studied genes in this functional category. A number of studies have revealed frequent mutations of *TNFRSF14* with a frequency ranging from 18–46 % [100, 101]. Additionally, the 1p36 genomic locus where *TNFRSF14* resides is targeted by recurrent deletions and aUPD. Whilst the presence of *TNFRSF14* mutations was initially linked to inferior outcome, this was not validated in a subsequent study and requires examination in larger patient cohorts. TNFRSF14, tumor necrosis factor super-receptor family 14 (also called herpes virus entry mediator, HVEM) is a surface molecule ubiquitously expressed by hematopoietic cells (B and T cells) belonging to the TNFR superfamily, with important roles in co-stimulation and co-inhibition. It performs a diverse and complex set of roles in controlling innate and adaptive immune responses, owing to its ability to interact with five different ligands: BTLA (B and T lymphocyte attenuator), LIGHT (also known as TNFSF14), CD160, lymphotoxin α and gD [102, 103]. Normally, it acts as an immune regulating molecular switch alternating between activatory and inhibitory signaling depending on which of its five ligands it engages and the stoichiometry of this engagement, with overriding evidence supporting co-stimulatory signaling through LIGHT binding and co-inhibitory signaling through BTLA. Its functional remit in FL initiation and development has not been fully explored.

Other genes involved in antigen recognition and immune modulation such as *B2M* (beta-2-microglobulin), *CD58* and *HLA* genes have been reported as frequent genetic events in DLBCL, however, the prevalence of these aberrations in FL has yet to be examined.

5.9.5 Alterations affecting NF-κB signaling

Deregulation and, in particular, constitutive activation of the anti-apoptotic NF-κB signaling pathway, caused by mutations affecting positive and negative regulators, is an established feature of a number of B-cell lymphomas especially activated B-cell-like subtype of DLBCL (ABC-DLBCL). In contrast, these mutations occur significantly less frequently in GC lymphomas such as Burkitt's and GC B-cell-like DLBCL (GCB DLBCL) [104–106].

Surprisingly, genetic aberrations affecting both positive and negative regulators of NF-κB and B-cell receptor signaling are frequent in FL. *CARD11* encoding a key scaffolding protein in the CBM (CARD11-BCL10-MALT1) signalosome complex that mediates NF-κB activation upon antigen receptor ligation in B cells is mutated in just over 10 % of cases, whilst *TNFAIP3* encoding the enzyme A20 that acts as a negative regulator of canonical NF-κB signaling, is mutated at a similar frequency. Other components along the signaling pathway including *CD79A*, *CD79B* and *BCL10* are mutated at frequencies of less than 5 %. The type and location of the detected aberrations are reminiscent of those seen in ABC DLBCLs, with the exception of *CD79B*. *CARD11* mutations occur within the coiled-coil domain and are known to be activating aberrations [106], whilst the majority of the *TNFAIP3* mutations were inactivating, usually combined with deletions of the second allele. Whereas the *CD79B* variants observed in FL were frameshift mutations affecting the immunoreceptor tyrosine-based activation motif (ITAM) domain, in contrast to predominantly missense mutations at the tyrosine 196 amino acid residue seen in ABC DLBCL [105].

Somatic mutations of *MYD88*, which encodes an immune receptor adaptor protein involved in Toll-like receptor (TIR) signaling, appear to occur predominantly during progression and transformation of FL. Recurrent oncogenic *MYD88* mutations were first identified in ABC DLBCLs [107] and subsequently in other lymphoma subtypes including Waldenstrom's macroglobulinemia, marginal zone lymphomas and infrequently CLL [107–112]. Mechanistically, the consequence of the mutation is activation of IL1-associated kinases (IRAK-1 and IRAK-4) that engage the NFκB and MAP kinase pathways leading to the production of IL6, IL10 and autocrine activation of the JAK pathway [105]. In these non-FL subtypes, the mutation affects a specific hotspot residue within the TIR domain, leading to the amino-acid substitution p.L265P. Conversely, in FL, the majority of mutations affect alternate residues within the TIR domain and are restricted to the transformation biopsies.

5.9.6 Alterations affecting JAK-STAT signaling

The Janus kinase (JAK)-signal transducer and activator of transcription (STAT) pathway mediates signaling by cytokine, chemokine and growth factor receptors and is critical in normal hematopoiesis. Many of these receptors transmit a plethora of cellular anti-apoptotic, proliferative and differentiation signals, and constitutive activation of the pathway has been implicated in several malignancies, including hematological cancers. Its role in FL development is still emerging. Mutations in *SOCS1*, a negative regulator of the JAK-STAT pathway that terminates the signaling by direct inhibition of the JAK tyrosine kinase, occur in less than 10 % of cases [80]. Aberrations in the STAT family of proteins are rare with the exception of *STAT6*, a cytoplasmic transcription factor, which is somatically mutated in approximately 11 % of cases [113, 114]. All the mutations were missense in nature and clustered within the DNA-binding domain, suggesting they play a role in disruption of this interaction. A recent study illustrated the mutations as activating in nature using *in vitro* transfected cell lines and primary tumor samples. Here, *STAT6* mutants were shown to preferentially localize to the nucleus and induce STAT6 target gene expression, particularly in the presence of IL-4 [114]. Similar mutations of *SOCS1* and *STAT6* have been observed in DLBCL, Hodgkin's and primary mediastinal B-cell lymphoma (PMBCL). Together, both mutations are thought to constitutively activate the signaling pathway and promote survival of tumor cells.

5.9.7 Alterations affecting genes involved in cell proliferation and cell cycle regulation

One of the key characteristics of tumor cells is their ability to proliferate uncontrolled and self-autonomously. To achieve this, tumor cells must circumvent the actions of several oncogenes and tumor suppressors. It is well documented in FL that several genes involved in the regulatory mechanisms of these processes are altered to give the tumor population a proliferative advantage. Alterations in the proto-oncogene, *MYC*, and its target genes correlate with FL transformation [115–117]. Mutations and deletions of the genomic loci encompassing *TP53*, a tumor suppressor gene, are frequently observed in FL linked with an adverse outcome [118, 119] and increase in the progression or transformation phase of the malignancy. In addition, certain cell cycle components and their regulators are also recurrently perturbed due to copy number aberrations. The genomic region 12q13–15 is subject to frequent copy number gains and interestingly this region encompasses several cell cycle regulatory components. Increased expression of MDM2, an ubiquitin-protein ligase that degrades TP53, is frequently linked to transformation [120, 121]. Likewise, cyclin-dependent kinases, *CDK2* and *CDK4*, that are essential for G1/S transition of the cell cycle, are frequently amplified and associated with high-risk FL. CDK activity is negatively regulated by an inhibitor CDKN2a/p16, thus preventing their interaction with cyclin D and their subsequent phosphorylation of the retinoblastoma (RB) proteins. The gene locus of *CDKN2A/p16* is subject to recurrent heterozygous and homozygous deletions, implicating a removal of its inhibitory effect allowing the potentially uncontrolled phosphorylation of RB proteins leading to the release of E2F transcription factors and continuous cell cycle progression. Cyclin D3 (CCND3), which encodes a binding partner of the CDKs, is infrequently mutated in FL, however co-occurs with CDK4 amplifications. Finally, the tumor suppressor, RB1 is also subject to deletions. There is a suggestion that these alterations occur in a mutually exclusive manner [122] and perturbation of this set of genes converge on both the p53 and RB axis, decreasing their activity, deregulating cell cycle and promoting tumor proliferation and survival.

5.9.8 Alterations in miscellaneous genes

Ephrin receptor A7 (EPHA7) was uncovered as a tumor suppressor gene within the frequently deleted 6q [123]. Whilst *EPHA7* is deleted in almost a quarter of cases, its protein expression is reduced or absent in about 70 % of specimens, thought to be a consequence of additional epigenetic silencing. EphA7 knockdown in a Bcl2 transgenic mouse model accelerated lymphoma onset with the lymphomas generated retaining key histological features of FL and demonstrated evidence of ongoing SHM like in human FL. Ephrin receptors are tyrosine kinases that dimerize and bind ephrin-A family ligands residing on adjacent cells, leading to bidirectional cell-cell interactions, particularly activating components of the ERK signaling pathway. Normally, GC B cells express only the truncated $EPHA7^{TR}$ isoform. In the absence of the full-length EPHA7 receptor, heterodimerization occurs with the EPHA2 receptor, thus inhibiting its homodimerization in a dominant-negative fashion. This maintains the EPHA2 receptor in an inactive unphosphorylated state. In FL, when $EPHA7^{TR}$ is deleted or epigenetically silenced, EPHA2 is able to homodimerize and activate oncogenic signaling pathways through ERC and STAT3. The administration of a recombinant EPHA7 ectodomain protein (EPHA7FC) was shown to inhibit EPHA2 phosphorylation, reduce ERK signaling and inhibit cell growth *in vitro*, promoting tumor regression in human lymphoma xenografts.

EBF1 (early B-cell factor 1) is a transcription factor responsible for lineage commitment in B-cell development and operates as a homodimer binding the consensus sequence 5'TCCCNNGGGA [124, 125]. It is mutated in about 10 % of cases, with more than half of the mutations affecting the DNA-binding domain (DBD) at evolutionarily conserved residues. Target gene binding by EBF1 can induce transcriptional activation, repression or H3K4 dimethylation. *EBF1* mutants appear to impair upregulation of known Ebf1 target genes *Igll1* and *Cd79b*, suggesting the mutations in FL are loss of function in nature [80].

5.10 Disease progression, transformation and clonal evolution

Examination of sequential FL tumors has provided the unique opportunity to study patterns of genomic evolution of FL through its progression and transformation of the disease. Earlier studies determined modes of evolution by studying SHM and CSR patterns, later advancing to implementing copy number and more recently, full genomic profiling approaches. As such, the majority of studies have provided overwhelming evidence for complex nonlinear modes of tumor evolution, rather than a process of progressive acquisition of genetic events. Having said that, later relapse and transformation disease events typically harbor many more genetic aberrations. This increase arises from the influence of a myriad of intrinsic and extrinsic variables including a presumably higher degree of genomic instability as the tumor evolves and the impact of prior cytotoxic chemotherapy in inducing further genomic damage as well as modulating differential clonal selection.

Moreover, by comparing diagnostic to relapse and transformation tumors, there is supportive evidence that traces the evolution of all biopsies from a putative ancestral clone, the so-called CPC [60, 80, 117, 126]. Phylogenetic trees have been employed to allow reconstruction of the clonal relationships between diagnostic and latter biopsies from the same patient. Typically, these have shown that the relapse or transformation biopsy arises by divergent evolution from this ancestral clone with the acquisition of distinct genetic aberrations that presumably confer a selective advantage and are selected to undergo clonal expansion (Fig. 5.4). Furthermore, the latter progression or transformed tumors can either display a high (referred to as a 'rich' CPC) or minimal (referred to as a 'sparse' CPC) degree of genetic semblance to the ancestral clone [80].

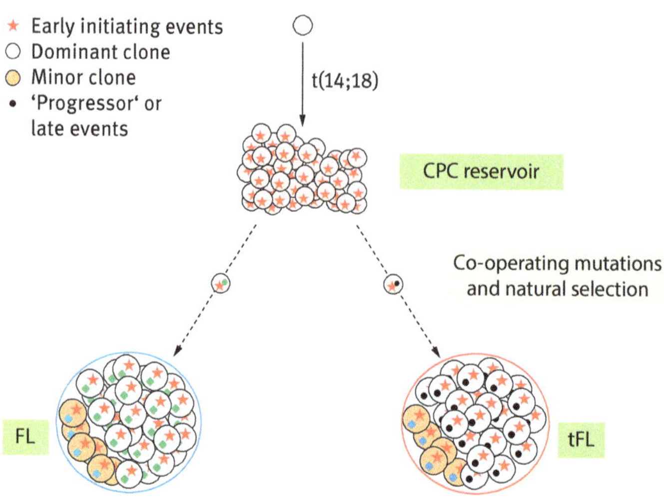

Fig. 5.4: **Tumor progression via divergent clonal evolution.** Following the acquisition of the t(14;18), acquisition of early initiating genetic events generates a reservoir population of common progenitor cells (CPC). These subsequently acquire a specific set of genetic events that are selectively advantageous, naturally selected out and undergo clonal expansion. The indolent and transformed FL tumors arise by divergent clonal evolution from this CPC population rather than sequential acquisition of genetic events in a linear evolution model.

Precise analyses of the clonal and genetic composition of temporally profiled tumors has brought into focus early, initiating versus progression- or transformation-specific genetic aberrations. A key attribute is the early occurrence of *BCL2* translocations with mutations in epigenetic regulators such as *CREBBP*, *EP300*, *EZH2* and *MLL2* that must occur within the ancestral population and likely represent initiating events that drive tumorigenesis [80, 117, 126]. These 'epimutations' are characteristically clonal, suggesting their presence within the entire tumor population, and are conserved during the progression of the disease from indolent FL to aggressive transformed FL. Conversely, aberrations that deregulate the cell cycle, promote DNA damage or activate signaling pathways that nurture cell survival (for example NF-κB) have typically been found to be late events that cultivate fitter tumor clones that drive progression and transformation [80, 117]. Gene expression profiling provides further evidence that oncogenic mechanisms which favored increased proliferation and NF-κB signaling were enriched at transformation [115, 127].

Overall, the prevalence of progression and transformation events during a patient's disease course suggests that our conventional measures are not specific enough to eradicate the tumor-initiating CPC population.

5.11 Other FL variants: Pediatric and t(14;18)-negative follicular lymphoma

Pediatric follicular lymphoma (PFL) is a variant of FL presenting as localized lymphadenopathy in the head and neck with a male predominance, and has been included as a new provisional entity in the WHO Lymphoma Classification 2008. Strict morphologic or immunophenotypic definitions have not been established and the demarcation with conventional FL is debatable. Clinically, in broad terms, PFL is distinct from FL with recurrence after therapy rare, PFLs lacking the t(14;18) signature lesion typically present with stage 1 disease and a high proliferation index (> 30 %) [128, 129]. While pediatric suggests a disease restricted to patients < 18 yrs, Louissant and colleagues [129] demonstrate PFL-like cases in adults into the 4th decade and have proposed "pediatric type" FL to represent this older age group. The similarity may not end there with striking overlap in features of PFL and t(14;18) negative cases of "adult" FL; these cases, which are thought to represent some 10 % of all FL patients frequently present with lower stage, high IPI scores (International Prognostic Index), with a portion staining positive for BCL2. Gene expression profiling hints at a distinct biology, with t(14;18)-negative profiles reminiscent of activated B-cell like tumors and enrichment for NF-κB signatures [130]. These changes do not translate into an overall survival advantage between "adult" FL with and without the t(14;18) but it is worth keeping in mind that the numbers of cases are few and historical in nature.

In comparison to FL [80], the mutational spectrum in PFL has not been explored in any significant depth. There is however a notably high frequency of mutations in *TNFRSF14* [131] suggesting that PFL may harbor an overlapping mutational repertoire with its incurable cousins. The best opportunity to assess the underlying biology and semblance between these lymphomas may ultimately rely on mutational profiling within PFL, "pediatric type" and "adult" FL where the *IgH-BCL2* status is available for comparison. This data may equally provide an explanation for the clinical heterogeneity observed across these groups.

5.12 Impact of the tumor microenvironment

FL cells exist immersed within a milieu of nonmalignant immune cells that compose the tumor microenvironment. In addition to the genetic alterations, there is now substantial evidence supporting the complementary crosstalk between tumor cells and its microenvironment, seen as pivotal in the disease development, maintenance and progression. A seminal study using gene expression profiling by Dave and colleagues identified that expression profiles derived from the nonneoplastic immune cells independently predict clinical outcome in FL, based on the T cell-macrophage composition [132]. A number of immunohistochemical studies were subsequently performed to recapitulate these observations and investigate the effect of specific cell populations of T cells, macrophages and stromal cell subsets on FL. These studies identified various associations between factors such as the number and/or spatial distribution of the infiltrating microenvironment immune cells and its impact on survival [133–137]. However, this approach produced differing outcomes that were not reproducible across the studies secondary to biases in patient selection criteria, treatment modalities and immunohistochemistry (IHC) scoring techniques.

Although the precise role of the cells that constitute the FL microenvironment is not fully elucidated, studies have proposed two overriding themes. Firstly, tumor cells can co-exist with their microenvironment as a consequence of signaling arising from the surrounding stroma that critically supports a tumor proliferative niche; and secondly, the tumor cells can modulate the immune cells within the microenvironment to be conducive for escaping immune surveillance. FL cells express high levels of CXCR4 and CXCR5 and are homed into follicles by chemokines such as CXCL12 and CXCL13 that are released by follicular T helper (FTH) cells or follicular dendritic cells (FDCs) [138, 139]. The role of stromal cells in FL pathogenesis has extended beyond just tumor cell homing as recent *in vitro* studies have shown that they also promote tumor B-cell survival, contribute to monocyte recruitment via tumor secretion of CCL2, and polarize macrophages toward the TAM (tumor associated macrophages)-like phenotype [140, 141].

Given the importance of this immune cell repertoire in normal B cell and FL maintenance, understanding how acquired genetic changes in FL modulate the interactions between the malignant B cell and the microenvironment populations may provide further insight into the mechanisms of the crosstalk and provide potential therapeutic avenues on how best to reverse these effects.

5.13 Susceptibility loci for FL

The FL diversity has not impeded genome-wide association efforts (GWAS) that have established the candidacy of the human leukocyte antigen (HLA) class I and II regions on 6p21.32-33 [142–144]. Indeed, in the latest study by Skibola and colleagues they amassed 4,523 case subjects and 13,344 controls as part of a large scale two-stage GWAS multicenter international initiative of European ancestry [145]. Their experiments have uncovered 5 non-HLA loci linked with FL risk on 11q23.3, 11q24.3, 3q28, 18q21.33 and 8q24.21. While functional validation of these variants is likely to prove challenging, these data highlight the complexity of FL and challenges faced in teasing apart the early initiating events in this indolent lymphoma.

5.14 Concluding remarks

Advances in the molecular characterization of FL have yielded important insights into the mechanisms underlying its initiation and progression. Whilst for decades the t(14;18) epitomized the first genetic hit in lymphomagenesis, it has become evident that epigenetic deregulation is as important a hallmark of the disease. Detailed functional studies are underway to decipher how each of these genetic attributes contribute independently and how they cooperate together in the development of FL. The identification of potential molecular targets and an understanding of the evolution mechanisms that drive progression will provide the impetus for development of novel targeted therapeutic approaches and may ultimately lead to a refinement in treatment algorithms towards a stratified or precision-led strategy in the future.

References

[1] Bosga-Bouwer AG, van Imhoff GW, Boonstra R, et al. Follicular lymphoma grade 3B includes 3 cytogenetically defined subgroups with primary t(14;18), 3q27, or other translocations: t(14;18) and 3q27 are mutually exclusive. Blood 2003;101:1149–1154.

[2] Horn H, Schmelter C, Leich E, et al. Follicular lymphoma grade 3B is a distinct neoplasm according to cytogenetic and immunohistochemical profiles. Haematologica 2011;96:1327–1334.

[3] Karube K, Guo Y, Suzumiya J, et al. CD10-MUM1+ follicular lymphoma lacks BCL2 gene translocation and shows characteristic biologic and clinical features. Blood 2007;109:3076-3079.

[4] Katzenberger T, Ott G, Klein T, Kalla J, Müller-Hermelink HK, Ott MM. Cytogenetic alterations affecting BCL6 are predominantly found in follicular lymphomas grade 3B with a diffuse large B-cell component. Am J Pathol 2004;165:481–490.

[5] Ott G, Katzenberger T, Lohr A, et al. Cytomorphologic, immunohistochemical, and cytogenetic profiles of follicular lymphoma: 2 types of follicular lymphoma grade 3. Blood 2002;99:3806–3812.

[6] Montoto S, Fitzgibbon J. Transformation of indolent B-cell lymphomas. J Clin Oncol 2011;29:1827–1834.

[7] Al-Tourah AJ, Gill KK, Chhanabhai M, et al. Population-based analysis of incidence and outcome of transformed non-Hodgkin's lymphoma. J Clin Oncol 2008;26:5165–5169.

[8] Bastion Y, Blay JY, Divine M, et al. Elderly patients with aggressive non-Hodgkin's lymphoma: disease presentation, response to treatment, and survival–a Groupe d'Etude des Lymphomes de l'Adulte study on 453 patients older than 69 years. J Clin Oncol 1997;15:2945–2953.

[9] Link BK, Maurer MJ, Nowakowski GS, et al. Rates and outcomes of follicular lymphoma transformation in the immunochemotherapy era: a report from the University of Iowa/Mayo Clinic specialized program of research excellence molecular epidemiology resource. J Clin Oncol 2013;31:3272–3278.

[10] Bodet-Milin C, Kraeber-Bodéré F, Moreau P, Campion L, Dupas B, Le Gouill S. Investigation of FDG-PET/CT imaging to guide biopsies in the detection of histological transformation of indolent lymphoma. Haematologica 2008;93:471–472.

[11] Noy A, Schöder H, Gönen M, et al. The majority of transformed lymphomas have high standardized uptake values (SUVs) on positron emission tomography (PET) scanning similar to diffuse large B-cell lymphoma (DLBCL). Ann Oncol 2009;20:508–512.

[12] Wondergem MJ, Rizvi SN, Jauw Y, et al. 18F-FDG or 3'-Deoxy-3'-18F-Fluorothymidine to detect transformation of follicular lymphoma. J Nucl Med 2015;56:216–221.

[13] De Jong D, Voetdijk BM, Beverstock GC, van Ommen GJ, Willemze R, Kluin PM. Activation of the c-myc oncogene in a precursor-B-cell blast crisis of follicular lymphoma, presenting as composite lymphoma. N Engl J Med 1988;318:1373–1378.

[14] Gauwerky CE, Hoxie J, Nowell PC, Croce CM. Pre-B-cell leukemia with a t(8; 14) and a t(14; 18) translocation is preceded by follicular lymphoma. Oncogene 1988;2:431–435.

[15] Kishimoto W, Shirase T, Chihara D, et al. Double-hit lymphoma with a feature of follicular lymphoma concurrent with clonally related B lymphoblastic leukemia: a preference of transformation for the bone marrow. J Clin Exp Hematop 2012;52:113–119.

[16] Xu X, Zhang L, Wang Y, Zhang Q, Sun B, Zhang Y. Double-hit and triple-hit lymphomas arising from follicular lymphoma following acquisition of MYC: report of two cases and literature review. Int J Clin Exp Pathol 2013;6:788–794.

[17] Young KH, Xie Q, Zhou G, et al. Transformation of follicular lymphoma to precursor B-cell lymphoblastic lymphoma with c-myc gene rearrangement as a critical event. Am J Clin Pathol 2008;129:157–166.

[18] Feldman AL, Arber DA, Pittaluga S, et al. Clonally related follicular lymphomas and histiocytic/dendritic cell sarcomas: evidence for transdifferentiation of the follicular lymphoma clone. Blood 2008;111:5433–5439.

[19] Wang E, Papalas J, Hutchinson CB, et al. Sequential development of histiocytic sarcoma and diffuse large b-cell lymphoma in a patient with a remote history of follicular lymphoma with genotypic evidence of a clonal relationship: a divergent (bilineal) neoplastic transformation of an indolent B-cell lymphoma in a single individual. Am J Surg Pathol 2011;35:457–463.

[20] Tsujimoto Y, Cossman J, Jaffe E, Croce CM. Involvement of the bcl-2 gene in human follicular lymphoma. Science 1985a;228:1440–1443.

[21] Weiss LM, Warnke RA, Sklar J, Cleary ML. Molecular analysis of the t(14;18) chromosomal translocation in malignant lymphomas. N Engl J Med 1987;317:1185–1189.

[22] Albinger-Hegyi A, Hochreutener B, Abdou MT, et al. High frequency of t(14;18)-translocation breakpoints outside of major breakpoint and minor cluster regions in follicular lymphomas: improved polymerase chain reaction protocols for their detection. Am J Pathol 2002;160:823–832.

[23] Buchonnet G, Jardin F, Jean N, et al. Distribution of BCL2 breakpoints in follicular lymphoma and correlation with clinical features: specific subtypes or same disease? Leukemia 2002;16:1852–1856.

[24] Cleary ML, Sklar J. Nucleotide sequence of a t(14;18) chromosomal breakpoint in follicular lymphoma and demonstration of a breakpoint-cluster region near a transcriptionally active locus on chromosome 18. Proc Natl Acad Sci USA 1985;82:7439–7443.

[25] Ngan BY, Nourse J, Cleary ML. Detection of chromosomal translocation t(14;18) within the minor cluster region of bcl-2 by polymerase chain reaction and direct genomic sequencing of the enzymatically amplified DNA in follicular lymphomas. Blood 1989;73:1759–1762.

[26] Tsai AG, Lu H, Raghavan SC, Muschen M, Hsieh CL, Lieber MR. Human chromosomal translocations at CpG sites and a theoretical basis for their lineage and stage specificity. Cell 2008;135:1130–1142.

[27] Tsujimoto Y, Gorham J, Cossman J, Jaffe E, Croce CM. The t(14;18) chromosome translocations involved in B-cell neoplasms result from mistakes in VDJ joining. Science 1985b;229:1390–1393.

[28] Hirt C, Dölken G, Janz S, Rabkin CS. Distribution of t(14;18)-positive, putative lymphoma precursor cells among B-cell subsets in healthy individuals. Br J Haematol 2007;138:349–353.

[29] Muramatsu M, Kinoshita K, Fagarasan S, Yamada S, Shinkai Y, Honjo T. Class switch recombination and hypermutation require activation-induced cytidine deaminase (AID), a potential RNA editing enzyme. Cell 2000;102:553–563.

[30] Hardianti MS, Tatsumi E, Syampurnawati M, et al. Activation-induced cytidine deaminase expression in follicular lymphoma: association between AID expression and ongoing mutation in FL. Leukemia 2004;18:826–831.

[31] Smit LA, Bende RJ, Aten J, Guikema JE, Aarts WM, van Noesel CJ. Expression of activation-induced cytidine deaminase is confined to B-cell non-Hodgkin's lymphomas of germinal-center phenotype. Cancer Res 2003;63:3894–3898.

[32] Pasqualucci L, Bhagat G, Jankovic M, et al. AID is required for germinal center-derived lymphomagenesis. Nat Genet 2008;40:108–112.

[33] Migliazza A, Martinotti S, Chen W, et al. Frequent somatic hypermutation of the 5' noncoding region of the BCL6 gene in B-cell lymphoma. Proc Natl Acad Sci USA 1995;92:12520–12524.

[34] Pasqualucci L, Migliazza A, Fracchiolla N, et al. BCL-6 mutations in normal germinal center B cells: evidence of somatic hypermutation acting outside Ig loci. Proc Natl Acad Sci USA 1998;95:11816–11821.

[35] Mottok A, Renné C, Seifert M, et al. Inactivating SOCS1 mutations are caused by aberrant somatic hypermutation and restricted to a subset of B-cell lymphoma entities. Blood 2009;114:4503–4506.

[36] Schuetz JM, Johnson NA, Morin RD, et al. BCL2 mutations in diffuse large B-cell lymphoma. Leukemia 2012;26:1383–1390.

[37] Wang Q, Oliveira T, Jankovic M, et al. Epigenetic targeting of activation-induced cytidine deaminase. Proc Natl Acad Sci USA 2014;111:18667–1872.

[38] Meng FL, Du Z, Federation A, et al. Convergent transcription at intragenic super-enhancers targets AID-initiated genomic instability. Cell 2014;159:1538–1548.

[39] Qian J, Wang Q, Dose M, et al. B cell super-enhancers and regulatory clusters recruit AID tumorigenic activity. Cell 2014;159:1524–1537.

[40] Dölken G, Illerhaus G, Hirt C, Mertelsmann R. BCL-2/JH rearrangements in circulating B cells of healthy blood donors and patients with nonmalignant diseases. J Clin Oncol 1996;14:1333–1344.
[41] Limpens J, Stad R, Vos C, et al. Lymphoma-associated translocation t(14;18) in blood B cells of normal individuals. Blood 1995;85:2528–2536.
[42] Summers KE, Goff LK, Wilson AG, Gupta RK, Lister TA, Fitzgibbon J. Frequency of the Bcl-2/IgH rearrangement in normal individuals: implications for the monitoring of disease in patients with follicular lymphoma. J Clin Oncol 2001;19:420–424.
[43] Baccarelli A, Hirt C, Pesatori AC, et al. t(14;18) translocations in lymphocytes of healthy dioxin-exposed individuals from Seveso, Italy. Carcinogenesis 2006;27:2001–2007.
[44] Schüler F, Dölken L, Hirt C, et al. Prevalence and frequency of circulating t(14;18)-MBR translocation carrying cells in healthy individuals. Int J Cancer 2009;124:958–963.
[45] Roulland S, Navarro JM, Grenot P, et al. Follicular lymphoma-like B cells in healthy individuals: a novel intermediate step in early lymphomagenesis. J Exp Med 2006;203:2425–2431.
[46] Katsumata M, Siegel RM, Louie DC, et al. Differential effects of Bcl-2 on T and B cells in transgenic mice. Proc Natl Acad Sci USA 1992;89:11376–11380.
[47] McDonnell TJ, Deane N, Platt FM, et al. bcl-2-immunoglobulin transgenic mice demonstrate extended B cell survival and follicular lymphoproliferation. Cell 1989;57:79–88.
[48] McDonnell TJ, Korsmeyer SJ. Progression from lymphoid hyperplasia to high-grade malignant lymphoma in mice transgenic for the t(14; 18). Nature 1991;349:254–256.
[49] Vogelstein B, Kinzler KW. The multistep nature of cancer. Trends Genet 1993;9:138–141.
[50] Cong P, Raffeld M, Teruya-Feldstein J, Sorbara L, Pittaluga S, Jaffe ES. In situ localization of follicular lymphoma: description and analysis by laser capture microdissection. Blood 2002;99:3376–3382.
[51] Henopp T, Quintanilla-Martínez L, Fend F, Adam P. Prevalence of follicular lymphoma in situ in consecutively analysed reactive lymph nodes. Histopathology 2011;59:139–142.
[52] Mamessier E, Song JY, Eberle FC, et al. Early lesions of follicular lymphoma: a genetic perspective. Haematologica 2014;99:481–488.
[53] Schmidt J, Salaverria I, Haake A, et al. Increasing genomic and epigenomic complexity in the clonal evolution from in situ to manifest t(14;18)-positive follicular lymphoma. Leukemia 2014;28:1103–1112.
[54] Roulland S, Kelly RS, Morgado E, et al. t(14;18) Translocation: A predictive blood biomarker for follicular lymphoma. J Clin Oncol 2014;32:1347–1355.
[55] Tellier J, Menard C, Roulland S, et al. Human t(14;18)positive germinal center B cells: a new step in follicular lymphoma pathogenesis? Blood 2014;123:3462–3465.
[56] Sungalee S, Mamessier E, Morgado E, et al. Germinal center reentries of BCL2-overexpressing B cells drive follicular lymphoma progression. J Clin Invest 2014;124:5337–5351.
[57] Bonnet D, Dick JE. Human acute myeloid leukemia is organized as a hierarchy that originates from a primitive hematopoietic cell. Nat Med 1997;3:730–737.
[58] Aarts WM, Bende RJ, Bossenbroek JG, Pals ST, van Noesel CJ. Variable heavy-chain gene analysis of follicular lymphomas: subclone selection rather than clonal evolution over time. Blood 2001;98:238–240.
[59] Carlotti E, Wrench D, Matthews J, et al. Transformation of follicular lymphoma to diffuse large B-cell lymphoma may occur by divergent evolution from a common progenitor cell or by direct evolution from the follicular lymphoma clone. Blood 2009;113:3553–3557.
[60] Eide MB, Liestøl K, Lingjaerde OC, et al. Genomic alterations reveal potential for higher grade transformation in follicular lymphoma and confirm parallel evolution of tumor cell clones. Blood 2010;116:1489–1497.

[61] Ruminy P, Jardin F, Picquenot JM, et al. S(mu) mutation patterns suggest different progression pathways in follicular lymphoma: early direct or late from FL progenitor cells. Blood 2008;112:1951–1959.
[62] Zelenetz AD, Chen TT, Levy R. Histologic transformation of follicular lymphoma to diffuse lymphoma represents tumor progression by a single malignant B cell. J Exp Med 1991;173:197–207.
[63] Weigert O, Kopp N, Lane AA, et al. Molecular ontogeny of donor-derived follicular lymphomas occurring after hematopoietic cell transplantation. Cancer Discov 2012;2:47–55.
[64] Fitzgibbon J, Iqbal S, Davies A, et al. Genome-wide detection of recurring sites of uniparental disomy in follicular and transformed follicular lymphoma. Leukemia 2007;21:1514–1520.
[65] Green MR, Alizadeh AA. Common progenitor cells in mature B-cell malignancies: implications for therapy. Curr Opin Hematol 2014;21:333–340.
[66] Bouska A, McKeithan TW, Deffenbacher KE, et al. Genome-wide copy-number analyses reveal genomic abnormalities involved in transformation of follicular lymphoma. Blood 2014;123:1681–1690.
[67] Cheung KJ, Shah SP, Steidl C, et al. Genome-wide profiling of follicular lymphoma by array comparative genomic hybridization reveals prognostically significant DNA copy number imbalances. Blood 2009;113:137–148.
[68] Horsman DE, Connors JM, Pantzar T, Gascoyne RD. Analysis of secondary chromosomal alterations in 165 cases of follicular lymphoma with t(14;18). Genes Chromosomes Cancer 2001;30:375–382.
[69] Horsman DE, Okamoto I, Ludkovski O, et al. Follicular lymphoma lacking the t(14;18)(q32;q21): identification of two disease subtypes. Br J Haematol 2003;120:424–433.
[70] Johnson NA, Al-Tourah A, Brown CJ, Connors JM, Gascoyne RD, Horsman DE. Prognostic significance of secondary cytogenetic alterations in follicular lymphomas. Genes Chromosomes Cancer 2008;47:1038–1048.
[71] Katzenberger T, Kalla J, Leich E, et al. A distinctive subtype of t(14;18)-negative nodal follicular non-Hodgkin lymphoma characterized by a predominantly diffuse growth pattern and deletions in the chromosomal region 1p36. Blood 2009;113:1053–1061.
[72] Schwaenen C, Viardot A, Berger H, et al. Microarray-based genomic profiling reveals novel genomic aberrations in follicular lymphoma which associate with patient survival and gene expression status. Genes Chromosomes Cancer 2009;48:39–54.
[73] Viardot A, Barth TF, Möller P, Döhner H, Bentz M. Cytogenetic evolution of follicular lymphoma. Semin Cancer Biol 2003;13:183–190.
[74] Fitzgibbon J, Smith LL, Raghavan M, et al. Association between acquired uniparental disomy and homozygous gene mutation in acute myeloid leukemias. Cancer Res 2005;65:9152–9154.
[75] Raghavan M, Lillington DM, Skoulakis S, et al. Genome-wide single nucleotide polymorphism analysis reveals frequent partial uniparental disomy due to somatic recombination in acute myeloid leukemias. Cancer Res 2005;65:375–378.
[76] Cheung KJ, Delaney A, Ben-Neriah S, et al. High resolution analysis of follicular lymphoma genomes reveals somatic recurrent sites of copy-neutral loss of heterozygosity and copy number alterations that target single genes. Genes Chromosomes Cancer 2010a;49:669–681.
[77] O'Shea D, O'Riain C, Gupta M, et al. Regions of acquired uniparental disomy at diagnosis of follicular lymphoma are associated with both overall survival and risk of transformation. Blood 2009;113:2298–2301.
[78] Ross CW, Ouillette PD, Saddler CM, Shedden KA, Malek SN. Comprehensive analysis of copy number and allele status identifies multiple chromosome defects underlying follicular lymphoma pathogenesis. Clin Cancer Res 2007;13:4777–4785.

[79] Morin RD, Mendez-Lago M, Mungall AJ, et al. Frequent mutation of histone-modifying genes in non-Hodgkin lymphoma. Nature 2011;476:298–303.
[80] Okosun J, Bödör C, Wang J, et al. Integrated genomic analysis identifies recurrent mutations and evolution patterns driving the initiation and progression of follicular lymphoma. Nat Genet 2014;46:176–181.
[81] Pasqualucci L, Dominguez-Sola D, Chiarenza A, et al. Inactivating mutations of acetyltransferase genes in B-cell lymphoma. Nature 2011;471:189–195.
[82] Glaser S, Schaft J, Lubitz S, et al. Multiple epigenetic maintenance factors implicated by the loss of Mll2 in mouse development. Development 2006;133:1423–1432.
[83] Ng SB, Bigham AW, Buckingham KJ, et al. Exome sequencing identifies MLL2 mutations as a cause of Kabuki syndrome. Nat Genet 2010;42:790–793.
[84] Bödör C, Grossmann V, Popov N, et al. EZH2 mutations are frequent and represent an early event in follicular lymphoma. Blood 2013;122:3165–3168.
[85] Bödör C, O'Riain C, Wrench D, et al. EZH2 Y641 mutations in follicular lymphoma. Leukemia 2011;25:726–729.
[86] Morin RD, Johnson NA, Severson TM, et al. Somatic mutations altering EZH2 (Tyr641) in follicular and diffuse large B-cell lymphomas of germinal-center origin. Nat Genet 2010;42:181–185.
[87] Ryan RJ, Nitta M, Borger D, et al. EZH2 codon 641 mutations are common in BCL2-rearranged germinal center B cell lymphomas. PLoS One 2011;6:e28585.
[88] Kirmizis A, Bartley SM, Kuzmichev A, et al. Silencing of human polycomb target genes is associated with methylation of histone H3 Lys 27. Genes Dev 2004;18:1592–1605.
[89] Viré E, Brenner C, Deplus R, et al. The Polycomb group protein EZH2 directly controls DNA methylation. Nature 2006;439:871–874.
[90] Majer CR, Jin L, Scott MP, et al. A687V EZH2 is a gain-of-function mutation found in lymphoma patients. FEBS Lett 2012;586:3448–3451.
[91] McCabe MT, Graves AP, Ganji G, et al. Mutation of A677 in histone methyltransferase EZH2 in human B-cell lymphoma promotes hypertrimethylation of histone H3 on lysine 27 (H3K27). Proc Natl Acad Sci USA 2012a;109:2989–2994.
[92] Sneeringer CJ, Scott MP, Kuntz KW, et al. Coordinated activities of wild-type plus mutant EZH2 drive tumor-associated hypertrimethylation of lysine 27 on histone H3 (H3K27) in human B-cell lymphomas. Proc Natl Acad Sci USA 2010;107:20980–20985.
[93] Yap DB, Chu J, Berg T, et al. Somatic mutations at EZH2 Y641 act dominantly through a mechanism of selectively altered PRC2 catalytic activity, to increase H3K27 trimethylation. Blood 2011;117:2451–2459.
[94] Velichutina I, Shaknovich R, Geng H, et al. EZH2-mediated epigenetic silencing in germinal center B cells contributes to proliferation and lymphomagenesis. Blood 2010;116:5247–5255.
[95] Béguelin W, Popovic R, Teater M, et al. EZH2 is required for germinal center formation and somatic EZH2 mutations promote lymphoid transformation. Cancer Cell 2013;23:677–692.
[96] Caganova M, Carrisi C, Varano G, et al. Germinal center dysregulation by histone methyltransferase EZH2 promotes lymphomagenesis. J Clin Invest 2013;123:5009–5022.
[97] Knutson SK, Wigle TJ, Warholic NM, et al. A selective inhibitor of EZH2 blocks H3K27 methylation and kills mutant lymphoma cells. Nat Chem Biol 2012;8:890–896.
[98] McCabe MT, Ott HM, Ganji G, et al. EZH2 inhibition as a therapeutic strategy for lymphoma with EZH2-activating mutations. Nature 2012b;492:108–112.
[99] Bararia D, Heide M, Romashova T, et al. ARID1A Is Recurrently and Significantly Mutated in Follicular Lymphoma (FL) and Impairs DNA Repair Efficiency. Blood 2014;124:73
[100] Cheung KJ, Johnson NA, Affleck JG, et al. Acquired TNFRSF14 mutations in follicular lymphoma are associated with worse prognosis. Cancer Res 2010b;70:9166–9174.

[101] Launay E, Pangault C, Bertrand P, et al. High rate of TNFRSF14 gene alterations related to 1p36 region in de novo follicular lymphoma and impact on prognosis. Leukemia 2011.
[102] Cai G, Freeman GJ. The CD160, BTLA, LIGHT/HVEM pathway: a bidirectional switch regulating T-cell activation. Immunol Rev 2009;229:244–258.
[103] Steinberg MW, Cheung TC, Ware CF. The signaling networks of the herpesvirus entry mediator (TNFRSF14) in immune regulation. Immunol Rev 2011;244:169–187.
[104] Compagno M, Lim WK, Grunn A, et al. Mutations of multiple genes cause deregulation of NF-kappaB in diffuse large B-cell lymphoma. Nature 2009;459:717–721.
[105] Davis RE, Ngo VN, Lenz G, et al. Chronic active B-cell-receptor signalling in diffuse large B-cell lymphoma. Nature 2010;463:88–92.
[106] Lenz G, Davis RE, Ngo VN, et al. Oncogenic CARD11 mutations in human diffuse large B cell lymphoma. Science 2008;319:1676–1679.
[107] Ngo VN, Young RM, Schmitz R, et al. Oncogenically active MYD88 mutations in human lymphoma. Nature 2011;470:115–119.
[108] Damm F, Mylonas E, Cosson A, et al. Acquired initiating mutations in early hematopoietic cells of CLL patients. Cancer Discov 2014;4:1088–1101.
[109] Landau DA, Carter SL, Stojanov P, et al. Evolution and impact of subclonal mutations in chronic lymphocytic leukemia. Cell 2013;152:714–726.
[110] Puente XS, Pinyol M, Quesada V, et al. Whole-genome sequencing identifies recurrent mutations in chronic lymphocytic leukaemia. Nature 2011;475:101–105.
[111] Rossi D, Trifonov V, Fangazio M, et al. The coding genome of splenic marginal zone lymphoma: activation of NOTCH2 and other pathways regulating marginal zone development. J Exp Med 2012;209:1537–1551.
[112] Treon SP, Xu L, Yang G, et al. MYD88 L265P somatic mutation in Waldenström's macroglobulinemia. N Engl J Med 2012;367:826–833.
[113] Yan Q, Huang Y, Watkins AJ, et al. BCR and TLR signaling pathways are recurrently targeted by genetic changes in splenic marginal zone lymphomas. Haematologica 2012;97:595–598.
[114] Yildiz M, Li H, Bernard D, et al. Activating STAT6 mutations in follicular lymphoma. Blood 2015;125:668–679.
[115] Davies AJ, Rosenwald A, Wright G, et al. Transformation of follicular lymphoma to diffuse large B-cell lymphoma proceeds by distinct oncogenic mechanisms. Br J Haematol 2007;136:286–293.
[116] Lossos IS, Alizadeh AA, Diehn M, et al. Transformation of follicular lymphoma to diffuse large-cell lymphoma: alternative patterns with increased or decreased expression of c-myc and its regulated genes. Proc Natl Acad Sci USA 2002;99:8886–8891.
[117] Pasqualucci L, Khiabanian H, Fangazio M, et al. Genetics of follicular lymphoma transformation. Cell Rep 2014;6:130–140.
[118] Lo Coco F, Gaidano G, Louie DC, Offit K, Chaganti RS, Dalla-Favera R. p53 mutations are associated with histologic transformation of follicular lymphoma. Blood 1993;82:2289–2295.
[119] O'Shea D, O'Riain C, Taylor C, et al. The presence of TP53 mutation at diagnosis of follicular lymphoma identifies a high-risk group of patients with shortened time to disease progression and poorer overall survival. Blood 2008;112:3126–3129.
[120] Møller MB, Nielsen O, Pedersen NT. Frequent alteration of MDM2 and p53 in the molecular progression of recurring non-Hodgkin's lymphoma. Histopathology 2002;41:322–330.
[121] Davies AJ, Lee AM, Taylor C, et al. A limited role for TP53 mutation in the transformation of follicular lymphoma to diffuse large B-cell lymphoma. Leukemia 2005;19:1459–1465.
[122] Oricchio E, Ciriello G, Jiang M, et al. Frequent disruption of the RB pathway in indolent follicular lymphoma suggests a new combination therapy. J Exp Med 2014;211:1379–1391.
[123] Oricchio E, Nanjangud G, Wolfe AL, et al. The Eph-receptor A7 is a soluble tumor suppressor for follicular lymphoma. Cell 2011;147:554–564.

[124] Hagman J, Belanger C, Travis A, Turck CW, Grosschedl R. Cloning and functional characterization of early B-cell factor, a regulator of lymphocyte-specific gene expression. Genes Dev 1993;7:760–773.
[125] Treiber N, Treiber T, Zocher G, Grosschedl R. Structure of an Ebf1:DNA complex reveals unusual DNA recognition and structural homology with Rel proteins. Genes Dev 2010;24:2270–2275.
[126] Green MR, Gentles AJ, Nair RV, et al. Hierarchy in somatic mutations arising during genomic evolution and progression of follicular lymphoma. Blood 2013;121:1604–1611.
[127] Brodtkorb M, Lingjærde OC, Huse K, et al. Whole-genome integrative analysis reveals expression signatures predicting transformation in follicular lymphoma. Blood 2014;123:1051–1054.
[128] Liu Q, Salaverria I, Pittaluga S, et al. Follicular lymphomas in children and young adults: a comparison of the pediatric variant with usual follicular lymphoma. Am J Surg Pathol 2013;37:333–343.
[129] Louissaint A, Ackerman AM, Dias-Santagata D, et al. Pediatric-type nodal follicular lymphoma: an indolent clonal proliferation in children and adults with high proliferation index and no BCL2 rearrangement. Blood 2012;120:2395–2404.
[130] Leich E, Salaverria I, Bea S, et al. Follicular lymphomas with and without translocation t(14;18) differ in gene expression profiles and genetic alterations. Blood 2009;114:826–834.
[131] Martin-Guerrero I, Salaverria I, Burkhardt B, et al. Recurrent loss of heterozygosity in 1p36 associated with TNFRSF14 mutations in IRF4 translocation negative pediatric follicular lymphomas. Haematologica 2013;98:1237–1241.
[132] Dave SS, Wright G, Tan B, et al. Prediction of survival in follicular lymphoma based on molecular features of tumor-infiltrating immune cells. N Engl J Med 2004;351:2159–2169.
[133] Carreras J, Lopez-Guillermo A, Roncador G, et al. High numbers of tumor-infiltrating programmed cell death 1-positive regulatory lymphocytes are associated with improved overall survival in follicular lymphoma. J Clin Oncol 2009;27:1470–1476.
[134] de Jong D, Koster A, Hagenbeek A, et al. Impact of the tumor microenvironment on prognosis in follicular lymphoma is dependent on specific treatment protocols. Haematologica 2009;94:70–77.
[135] Farinha P, Masoudi H, Skinnider BF, et al. Analysis of multiple biomarkers shows that lymphoma-associated macrophage (LAM) content is an independent predictor of survival in follicular lymphoma (FL). Blood 2005;106:2169–2174.
[136] Lee AM, Clear AJ, Calaminici M, et al. Number of CD4+ cells and location of forkhead box protein P3-positive cells in diagnostic follicular lymphoma tissue microarrays correlates with outcome. J Clin Oncol 2006;24:5052–5059.
[137] Wahlin BE, Aggarwal M, Montes-Moreno S, et al. A unifying microenvironment model in follicular lymphoma: outcome is predicted by programmed death-1–positive, regulatory, cytotoxic, and helper T cells and macrophages. Clin Cancer Res 2010;16:637–650.
[138] Amé-Thomas P, Le Priol J, Yssel H, et al. Characterization of intratumoral follicular helper T cells in follicular lymphoma: role in the survival of malignant B cells. Leukemia 2012;26:1053–1063.
[139] López-Giral S, Quintana NE, Cabrerizo M, et al. Chemokine receptors that mediate B cell homing to secondary lymphoid tissues are highly expressed in B cell chronic lymphocytic leukemia and non-Hodgkin lymphomas with widespread nodular dissemination. J Leukoc Biol 2004;76:462–471.
[140] Amé-Thomas P, Maby-El Hajjami H, Monvoisin C, et al. Human mesenchymal stem cells isolated from bone marrow and lymphoid organs support tumor B-cell growth: role of stromal cells in follicular lymphoma pathogenesis. Blood 2007;109:693–702.

[141] Guilloton F, Caron G, Ménard C, et al. Mesenchymal stromal cells orchestrate follicular lymphoma cell niche through the CCL2-dependent recruitment and polarization of monocytes. Blood 2012;119:2556–2567.
[142] Conde L, Halperin E, Akers NK, et al. Genome-wide association study of follicular lymphoma identifies a risk locus at 6p21.32. Nat Genet 2010;42:661–664.
[143] Skibola CF, Bracci PM, Halperin E, et al. Genetic variants at 6p21.33 are associated with susceptibility to follicular lymphoma. Nat Genet 2009;41:873–875.
[144] Smedby KE, Foo JN, Skibola CF, et al. GWAS of follicular lymphoma reveals allelic heterogeneity at 6p21.32 and suggests shared genetic susceptibility with diffuse large B-cell lymphoma. PLoS Genet 2011;7:e1001378.
[145] Skibola CF, Berndt SI, Vijai J, et al. Genome-wide association study identifies five susceptibility loci for follicular lymphoma outside the HLA region. Am J Hum Genet 2014;95:462–471.

Ming-Qing Du

6 Genetics and molecular pathogenesis of marginal zone lymphoma

6.1 Introduction

There are three indolent B-cell lymphomas that are thought to derive from marginal zone B cells and they include extranodal marginal zone lymphoma of mucosa-associated lymphoid tissue (MALT lymphoma), splenic marginal zone lymphoma (SMZL), and nodal marginal zone lymphoma. Significant advances have been made in the genetic characterization of MALT lymphoma and SMZL, and these advances have provided major insights into the molecular pathogenesis of these lymphoma subtypes. This chapter focuses on the recent advances in the genetic and molecular characterization of MALT lymphoma and SMZL.

6.2 Extranodal marginal zone lymphoma of mucosa-associated lymphoid tissue

6.2.1 Etiology of MALT lymphoma

MALT lymphoma arises from the acquired MALT that is triggered by chronic inflammatory disorders at diverse mucosal sites. Although not fully characterized, there is a distinct association between MALT lymphoma of different mucosal sites and etiological factors. For example, most of gastric MALT lymphomas are associated with *Helicobacter pylori* infection, while those from the skin, ocular adnexa, and immunoproliferative small intestine disease are variably linked to infection with *Borrelia burgdorferi* [1–6], *Chlamydia psittaci* [7–10] and *Campylobacter jejuni* [11], respectively. Furthermore, MALT lymphomas of the salivary gland and thyroid are closely associated with lymphoepithelial sialadenitis and Hashimoto thyroiditis, respectively [12, 13]. These etiological factors not only trigger acquired MALT, a setting for malignant transformation, but also generate immunological stimulations critical for the clonal expansion and survival of transformed cells. The critical role of these etiological factors in MALT lymphoma development is unequivocally demonstrated by responses of MALT lymphoma to antibiotic therapies in several independent settings, albeit understanding of their molecular mechanisms is incomplete.

6.2.2 Role of immunological stimulation in the pathogenesis of MALT lymphoma

Among the various etiological factors implicated in MALT lymphoma, the molecular mechanisms underlying their pathogenic role are best understood in the setting of gastric MALT lymphoma. The findings to date support that *H. pylori* infection triggers immunological responses that stimulate the lymphoma cells, and these stimulations include both T-cell dependent help and BCR signaling.

6.2.2.1 T-cell help

Early *in vitro* studies showed that the growth of gastric MALT lymphoma B cells responds to *H. pylori* antigens, but this critically depends on tumor infiltrating T cells and requires cognate interaction between B and T cells involving CD40 and CD40L co-stimulatory molecules [14–16]. *H. pylori* antigens do not directly stimulate MALT lymphoma B cells, but rather activate tumor infiltrating T cells and, through these, *H. pylori* specific T cells to promote the proliferation of malignant B cells [17, 18]. It is believed that *H. pylori* infection triggers T-cell dependent responses through the classical germinal center reaction, and generates a background of polyclonal reactive B and T cells. These *H. pylori* specific T cells may spread to the tumor area and stimulate malignant B cells in a manner similar to bystander T-cell help to memory B cells [19]. Additionally, the soluble ligands and cytokines, such as CD40L and BAFF generated by *H. pylori* mediated immune responses, may also help to promote the survival and expansion of malignant B cells.

The above model is supported by the invariable presence of reactive B-cell follicles and by an active immune response to *H. pylori* infection, as shown by the enriched expression of pro-inflammatory cytokines such as IL8 and IL1β, and molecules involved in B- and T-cell interaction such as CD86, CD28 and ICOS in gastric MALT lymphoma, particularly those without chromosomal translocations [20]. It remains to be investigated whether this model also applies to other MALT lymphomas that are associated with infection of microbial agents.

6.2.2.2 BCR stimulation

There are several lines of evidence suggesting that direct BCR stimulation plays an important role in the development of MALT lymphoma. First, MALT lymphoma cells, although originating from post-germinal center memory B cells of the acquired MALT, frequently re-visit B-cell follicles, known as follicular colonization, and show ongoing intraclonal sequence variations in their rearranged immunoglobulin (*IG*) genes [21–25]. Second, although not yet fully characterized, findings to date suggest that at least a proportion of MALT lymphoma derived immunoglobulins might be potentially autoreactive. Evidence to support this includes their biased usage of certain *IG* germline family members, including *IGVH4-34* and *IGVH1-69*, that are commonly employed by autoantibodies, and clinical presentation of autoimmunity in patients with MALT lymphoma. In general, these features are more evident in MALT lymphoma of the ocular adnexa, salivary gland and thyroid, although they have also been reported in cases of gastric MALT lymphoma [17, 25–33]. Together, these observations suggest a role of direct antigenic stimulation in the pathogenesis of MALT lymphoma (Fig. 6.1). This

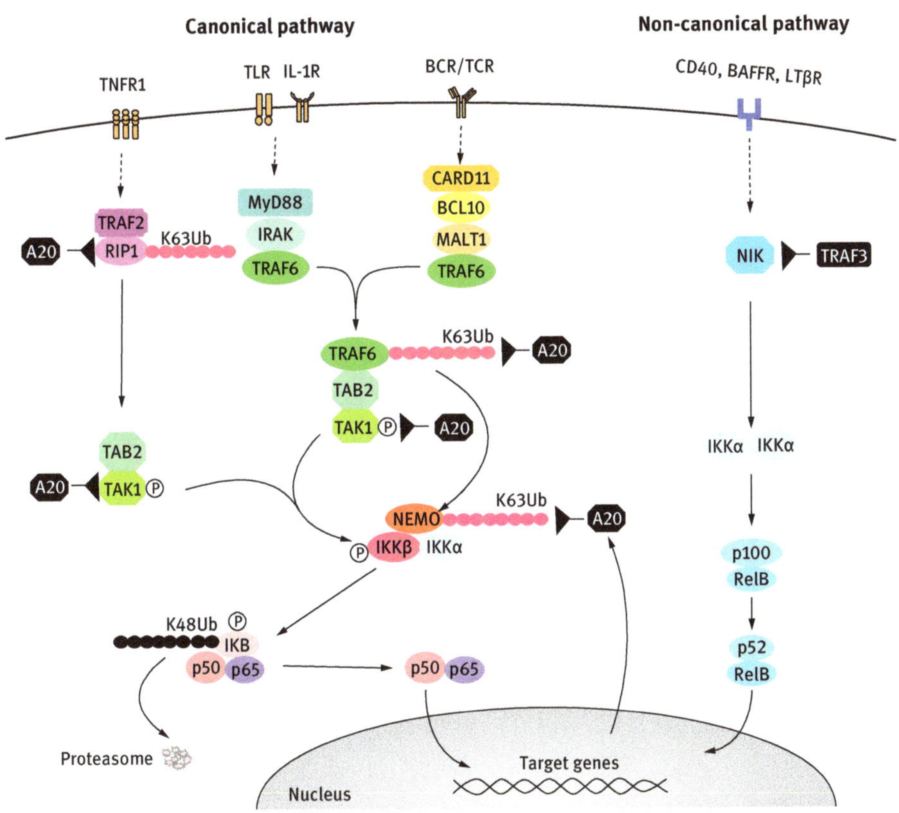

Fig. 6.1: The NF-κB pathway and the major players involved in MALT lymphoma pathogenesis. Signaling from the TRAFR, TLR, IL-1R, and antigen receptor activate the canonical NF-κB pathway that is characterized by activation of the IKK complex, phosphorylation and degradation of IκB. The signaling from CD40, BAFFR and LTβR activates the noncanonical NF-κB pathway, which is featured by activation of NIK, p100 processing and generation of functional active p52. Both canonical and noncanonical pathways are governed by several negative regulators, of which A20 (TNFAIP3) and TRAF3 are typical representatives. Reproduced with permission from Du MQ, Histopathology 2011 [44]. Abbreviations: TNFR: tumor necrosis factor receptor; TLR: Toll-like receptor; IL-1R: interleukin 1 receptor; BCR: B-cell receptor; TCR: T-cell receptor; TRAF: TNF associated factor; RIP1: receptor interacting protein 1; TAK1: transforming growth factor β activating kinase; TAB: TAK binding protein; IKK: inhibitor of NF-κB kinase; NEMO: NF-κB essential modulator; IκB: inhibitor of NF-κB; BAFFR: B-cell activating factor receptor; LTβR: lymphotoxin β receptor; NIK: NF-κB inducing kinase; K63Ub: K63 linked ubiquitin chain; K48Ub: K48 linked ubiquitin chain.

notion is further reinforced by the identification in these tumors of diverse genetic abnormalities that augment the signaling pathways downstream of BCR.

6.2.3 Genetic abnormalities in MALT lymphoma

A number of recurrent genetic abnormalities have been identified in MALT lymphoma by conventional cytogenetic studies or candidate gene mutation analyses. Interestingly, these genetic changes are largely mutually exclusive and occur at variable frequencies in MALT lymphoma of different mucosal sites, but nonetheless affect a common or overlapping molecular pathway (Fig. 6.1).

6.2.3.1 t(1;14)(p22;q32)

This translocation occurs primarily in MALT lymphoma of the lung (9 %) and stomach (4 %), but not or rarely in those of the ocular adnexa, salivary gland, thyroid and skin (Fig. 6.2) [34]. Gastric MALT lymphomas with t(1;14)(p22;q32) are typically of advanced stage, unlikely to respond to *H. pylori* eradication [35]. The translocation juxtaposes the *BCL10* gene under the regulatory control of the *IG* gene enhancer, leading to BCL10 overexpression [36, 37].

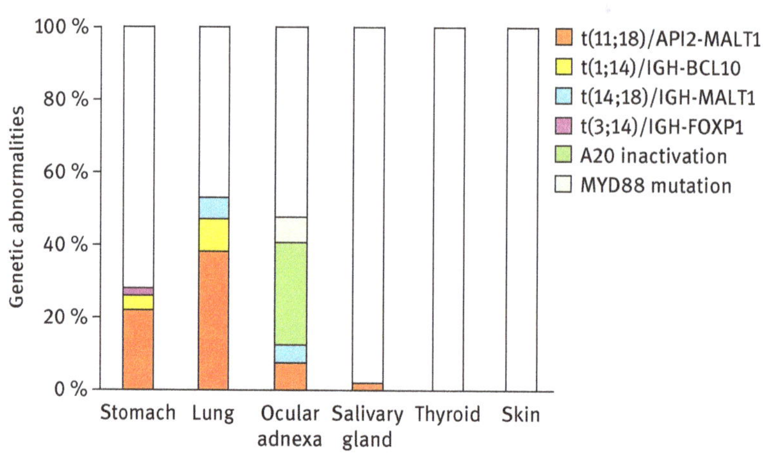

Fig. 6.2: **Frequencies of genetic abnormalities in MALT lymphoma of different sites.** The data are based on our previous work [34, 38–43].

BCL10 contains a CARD domain in its N-terminal region and is an essential part of the CARD11 (CARMA1)/BCL10/MALT1 signalosome complex that links antigen receptor signaling to canonical NF-κB pathway activation (Figs. 6.1 and 6.3) [45, 46]. Upon BCR binding by the antigen, CARD11 is recruited to the immunological synapse,

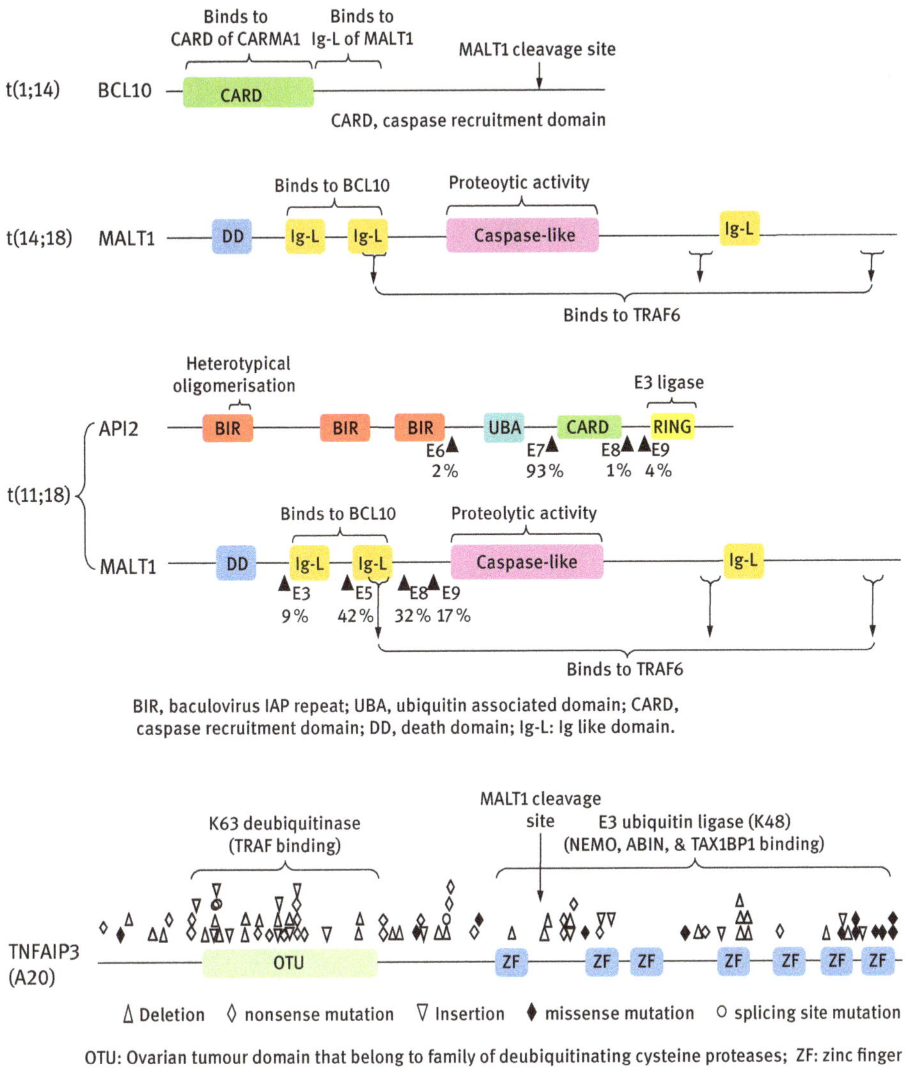

Fig. 6.3: **Key features of MALT lymphoma associated oncogenes or tumor suppresser genes.** t(1;14)(p22;q32)/*BCL10-IGH* and t(14;18)(q32;a21)/*IGH-MALT1* cause overexpression of BCL10 and MALT1 respectively, while t(11;18)(q21;q21)/*BIRC3*(API2)-*MALT1* fuses the N-terminal API2 to the C-terminal MALT1 and generates a chimeric fusion product. Various breakpoints in API2 and MALT1 and their frequencies are indicated. *TNFAIP3* (*A20*) is commonly inactivated by deleterious mutations and deletions. Reproduced with permission from Du MQ, Histopathology 2011 [44].

and is phosphorylated in its linker region by PKCβ, which induces a conformational change, releasing its auto-inhibition. Activated CARD11 recruits BCL10 and MALT1, and promotes BCL10 filamentous assembly and MALT1 oligomerization and activation [47, 48]. MALT1 then serves as a further scaffold molecule to recruit downstream signaling components such as TRAF6 and trigger nondegradative K63-linked polyubiquitination of TRAF6 and NEMO, the regulatory subunit of the IKK complex, consequently leading to activation of the IKK complex and hence of the canonical NF-κB pathway.

In line with the role of BCL10 in linking BCR signaling to the canonical NF-κB pathway activation, the protein is expressed in the cytoplasm of normal B cells including marginal zone B cells [49], the normal cell counterpart of MALT lymphoma. Overexpression of BCL10 by chromosomal translocation leads to constitutive activation through oligomerization via homotypic CARD/CARD interaction, thus enhancing NF-κB activity. Intriguingly, the BCL10 protein is expressed strongly in the nuclei of lymphoma cells with t(1;14)(q22;q32)/*BCL10-IGH* and also moderately in those with t(11;18)(q21;q21)/*BIRC3 (API2)-MALT1* [34, 49–51]. Strong BCL10 nuclear expression is also a feature of the marginal zone B cells of Eu-BCL10 mice [52]. These observations point to a role of nuclear BCL10 in the pathogenesis of MALT lymphoma. Nonetheless, the molecular mechanisms underlying BCL10 nuclear localization and function of nuclear BCL10 remain to be investigated.

6.2.3.2 t(14;18)(q32;q21)

This translocation occurs recurrently in MALT lymphoma of ocular adnexa (7 %) and the lung (6 %), but not or rarely in those of the stomach, salivary gland, thyroid and skin (Fig. 6.2) [39, 53]. The translocation brings the entire *MALT1* gene under the regulatory control of the *IGH* gene and hence causes its overexpression [54].

MALT1 contains several functional domains, including an N-terminal death domain, three immunoglobulin-like domains and a proteolytically active paracaspase domain (Fig. 6.3). As mentioned above, MALT1 is a critical effector molecule of the CARD11/BCL10/MALT1 signalosome complex and functions as a scaffold protein to recruit downstream molecules and activate the canonical NF-κB pathway [55, 56]. The ability of MALT1 to activate the canonical NF-κB pathway depends on its interaction with BCL10 as BCL10 binds to the two N-terminal Ig-like domains of MALT1, mediating its oligomerization and its activation [57, 58]. In line with this, both MALT1 and BCL10 are strongly expressed in cytoplasm in the MALT lymphoma cells carrying t(14;18)(q32;q21) [39].

Apart from its direct role in signal transduction, MALT1 also regulates NF-κB activation through its protease activities, which are dispensable for its ability to activate the canonical NF-κB pathway [59, 60]. Upon activation, MALT1 can specifically cleave several NF-κB regulators including A20 (also known as TNFAIP3), CYLD (cylindromatosis), RelB and BCL10, triggering their degradation [60–63]. A20, a target of NF-κB, serves as a negative feedback regulator to attenuate NF-κB activity through inactivation of NEMO, TRAF6, RIP1, and TAK1, thus negatively regulating several cellular signalings that activate the canonical NF-κB pathway (Fig. 6.1) [64–67]. CYLD is another global NF-κB negative regulator that also inactivates NEMO, TRAF6, RIP1 and TAK1 [63]. RelB is a newly recognized inhibitor of the canonical NF-κB pathway and MALT1 mediated RelB cleavage, and degradation enhances DNA binding of RelA or c-Rel-containing NF-κB complex [68]. In MALT lymphoma cells with t(1;14)(p22;q32)/*BCL10-IGH* or t(14;18)(q32;q21)/*IGH- MALT1*, MALT1 is most likely constitutively activated and may cleave the above NF-κB negative regulators, abolishing the physiological auto-negative feedback and causing constitutive NF-κB activation.

In addition, MALT1 can directly interact and activate caspase-8 in a protease independent manner [69, 70]. Finally, MALT1 is also involved in noncanonical NF-κB pathway activation, as its deficiency causes impaired activation of this signaling pathway induced by BAFF [71]. These functional properties of MALT1 may also play a role in the pathogenesis of MALT lymphoma.

6.2.3.3 t(11;18)(q21;q21)

This translocation occurs frequently in MALT lymphoma of the stomach (24 %) and lung (38 %), but infrequently or not in those of the ocular adnexa, salivary gland, thyroid and skin (Fig. 6.2) [39, 53]. The majority of gastric MALT lymphomas with t(11;18)(q21;q21) do not respond to *H. pylori* eradication [38, 72, 73], suggesting that their growth is no longer critically dependent on *H. pylori* mediated immunological stimulations. The translocation fuses the N-terminal region of the *BIRC3* (*API2)* to the C-terminal region of the *MALT1* and generates a functional chimeric fusion product (Fig. 6.3) [75, 76].

The API2-MALT1 fusion product is able to activate both the canonical and the non-canonical NF-κB pathways. In the absence of any upstream signaling, the API2-MALT1 fusion products are capable of forming oligomers through heterotypic interaction between the BIR1 of the API2 moiety and the C-terminal region of MALT1 (Fig. 6.3) [77, 78]. This auto-oligomerization is believed to recapitulate the oligomerization of wild-type MALT1, resulting from assembly of the CARD11/BCL10/MALT1 signalosome complex, and activates the canonical NF-κB pathway. Nonetheless, the optimal activation of the canonical NF-κB pathway by the API2-MALT1 fusion product depends on its recruitment of TRAF2 and RIP1, and subsequent RIP1 polyubiquitination [77, 79].

The API2-MALT1 fusion products invariably contain the intact caspase-like domain of the C-terminal MALT1 and thus possess its protease activities. Similar to MALT1, API2-MALT1 can also cleave A20 and CYLD [61, 80]. By proteolytic inactivation of these NF-κB regulators, expression of the API2-MALT1 fusion product by chromosomal translocation would cause constitutive activation of the canonical NF-κB pathway. In addition, the API2-MALT1 induced NF-κB activation may enhance its own expression, potentially forming a positive feedback loop since API2 is a target of NF-κB [81]. This notion is supported by the observation of high levels of polyubiquitination of NEMO in both MALT lymphomas with t(11;18)(q21;q21)/*API2-MALT1* and splenic marginal zone B cells of Eμ-API2-MALT1 mice [78, 82].

Through the concerted actions of both API2 and MALT1 moieties, the API2-MALT1 fusion product also gains ability to cleave novel paracaspase targets, and remarkably generates novel oncogenic products. Rosebeck and colleagues have shown that the API2-MALT1 fusion product induces proteolytic cleavage of NF-κB-inducing kinase (NIK), which leads to activation of the noncanonical NF-κB pathway (Fig. 6.4) [83]. The API2 moiety of the fusion product recruits NIK and places it in close proximity of the activated MALT1 protease domain, leading to cleavage of NIK at arginine 325. The cleavage generates a C-terminal NIK fragment that retains kinase activity, but is resistant to proteasomal degradation, and consequently results in constitutive noncanonical NF-κB signaling, enhanced B-cell adhesion, and apoptosis resistance [83].

Recently, Nie and colleagues have demonstrated that the API2-MALT1 fusion product also induces proteolytic cleavage of the tumor suppressor protein LIM domain and actin-binding protein 1 (LIMA1), and this leads to the generation of an oncogenic LIM domain-only (LMO) fragment (Fig. 6.4) [84]. Similarly, the API2 moiety and the MALT1 caspase-like domain mediate the recruitment of LIMA1 and its proteolytic cleavage, respectively. Expression of the novel LMO fragment promotes survival and proliferation of primary B cells *in vitro* and tumor formation in xenograft models, while expression of cleavage resistant LIMA1 suppresses API2-MALT1 induced lymphomagenesis [84]. In this context, the API2-MALT1 fusion product causes potentially a dual effect: inactivating the tumor suppressor LIMA1 and generating a novel oncogenic product.

6.2.3.4 Other infrequent chromosome translocations

Several novel chromosome translocations have also been described in MALT lymphoma and these include t(3;14)(p13;q32)/*FOXP1-IGH* [85–87], t(1;14)(p21;q32)/*CNN3-IGH*, t(5;14)(q34;q32)/ODZ2-IGH, t(9;14)(p24;q32)/*JMJD2C-IGH* [88] and t(X;14)(p11.4;q32)/*GPR34-IGH* [89, 90]. Like classical chromosomal translocations involving the *IG* gene loci, these aberrations typically cause overexpression of the involved oncogene.

6.2.3.5 TNFAIP3/A20 inactivation

The majority of MALT lymphomas, particularly those of the ocular adnexa, salivary gland, thyroid and skin, do not carry the above mentioned chromosomal translocations, but show frequent copy number alterations such as trisomy 3, 12 and/or 18. By array comparative genomic (CGH) analyses of ocular adnexal MALT lymphoma, *TNFAIP3* (*A20*) was identified as the target of 6q23 deletion [41, 91, 92]. Subsequent studies confirm that *TNFAIP3* (A20) is also inactivated by deleterious mutations (Fig. 6.3) [41, 42, 93]. Interestingly, *TNFAIP3* (A20) deletions and mutations are frequently seen in MALT lymphoma of the ocular adnexa, salivary gland and thyroid, but rarely in those of stomach, lung and skin (Fig. 6.2) [41–43, 93, 94]. Furthermore, unlike the above described chromosomal translocations that are specifically associated with MALT lymphoma, *TNFAIP3/A20* deletions and mutations are also frequently seen in other B-cell lymphoma entities characterized by constitutive NF-κB activation, such as

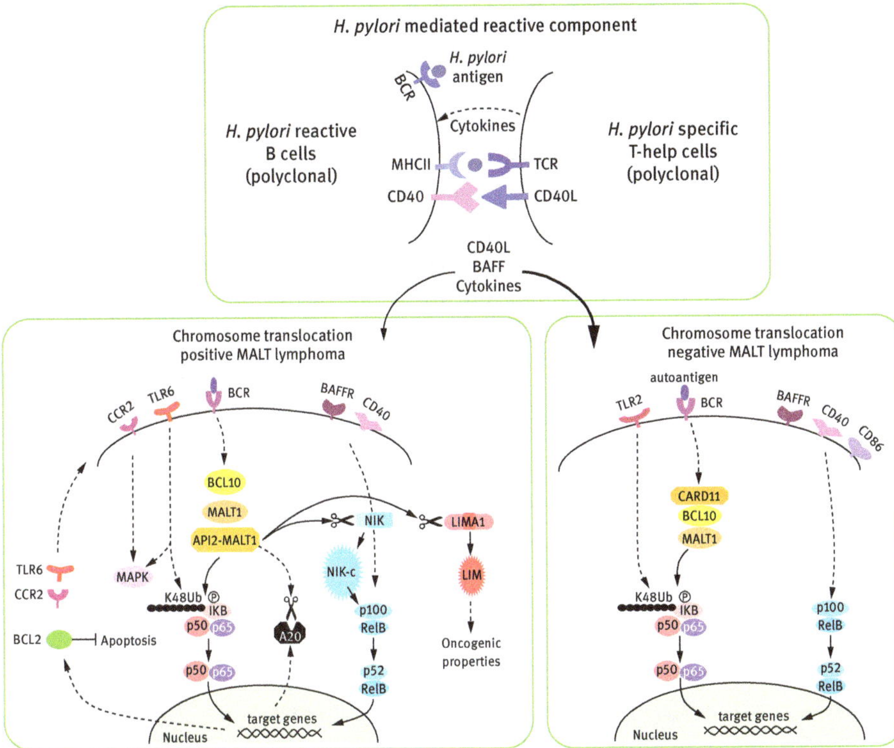

Fig. 6.4: The proposed model of the molecular pathogenesis of gastric MALT lymphoma with and without chromosomal translocations. The oncogenic products of t(1;14)(p22;q32)/*BCL10-IGH*, t(14;18)(q32;21)/*IGH-MALT1* and t(11;18)(q21;q21)/*API2-MALT1* are potent activators of the canonical NF-κB pathway. They may further augment the mediated NF-κB activation by enhancing expression of surface receptors TLR6 and CCR2, as well as by proteolytic cleavage of the negative inhibitor A20. These oncogenic products may also potentially cooperate with the signaling from BCR, BAFFR and CD40, via bystander T cells generated in the *H. pylori* mediated reactive component. In addition, the API2-MALT1 fusion product has the ability to cleave NIK and generate a stable NIK C-terminal fragment, capable of activating the noncanonical NF-κB pathway, and also to cleave LIMA1 and generate a LIM domain-only (LMO) fragment, conferring oncogenic properties. Growth of translocation-negative MALT lymphomas is largely driven by *H. pylori* generated immune responses including signaling from CD40 and CD86 through bystander T cells, and direct triggering of TLR and BCR by *H. pylori* associated lipopolysaccharides and autoantigen respectively. This underscores that most of translocation negative gastric MALT lymphomas can be cured by *H. pylori* eradication. Modified with permission from Du MQ, Histopathology 2011 [44].
Abbreviations: TLR: Toll-like receptor; BCR: B-cell receptor; MAPK: MAP kinase; IκB: inhibitor of NF-κB; BAFFR: B-cell activating factor receptor; K48Ub: K48 linked ubiquitin chain. NIK: NF-κB inducible kinase; LIMA1: LIM domain and actin-binding protein 1.

activated B-cell-like diffuse large B-cell lymphoma, Hodgkin lymphoma and primary mediastinal B-cell lymphoma [42, 93, 95–98].

As mentioned above, TNFAIP3 (A20) is a target of NF-κB and serves as an auto-negative feedback regulator to attenuate NF-κB activities triggered by several surface receptors, including BCR, TNFR and TLR (Fig. 6.1) [65–67]. TNFAIP3 (A20) contains an N-terminal OTU domain that possesses deubiquitinating cysteine protease activity, and 7 zinc finger domains in its C-terminus, which confer the E3 ubiquitin ligase activity [64, 99]. Through these ubiquitin editing enzyme activities, i.e. removing the K63-linked ubiquitin chain and catalyzing the K48-linked polyubiquitination, A20 can inactivate several NF-κB positive regulators including RIP1/2, TRAF6, NEMO and TAK1 [65–67]. Thus, TNFAIP3 (A20) inactivation by mutation and/or deletion can potentially augment NF-κB activation by signaling from multiple surface receptors.

6.2.3.6 Somatic mutations affecting other NF-κB regulators

Given that MALT lymphoma is characterized by several genetic changes of NF-κB regulators, a number of studies investigated whether the somatic mutations found in other B-cell lymphomas characterized by constitutive NF-κB activation also occur in MALT lymphoma [94, 100–104]. In general, mutations affecting other NF-κB regulators including *MYD88*, *CD79A*, *CD79B*, *CARD11* and *TRAF3* are infrequent in MALT lymphoma. These observations further highlight the difference in genetics between MALT lymphoma and other B-cell lymphomas characterized by constitutive NF-κB activation despite their common involvement of TNFAIP3 (A20) inactivation, and also suggest the presence of unknown genetic abnormalities that await to be discovered in MALT lymphoma.

6.2.4 Oncogenic cooperation between genetic abnormalities and antigenic stimulation

Chromosomal translocations and *TNFAIP3* (*A20*) deletions/mutations are the primary genetic events in MALT lymphoma development. Nevertheless, none of these genetic aberrations alone is sufficient for malignant transformation. Both Eμ-*BCL10* and Eμ-*API2-MALT1* mice develop splenic marginal zone hyperplasia but not lymphoma [52, 82], while TNFAIP3 (A20) deficiency in B cells enhances B-cell proliferation and results in the generation of autoantibodies [105]. Thus, additional factors are required to cooperate with these genetic events in MALT lymphoma development. On the one hand, it is possible that there are additional genetic defects not yet identified, while on the other hand, there is likely a close oncogenic cooperation between these genetic abnormalities and immunological stimulation in MALT lymphoma development.

There are several lines of evidence suggesting cooperation between chromosomal translocations and surface receptor stimulation (Fig. 6.4). First, NF-κB activation by expression of API2-MALT1, MALT1 and BCL10 can be enhanced by stimulation of surface immune receptors such as CD40 [106, 107]. Second, gene expression profiling analyses show increased expression of surface receptors such as TLR6 and CCR2, and the anti-apoptotic protein BCL2 in translocation positive MALT lymphoma [20]. Expression of TLR6, in the presence of its heterodimer partner TLR2, can enhance both API2-MALT1 and BCL10 mediated NF-κB activation *in vitro* [20]. As these genes themselves are the transcriptional targets of NF-κB, their enhanced expression by chromosomal translocation may act as a positive feedback to NF-κB activation. Finally, stimulation of Eμ-API2-MALT1 mice with Freund's complete adjuvant leads to development of splenic marginal zone lymphoma-like lesion [108].

There is also a possibility of cooperation between TNFAIP3 (A20) inactivation and immunological stimulation in MALT lymphoma development. Given that TNFAIP3 (A20) negatively regulates NF-κB activation by signaling from several surface receptors including BCR, TNFR and TLR [65–67], a functional cooperation between TNFAIP3 (A20) inactivation and these receptor signaling pathways is expected. As *TNFAIP3* (*A20*) deletions and mutations are frequently seen in MALT lymphoma of the ocular adnexa, salivary gland and thyroid, which show the strongest association with autoimmunity among various MALT lymphoma sites, it is pertinent to speculate that TNFAIP3 (A20) inactivation may potentiate the effect of BCR signaling by autoantigen. In addition, TNFAIP3 (A20) inactivation could augment the signaling by TNFR and TLR, triggered by the respective stimuli present in the chronic inflammatory microenvironment.

6.3 Splenic marginal zone lymphoma

6.3.1 Etiology of SMZL

There are several lines of evidence suggesting an infectious etiology in at least of a proportion of SMZL. Epidemiological studies and meta-analyses show a strong association among chronic HCV infection, type-II mixed cryoglobulinemia, and several non-Hodgkin's lymphomas including SMZL [109–111]. Such an association is particularly strong in geographical regions such as Japan and Italy, where HCV infection is highly prevalent and is considered endemic in certain areas [112, 113]. In support of a causative role of HCV infection in SMZL, antiviral therapy with interferon (with or without ribavirin) leads to complete or partial lymphoma regression in a significant proportion of cases with HCV-associated SMZL [114, 115]. Lymphoma regression is most likely the result of HCV clearance rather than an antiproliferative effect of interferon as a rapid regression of HCV-related SMZL is also seen following an interferon-free, HCV NS3/NS4A inhibitor-based treatment [116].

6.3.2 Role of antigenic drive in pathogenesis of SMZL

The regression of HCV-associated SMZL following antiviral therapy strongly indicates an active role of antigenic stimulation in the lymphoma pathogenesis. There are several potential possibilities that chronic HCV infection may promote the survival and proliferation of malignant B cells although the exact role of the viral infection in B-cell lymphomagenesis is unknown. The envelope protein E2 of HCV can interact with the surface protein CD81 in B cells and stimulates their proliferation, particularly in the presence of BCR engagement [117]. In addition, HCV infection induces the release of cytokines and growth factors such as BAFF and interleukin-6 [118, 119], which are known stimulators of B cells. Finally, HCV infection also induces TLR4 expression in B cells [120], thus potentially enhancing TLR signaling.

A potential role of antigenic drive in the pathogenesis of SMZL is also suggested by the genetic features of the rearranged immunoglobulin genes. SMZLs show a biased usage of both immunoglobulin heavy and light chain genes, with *IGHV1-2, IGKV3-20* and *IGLV2-14* being considerably overrepresented [121–124]. The majority of the rearranged immunoglobulin variable genes show somatic mutations, although at considerably variable frequencies. The *IGHV1-2* rearrangements in SMZL are characterized by minimal somatic mutations and longer complementarity determining region-3 (CDR3) sequences with common motifs, suggesting possible selection by superantigens [121, 122]. In line with this speculation, the majority of SMZL with *IGHV1-2* rearrangements show clinical presentation of autoimmunity [124]. Thus, *IGHV1-2* expressing BCR may potentially confer an active BCR signaling although the antigen recognized by the BCR remains to be characterized.

6.3.3 Genetic abnormalities in SMZL

SMZL lacks any specific and recurrent chromosomal translocations, although translocations involving the *CCND1*, *CCND3*, *BCL6*, *MUM1* or *PAX5* loci were reported in isolated cases [125, 126]. CGH analyses identify several recurrent chromosomal gains and losses, with 7q deletion being characteristically seen in SMZL [127–132].

6.3.3.1 7q deletion in SMZL

Array CGH analyses mapped the minimal common region deleted in 7q to a ~ 3Mb region at 7q32.1-32.2 [131–133]. Extensive screening of the 7q32 deletion by interphase fluorescence in situ hybridization analysis identified this deletion in 34 % of SMZL and in ~ 33 % of splenic B-cell lymphoma/leukemia unclassifiable (Fig. 6.5), but not or rarely in other lymphomas involving the spleen, including chronic lymphocytic leukemia and hairy cell leukemia [131, 132].

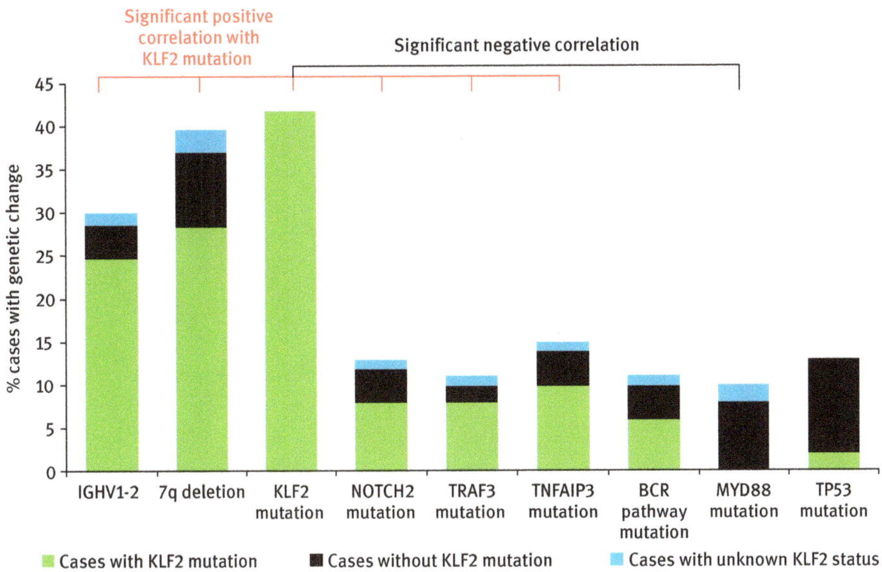

Fig. 6.5: **Frequencies of genetic abnormalities in SMZL.** The data is based on the study by Clipson et al. (reproduced with permission) [140] and show a significant positive correlation between *KLF2* mutation and *IGHV1-2* rearrangement, 7q32 deletion, *NOTCH2*, *TRAF3* and *TNFAIP3* mutations as well as a significant negative correlation between *KLF2* and *MYD88* mutations.

The genetic target(s) of 7q32 deletion in SMZL is unknown despite extensive research for potential tumor suppresser gene(s). There is no homozygous deletion, nor recurrent breakpoints at 7q32 in SMZL [131–133]. There are a number of coding genes and several noncoding microRNAs, which could be the potential targets of the deletion. Sequencing of these candidate genes to date did not identify any recurrent pathogenic mutations [131–133]. Nevertheless, the expression of many of these candidate genes was reduced in SMZL with 7q32 deletion in comparison with those without the deletion [133]. It remains to be investigated whether a haplodeficiency of one or more of these candidate genes may have a role in the pathogenesis of SMZL.

6.3.3.2 Mutation in the NOTCH pathway

Recent studies using whole exome sequencing (WES) approaches together with candidate gene sequencing identified a plethora of somatic mutations in SMZL, which involve the NOTCH, KLF2, NF-κB, BCR and TLR pathways, as well as histone modifiers and transcriptional regulators [134–140].

NOTCH2 mutations are detectable in 6.5–25 % of SMZL (Figs. 6.5 and 6.6), but not or rarely in other low grade B-cell lymphomas, including chronic lymphocytic leukemia and hairy cell leukemia [134, 135]. The vast majority of *NOTCH2* mutations are frameshift indels or nonsense mutations, and clustered in a hotspot region up-

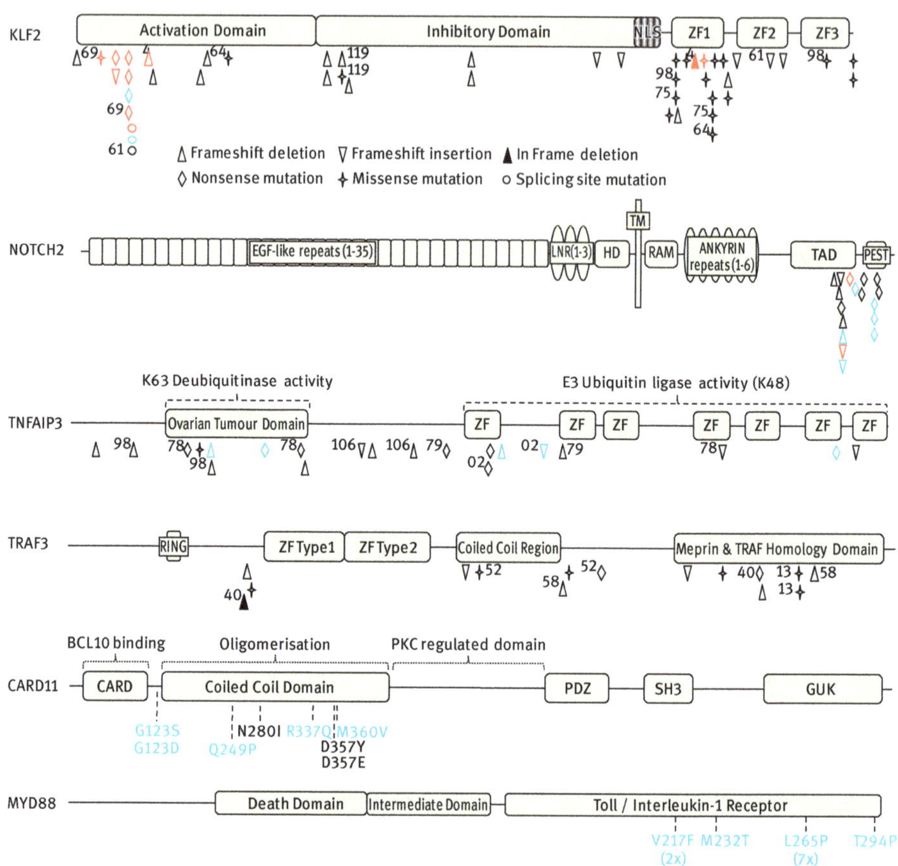

Fig. 6.6: **Key features of SMZL associated oncogenes or tumor suppresser genes and their mutations.** The mutation data are from the study by Clipson et al. (reproduced with permission) [140]. Abbreviations: LNR: LIN-12/NOTCH repeats; HD: heterodimerization; TM: transmembrane; RAM: regulation of amino acid metabolism; TAD: transactivation domain; PEST: Proline, glutamic acid, serine and threonine rich domain; ZF: zinc finger; CARD: caspase recruitment domain; PDZ: (PSD95, DLG and ZO1 homology) domain; SH3: Src homology motif; GUK: guanylate kinase domain

stream or within the C-terminal PEST domain that is critical for NOTCH2 proteasomal degradation [134–137, 140, 141]. These mutations eliminate or truncate the C-terminal PEST domain and thus potentially enhance the stability of the active NOTCH2 intracellular domain, and its function [134–137, 140, 141].

In addition to *NOTCH2*, several other NOTCH signaling regulators including *NOTCH1*, *SPEN* and *DTX1* are also targeted by somatic mutations but at much lower frequencies. Interestingly, mutations in these genes appear to be largely mutually exclusive [134, 135].

NOTCH2 signaling is critical for the generation of marginal zone B cells and their retention in the splenic marginal zone [142–145], but has little impact on cell proliferation and survival [144, 145]. *NOTCH2* activation by mutation may thus underscore the phenotype of SMZL.

6.3.3.3 Mutations of Krüppel-like factor 2 (KLF2)

KLF2 mutations are found in 20–42% of SMZL (Figs. 6.5 and 6.6), and are also variably seen in hairy cell leukemia (10–16%), diffuse large B-cell lymphoma (up to 11%), nodal marginal zone lymphoma (9%) and MALT lymphoma (4–8%), but not in chronic lymphocytic leukemia [140, 146]. The majority of *KLF2* mutations are frameshift indels, nonsense mutations, or substitutions affecting the essential splice site, and localized toward the N-terminal activation domain (AD) and middle inhibitory domain (ID), thus potentially resulting in a truncated protein product. The remaining mutations are missense substitutions and cluster in the C-terminal zinc finger (ZF) 1, with mutational hotspots on conserved amino acids residues.

KLF2 is a member of the KLF family of transcription factors, and has been shown to be a negative regulator of inflammation and NF-κB activation [147–150]. KLF2 deficiency in B cells in mice appears to promote follicular B cells to gain a marginal zone like phenotype and migrate to the splenic marginal zone, but have little impact on their proliferation [151–153]. In line with the above observations, *KLF2* mutations abrogated the ability of KLF2 to suppress NF-κB activation by TLR, BCR, BAFFR and TNFR signaling *in vitro* [140]. *KLF2* inactivation by mutation may deregulate gene expression through the modulation of NF-κB activities and other currently unknown mechanisms, thereby promoting B cells homing to the splenic marginal zone.

6.3.3.4 Mutations affecting the NF-κB pathway

Although not yet comprehensively investigated, studies to date have shown somatic mutations affecting several NF-κB regulators in SMZL. These included activating mutations in positive regulators such as *CARD11* (~7%), *MYD88* (5–13%) and *IKBKB* (7%), and inactivating mutations in negative regulators such as *TNFAIP3* (*A20*) (7–13%), *BIRC3* (*API2*) (6.3%) and *TRAF3* (5%) (Figs. 6.5 and 6.6) [134, 138–140]. Interestingly, these mutations are largely mutually exclusive [134, 138, 140]. Mutations in *CARD11*, *MYD88* and *IKBKB* most likely enhance the activation of the canonical NF-κB pathway, while those in *BIRC3* and *TRAF3* likely argument the signaling of the noncanonical NF-κB pathway. *TNFAIP3* (*A20*) inactivation by mutation may promote signaling from several surface receptors including TNFR, TLR/IL1-R and BCR [154]. Through the enhanced signaling of these pathways, these mutations may contribute to both the development and survival of marginal zone B cells although their precise role remains to be dissected.

6.3.3.5 Mutations in epigenetic regulators

Apart from mutations seen in the above described molecular pathways, that are important for marginal zone B-cell development, whole exome or genome sequencing has also identified mutations in several epigenetic regulators [134–137]. These genes include *MLL2* (a histone methyltransferase), *ARID1A* (a member of the SWI/SNF adenosine triphosphate-dependent chromatin-remodeling complexes), *EP300* (a histone and non-histone acetyltransferase), *SIN3A* (a core component of the SIN3-HDAC1/2 histone deacetylase complex), and *TBL1XR1* (an intrinsic component of the SMRT-N-CoR transcription co-repressor machinery). Larger cohorts of patients will need to be investigated in order to assess the true incidence of these epigenetic regulator mutations and their role in the pathogenesis of SMZL.

Finally, *TP53* mutations are seen in ~ 12 % of SMZL and these mutations likely inactivate TP53 function [140]. Like in other lymphomas, *TP53* mutations in SMZL are likely to be a secondary genetic event. Among the various genetic changes reported in SMZL, only TP53 inactivation has been consistently shown to be associated with poor prognosis [132, 140, 155, 156].

6.3.4 Oncogenic cooperation between genetic changes and antigenic stimulation

As shown above, a diverse spectrum of genetic changes has been identified in SMZL. With the exception of 7q deletion, *KLF2* and *NOTCH2* mutation, all other mutations occur in less than 10 % of cases (Fig. 6.5). None of these mutations alone is sufficient for malignant transformation. However, several of these genetic changes show a significant association, suggesting potential oncogenic cooperation (Fig. 6.5) [140, 146].

The majority of SMZL with *KLF2* mutation have both 7q32 deletion and *IGHV1-2* rearrangement, and additional mutations in *NOTCH2*, or *TNFAIP3* (*A20*), or *TRAF3*. The genes targeted by 7q32 deletion and their role in SMZL pathogenesis are unclear [132, 133]. Nonetheless, potential cooperation may be expected among other concurrent genetic changes, based on their biological properties (Fig. 6.7). For example, the activation of the canonical NF-κB pathway by *IGHV1-2* expressing BCR could be enhanced by KLF2 and TNFAIP3 (A20) inactivation. Similarly, the activation of the non-canonical NF-κB pathway by BAFF or CD40L could be augmented by KLF2 and TRAF3 inactivation. Clearly, our current understanding of the molecular mechanisms sustained by these genetic changes is rather limited, and their precise oncogenic cooperation and impact on marginal zone B-cell differentiation, cell migration, retention and clonal expansion in the splenic marginal zone remain to be investigated. In addition, a high proportion of SMZL, particularly those negative for *KLF2* mutations, show paucity of pathogenic mutations [140] and their genetic bases may need to be better characterized in future studies.

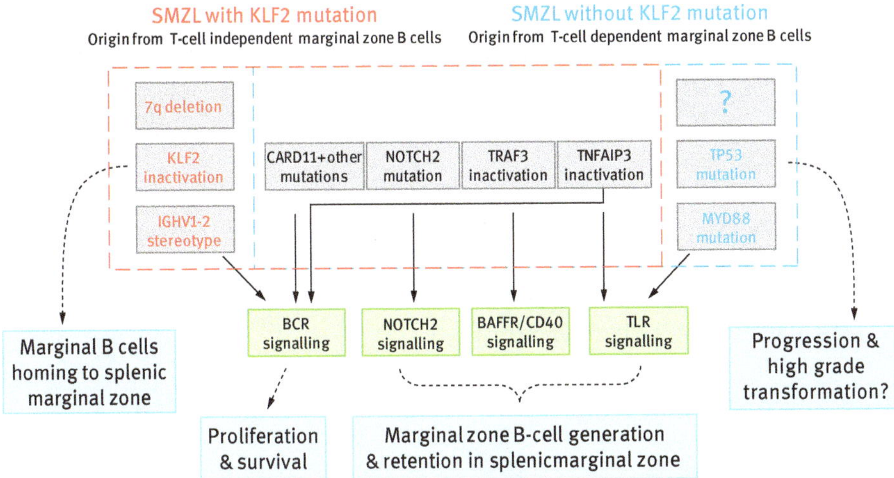

Fig. 6.7: **Summary of genetic aberrations and their possible biological impact in the biology of SMZL [140].** The majority of SMZLs with *KLF2* mutation have the rearranged *IGHV1-2* that carries minimal somatic mutations, suggesting derivation of these lymphoma cells from T-cell independent marginal zone B cells. The biased usage of *IGHV1-2* indicates a possible antigenic drive by a superantigen. *KLF2* inactivation by mutations may facilitate marginal zone B-cell differentiation and their homing to the splenic marginal zone. *CARD11*, *NOTCH2*, *TRAF3* and *TNFAIP3* mutations are found in SMZL with and without *KLF2* mutation. Mutations in *CARD11* and *CD79A/B*, *BCL10* may lead to active BCR signaling, thereby promoting cell proliferation and survival. Activation of NOTCH2, BAFFR/CD40 and TLR signaling by *NOTCH2*, *TRAF3* and *TNFAIP3* mutations may primarily contribute to marginal zone B-cell generation and their retention in the splenic marginal zone. The majority of SMZLs without *KLF2* mutations have heterogeneous usage of *IGHV* that carries high loads of somatic mutation, suggesting an origin of these lymphomas from T-cell dependent marginal zone B cells. *MYD88* and *TP53* mutations are virtually exclusively seen in cases without *KLF2* mutation. *MYD88* mutations cause constitutive TLR signaling, while *TP53* mutations inactivate its tumor suppressor function and may promote disease progression.

6.4 Concluding remarks

Although the genetics of MALT lymphoma and SMZL and their molecular mechanisms are not yet fully characterized, the discoveries to date have provided significant insights into the pathogenesis of these low-grade B-cell lymphomas. There are several important common features emerging in both MALT lymphoma and SMZL, despite the fact that they are distinct lymphoma entities. First, there are diverse genetic changes in both MALT lymphoma and SMZL, and many of these different genetic changes appear to affect a common or related molecular pathway. Second, there is a potential cooperation among concurrent genetic changes, particularly in SMZL, as these genetic changes likely confer complementary oncogenic properties. Third, there is also a potential cooperation between genetic changes and antigenic stimulations, which may involve multiple immune receptors including BCR. Our current knowledge on the biology of MALT lymphoma and SMZL not only guides future investigations of these lymphoma entities, but also provides a paradigm for studies of other B-cell lymphomas.

Acknowledgements

The studies described from Professor Ming-Qing Du's laboratory were supported by research grants from the Leukaemia Lymphoma Research, U.K., Kay Kendall Leukaemia Fund, the Elimination of Leukemia Fund, U.K., the Lady Tata Memorial Trust, U.K., and the Addenbrooke's Charitable Trust.

References

[1] Garbe C, Stein H, Dienemann D, Orfanos CE. Borrelia burgdorferi-associated cutaneous B cell lymphoma: clinical and immunohistologic characterization of four cases. J Am Acad Dermatol 1991;24:584–590.
[2] Cerroni L, Zochling N, Putz B, Kerl H. Infection by Borrelia burgdorferi and cutaneous B-cell lymphoma. J Cutan Pathol 1997;24:457–461.
[3] Sonck CE, Viljanen M, Hirsimaki P, Soderstrom KO, Ekfors TO. Borrelial lymphocytoma – a historical case. APMIS 1998;106:947–952.
[4] Goodlad JR, Davidson MM, Hollowood K, et al. Primary cutaneous B-cell lymphoma and Borrelia burgdorferi infection in patients from the Highlands of Scotland. Am J Surg Pathol 2000;24:1279–1285.
[5] Grange F, Wechsler J, Guillaume JC, et al. Borrelia burgdorferi-associated lymphocytoma cutis simulating a primary cutaneous large B-cell lymphoma. J Am Acad Dermatol 2002;47:530-4.
[6] Bogle MA, Riddle CC, Triana EM, Jones D, Duvic M. Primary cutaneous B-cell lymphoma. J Am Acad Dermatol 2005;53:479–484.
[7] Ferreri AJ, Guidoboni M, Ponzoni M, et al. Evidence for an association between Chlamydia psittaci and ocular adnexal lymphomas. J Natl Cancer Inst 2004;96:586–594.
[8] Rosado MF, Byrne GE, Jr., Ding F, et al. Ocular adnexal lymphoma: a clinicopathologic study of a large cohort of patients with no evidence for an association with Chlamydia psittaci. Blood 2006;107:467–472.
[9] Vargas RL, Fallone E, Felgar RE, et al. Is there an association between ocular adnexal lymphoma and infection with Chlamydia psittaci? The University of Rochester experience. Leuk Res 2006;30:547–551.
[10] Chanudet E, Zhou Y, Bacon CM, et al. Chlamydia psittaci is variably associated with ocular adnexal MALT lymphoma in different geographical regions. J Pathol 2006;209:344–351.
[11] Lecuit M, Abachin E, Martin A, et al. Immunoproliferative small intestinal disease associated with Campylobacter jejuni. N Engl J Med 2004;350:239–248.
[12] Voulgarelis M, Moutsopoulos HM. Mucosa-associated lymphoid tissue lymphoma in Sjogren's syndrome: risks, management, and prognosis. Rheum Dis Clin North Am 2008;34:921–933, viii.
[13] Watanabe N, Noh JY, Narimatsu H, et al. Clinicopathological features of 171 cases of primary thyroid lymphoma: a long-term study involving 24553 patients with Hashimoto's disease. Br J Haematol 2011;153:236–243.

[14] Hussell T, Isaacson PG, Crabtree JE, Spencer J. The response of cells from low-grade B-cell gastric lymphomas of mucosa-associated lymphoid tissue to Helicobacter pylori. Lancet 1993;342:571–574.

[15] Hussell T, Isaacson PG, Crabtree JE, Spencer J. Helicobacter pylori-specific tumour-infiltrating T cells provide contact dependent help for the growth of malignant B cells in low- grade gastric lymphoma of mucosa-associated lymphoid tissue. J Pathol 1996;178:122–127.

[16] Greiner A, Knorr C, Qin Y, et al. Low-grade B cell lymphomas of mucosa-associated lymphoid tissue (MALT-type) require CD40-mediated signaling and Th2-type cytokines for in vitro growth and differentiation. Am J Pathol 1997;150:1583–1593.

[17] Hussell T, Isaacson PG, Crabtree JE, Dogan A, Spencer J. Immunoglobulin specificity of low grade B cell gastrointestinal lymphoma of mucosa-associated lymphoid tissue (MALT) type. Am J Pathol 1993;142:285–292.

[18] D'Elios MM, Amedei A, Manghetti M, et al. Impaired T-cell regulation of B-cell growth in Helicobacter pylori – related gastric low-grade MALT lymphoma. GASTROENTEROLOGY 1999;117:1105–1112.

[19] Bernasconi NL, Traggiai E, Lanzavecchia A. Maintenance of serological memory by polyclonal activation of human memory B cells. Science 2002;298:2199–2202.

[20] Hamoudi RA, Appert A, Ye H, et al. Differential expression of NF-kappaB target genes in MALT lymphoma with and without chromosome translocation: insights into molecular mechanism. Leukemia 2010;24:1487–1497.

[21] Du MQ, Xu CF, Diss TC, et al. Intestinal dissemination of gastric mucosa-associated lymphoid tissue lymphoma. Blood 1996;88:4445–4451.

[22] Du M, Diss TC, Xu C, Peng H, Isaacson PG, Pan L. Ongoing mutation in MALT lymphoma immunoglobulin gene suggests that antigen stimulation plays a role in the clonal expansion. Leukemia 1996;10:1190–1197.

[23] Qin Y, Greiner A, Hallas C, Haedicke W, Muller Hermelink HK. Intraclonal offspring expansion of gastric low-grade MALT-type lymphoma: evidence for the role of antigen-driven high-affinity mutation in lymphomagenesis. Lab Invest 1997;76:477–485.

[24] Thiede C, Alpen B, Morgner A et al. Ongoing somatic mutations and clonal expansions after cure of Helicobacter pylori infection in gastric mucosa-associated lymphoid tissue B-cell lymphoma. J Clin Oncol 1998;16:3822–3831.

[25] Zhu D, Lossos C, Chapman-Fredricks JR, et al. Biased use of the IGHV4 family and evidence for antigen selection in Chlamydophila psittaci-negative ocular adnexal extranodal marginal zone lymphomas. PLoS One 2011;6:e29114.

[26] van Maldegem F, Wormhoudt TA, Mulder MM, et al. Chlamydia psittaci-negative ocular adnexal marginal zone B-cell lymphomas have biased VH4-34 immunoglobulin gene expression and proliferate in a distinct inflammatory environment. Leukemia 2012;26:1647–1653.

[27] Perez M, Pacchiarotti A, Frontani M, et al. Primary cutaneous B-cell lymphoma is associated with somatically hypermutated immunoglobulin variable genes and frequent use of VH1-69 and VH4-59 segments. Br J Dermatol 2010;162:611–618.

[28] Hansen A, Reiter K, Pruss A, et al. Dissemination of a Sjogren's syndrome-associated extranodal marginal-zone B cell lymphoma: circulating lymphoma cells and invariant mutation pattern of nodal Ig heavy- and light-chain variable-region gene rearrangements. Arthritis Rheum 2006;54:127–137.

[29] Miklos JA, Swerdlow SH, Bahler DW. Salivary gland mucosa-associated lymphoid tissue lymphoma immunoglobulin V(H) genes show frequent use of V1-69 with distinctive CDR3 features. Blood 2000;95:3878–3884.

[30] Diss TC, Peng H, Wotherspoon AC, Pan L, Speight PM, Isaacson PG. Brief report: a single neoplastic clone in sequential biopsy specimens from a patient with primary gastric-mucosa-

associated lymphoid-tissue lymphoma and Sjogren's syndrome. N Engl J Med 1993;329:172–175.

[31] Raderer M, Osterreicher C, Machold K, et al. Impaired response of gastric MALT-lymphoma to Helicobacter pylori eradication in patients with autoimmune disease. Ann Oncol 2001;12:937–939.

[32] Folli F, Perego L, Ponzoni M, et al. Autoantibodies against a 72-kDa ductal cell membrane glycoprotein in a patient affected by Sjogren's syndrome and gastric MALT lymphoma. Ann Hematol 2002;81:597–602.

[33] Moshynska OV, Saxena A. Clonal relationship between Hashimoto thyroiditis and thyroid lymphoma. J Clin Pathol 2008;61:438–444.

[34] Ye H, Liu H, Attygalle A, et al. Variable frequencies of t(11;18)(q21;q21) in MALT lymphomas of different sites: significant association with CagA strains of H pylori in gastric MALT lymphoma. Blood 2003;102:1012–1018.

[35] Ye H, Gong L, Liu H, et al. Strong BCL10 nuclear expression identifies gastric MALT lymphomas that do not respond to H pylori eradication. Gut 2006;55:137–138.

[36] Willis TG, Jadayel DM, Du MQ, et al. Bcl10 is involved in t(1;14)(p22;q32) of MALT B cell lymphoma and mutated in multiple tumor types. Cell 1999;96:35–45.

[37] Zhang Q, Siebert R, Yan M, et al. Inactivating mutations and overexpression of BCL10, a caspase recruitment domain-containing gene, in MALT lymphoma with t(1;14)(p22;q32). Nat Genet 1999;22:63–68.

[38] Liu H, Ye H, Ruskone-Fourmestraux A, et al. T(11;18) is a marker for all stage gastric MALT lymphomas that will not respond to H. pylori eradication. Gastroenterology 2002;122:1286–1294.

[39] Ye H, Gong L, Liu H, et al. MALT lymphoma with t(14;18)(q32;q21)/IGH-MALT1 is characterized by strong cytoplasmic MALT1 and BCL10 expression. J Pathol 2005;205:293-301.

[40] Goatly A, Bacon CM, Nakamura S et al. FOXP1 abnormalities in lymphoma: translocation breakpoint mapping reveals insights into deregulated transcriptional control. Mod Pathol 2008;21:902–911.

[41] Chanudet E, Ye H, Ferry J, et al. A20 deletion is associated with copy number gain at the TNFA/B/C locus and occurs preferentially in translocation-negative MALT lymphoma of the ocular adnexa and salivary glands. J Pathol 2009;217:420–430.

[42] Chanudet E, Huang Y, Ichimura K, et al. A20 is targeted by promoter methylation, deletion and inactivating mutation in MALT lymphoma. Leukemia 2010;24:483–487.

[43] Bi Y, Zeng N, Chanudet E, et al. A20 inactivation in ocular adnexal MALT lymphoma. Haematologica 2012;97:926–930.

[44] Du MQ. MALT lymphoma: many roads lead to nuclear factor-kappab activation. Histopathology 2011;58:26–38.

[45] Thome M, Weil R. Post-translational modifications regulate distinct functions of CARMA1 and BCL10. Trends Immunol 2007;28:281–288.

[46] Turvey SE, Durandy A, Fischer A, et al. The CARD11-BCL10-MALT1 (CBM) signalosome complex: Stepping into the limelight of human primary immunodeficiency. J Allergy Clin Immunol 2014;134:276–284.

[47] Thome M, Charton JE, Pelzer C, Hailfinger S. Antigen receptor signaling to NF-kappaB via CARMA1, BCL10, and MALT1. Cold Spring Harb Perspect Biol 2010;2:a003004.

[48] Qiao Q, Yang C, Zheng C et al. Structural architecture of the CARMA1/Bcl10/MALT1 signalosome: nucleation-induced filamentous assembly. Mol Cell 2013;51:766–779.

[49] Ye H, Dogan A, Karran L, et al. BCL10 expression in normal and neoplastic lymphoid tissue: nuclear localization in MALT lymphoma. Am J Pathol 2000;157:1147–154.

[50] Liu H, Ye H, Dogan A, et al. T(11;18)(q21;q21) is associated with advanced mucosa-associated lymphoid tissue lymphoma that expresses nuclear BCL10. Blood 2001;98:1182–1187.

[51] Maes B, Demunter A, Peeters B, Wolf-Peeters C. BCL10 mutation does not represent an important pathogenic mechanism in gastric MALT-type lymphoma, and the presence of the API2-MLT fusion is associated with aberrant nuclear BCL10 expression. Blood 2002;99:1398–1404.

[52] Li Z, Wang H, Xue L, et al. Emu-BCL10 mice exhibit constitutive activation of both canonical and noncanonical NF-kappaB pathways generating marginal zone (MZ) B-cell expansion as a precursor to splenic MZ lymphoma. Blood 2009;114:4158–4168.

[53] Streubel B, Simonitsch-Klupp I, Mullauer L, et al. Variable frequencies of MALT lymphoma-associated genetic aberrations in MALT lymphomas of different sites. Leukemia 2004;18:1722–1726.

[54] Streubel B, Lamprecht A, Dierlamm J, et al. T(14;18)(q32;q21) involving IGH and MALT1 is a frequent chromosomal aberration in MALT lymphoma. Blood 2003;101:2335–2359.

[55] Ruefli-Brasse AA, French DM, Dixit VM. Regulation of NF-kappaB-dependent lymphocyte activation and development by paracaspase. Science 2003;302:1581–1584.

[56] Ruland J, Duncan GS, Wakeham A, Mak TW. Differential requirement for Malt1 in T and B cell antigen receptor signaling. Immunity 2003;19:749–758.

[57] Uren GA, O'Rourke K, Aravind L, et al. Identification of paracaspases and metacaspases: two ancient families of caspase-like proteins, one of which plays a key role in MALT lymphoma. Mol Cell 2000;6:961–967.

[58] Lucas PC, Yonezumi M, Inohara N, et al. Bcl10 and MALT1, independent targets of chromosomal translocation in malt lymphoma, cooperate in a novel NF-kappa B signaling pathway. J Biol Chem 2001;276:19012–19019.

[59] Rebeaud F, Hailfinger S, Posevitz-Fejfar A, et al. The proteolytic activity of the paracaspase MALT1 is key in T cell activation. Nat Immunol 2008;9:272–281.

[60] Duwel M, Welteke V, Oeckinghaus A, et al. A20 negatively regulates T cell receptor signaling to NF-kappaB by cleaving Malt1 ubiquitin chains. J Immunol 2009;182:7718–7728.

[61] Coornaert B, Baens M, Heyninck K, et al. T cell antigen receptor stimulation induces MALT1 paracaspase-mediated cleavage of the NF-kappaB inhibitor A20. Nat Immunol 2008;9:263–271.

[62] Kirchhofer D, Vucic D. Protease activity of MALT1: a mystery unravelled. Biochem J 2012;444:e3–e5.

[63] Bartuzi P, Hofker MH, van de Sluis B. Tuning NF-kappaB activity: a touch of COMMD proteins. Biochim Biophys Acta 2013;1832:2315–2321.

[64] Coornaert B, Carpentier I, Beyaert R. A20: central gatekeeper in inflammation and immunity. J Biol Chem 2009;284:8217–8221.

[65] Vereecke L, Beyaert R, Van Loo G. The ubiquitin-editing enzyme A20 (TNFAIP3) is a central regulator of immunopathology. Trends Immunol 2009;30:383–391.

[66] Boone DL, Turer EE, Lee EG, et al. The ubiquitin-modifying enzyme A20 is required for termination of Toll-like receptor responses. Nat Immunol 2004;5:1052–1060.

[67] Thome M, Tschopp J. TCR-induced NF-kappaB activation: a crucial role for Carma1, Bcl10 and MALT1. Trends Immunol 2003;24:419–424.

[68] Hailfinger S, Nogai H, Pelzer C, et al. Malt1-dependent RelB cleavage promotes canonical NF-kappaB activation in lymphocytes and lymphoma cell lines. Proc Natl Acad Sci USA 2011;108:14596–14601.

[69] Kawadler H, Gantz MA, Riley JL, Yang X. The paracaspase MALT1 controls caspase-8 activation during lymphocyte proliferation. Mol Cell 2008;31:415–421.

[70] Kingeter LM, Schaefer BC. Malt1 and cIAP2-Malt1 as effectors of NF-kappaB activation: kissing cousins or distant relatives? Cell Signal 2010;22:9–22.

[71] Tusche MW, Ward LA, Vu F, et al. Differential requirement of MALT1 for BAFF-induced outcomes in B cell subsets. J Exp Med 2009;206:2671–2683.

[72] Liu H, Ruskone Fourmestraux A, Lavergne-Slove A, et al. Resistance of t(11;18) positive gastric mucosa-associated lymphoid tissue lymphoma to *Helicobacter pylori* eradication therapy. Lancet 2000:357:39–40.

[73] Ruskone-Fourmestraux A, Fischbach W, Aleman BM, et al. EGILS consensus report. Gastric extranodal marginal zone B-cell lymphoma of MALT. Gut 2011;60:747–758.

[74] Dierlamm J, Baens M, Wlodarska I, et al. The apoptosis inhibitor gene API2 and a novel 18q gene, MLT, are recurrently rearranged in the t(11;18)(q21;q21) associated with mucosa-associated lymphoid tissue lymphomas. Blood 1999;93:3601–3609.

[75] Akagi T, Motegi M, Tamura A, et al. A novel gene, MALT1 at 18q21, is involved in t(11;18)(q21;q21) found in low-grade B-cell lymphoma of mucosa-associated lymphoid tissue. Oncogene 1999;18:5785–5794.

[76] Morgan JA, Yin Y, Borowsky AD, et al. Breakpoints of the t(11;18)(q21;q21) in mucosa-associated lymphoid tissue (MALT) lymphoma lie within or near the previously undescribed gene MALT1 in chromosome 18. Cancer Res 1999;59:6205–6213.

[77] Lucas PC, Kuffa P, Gu S, et al. A dual role for the API2 moiety in API2-MALT1-dependent NF-kappaB activation: heterotypic oligomerization and TRAF2 recruitment. Oncogene 2007;26:5643–5654.

[78] Zhou H, Du MQ, Dixit VM. Constitutive NF-kappaB activation by the t(11;18)(q21;q21) product in MALT lymphoma is linked to deregulated ubiquitin ligase activity. Cancer Cell 2005;7:425–431.

[79] Rosebeck S, Rehman AO, Apel IJ, et al. The API2-MALT1 fusion exploits TNFR pathway-associated RIP1 ubiquitination to promote oncogenic NF-kappaB signaling. Oncogene 2014;33:2520–2530.

[80] Staal J, Driege Y, Bekaert T, et al. T-cell receptor-induced JNK activation requires proteolytic inactivation of CYLD by MALT1. EMBO J 2011;30:1742–1752.

[81] Hosokawa Y, Suzuki H, Nakagawa M, Lee TH, Seto M. API2-MALT1 fusion protein induces transcriptional activation of the API2 gene through NF-kappaB binding elements: evidence for a positive feed-back loop pathway resulting in unremitting NF-kappaB activation. Biochem Biophys Res Commun 2005;334:51–60.

[82] Baens M, Fevery S, Sagaert X, et al. Selective expansion of marginal zone B cells in Emicro-API2-MALT1 mice is linked to enhanced IkappaB kinase gamma polyubiquitination. Cancer Res 2006;66:5270–5277.

[83] Rosebeck S, Madden L, Jin X, et al. Cleavage of NIK by the API2-MALT1 fusion oncoprotein leads to noncanonical NF-kappaB activation. Science 2011;331:468–472.

[84] Nie Z, Du MQ, McAllister-Lucas LM, et al. Conversion of the LIMA1 tumor suppressor into an oncogenic LMO-like protein by API2-MALT1 in MALT lymphoma. Nat Commun 2015;6:5908.

[85] Streubel B, Vinatzer U, Lamprecht A, Raderer M, Chott A. T(3;14)(p14.1;q32) involving IGH and FOXP1 is a novel recurrent chromosomal aberration in MALT lymphoma. Leukemia 2005;19:652–658.

[86] Wlodarska I, Veyt E, De Paepe P, et al. FOXP1, a gene highly expressed in a subset of diffuse large B-cell lymphoma, is recurrently targeted by genomic aberrations. Leukemia 2005;19:1299–1305.

[87] Fenton JA, Schuuring E, Barrans SL, et al. t(3;14)(p14;q32) results in aberrant expression of FOXP1 in a case of diffuse large B-cell lymphoma. Genes Chromosomes Cancer 2006;45:164–168.

[88] Vinatzer U, Gollinger M, Mullauer L, Raderer M, Chott A, Streubel B. Mucosa-associated lymphoid tissue lymphoma: novel translocations including rearrangements of ODZ2, JMJD2C, and CNN3. Clin Cancer Res 2008;14:6426–6431.
[89] Ansell SM, Akasaka T, McPhail E, et al. t(X;14)(p11;q32) in MALT lymphoma involving GPR34 reveals a role for GPR34 in tumor cell growth. Blood 2012;120:3949–3957.
[90] Baens M, Finalet FJ, Tousseyn T, et al. t(X;14)(p11.4;q32.33) is recurrent in marginal zone lymphoma and up-regulates GPR34. Haematologica 2012;97:184–188.
[91] Kim WS, Honma K, Karnan S, et al. Genome-wide array-based comparative genomic hybridization of ocular marginal zone B cell lymphoma: comparison with pulmonary and nodal marginal zone B cell lymphoma. Genes Chromosomes Cancer 2007;46:776–783.
[92] Honma K, Tsuzuki S, Nakagawa M, et al. TNFAIP3 is the target gene of chromosome band 6q23.3-q24.1 loss in ocular adnexal marginal zone B cell lymphoma. Genes Chromosomes Cancer 2008;47:1–7.
[93] Honma K, Tsuzuki S, Nakagawa M, et al. TNFAIP3/A20 functions as a novel tumor suppressor gene in several subtypes of non-Hodgkin lymphomas. Blood 2009;114:2467–2475.
[94] Moody S, Escudero-Ibarz L, Wang M, et al. Distinct mutation profile of MALT lymphoma compared to otehr B-cell lymphoma characterized by NF-kB activation. 17th meeting of the European Association for Haematopathology 2014: PP-Lymph-002.
[95] Novak U, Rinaldi A, Kwee I, et al. The NF-{kappa}B negative regulator TNFAIP3 (A20) is inactivated by somatic mutations and genomic deletions in marginal zone lymphomas. Blood 2009;113:4918–4921.
[96] Kato M, Sanada M, Kato I, et al. Frequent inactivation of A20 in B-cell lymphomas. Nature 2009;459:712–716.
[97] Compagno M, Lim WK, Grunn A, et al. Mutations of multiple genes cause deregulation of NF-kappaB in diffuse large B-cell lymphoma. Nature 2009;459:717–721.
[98] Schmitz R, Hansmann ML, Bohle V, et al. TNFAIP3 (A20) is a tumor suppressor gene in Hodgkin lymphoma and primary mediastinal B cell lymphoma. J Exp Med 2009;206:981–989.
[99] Sun SC. Deubiquitylation and regulation of the immune response. Nat Rev Immunol 2008;8:501–511.
[100] Li ZM, Rinaldi A, Cavalli A, et al. MYD88 somatic mutations in MALT lymphomas. Br J Haematol 2012;158:662–664.
[101] Liu F, Karube K, Kato H, et al. Mutation analysis of NF-kappaB signal pathway-related genes in ocular MALT lymphoma. Int J Clin Exp Pathol 2012;5:436–441.
[102] Yan Q, Wang M, Moody S, et al. Distinct involvement of NF-kappaB regulators by somatic mutation in ocular adnexal malt lymphoma. Br J Haematol 2013;160:851–854.
[103] Gachard N, Parrens M, Soubeyran I, et al. IGHV gene features and MYD88 L265P mutation separate the three marginal zone lymphoma entities and Waldenstrom macroglobulinemia/lymphoplasmacytic lymphomas. Leukemia 2013;27:183–189.
[104] Zhu D, Ikpatt OF, Dubovy SR, et al. Molecular and genomic aberrations in Chlamydophila psittaci negative ocular adnexal marginal zone lymphomas. Am J Hematol 2013;88:730–735.
[105] Hovelmeyer N, Reissig S, Xuan NT, et al. A20 deficiency in B cells enhances B-cell proliferation and results in the development of autoantibodies. Eur J Immunol 2011;41:595–601.
[106] Ho L, Davis RE, Conne B, et al. MALT1 and the API2-MALT1 fusion act between CD40 and IKK and confer NF-kappa B-dependent proliferative advantage and resistance against FAS-induced cell death in B cells. Blood 2005;105:2891–2899.
[107] Tian MT, Gonzalez G, Scheer B, DeFranco AL. Bcl10 can promote survival of antigen-stimulated B lymphocytes. Blood 2005.

[108] Sagaert X, Theys T, Wolf-Peeters C, Marynen P, Baens M. Splenic marginal zone lymphoma-like features in API2-MALT1 transgenic mice that are exposed to antigenic stimulation. Haematologica 2006;91:1693–1696.
[109] Viswanatha DS, Dogan A. Hepatitis C virus and lymphoma. J Clin Pathol 2007;60:1378-83.
[110] Ferreri AJ, Govi S, Ponzoni M. Marginal zone lymphomas and infectious agents. Semin Cancer Biol 2013;23:431–440.
[111] Sulyok M, Makara M, Ujhelyi E, Valyi-Nagy I. Non-Hodgkin Lymphoma and Hepatitis C: Where We are and What Next? Pathol Oncol Res 2015;21:1–7.
[112] Arcaini L, Lazzarino M, Colombo N, et al. Splenic marginal zone lymphoma: a prognostic model for clinical use. Blood 2006;107:4643–4649.
[113] Negri E, Little D, Boiocchi M, La Vecchia C, Franceschi S. B-cell non-Hodgkin's lymphoma and hepatitis C virus infection: a systematic review. Int J Cancer 2004;111:1–8.
[114] Arcaini L, Bruno R. Hepatitis C virus infection and antiviral treatment in marginal zone lymphomas. Curr Clin Pharmacol 2010;5:74–81.
[115] Arcaini L, Vallisa D, Rattotti S, et al. Antiviral treatment in patients with indolent B-cell lymphomas associated with HCV infection: a study of the Fondazione Italiana Linfomi. Ann Oncol 2014;25:1404–1410.
[116] Rossotti R, Travi G, Pazzi A, Baiguera C, Morra E, Puoti M. Rapid clearance of HCV-related splenic marginal zone lymphoma under an interferon-free, NS3/NS4A inhibitor-based treatment. A case report. J Hepatol 2015;62:234–237.
[117] Rosa D, Saletti G, De Gregorio E, et al. Activation of naive B lymphocytes via CD81, a pathogenetic mechanism for hepatitis C virus-associated B lymphocyte disorders. Proc Natl Acad Sci USA 2005;102:18544–18549.
[118] Fabris M, Quartuccio L, Sacco S, et al. B-Lymphocyte stimulator (BLyS) up-regulation in mixed cryoglobulinaemia syndrome and hepatitis-C virus infection. Rheumatology (Oxford) 2007;46:37–43.
[119] Feldmann G, Nischalke HD, Nattermann J, et al. Induction of interleukin-6 by hepatitis C virus core protein in hepatitis C-associated mixed cryoglobulinemia and B-cell non-Hodgkin's lymphoma. Clin Cancer Res 2006;12:4491–4498.
[120] Machida K, Cheng KT, Sung VM, Levine AM, Foung S, Lai MM. Hepatitis C virus induces toll-like receptor 4 expression, leading to enhanced production of beta interferon and interleukin-6. J Virol 2006;80:866–874.
[121] Zibellini S, Capello D, Forconi F, et al. Stereotyped patterns of B-cell receptor in splenic marginal zone lymphoma. Haematologica 2010;95:1792–1796.
[122] Bikos V, Darzentas N, Hadzidimitriou A, et al. Over 30 % of patients with splenic marginal zone lymphoma express the same immunoglobulin heavy variable gene: ontogenetic implications. Leukemia 2012;26:1638–1646.
[123] Bikos V, Stalika E, Baliakas P, et al. Selection of antigen receptors in splenic marginal-zone lymphoma: further support from the analysis of the immunoglobulin light-chain gene repertoire. Leukemia 2012;26:2567–2569.
[124] Brisou G, Verney A, Wenner T, et al. A restricted IGHV gene repertoire in splenic marginal zone lymphoma is associated with autoimmune disorders. Haematologica 2014;99:e197–e198.
[125] Oscier D, Owen R, Johnson S. Splenic marginal zone lymphoma. Blood Rev 2005;19:39–51.
[126] Remstein ED, Law M, Mollejo M, Piris MA, Kurtin P, Dogan A. IGH translocation and 7 deletions are frequent genetic abnormalities in splenic marginal zone lymphoma and are present in mutually exclusive subjects. XIII meeting of the European Association for Haematopathology 2006:2006:LO8.

[127] Dierlamm J, Rosenberg C, Stul M, et al. Characteristic pattern of chromosomal gains and losses in marginal zone B cell lymphoma detected by comparative genomic hybridization. Leukemia 1997;11:747–758.

[128] Hernandez JM, Garcia JL, Gutierrez NC, et al. Novel genomic imbalances in B-cell splenic marginal zone lymphomas revealed by comparative genomic hybridization and cytogenetics. Am J Pathol 2001;158:1843–1850.

[129] Andersen CL, Gruszka-Westwood A, Atkinson S, et al. Recurrent genomic imbalances in B-cell splenic marginal-zone lymphoma revealed by comparative genomic hybridization. Cancer Genet Cytogenet 2005;156:122–128.

[130] Boonstra R, Bosga-Bouwer A, van Imhoff GW, et al. Splenic marginal zone lymphomas presenting with splenomegaly and typical immunophenotype are characterized by allelic loss in 7q31-32. Mod Pathol 2003;16:1210–1217.

[131] Watkins AJ, Huang Y, Ye H, et al. Splenic marginal zone lymphoma: characterization of 7q deletion and its value in diagnosis. J Pathol 2010;220:461–474.

[132] Fresquet V, Robles EF, Parker A, et al. High-throughput sequencing analysis of the chromosome 7q32 deletion reveals IRF5 as a potential tumour suppressor in splenic marginal-zone lymphoma. Br J Haematol 2012;158:712–726.

[133] Watkins AJ, Hamoudi RA, Zeng N, et al. An integrated genomic and expression analysis of 7q deletion in splenic marginal zone lymphoma. PLoS One 2012;7:e44997.

[134] Rossi D, Trifonov V, Fangazio M, et al. The coding genome of splenic marginal zone lymphoma: activation of NOTCH2 and other pathways regulating marginal zone development. J Exp Med 2012;209:1537–1551.

[135] Kiel MJ, Velusamy T, Betz BL, et al. Whole-genome sequencing identifies recurrent somatic NOTCH2 mutations in splenic marginal zone lymphoma. J Exp Med 2012;209:1553–1565.

[136] Martinez N, Almaraz C, Vaque JP, et al. Whole-exome sequencing in splenic marginal zone lymphoma reveals mutations in genes involved in marginal zone differentiation. Leukemia 2014;28:1334–1340.

[137] Parry M, Rose-Zerilli MJ, Gibson J, et al. Whole exome sequencing identifies novel recurrently mutated genes in patients with splenic marginal zone lymphoma. PLoS One 2013;8:e83244.

[138] Yan Q, Huang Y, Watkins AJ, et al. BCR and TLR signaling pathways are recurrently targeted by genetic changes in splenic marginal zone lymphomas. Haematologica 2012;97:595–598.

[139] Rossi D, Deaglio S, Dominguez-Sola D et al. Alteration of BIRC3 and multiple other NF-{kappa}B pathway genes in splenic marginal zone lymphoma. Blood 2011.

[140] Clipson A, Wang M, de Leval L, et al. KLF2 mutation is the most frequent somatic change in splenic marginal zone lymphoma and identifies a subset with distinct genotype. Leukemia 2014. doi: 10.1038/leu.2014.330. [Epub ahead of print]

[141] Troen G, Wlodarska I, Warsame A, Hernandez LS, Wolf-Peeters C, Delabie J. NOTCH2 mutations in marginal zone lymphoma. Haematologica 2008;93:1107–1109.

[142] Tanigaki K, Han H, Yamamoto N, et al. Notch-RBP-J signaling is involved in cell fate determination of marginal zone B cells. Nat Immunol 2002;3:443–450.

[143] Saito T, Chiba S, Ichikawa M, et al. Notch2 is preferentially expressed in mature B cells and indispensable for marginal zone B lineage development. Immunity 2003;18:675–685.

[144] Hampel F, Ehrenberg S, Hojer C, et al. CD19-independent instruction of murine marginal zone B-cell development by constitutive Notch2 signaling. Blood 2011;118:6321–6331.

[145] Simonetti G, Carette A, Silva K, et al. IRF4 controls the positioning of mature B cells in the lymphoid microenvironments by regulating NOTCH2 expression and activity. J Exp Med 2013;210:2887–2902.

[146] Piva R, Deaglio S, Fama R, et al. The Krüppel-like factor 2 transcription factor gene is recurrently mutated in splenic marginal zone lymphoma. Leukemia 2014: doi: 10.1038/leu.2014.294. [Epub ahead of print]
[147] Mahabeleshwar GH, Kawanami D, Sharma N, et al. The myeloid transcription factor KLF2 regulates the host response to polymicrobial infection and endotoxic shock. Immunity 2011;34:715–728.
[148] Heo KS, Chang E, Takei Y, et al. Phosphorylation of protein inhibitor of activated STAT1 (PIAS1) by MAPK-activated protein kinase-2 inhibits endothelial inflammation via increasing both PIAS1 transrepression and SUMO E3 ligase activity. Arterioscler Thromb Vasc Biol 2013;33:321–329.
[149] Zahlten J, Steinicke R, Opitz B, et al. TLR2- and nucleotide-binding oligomerization domain 2-dependent Krüppel-like factor 2 expression down-regulates NF-kappa B-related gene expression. J Immunol 2010;185:597–604.
[150] Nayak L, Goduni L, Takami Y, et al. Kruppel-like factor 2 is a transcriptional regulator of chronic and acute inflammation. Am J Pathol 2013;182:1696–1704.
[151] Hoek KL, Gordy LE, Collins PL, et al. Follicular B cell trafficking within the spleen actively restricts humoral immune responses. Immunity 2010;33:254–265.
[152] Winkelmann R, Sandrock L, Porstner M, et al. B cell homeostasis and plasma cell homing controlled by Kruppel-like factor 2. Proc Natl Acad Sci USA 2011;108:710–715.
[153] Hart GT, Wang X, Hogquist KA, Jameson SC. Kruppel-like factor 2 (KLF2) regulates B-cell reactivity, subset differentiation, and trafficking molecule expression. Proc Natl Acad Sci USA 2011;108:716–721.
[154] Catrysse L, Vereecke L, Beyaert R, Van Loo G. A20 in inflammation and autoimmunity. Trends Immunol 2014;35:22–31.
[155] Sol MM, Mollejo M, Villuendas R, et al. Analysis of the frequency of microsatellite instability and p53 gene mutation in splenic marginal zone and MALT lymphomas. Mol Pathol 1998;51:262–267.
[156] Gruszka-Westwood AM, Hamoudi RA, Matutes E, Tuset E, Catovsky D. p53 abnormalities in splenic lymphoma with villous lymphocytes. Blood 2001;97:3552–3558.

David Dominguez-Sola and Riccardo Dalla-Favera

7 Burkitt lymphoma

7.1 Introduction

Burkitt Lymphoma (BL) is a malignancy of mature B cells that can occur at any age, although its peak incidence is in the first decade of life. The unusual epidemiological distribution of BL has suggested complex biological interactions with the host and the environment [1]. Specifically, the incidence of this lymphoma varies significantly throughout the world, reflecting that of its major epidemiologically defined subtypes: (i) sporadic BL (sBL), occurring with an approximate frequency of 1–3 cases per million worldwide; (ii) endemic BL (eBL), occurring in equatorial regions of Africa and South America, often in association with endemic malaria, at a frequency of 5–10 cases per million (i.e. in Africa); and (iii) HIV-associated BL (AIDS-BL), occurring in the context of the AIDS pandemic. These three BL subtypes have variable association with EBV infection of tumor cells, ranging from 10–30 % of cases for sBL and AIDS-BL, to virtually 100 % of cases for eBL. The contribution of EBV to BL pathogenesis is still a matter of debate and active research [1] (see below). Analogously, the association of eBL with endemic malaria is supported by epidemiological data and animal models of plasmodium infection [2], but the precise biological mechanisms underlying this association are still unclear.

All three BL subtypes display similar morphologic features [3], including a diffuse architecture with characteristic "starry-sky" pattern due to the abundant presence of phagocytic histiocytes engulfing apoptotic cell debris. In the Western countries, BLs are curable in a large majority of cases when treated with specific intensive chemotherapeutic regimes. The cure rate is, however, much lower for eBL in Africa, due to the unmanageable toxicity derived from these treatments [4]. This chapter will focus on the pathogenic mechanisms leading to BL lymphomagenesis, with emphasis on the genetic lesions that play a role in this process. We refer to other reviews for more extensive descriptions of the epidemiological, pathologic and clinical features of BL [4–6].

7.2 Cellular origin of Burkitt lymphoma

The origin of BL was initially established to be from B cells within germinal centers (GC) of secondary follicles based on morphologic and immunophenotypic evidence [7–9]. This notion was conclusively demonstrated by showing that BL cases display hypermutated immunoglobulin variable (*IGV*) gene sequences, which supports their derivation from cells blocked within the GC or that have transited through this lymphoid compartment [10, 11]. Further refinement on the cell of origin of BL has been obtained from the analysis of gene expression profiles comparing BL cells to normal human B-cell subpopulations, which confirmed that the transcriptional program of BL is more similar to GC B cells than to any other mature B-cell subpopulation, including naïve B cells and memory B cells [10]. Notably, more detailed gene expression profile analyses have revealed that BL displays a transcriptional program similar to GC dark-zone cells (centroblasts), which represent the proliferative zone of the GC, where somatic hypermutation (SHM) of the immunoglobulin genes occurs [12–14]. This is different from all other GC-derived non-Hodgkin lymphomas, including GCB-type Diffuse Large B Cell Lymphomas (GCB-DLBCL), which appear to be more related to light-zone germinal center B cells (centrocytes) or to later stages of post-germinal center differentiation [13]. A derivation from GC centroblasts seems to be common to all BL subtypes, and is independent of their infection by EBV.

7.3 Genetic lesions involved in BL pathogenesis

7.3.1 Chromosomal translocations involving the c-MYC proto-oncogene

All tumor cells in virtually all cases of BL harbor reciprocal balanced chromosomal translocations involving the c-*MYC (MYC)* locus on chromosome 8q24 and one of three alternative chromosomal regions containing an immunoglobulin gene locus: IGH, IGκ, or IGλ [15–20]. A majority (~ 80 %) of cases involve the IGH locus at chromosome 14 (t(8;14)(q24;q32)), while the remaining cases carry a t(2;8)(p11;q24), involving the IGκ locus (15 % of cases); or a t(8;22)(q24;q11), involving the IGλ locus (5 % of cases). These BL-associated translocations are very heterogeneous at the molecular level. The t(8;14) breakpoints are located 5′ and centromeric to *MYC*, whereas, in t(2;8) and t(8;22), they map 3′ to *MYC* [15–20]. Further molecular heterogeneity derives from the exact breakpoint sites on chromosomes 8 and 14 in t(8;14) (Fig. 7.1). In eBL the *IGH-MYC* translocation generally involves sequences on chromosome 8 that are located at undefined distances (more than 100 kb) 5′ to *MYC*, and sequences on chromosome 14 within or in proximity of the IGJ_{H} region [21, 22]. Conversely, in sBL t(8;14) preferentially involves sequences located within or immediately 5′ (less than 3 kb) to *MYC* on chromosome 8 and sequences on chromosome 14 mapping within the IG switch regions [21, 22] (Fig. 7.1). These differences suggest that these gene rearrangements are driven by alternative molecular mechanisms, which are only partially understood at present. Prevalent theories attribute these genetic lesions to mistakes in the DNA remodeling processes that take place in the germinal center. Translocations associated with eBL are probably due to undesired consequences of the SHM mechanism, which almost exclusively occurs in the germinal center, while translocations associated with sBL and AIDS-BL appear to involve mistakes in immunoglobulin class-switch recombination (CSR) events [23]. Both of the above models are consistent with more recent data suggesting that AID, the enzyme mediating SHM, affects a discrete number of loci in addition to the Ig genes, acting preferentially on

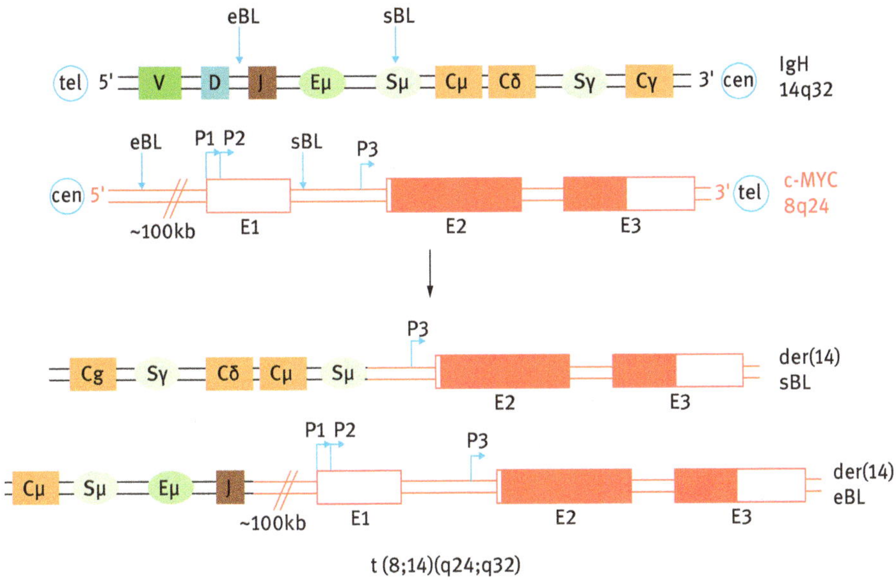

Fig. 7.1: **Architecture of translocations between MYC and the immunoglobulin gene loci in human Burkitt lymphoma.** The top panel shows a schematic view of the human immunoglobulin heavy chain (IGH) and MYC loci. Exons are represented as boxes, with coding exons filled in colors. Arrows point to the location of chromosomal breakpoints caused by AID targeting during the germinal center reaction. Note the different location of these lesions in eBL and sBL. P1, P2 and P3 are three different 5′ regulatory regions alternatively driving transcription of the MYC locus in B cells. The activity of these promoters is highly context dependent, and in normal conditions, P2 is the most prominent (cen, centromere; tel, telomere). The bottom panel shows the resulting architecture of the chromosomal translocations most commonly found in sBL and eBL. Note that the identity of IGH regulatory elements juxtaposed to the coding exons of MYC, and thus driving MYC expression, is different in sBL and eBL. The implications of these differences in the pathogenesis of BL, i.e. the pattern of MYC expression deregulation, are poorly understood. (Reprinted from [24])

sequences around the transcription start sites of actively transcribed genes, including MYC [25, 26]. AID targeting initiates DNA double strand breaks that can occasionally be resolved into aberrant recombination events with the immunoglobulin loci, leading to structural gene rearrangements equivalent to those found in BL and other germinal center-derived lymphomas [25–27]. Since the recombining translocation partners are preferentially selected based on their proximity during interphase and their transcriptional activation during B-cell development [28–31], the close spatial proximity between *MYC* and the immunoglobulin heavy chain locus (*IGH*) within transcription factories is believed to facilitate the appearance of *MYC-IG* translocations [28, 32, 33].

MYC-IG translocations in BL are expected to disrupt a strict cycle that controls MYC expression in normal germinal center B cells (Fig. 7.2). MYC expression is induced in the founding B cells of the germinal center, shortly after T-cell dependent antigen activation, but is then transcriptionally repressed by BCL6 in the bulk of the hyperproliferative dark zone of the germinal center. MYC expression is later re-induced in affinity-selected antigen-specific B cells in the light zone, which re-enter the cell cycle and transit back to the dark-zone for further rounds of selection based on antigen specificity, a process known as "cyclic re-entry" that is key to sustaining affinity maturation during adaptive immune responses [34–36] (Fig. 7.2). Thus, juxtaposition of *MYC* coding regions to heterologous immunoglobulin enhancers causes the constitutive and ectopic expression of MYC throughout the germinal center reaction (Fig. 7.2).

In addition to deregulated transcription, oncogenic conversion of *MYC* may also be facilitated by acquired mutations in its coding region (specifically, in exon 2). These mutations are likely caused by the juxtaposition of *MYC* to the *IG* loci, which would abnormally recruit the SHM activity to *MYC* coding sequences [37, 38]. While the functional significance of these mutations is suggested by their presence also in the viral v-myc oncogene [39], the selective advantage that these MYC somatic mutations confer to Burkitt lymphoma cells and lymphoma progression is only partially understood [40].

The biological consequences of MYC deregulation during lymphomagenesis are most probably an exacerbation of its physiologic activity in proliferating cells. MYC is a basic-helix-loop-helix (bHLH), sequence-specific DNA-binding protein. It requires its heterodimerization with another bHLH protein, MAX, to exert most if not all of its known biological activities [41, 42]. Binding of MYC/MAX heterodimers to DNA has wide and general effects on gene transcription: consistently, mammalian cells show important changes in gene expression programs upon MYC activation, with net effects of cell growth and proliferation [41, 43]. The molecular nature of MYC-dependent global effects on transcription is, however, controversial. While some studies refer to MYC as a global, nonspecific transcriptional amplifier [44, 45], other results suggest the existence of a core of MYC target genes common to all cell types [46].

Among the repertoire of MYC target genes are those involved in energy metabolism, protein and RNA biogenesis, and telomere maintenance [47–53]. A MYC gene transcriptional signature can be identified in all BL cases, and the presence of a *MYC/IG* translocation is strongly correlated with this signature even in other B-NHL, such as GCB-DLBCL, which carries this translocation in a subset (5–14 %) of cases [54, 55]. These observations indicate that the control of core cellular functions is key to MYC-dependent lymphomagenesis, and reveal specific vulnerabilities in MYC-driven cancer cells that could be exploited for the therapeutic targeting of MYC [56].

MYC has also a role in DNA replication, a function that was initially seen as a consequence of its transcriptional activity [57]. However, more recent studies have documented the ability of MYC to control DNA replication initiation in a transcription-independent fashion [58, 59]. This biological function could be relevant for lymphomagenesis since MYC expression induces replication stress and DNA damage, which could contribute to the selection of genetically unstable, transformation-prone

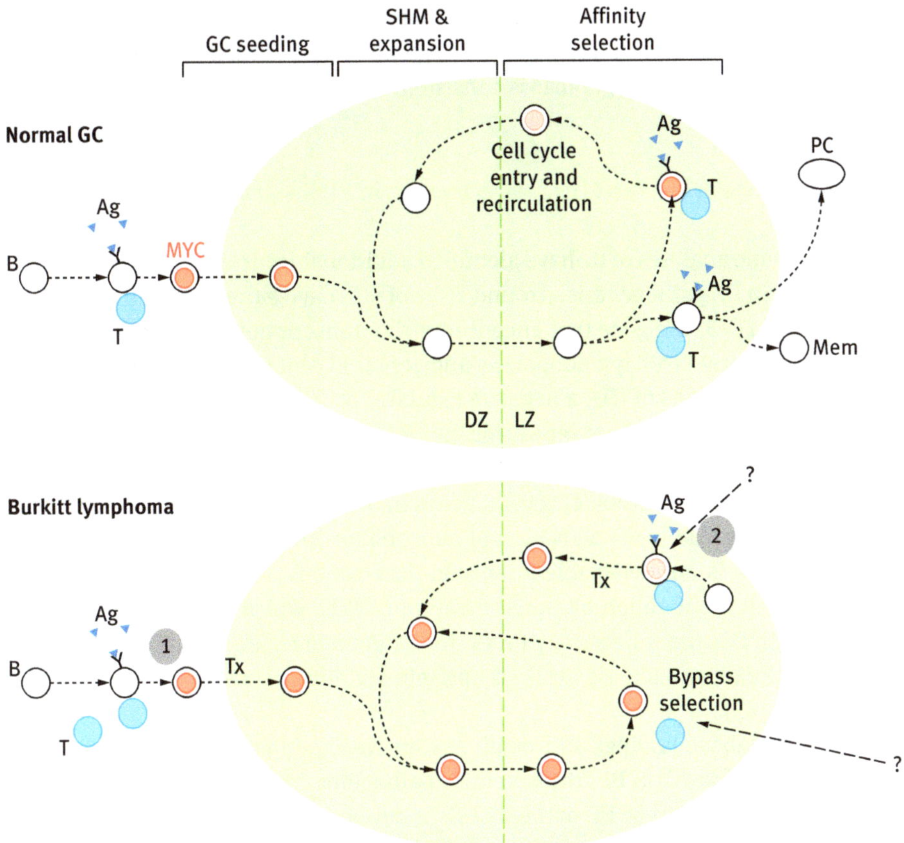

Fig. 7.2: Normal dynamics of MYC expression during the germinal center reaction and its deregulation in Burkitt lymphoma. Top panel: Dynamics of MYC expression in normal germinal center (GC) B cells. MYC expression is induced in two waves during a normal germinal center reaction (cells with red nuclei), first during germinal center entry, and later, once germinal centers reach maturity. In this structure, MYC expression is confined to a few affinity-selected B cells found in the light zone (LZ), which will transit to the dark zone (DZ) compartment for further expansion and a new cycle of somatic hypermutation of their immunoglobulin genes. In both cases, MYC activation precedes the proliferative expansion of these cells, and correlates with active T-cell dependent selection. Coupling of MYC expression to affinity selection is critical for the maintenance of the germinal center reaction. Bottom panel: Models for the generation of MYC translocations in BL. Translocations involving the IG genes and the MYC gene are thought to derive from mistakes occurring during AID-mediated SHM or CSR in the germinal center. Since AID activity requires gene transcription, the introduction of DNA double strand breaks that precedes the appearance of chromosomal rearrangements might occur in the small subset of MYC^+ germinal center B cells, either at the beginning of the germinal center reaction (1), or in light zone germinal center B cells (2). Whether these two cellular "origins" could give rise to specific variants of BL is not known. MYC/IG translocations would deregulate MYC expression in otherwise normal germinal center B cells, and uncouple MYC activation from affinity selection. The consequences of these events in germinal center homeostasis, affinity maturation, and the survival of germinal center B cells remains a matter of active investigation.

cells during precancerous stages [60]. The observation that activated B cells exposed to replication stress from different sources accumulate genetic lesions similar to those found in human B-cell lymphomas is consistent with this notion [61].

7.3.2 Mutations in TCF3 and ID3, and activation of PI3K signaling

Recent genomic analyses of BL have identified additional recurrent lesions of possible significance in BL pathogenesis. Around 70 % of sBL have either recurrent activating mutations of *TCF3*, the gene that encodes for the transcription factor E2A and plays major roles in early B-cell lymphopoiesis and germinal center development [62, 63], or inactivating mutations of ID3, a negative regulator of TCF3 [64–66]. These mutations cause the escape of E2A from repression by ID3 and thus result in its deregulation, which appears to promote antigen-independent "tonic" B cell receptor (BCR) signaling by altering the expression of several genes in the BCR pathway [66]. In particular, these alterations lead to the activation of phosphoinositide 3-kinase (PI3K) signaling (Fig. 7.3), which is not detectable in normal dark zone B cells [67, 68]. In addition, E2A affects cell proliferation by transactivating *CCND3*, which encodes for the G1–S-specific cyclin D3, and is necessary for GC formation and expansion [69, 70]. Interestingly, mutations that increase cyclin D3 protein stability are also found in ~ 40 % of sBL [66] (Fig. 7.3).

While the above findings still require extensive experimental validation and extension to eBL and AIDS-BL, the potential pathogenic relevance of combined MYC and PI3K dysregulation in BL pathogenesis is supported by the fact that transgenic mice engineered to activate both pathways in this B-cell type develop lymphomas that largely resemble human BL, including the sporadic acquisition of *CCND3* mutations (see below) [71].

7.3.3 Additional recurrent genetic lesions

Since MYC deregulation and PI3K activation in mice results in the development of B cell lymphomas clonal in nature, it is likely that additional lesions are necessary for effective lymphomagenesis. Accordingly, in addition to the ~ 40 % of cases displaying cyclin D3 mutations, one-third of human BL cases have inactivating mutations of several tumor suppressors including *TP53 (*which is inactivated at variable frequencies in the different BL subtypes [72]), *PTEN* and *CDKN2A* [74]. Moreover, almost 60 % of BL cases present non-synonymous mutations in the S1PR2, GNA13 and P2RY8 genes, which encode for components of the Gα13 signaling pathway involved in growth regulation and confinement of GC B cells [73]. The relative contribution of these genetic lesions to lymphomagenesis needs to be addressed experimentally, and their possible crosstalk with MYC deregulation further evaluated.

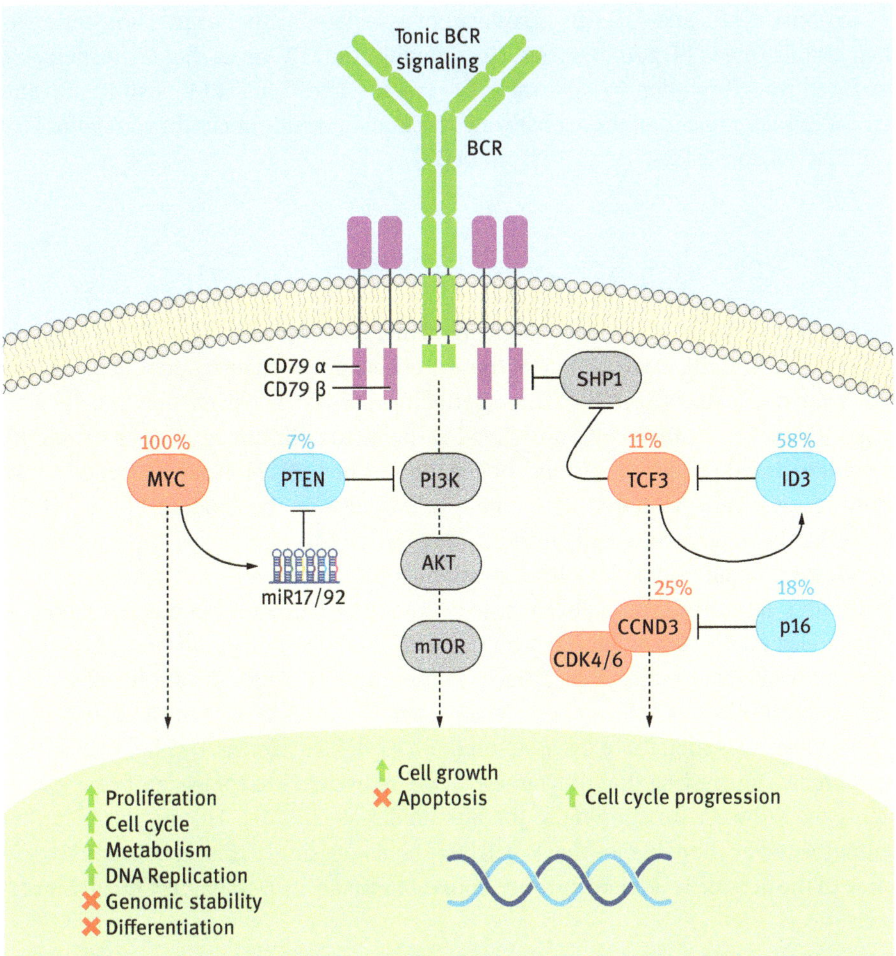

Fig. 7.3: **Summary of pathways and genetic lesions recurrently altered in human Burkitt lymphoma, and their connection to B-cell receptor signaling.** The scheme summarizes the results of diverse recent sequencing efforts aimed at identifying the mutational repertoire of human BL (see text). Oncogenes (red) and tumor suppressors (blue) are highlighted. The percentages indicate the fraction of BL cases in which these genes have been found mutated. Black arrows (positive) and broken lines (negative) highlight specific regulatory relationships among mutated genes. Most of these genetic lesions seem to converge in the deregulation of three main biological outcomes, namely cell cycle control, B cell receptor signaling and MYC-dependent activities. (Reprinted from [24])

Role of EBV infection. As already mentioned, the role of EBV infection in BL pathogenesis is unclear. In favor of its contribution to lymphomagenesis are its clear immortalizing role on human B cells *in vitro* and the presence of the viral genome in virtually all eBL cases and in a significant fraction of sBL and AIDS-BL cases. On the other hand, EBV infection is present in the majority of the population worldwide, and the majority of sBL and AIDS-BL cases are not associated with the virus, indicating its dispensable role in BL pathogenesis. Most notably, the EBV genes that are responsible for B-cell transformation *in vitro* are not expressed in eBL and EBV-positive sBL and AIDS-BL. Taken together, these observations cannot exclude or confirm a role for EBV in BL pathogenesis (discussed in detail in [74]).

7.4 Mouse models of Burkitt lymphoma

Several mouse models have been engineered in an attempt to recapitulate the pathogenesis of human BL *in vivo*. Some of these models have been instrumental to demonstrate the oncogenic effect of MYC deregulation [75–78]; however, they mostly activated MYC in immature B cells, and therefore failed to faithfully reproduce human BL in cellular, immunophenotypic and/or genetic features [79]. Moreover, the majority of these models were generated with limited knowledge on the genetics of human BL, which has been uncovered only recently [64–66].

However, a mouse model with clinical, histopathological, phenotypic, molecular and genetic features very similar to those of human BL has been recently reported [71]. This mouse model is characterized by the conditional co-deregulation of MYC and a constitutive active form of the P110alpha subunit of PI3K specifically in GC B cells. Although these two specific genetic lesions rarely coexist in human BL, constitutive PI3K activation is thought to be common in BL as well as in other B-cell non-Hodgkin lymphomas, suggesting that this mouse model reproduces to some significant extent signaling pathways functioning in the human disease [66, 71]. This notion is further strengthened by the evidence that B-cell lymphomas arising in this mouse model carry some of the genetic lesions recurrently found in human BL [71], namely CCND3 mutations.

7.5 Conclusions and perspectives

Despite being one of the most studied and curable lymphomas, there are still some important unresolved questions related to BL pathogenesis. A great deal is known about the mechanisms by which MYC controls normal cell physiology; however, the specific contribution of each of its biological activities to its oncogenic activity, including its role in transcription versus DNA replication, has not yet been addressed. Despite our knowledge of an extensive network of MYC transcriptional targets, we do not precisely know which parts of this network are essential for its tumorigenic activity. Finally, the role of MYC-induced replication stress in the generation the genomic instability favoring tumor progression remains to be investigated.

The accumulated knowledge should be translated into therapeutic advances allowing similar or higher efficacy than that of chemotherapy-based regimens, but with much less toxicity, an issue particularly relevant for eBL. The compelling evidence in humans and mice that BL may be sustained by the combined action of constitutive MYC expression and PI3K activation should stimulate the clinical testing of compounds targeting these two pathways.

References

[1] Magrath I. Epidemiology: clues to the pathogenesis of Burkitt lymphoma. Br J Haematol 2012;156:744–756.

[2] Robbiani DF et al. Plasmodium Infection Promotes Genomic Instability and AID-Dependent B Cell Lymphoma. Cell 2015;162:727–737.

[3] Molyneux EM et al. Burkitt's lymphoma. Lancet 2012;379:1234–1244.

[4] Campo E et al. The 2008 WHO classification of lymphoid neoplasms and beyond: evolving concepts and practical applications. Blood 2011;117:5019–5032.

[5] De Vita VT, Lawrence TS & Rosenberg SA (eds.). Cancer: Principles and Practice of Oncology. 10th edition, LWW, 2014.

[6] Cancer, I.A.f.R.o. WHO Classification of Tumours of Haematopoietic and Lymphoid Tissue, World Health Organization, 2008.

[7] Gregory CD et al. Identification of a subset of normal B cells with a Burkitt's lymphoma (BL)-like phenotype. J Immunol 1987;139:313–318.

[8] Mann RB et al. Non-endemic Burkitts's lymphoma. A B-cell tumor related to germinal centers. N Engl J Med 1976;295:685–691.

[9] van der Valk P, Jansen J, Daha MR & Meijer CJ. Characterization of B-cell non-Hodgkin's lymphomas. A study using a panel of monoclonal and heterologous antibodies. Virchows Arch A Pathol Anat Histopathol 1983;401:289–305.

[10] Klein U & Dalla-Favera R. Germinal centres: role in B-cell physiology and malignancy. Nat Rev Immunol 2008;8:22–33.

[11] Kuppers R, Klein U, Hansmann ML & Rajewsky K. Cellular origin of human B-cell lymphomas. N Engl J Med 1999;341:1520–1529.

[12] De Silva NS & Klein U. Dynamics of B cells in germinal centres. Nat Rev Immunol 2015;15:137–148.

[13] Victora GD et al. Identification of human germinal center light and dark zone cells and their relationship to human B-cell lymphomas. Blood 2012;120:2240–2248.

[14] Victora GD & Nussenzweig MC. Germinal centers. Annu Rev Immunol 2012;30:429–457.

[15] Dalla-Favera R. Chromosomal translocations involving the c-myc oncogene in lymphoid neoplasia. In: Kirsch IR (ed.). The causes and consequences of chromosomal aberrations. Boca Raton: CRC Press, 1993:312.

[16] Dalla-Favera R. et al. Human c-myc onc gene is located on the region of chromosome 8 that is translocated in Burkitt lymphoma cells. Proc Natl Acad Sci U S A 1982;79:7824–7827.

[17] Dalla-Favera R, Martinotti S, Gallo RC, Erikson J & Croce CM. Translocation and rearrangements of the c-myc oncogene locus in human undifferentiated B-cell lymphomas. Science 1983;219:963–967.

[18] Taub R. et al. Translocation of the c-myc gene into the immunoglobulin heavy chain locus in human Burkitt lymphoma and murine plasmacytoma cells. Proc Natl Acad Sci USA 1982;79:7837–7841.

[19] Davis M, Malcolm S & Rabbitts TH. Chromosome translocation can occur on either side of the c-myc oncogene in Burkitt lymphoma cells. Nature 1984;308:286–288.

[20] Hollis GF et al. A variant translocation places the lambda immunoglobulin genes 3′ to the c-myc oncogene in Burkitt's lymphoma. Nature 1984;307:752–755.
[21] Neri A, Barriga F, Knowles DM, Magrath IT & Dalla-Favera R. Different regions of the immunoglobulin heavy-chain locus are involved in chromosomal translocations in distinct pathogenetic forms of Burkitt lymphoma. Proc Natl Acad Sci USA 1988;85:2748–2752.
[22] Pelicci PG, Knowles DM 2nd, Magrath I & Dalla-Favera R. Chromosomal breakpoints and structural alterations of the c-myc locus differ in endemic and sporadic forms of Burkitt lymphoma. Proc Natl Acad Sci USA 1986;83:2984–2988.
[23] Kuppers R & Dalla-Favera R. Mechanisms of chromosomal translocations in B cell lymphomas. Oncogene 2001;20:5580–5594.
[24] Fangazio M, Pasqualucci L, and Dalla-Favera R: Chromosomal Translocations in B Cell Lymphoma. In J.D. Rowley et al. (eds), Chromosomal Translocations and Genome Rearrangements in Cancer, Springer International Publishing, Switzerland 2015.
[25] Chiarle R et al. Genome-wide translocation sequencing reveals mechanisms of chromosome breaks and rearrangements in B cells. Cell 2011;147:107–119.
[26] Klein IA et al. Translocation-capture sequencing reveals the extent and nature of chromosomal rearrangements in B lymphocytes. Cell 2011;147:95–106.
[27] Robbiani DF et al. AID produces DNA double-strand breaks in non-Ig genes and mature B cell lymphomas with reciprocal chromosome translocations. Mol Cell 2009;36:631–641.
[28] Hakim O et al. DNA damage defines sites of recurrent chromosomal translocations in B lymphocytes. Nature 2012;484:69–74.
[29] Kosak ST et al. Subnuclear compartmentalization of immunoglobulin loci during lymphocyte development. Science 2002;296:158–162.
[30] Parada LA, McQueen PG, Munson PJ & Misteli T. Conservation of relative chromosome positioning in normal and cancer cells. Curr Biol 2002;12:1692–1697.
[31] Roukos V et al. Spatial dynamics of chromosome translocations in living cells. Science 2013;341:660–664.
[32] Osborne CS et al. Myc dynamically and preferentially relocates to a transcription factory occupied by Igh. PLoS Biol 2007;5:e192.
[33] Roix JJ, McQueen PG, Munson PJ, Parada LA & Misteli T. Spatial proximity of translocation-prone gene loci in human lymphomas. Nat Genet 2003;34:287–291.
[34] Calado DP et al. The cell-cycle regulator c-Myc is essential for the formation and maintenance of germinal centers. Nat Immunol 2012;13:1092–1100.
[35] Dominguez-Sola D. et al. The proto-oncogene MYC is required for selection in the germinal center and cyclic reentry. Nat Immunol 2012;13:1083–1091.
[36] Victora GD et al. Germinal center dynamics revealed by multiphoton microscopy with a photoactivatable fluorescent reporter. Cell 2010;143:592–605.
[37] Bhatia K et al. Point mutations in the c-Myc transactivation domain are common in Burkitt's lymphoma and mouse plasmacytomas. Nat Genet 1993;5:56–61.
[38] Bhatia K et al. Mutations in the coding region of c-myc occur frequently in acquired immunodeficiency syndrome-associated lymphomas. Blood 1994;84:883–888.
[39] Lee CM & Reddy EP. The v-myc oncogene. Oncogene 1999;18:2997–3003.
[40] Hemann MT et al. Evasion of the p53 tumor surveillance network by tumor-derived MYC mutants. Nature 2005;436:807–811.
[41] Meyer N & Penn LZ. Reflecting on 25 years with MYC. Nat Rev Cancer 2008;8:976–990.
[42] Steiger D, Furrer M, Schwinkendorf D & Gallant P. Max-independent functions of Myc in Drosophila melanogaster. Nat Genet 2008;40:1084–1091.
[43] Kenneth NS & White RJ. Regulation by c-Myc of ncRNA expression. Curr Opin Genet Dev 2009;19:38–43.

[44] Lin CY et al. Transcriptional amplification in tumor cells with elevated c-Myc. Cell 2012;151:56–67.

[45] Nie Z et al. c-Myc is a universal amplifier of expressed genes in lymphocytes and embryonic stem cells. Cell 2012;151:68–79.

[46] Sabo A et al. Selective transcriptional regulation by Myc in cellular growth control and lymphomagenesis. Nature 2014;511:488–492.

[47] Dang CV. c-Myc target genes involved in cell growth, apoptosis, and metabolism. Mol Cell Biol 1999;19:1–11.

[48] Coller HA et al. Expression analysis with oligonucleotide microarrays reveals that MYC regulates genes involved in growth, cell cycle, signaling, and adhesion. Proc Natl Acad Sci USA 2000;97:3260–3265.

[49] Barna M et al. Suppression of Myc oncogenic activity by ribosomal protein haploinsufficiency. Nature 2008;456:971–975.

[50] Hsu TY et al. The spliceosome is a therapeutic vulnerability in MYC-driven cancer. Nature 2015;525:384–388.

[51] Koh CM et al. MYC regulates the core pre-mRNA splicing machinery as an essential step in lymphomagenesis. Nature 2015;523:96–100.

[52] Pourdehnad M et al. Myc and mTOR converge on a common node in protein synthesis control that confers synthetic lethality in Myc-driven cancers. Proc Natl Acad Sci USA 2013;110:11988–11993.

[53] van Riggelen J, Yetil A & Felsher DW. MYC as a regulator of ribosome biogenesis and protein synthesis. Nat Rev Cancer 2010;10:301–309.

[54] Dave SS et al. Molecular diagnosis of Burkitt's lymphoma. N Engl J Med 2006;354:2431–2442.

[55] Hummel M et al. A biologic definition of Burkitt's lymphoma from transcriptional and genomic profiling. N Engl J Med 2006;354:2419–2430.

[56] Cermelli S, Jang IS, Bernard B & Grandori C. Synthetic lethal screens as a means to understand and treat MYC-driven cancers. Cold Spring Harb Perspect Med 2014;4(3):387–404.

[57] Obaya AJ, Mateyak MK & Sedivy JM. Mysterious liaisons: the relationship between c-Myc and the cell cycle. Oncogene 1999;18:2934–2941.

[58] Dominguez-Sola D et al. Non-transcriptional control of DNA replication by c-Myc. Nature 2007;448:445–451.

[59] Srinivasan SV, Dominguez-Sola D, Wang LC, Hyrien O & Gautier J. Cdc45 is a critical effector of myc-dependent DNA replication stress. Cell Rep 2013;3:1629–1639.

[60] Halazonetis TD, Gorgoulis VG & Bartek J. An oncogene-induced DNA damage model for cancer development. Science 2008;319:1352–1355.

[61] Barlow JH et al. Identification of early replicating fragile sites that contribute to genome instability. Cell 2013;152:620–632.

[62] Bain G et al. E2A proteins are required for proper B cell development and initiation of immunoglobulin gene rearrangements. Cell 1994;79:885–892.

[63] Kwon K et al. Instructive role of the transcription factor E2A in early B lymphopoiesis and germinal center B cell development. Immunity 2008;28:751–762.

[64] Love C et al. The genetic landscape of mutations in Burkitt lymphoma. Nat Genet 2012;44:1321–1325.

[65] Richter J et al. Recurrent mutation of the ID3 gene in Burkitt lymphoma identified by integrated genome, exome and transcriptome sequencing. Nat Genet 2012;44:1316–1320.

[66] Schmitz R et al. Burkitt lymphoma pathogenesis and therapeutic targets from structural and functional genomics. Nature 2012;490:116–120.

[67] Dominguez-Sola D et al. The FOXO1 transcription factor instructs the germinal center dark zone program. Immunity 2015; doi:10.106/jimmuni.2015.10-015.
[68] Sander et al. PI3 Kinase and Foxo1 Transcription Factor Activity Differentially Control B Cells in the Germinal Center Light and Dark Zones, Immunity 2015; doi: 10.106/jimmuni.2015.10.021
[69] Cato MH, Chintalapati SK, Yau IW, Omori SA & Rickert RC. Cyclin D3 is selectively required for proliferative expansion of germinal center B cells. Mol Cell Biol 2011;31:127–137.
[70] Peled JU et al. Requirement for cyclin D3 in germinal center formation and function. Cell Res 2010;20:631–646.
[71] Sander S et al. Synergy between PI3K signaling and MYC in Burkitt lymphomagenesis. Cancer Cell 2012;22:167–179.
[72] Gaidano G et al. p53 mutations in human lymphoid malignancies: association with Burkitt lymphoma and chronic lymphocytic leukemia. Proc Natl Acad Sci U S A 1991;88:5413–5417.
[73] Bornkamm et al., Epstein-Barr virus and its role in the pathogenesis of Burkitt's lymphoma: an unresolved issue. Semin Cancer Biol. 2009;19:351–365.
[74] Muppidi et al., Loss of signalling via Gα13 in germinal centre B-cell-derived lymphoma. Nature 2014;51:254–258.
[75] Adams JM et al. The c-myc oncogene driven by immunoglobulin enhancers induces lymphoid malignancy in transgenic mice. Nature 1985;318:533–538.
[76] Kovalchuk AL et al. Burkitt lymphoma in the mouse. J Exp Med 2000;192:1183–11890.
[77] Klein G. Specific chromosomal translocations and the genesis of B-cell-derived tumors in mice and men. Cell 1983;32:311–315.
[78] Adams JM & Cory S. Oncogene co-operation in leukaemogenesis. Cancer Surv 1992;15:119–141.
[79] Zhu D, Qi CF, Morse HC 3rd, Janz S & Stevenson FK. Deregulated expression of the Myc cellular oncogene drives development of mouse "Burkitt-like" lymphomas from naive B cells. Blood 2005;105:2135–2137.

Laura Pasqualucci

8 Diffuse large B-cell lymphoma

8.1 Introduction

Diffuse large B-cell lymphoma (DLBCL) is a heterogeneous category of aggressive malignancies and represents the most common type of B-cell non-Hodgkin lymphoma (B-NHL) worldwide, accounting for 30 to 40 % of all new diagnoses and approximately 20,000 new cases each year [1, 2]. These tumors can affect patients of all ages but are predominantly diagnosed in older individuals (median age at presentation, 70 years), and arise *de novo* or from the histologic transformation of more indolent lymphomas – most commonly, follicular lymphoma (FL) [3] and chronic lymphocytic leukemia (CLL) [4].

Over the past decades, considerable progress has been made in the biological understanding of DLBCL as a clonal malignant expansion of B cells that are trapped at distinct stages of germinal center (GC) B-cell differentiation. Nonetheless, DLBCL remains a challenging clinical problem and up to one third of patients are not cured by standard first-line therapeutic approaches consisting of rituximab-based combination immuno-chemotherapy [5]. Current limits to effective treatment are explained in part by the remarkable heterogeneity of this disease, which can be recognized at multiple levels. Indeed, molecular analysis of DLBCL by gene expression profiling (GEP) has revealed that this morphologically indistinguishable disease is not a single homogeneous entity, but comprises multiple subtypes characterized by distinct transcriptional features that, in turn, influence clinical behavior. More recently, advances in the genomic and functional characterization of these tumors uncovered a multitude of cellular pathways that are targeted by genetic lesions and play key roles in tumor development and maintenance, as well as in the response to therapy. These discoveries are set to provide a molecular framework for the development of improved diagnostic and prognostic markers allowing the stratification of patients according to risk, and the design of more effective therapeutic approaches aimed at targeting oncogenic addictions specific to distinct lymphoma subtypes.

This chapter will focus on the molecular pathogenesis of the two most common subtypes of DLBCL, with emphasis on the mechanisms of genetic lesion and the nature of recurrently involved genes/pathways, in relationship to the normal biology of lymphocytes.

8.2 Cell of origin and classification

As most other B-NHLs, DLBCL derives from the malignant transformation of mature B cells that have experienced the GC reaction [6]. This micro-anatomical compartment forms in secondary lymphoid organs following encounter of a naive B cell with a T-cell dependent antigen [7], and fosters unique biological and genetic programs aimed at the generation and selection of cells that produce highly efficient neutralizing antibodies of different isotype classes [8, 9]. The above processes are compartmentalized within two anatomically distinct areas where B cells recycle bidirectionally: the dark zone (DZ), mostly composed of proliferating B cells that diversify their antibody genes through the process of somatic hypermutation (SHM), and the light zone (LZ), where lymphocytes capture antigen displayed on the surface of follicular dendritic cells and are selected based on their affinity for the antigen (Fig. 8.1) [10–14]. The initiation of

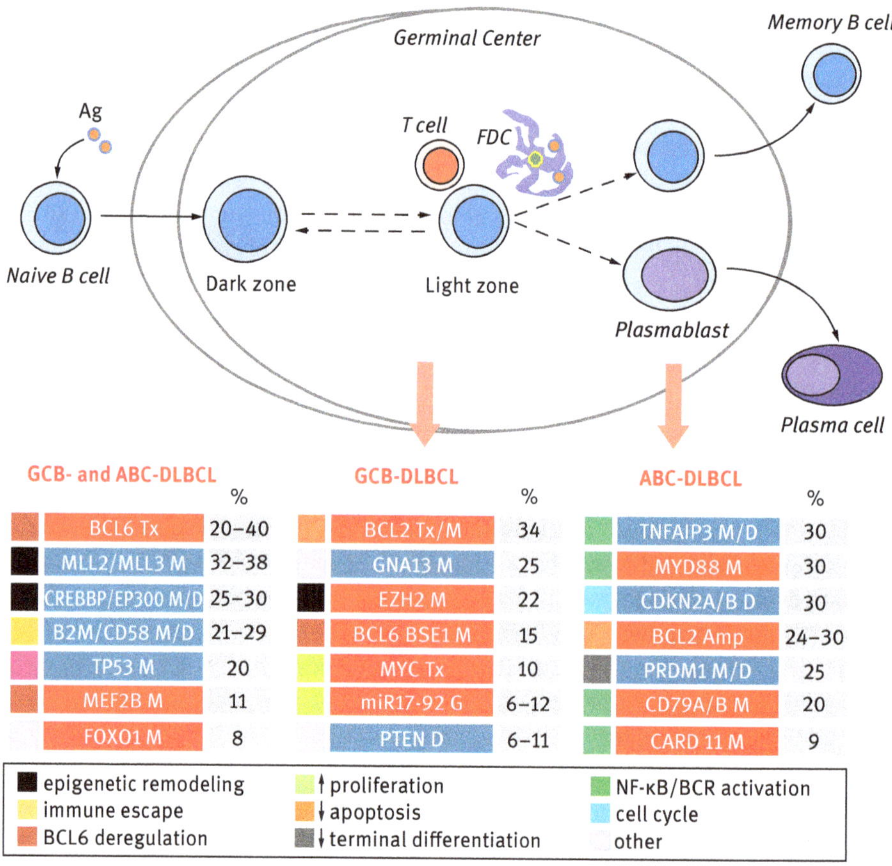

Fig. 8.1: **Postulated cellular origin of major DLBCL subtypes.** Schematic representation of the GC reaction and its relationship with the two major molecular subtypes of DLBCL, germinal center B-cell-like (GCB) and activated B-cell-like (ABC) DLBCL. The most common genetic alterations associated with this disease (including those shared across different subtypes and those subtype-specific) are shown, with color codes indicating the subverted biological pathway. Blue, loss-of-function; red, gain-of-function (modified from [182]). Tx, chromosomal translocation; M, point mutation; D, deletion; G, gain; Amp, high copy number amplification.

the GC reaction, the transition between these two phases, and the final exit from the GC to become a differentiated plasma cell or a memory B cell are finely regulated by a network of transcription factors (e.g. the GC master regulator BCL6) [15] that enable the establishment of well-defined transcriptional programs in response to a variety of signals provided by T cells and accessory cells [8, 9].

While we refer to Chapter 2 for a detailed description of the GC reaction, the relationship between the physiologic circuitries that operate within its compartments and B-NHL represents a fundamental concept for the correct understanding of lymphomagenesis, because it is the dysregulation of these same mechanisms and regulatory networks, mostly through genetic lesions, that drives the development of these tumors.

The central role of the GC reaction as the target structure of malignant transformation in B-NHL, including DLBCL is evidenced by multiple observations: first, most tumors harbor somatically hypermutated immunoglobulin (*IG*) genes, an irreversible hallmark of transit through the GC [16]; second, two major mechanisms of oncogenic transformation in DLBCL – namely chromosomal translocations and aberrant somatic hypermutation (ASHM) – result from mistakes in the AID-dependent machinery that normally diversifies the *IG* genes during B-lymphocyte development through SHM and class switch recombination (CSR) [17, 18]; third, the same transcriptional programs that reflect discrete functional phases during B-cell transit through the GC can be recognized at least in part in the phenotype of the two major molecular subtypes of DLBCL [19]; and, finally, a substantial number of genetic alterations associated with DLBCL converge on signaling pathways (e.g. NF-κB, B-cell receptor [BCR]) [20, 21] and transcription factors (e.g. BCL6, BLIMP1) [22–28] with key roles in the control of the complex GC dynamics.

8.2.1 The "cell-of-origin" (COO) classification

In 2001, the early application of microarray technologies to human patient biopsies for the genome-wide analysis of gene expression profiles (GEP) led to the identification of multiple phenotypic subgroups of DLBCL, reflecting the derivation from B cells arrested at various functional stages of differentiation. Based on similarities to the putative cell of origin (COO), DLBCL can be subdivided into three well-characterized subtypes, termed germinal center B-cell-like (GCB) DLBCL, activated B-cell-like (ABC) DLBCL, and primary mediastinal large B-cell lymphoma (PMBCL) [19, 29, 30]. An additional 10–15 % of cases exhibit transcriptional signatures that are intermediate between GCB- and ABC-DLBCL, and are thus referred to as unclassified [19, 31].

Although both GCB- and ABC-DLBCL appear more related to light zone B cells [32], the expression profile of GCB-DLBCL is enriched in genes that are highly transcribed in proliferating GC B cells, such as BCL6, and lacks post-GC differentiation markers [19]. These tumors show evidence of ongoing SHM in their antigen receptor genes [33] and, in most cases, express IgG on the cell surface, indicative of successful CSR [19].

In contrast, ABC-DLBCL shares its transcriptional profile with the one induced in BCR-activated B cells and also observed in a small subset of lymphocytes that are located in the LZ of the GC and have initiated the pathway – or are poised – to plasma cell differentiation (plasma blasts) [19]. These tumors have downregulated the expression of GC-specific markers and upregulated genes that are necessary for terminal B-cell differentiation, such as IRF4 and XBP1 [34, 35], together with numerous targets of the NF-κB transcription complex [36], indicating constitutive activation of this signaling pathway. Also consistent with the derivation from a post-GC cell, ABC-DLBCLs lack the footprint of ongoing SHM [33]. However, as will be discussed later, these tumors are precluded from completing the process of plasmacytic differentiation due to genetic lesions that disrupt this signaling cascade, exemplified by the loss of *PRDM1* and the translocation of *BCL6* [26–28]. Moreover, while most GC B cells switch their antigen receptor from IgM to IgG, ABC-DLBCL retain expression of IgM-BCR and show evidence of illegitimate CSR in the productive *IGH* allele, a finding attributed to abnormalities in the regulation of this process [37].

The third subtype of DLBCL, PMBCL, is postulated to arise from thymic medullary B cells or from post-GC B cells that migrated to the thymus [29, 30]. This clinicopathological entity, which typically involves the mediastinum and is most commonly observed in young female adults, shares significant histological, molecular and clinical characteristics with nodular sclerosis Hodgkin lymphoma (HL), including the diminished expression of BCR signaling pathway components, the evidence of constitutive NF-κB activation, the presence of an immune/inflammatory infiltrate, and a distinctive cytokine profile [29, 30, 38]. PMBCL has been recognized as a separate disease from systemic DLBCL in the revised 2008 World Health Organization Classification of Tumors of Hematopoietic and Lymphoid Tissues [2], and is discussed in detail in Chapter 9.

Of note, several studies have shown that DLBCL risk stratification according to GEP has prognostic value even after integration of the drug rituximab into standard combination therapies, with GCB-DLBCL showing a more favorable course and ABC-DLBCL representing the most aggressive form of the disease [39–43]. Nonetheless, the requirement of tools (i.e. GEP) that are not routinely available in every diagnostic lab for the correct classification of these tumors, and the imperfect reproducibility of immunohistochemistry-based classification trees [44, 45] have contributed to a discordant literature on the prognostic significance of COO subtypes [46]. As a consequence, the distinction in GCB- and ABC-DLBCL has not yet been officially incorporated into the WHO classification. This hurdle may be solved in the near future thanks to the development of novel technologies such as the gene expression-based Nanostring platform; in particular, a recent study showed that assignment of DLBCL COO by the Lymph2Cx assay using formalin-fixed paraffin-embedded tissue biopsies was able to successfully identify patient groups with significantly different clinical outcomes after R-CHOP, independent of the IPI score [47].

8.2.2 The "comprehensive-consensus-cluster" (CCC) classification

Similarities with the presumed cell of origin are not the only aspect of DLBCL that can be captured by GEP. A separate classification scheme built on comprehensive consensus clustering (CCC) methods allowed the recognition of three highly reproducible DLBCL subgroups characterized by shared molecular features within the tumor and termed OxPhos, BCR and host-response (HR) [48]. The OxPhos group is defined by the expression of genes involved in oxidative phosphorylation and mitochondrial metabolism; the BCR group is characterized by the increased expression of multiple components of the BCR/proliferation signaling pathway; and the HR group is enriched in genes associated with the tumor microenvironment/host inflammatory response [48]. Analogous to the COO classification, the CCC classification provided important insights into disease pathogenesis, suggesting potential tumor cell dependencies and rational therapeutic targets. Specifically, BCR-DLBCLs were shown to rely upon SYK/BCR-mediated survival signals [49], while OxPhos tumors are unresponsive to BCR inhibitors but appear to be uniquely sensitive to perturbations of the fatty acid oxidation program and glutathione synthesis [50]; BCR-, different from OxPhos-DLBCLs, were also shown to rely on BCL6 signaling [51, 52].

8.3 Mechanisms of genetic lesion in DLBCL

Analogous to other tumors, the pathogenesis of DLBCL represents a multistep process entailing the accumulation of multiple genetic lesions that alter the structure and/or the expression pattern of proto-oncogenes and tumor suppressor genes. Several mechanisms contribute to oncogenic dysregulation in DLBCL, including somatically acquired nonsilent point mutations and gene copy number changes, with amplifications being generally less common than deletions. In addition, DLBCL takes advantage of two mechanisms of genetic damage that are intimately connected to the physiologic *IG* DNA remodeling processes (VDJ recombination, SHM and CSR) operating in the B lymphocyte as it progresses down its differentiation pathway: (i) chromosomal translocations, and (ii) aberrant somatic hypermutation (ASHM) [17, 18].

The GC reaction and the above mentioned DNA modification mechanisms are essential components for the execution of effective immune responses, and individuals that are unable to accomplish these functions suffer from severe forms of immunedeficiencies [53, 54]. However, these same reactions represent a major source of disease, as they expose the genome of GC B cells to a constant danger through their ability to introduce single and double stranded DNA breaks. The function of SHM and CSR is particularly dangerous when considering that these events take place in rapidly proliferating cells (GC lymphocytes are among the fastest replicating mammalian cells, with a cell cycle period estimated between 6 and 12 hours) [10, 12, 55], where DNA damage checkpoints are silenced as the result of the BCL6 transrepression activity on genes such as TP53, ATR and CHEK1 [56–59]. Therefore, it does not come as a surprise that a large group of genetic lesions implicated in DLBCL development are generated through mistakes occurring during CSR and SHM. The importance of the GC and its associated DNA remodeling events in the pathogenesis of lymphoma has been experimentally demonstrated by studies showing that the deletion of Aicda, the enzyme required for CSR and SHM [53, 60], in lymphoma-prone mouse models is sufficient to prevent the occurrence of *MYC-IGH* translocations and the development of GC-derived lymphomas [61–63].

8.3.1 Chromosomal translocations

DLBCL-associated chromosomal translocations represent reciprocal and balanced recombination events that promote the exchange of genetic material between two distinct chromosomes [64]. The common denominator of all NHL-associated chromosomal translocations is the presence of a proto-oncogene in proximity to the recombination sites, with the *IG* loci being most frequently involved in the partner chromosome [17]. At variance with what is commonly observed in acute leukemias, and with few exceptions, chromosomal translocations in lymphoma rarely affect the coding domain of the involved proto-oncogene; instead, they place heterologous regulatory sequences (promoters or enhancers) derived from the partner chromosome in the proximity of an intact coding sequence, causing deregulated expression of a normal protein [64]. In most cases, the heterologous regulatory sequences that are juxtaposed to the involved proto-oncogene and are responsible for its deregulation derive from antigen receptor loci, which are expressed at high levels in GC/post-GC cells, although multiple different partners can exchange material with certain proto-oncogenes in so-called "promiscuous" translocations (typically, those involving *BCL6*) [65–67]. As a consequence, proto-oncogenes that are tightly regulated in the normal precursor cell may become constitutively expressed in the lymphoma cell ("homotopic deregulation"; e.g. translocations of *BCL6*), and proto-oncogenes that are normally not expressed in the pre-neoplastic counterpart may undergo ectopic expression in the lymphoma ("heterotopic deregulation"; e.g. translocations of *MYC* and *BCL2*, which encode for two proteins normally not expressed in GC cells) [68, 69].

8.3.2 Aberrant somatic hypermutation

The term aberrant somatic hypermutation (ASHM) was coined to define a mechanism of genetic damage that is due to the abnormal functioning of the physiologic *IG*-associated SHM process, resulting in the accumulation of multiple mutations around the 5′ sequences of genes that are normally protected from its activity in the GC [18]. In this unique microenvironment, SHM is tightly regulated both spatially and temporally, to guarantee that mutations are only introduced in the rearranged *IG* variable region genes [70] and in a few other genes, including *BCL6* [71–73] (though the function of *BCL6* mutations in normal GC cells remains unclear). Such restricted mutational activity relies on high-accuracy repair mechanisms [74] that counteract that ability of AID to bind multiple DNA sequences across the genome [75, 76]. In contrast, over half of newly diagnosed DLBCL samples, irrespective of their molecular subtype (and, at lower frequencies, a few other B-NHLs) [77–81] show evidence of ASHM in genes outside the *IG* loci, including among others the well-known proto-oncogenes *PIM1* and *MYC* [18]. These mutations are frequently multiple within the same allele and display typical features of the SHM process, including the distribution within approximately 2 kb from the transcription initiation site, the requirement for active transcription, a predominance of transversions over transitions, the preferential targeting of hotspot motifs, and frequencies that are several orders greater than the background mutation rate in mammalian cells [82]. Because, depending on the genomic configuration of the target gene, both coding and noncoding regions can be affected, ASHM is considered a powerful mechanism of transformation that may cause perturbations in gene expression as well as changes in structural and functional properties of numerous oncogenes or tumor suppressor genes. The number and identity of the genes that accumulate mutations due to this mechanism varies in different DLBCL cases [18, 83] but, although this phenomenon was first reported over 10 years ago, a comprehensive characterization of the potentially extensive genetic damage caused by ASHM is still lacking, and so is a precise understanding of its underlying molecular events. Interestingly, recent work demonstrated that the sequences targeted by ASHM and chromosomal translocations are not randomly distributed across the genome, but are predominantly grouped within super-enhancers or regulatory clusters in which sense and anti-sense transcription converge, demonstrating that these elements are key mediators of AID recruitment [84, 85]. ASHM may therefore contribute to the heterogeneity of DLBCL via the alteration of different cellular pathways in different cases.

8.4 The genetic landscape of DLBCL

In recent years, we have witnessed tremendous advances in our understanding of the molecular pathogenesis of DLBCL, fueled by the rapid expansion of next-generation sequencing technologies (whole-genome, whole-exome and transcriptome based) and by their integration with complementary approaches such as high-density SNP arrays (for the detection of copy number changes) and functional shRNA screens (for the identification of tumor dependencies). The application of these tools to large cohorts of patient samples revealed a remarkable complexity in the DLBCL coding genome, which harbors on average between 30 and > 100 lesions/case, with great inter-individual variability [86–89]. Sequencing of the DLBCL exome also uncovered several previously unappreciated genes/pathways that are subverted by genetic lesions, and thus presumably play important roles in the growth and survival of the lymphoma cell. Importantly, the numbers reported above likely represent an underestimate of the DLBCL mutation load, since the sequencing studies published to date did not interrogate noncoding portions of the genome, and thus cannot inform on the activity of ASHM, nor on the status of long noncoding RNAs, microRNAs and other genomic regions with key regulatory functions. Supporting this view, recent whole genome sequencing studies uncovered clusters of mutations at enhancers [84, 85], and several mutation hotspots were found in untranslated regions by a targeted approach integrating bioinformatics tools with the sequencing of actively transcribed regions marked by H3K4me3 [90]. While a detailed functional characterization has only been obtained for a subset of the identified lesions, these studies provided an incremental gain in our knowledge of DLBCL biology, leading to the discovery of potential targets for therapy.

8.5 Dysregulated pathways in DLBCL

The recent progresses spurred by genomic and functional efforts have led to a better definition of the oncogenic pathways that need to be disrupted to favor the initiation and expansion of DLBCL. These studies revealed that genetic aberrations in DLBCL often converge on signaling cascades and transcription factor networks that are important for GC B-cell physiology, and are co-opted by the tumor cell during the malignant transformation process that leads to lymphomagenesis. While a number of lesions are common to all DLBCL subgroups, likely reflecting general mechanisms of transformation, others are found preferentially or even exclusively associated with individual DLBCL subtypes, indicating that these tumors are addicted to distinct oncogenic pathways, and uncovering vulnerabilities that may in turn influence treatment outcome. The following section focuses on functionally well-characterized and historically relevant genetic lesions, as related to the two major subtypes of systemic DLBCL, defined by cell of origin.

8.5.1 Programs dysregulated in both GCB-DLBCL and ABC-DLBCL

8.5.1.1 Histone/chromatin modifications

Modulation of chromatin accessibility through histone modification is a key step in the regulation of gene transcription, and has been implicated in multiple types of cancers, including DLBCL. Indeed, the discovery of recurrent mutations in histone/chromatin modifier genes, including methyltransferases, acetyltransferases and histones themselves, has been a consistent theme in latest DLBCL sequencing studies, indicating a key role for epigenetic remodeling in lymphomagenesis (Fig. 8.2) [86–88]. Additionally, investigations aimed at reconstructing the history of clonal evolution during FL transformation to DLBCL suggest that mutations of epigenetic modifiers (namely,

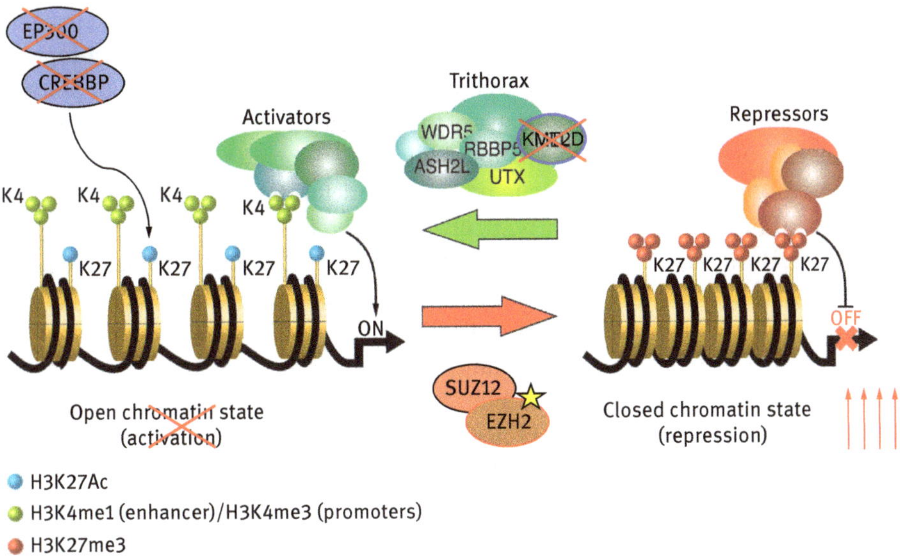

Fig. 8.2: **Genetic lesions in histone/chromatin modification enzymes.** The histone methylation pattern at a specific gene is determined by the equilibrium between the activities of two opposing complexes, exemplified here by the Trithorax-related KMT2D/KMT2C complex, which favors an active chromatin conformation via depositing H3K4me1 and H3K4me3 marks at enhancers and promoters, respectively, and the Polycomb Repressive Complex, which catalyzes the trimethylation of H3K27 (H3K27me3) associated with transcriptionally silenced chromatin. In DLBCL, a large number of genetic aberrations target histone/epigenetic modifiers, strongly suggesting a key role for epigenetic reprogramming during malignant transformation. Crosses depict loss-of-function genetic alterations and stars depict gain-of-function mutations.

CREBBP and *KMT2D*) represent early events introduced in a common ancestral clone before divergent evolution to FL or transformed FL (tFL) [91–94], suggesting that these lesions contribute to the initial phases of the malignant transformation process.

Inactivation of acetyltransferase genes. Nearly 30 % of all DLBCL samples, with some preference for the GCB-DLBCL subtype, harbor mutations and/or deletions in the acetyltransferase genes *CREBBP* and, less frequently, *EP300* [95]. These genes encode for two ubiquitously expressed enzymes that catalyze the addition of acetyl groups to the lysine residues of multiple histone and nonhistone proteins implicated in diverse signaling and developmental pathways [96]. Mutations include prototypical loss-of-function events that remove the C-terminal histone acetyltransferase (HAT) domain, as well as amino acid changes that also cluster in the HAT domain, leading to impaired enzymatic activity via diminished affinity for Acetyl-CoA [95]. Except for a few cases, they are observed in heterozygosis and are accompanied by expression of the residual wild-type allele, suggesting that CREBBP/EP300 function as haploinsufficient tumor suppressors. In support to a dose-dependent pathogenic effect of these proteins, loss of a single *CREBBP (or EP300)* allele is the causative genetic event in a rare congenital disorder known as Rubinstein–Taybi syndrome [97]. Given the widespread activity of these acetyltransferases on chromatin and the broad number of proteins that have been recognized as substrates of their enzymatic activity, it is expected that the consequences of CREBBP/EP300 inactivation will be broad on transcriptional regulation, and efforts are ongoing to comprehensively define the network of genes modulated by CREBBP/EP300 in normal GC B cells and disrupted in DLBCL patients carrying mutations. However, it has been shown that mutations of CREBBP have major impacts on the regulation and activity of two well-characterized substrates with central functions in GC physiology and pathology [95]: the onco-protein BCL6, where acetylation serves as a negative modulator of its transrepression activity [98], and the tumor suppressor TP53, which requires acetylation for its transcriptional activation [99, 100]. The balanced activity of these two proteins is fundamental for regulating DNA damage checkpoints during *IG* remodeling in the GC [56]; thus, one consequence of BCL6 activity overriding p53 would be an increased tolerance for DNA damage in the absence of proper apoptotic responses. The discovery of mutations in *CREBBP* and *EP300* has clinical implications in view of available drugs (i.e. histone deacetylase inhibitors) that could provide therapeutic benefits by re-establishing physiologic acetylation in patients stratified based on genetic status.

Inactivating mutations of KMT2D. The *KMT2D* gene (also known as *MLL2*) encodes a conserved member of the SET1 family of histone methyltransferases that controls epigenetic transcriptional regulation by mono-, di- and tri-methylating the lysine 4 position of histone H3, leading to an active chromatin conformation [101]. In GC B cells, KMT2D was shown to occupy chromatin domains at enhancer and promoter regions of genes with key roles in B-cell physiology [102, 103].

Somatic mutations of *KMT2D* are found in approximately one third of DLBCL cases and > 80 % of FL cases, representing one of the most frequent genetic aberrations associated with B-NHL [86, 87]. These events abrogate the enzymatic activity of KMT2D by either removing the C-terminal cluster of conserved domains, including the SET domain (truncating mutations) [86, 87], or suppressing its ability to catalyze H3K4 methylation *in vitro* (missense mutations in the carboxy terminal portion of the KMT2D protein) [103]. In support to the “common precursor” model, conditional B-cell specific deletion of *Kmt2d* before GC formation leads to an increased percentage of GC B cells featuring a distinct transcriptional program that is enriched in anti-apoptotic and cell cycle regulatory genes [103]. The tumor suppressor role of KMT2D has been demonstrated in mice where its genetic loss cooperates with BCL2 deregulation to induce the development of lymphoid malignancies recapitulating the spectrum of phenotypes observed during human FL to DLBCL transition [102, 103].

Collectively, the findings accumulated to date suggest that alterations in epigenetic modifiers may act differently from classical tumor suppressor genes, and facilitate the initial stages of transformation by creating a permissive environment for the proliferation and survival of the cancer clone, in cooperation with other genetic lesions (e.g. BCL2 translocations).

8.5.1.2 Deregulation of BCL6 activity

BCL6 is a master regulator of the GC reaction [15, 104] and deregulation of its activity constitutes a major mechanism of B-cell malignant transformation, by virtue of its involvement in multiple biological programs that include, among others, the establishment of a proliferative phenotype, the attenuation of DNA damage sensing and replication checkpoints [56–59], the protection from programmed cell death [105, 106], and the block of terminal differentiation [107]. BCL6 deregulation in DLBCL cells is achieved through multiple genetic aberrations, either directly or indirectly (Fig. 8.3). As many as 35 % of DLBCL patients – with a 2:1 ratio in ABC vs GCB-DLBCL [108] – carry a chromosomal translocation that juxtaposes the intact coding sequences of *BCL6* on chromosomal band 3q27 downstream and in the same transcriptional orientation to heterologous sequences bestowed by over 20 chromosomal partners (most commonly, the *IG* heavy and light chain loci) [22–25], leading to its "promoter substitution" [109]. More rarely (~ 6 % of cases), translocations cluster at an alternative breakpoint region, which may juxtapose BCL6 to distantly acting transcriptional regulatory elements [110]. The common feature of these promoters is a broader spectrum of activity throughout B-cell development, and particularly in the post-GC differentiation stage. As a consequence, BCL6 translocations prevent the downregulation of BCL6 transcription that normally licenses differentiation into post-GC B cells.

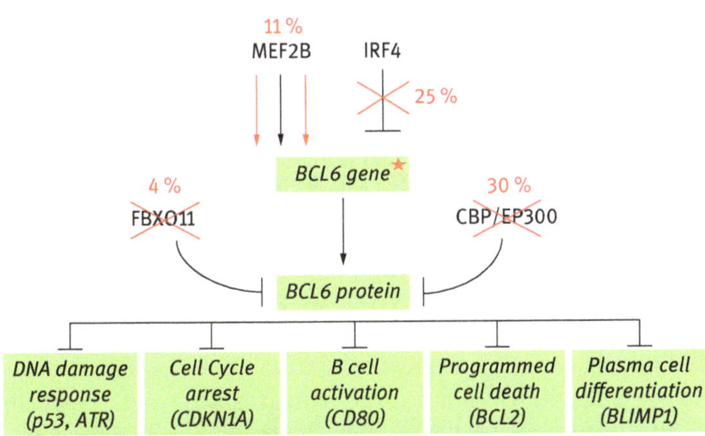

Fig. 8.3: **Deregulation of BCL6 activity by genetic lesions in DLBCL.** Recurrent genetic alterations deregulating the function of BCL6 in DLBCL, either directly or indirectly. Representative biological programs modulated by BCL6 in the GC, and disrupted as a consequence of these lesions are shown at the bottom. Symbols depict loss-of-function and gain-of-function genetic alterations (reprinted with permission from [182]).

BCL6 is also altered by multiple somatically acquired mutations that are distributed within a ~ 2 Kb region downstream to its TSS, encompassing the first noncoding exon (~ 75 % of cases). While most of these events reflect the physiologic activity of SHM in GC B cells [71, 72], a cluster of nucleotide changes within the untranslated exon 1 were found to be restricted to lymphoma (particularly GCB-DLBCL, 10 % of cases) [108, 111]. These mutations abrogate a negative autoregulatory loop by which the BCL6 protein controls its own transcription [111, 112] or prevent IRF4-mediated transcriptional repression induced by CD40 signaling in the GC LZ [113]. Overall, as many as 35 % of DLBCLs carry *BCL6* genetic abnormalities of proven functional consequences, and frequencies are expected to increase as we continue to dissect the full extent of functionally disruptive BCL6 mutations.

In addition to alterations directly affecting the *BCL6* gene, DLBCL have devised a number of ways to deregulate the BCL6 function indirectly. Besides the inactivating mutations of *CREBBP/EP300* described above, 10–15 % of patients harbor gain-of-function somatic mutations in the MEF2B transcription factor, a protein highly expressed during the GC reaction and involved in the transcriptional activation of BCL6 [86, 87, 114]. *MEF2B* mutations enhance its function by at least two mechanisms, including an increased transcriptional activation (mutations in the N-terminal MEF/MAD domain) or the loss of phosphorylation- and sumoylation-mediated negative regulatory motifs (mutations in the C-terminal portion of the protein) [114]. In another 5 % of cases, loss-of-function mutations and/or deletions of *FBXO11* impair the ubiquitylation and proteasomal-mediated degradation of BCL6, which is controlled by this ubiquitin adaptor protein [115]. Consistent with a prominent oncogenic role, deregulated BCL6 expression in a mouse model mimicking the *BCL6-IGH* rearrangement leads to the development of clonal lymphoproliferative disorders recapitulating genetic and phenotypic features of the human disease [116].

8.5.1.3 Loss of immune surveillance mechanisms

A large proportion of DLBCL cases have evolved means to deceive the immune system by interfering with the function of the major histocompatibility complex (MHC) class I, a class of molecules that is normally present on the surface of nearly all nucleated cells and is required for the proper recognition of endogenous peptides by cytotoxic T lymphocytes (CTL) [117].

Over 60 % of DLBCL samples fail to express MHC class I due to a variety of genetic and epigenetic mechanisms, which include: (i) structurally disruptive biallelic mutations and/or focal homozygous deletions of the *B2M* locus, encoding for the β2-microglobulin invariant subunit (29 % of cases) [118]; (ii) point mutations and genomic loss of the *HLA-A*, *-B*, and *-C* loci [119]; and (iii) lack of expression or aberrant cytoplasmic localization of the B2M protein, implying defects in the assembly and transport from the endoplasmic reticulum to the cell surface (30–45 % of cases) [118]. Supporting its relevance in DLBCL pathogenesis, the loss of HLA-I expression is infrequently observed in other B-NHL types and, in cases derived from the clinical evolution of FL, it is selectively acquired at transformation, concurrent with *B2M* genetic lesions [94]. Thus, immune escape may represent an important contributor to the more aggressive phenotype associated with this disease (Fig. 8.4).

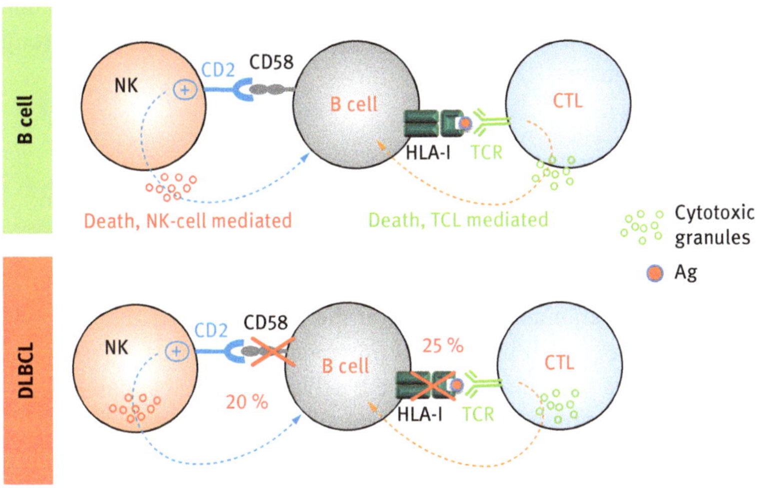

Fig. 8.4: **Genetic mechanisms of immune evasion in DLBCL.** Postulated mechanism of immune escape driven by genetic lesions in DLBCL. Expression of the MHC class I complex on the surface of nucleated cells is required for presentation of endogenous peptides to cytotoxic T lymphocytes (CTL), which will recognize non-self molecules (e.g. tumor antigens) and subsequently kill the presenting cell. HLA-I also acts as a ligand for the NK cell inhibitory receptors, which transduce signals to counterbalance the activating cues delivered through interaction of CD2 with CD58. Thus, tumor cells devoid of HLA-I are vulnerable to NK cell mediated attacks due to their activation by CD58 (top). In DLBCL (bottom), concurrent loss of HLA-I and CD58 by disruptive mutations and focal deletions may confer protection from both CTL and NK cell mediated lysis, leading to immune evasion of the tumor cell.

DLBCL also harbors recurrent genetic alterations inactivating the *CD58* gene, a member of the immunoglobulin superfamily that functions as the ligand of the CD2 receptor expressed on T cells and natural killer (NK) cells, facilitating antigen presentation [120]. Of note, loss of HLA-I and CD58 expression are often concurrent in the same cases [118], suggesting that they have been co-selected for their ability to interfere with both CTL-mediated and NK cell-mediated immune surveillance mechanisms.

Deleterious mutations have been reported in a number of other genes that are implicated in the regulation of immune responses or in the crosstalk between the tumor and the microenvironment, as exemplified by *TNFRSF14* (9–22 % of cases) [86, 88] and *CD70* (~ 10–22 % of cases) [86, 88, 121]. Genetic aberrations of *TNFRSF14* are found almost exclusively in GCB-DLBCL but they have not been functionally characterized yet; as such, their precise role in lymphomagenesis remains unclear.

Other lesions. Genetic disruption of the ***TP53*** gene is observed in ~20 % of DLBCL patients at diagnosis, regardless of COO subtype [86–88]. This frequency is relatively low compared to other lymphoid malignancies, and may in part reflect the preferential association of *TP53* aberrations with cases derived from FL transformation [122]. More recently, multiple low-frequency copy number aberrations were found in upstream regulators of a pathway that converges to deregulate p53 and various cell cycle components, resulting in increased expression of E2F1 target genes and identifying a group of patients with inferior outcome [123].

8.5.2 Genetic lesions associated with GCB-DLBCL

8.5.2.1 Chromosomal translocations of BCL2

The *BCL2* proto-oncogene encodes for the founding member of a protein family that governs commitment to programmed cell death at the mitochondrion [124]. BCL2 is also a direct target of BCL6, which binds to its promoter via the transcriptional co-activator Miz1 and prevents its expression in GC B cells, ensuring the maintenance of a default pro-apoptotic program [69, 106]. This regulatory axis is disrupted in 30–40 % of GCB-DLBCL due to the t(14;18) translocation, which brings the *BCL2* coding exons under the control of potent regulatory elements from the *IG* locus, resulting in its ectopic expression [125]. Moreover, BCL2 is one of the most highly mutated genes in this lymphoma subtype, reflecting the aberrant activity of SHM redirected by the juxtaposed *IG* enhancer to its promoter and coding sequences [69, 126]. In line with its ability to confer survival advantage, deregulation of BCL2 has been associated with an inferior outcome [127], particularly when coupled with MYC deregulation (see paragraph on “double-hit lymphomas”).

8.5.2.2 Chromosomal translocations of MYC

The MYC gene encodes for a sequence-specific transcription factor that acts as an amplifier of transcriptional programs associated with numerous biological functions, including proliferation, cell growth, energy metabolism, differentiation and apoptosis [128], as well as DNA replication independently of its transcriptional activity [129]. MYC is ectopically and constitutively expressed in 10–14 % of GCB-DLBCLs as the result of the t(8;14) translocation, which joins its intact coding exons to the *IG* heavy or light chains loci [130–133]. One consequence of the translocation is the loss of regulatory sequences by which BCL6 negatively modulates MYC expression in the GC, where its pattern is tightly regulated [68]. Analogous to BCL2, the presence of MYC translocations has been linked with worse prognosis in DLBCL [134].

8.5.2.3 Mutations of the EZH2 methyltransferase

EZH2 encodes for a SET-domain containing histone methyltransferase that is responsible for trimethylating the lysine 27 residue of histone H3 (H3K27me3), an epigenetic mark associated with transcriptional repression [135–137]. As a component of the polycomb repressive complex-2 (PRC2), EZH2 plays important functions during normal development by controlling the expression patterning in embryonic stem cells [138]. EZH2 is also a master regulator of the GC, where it is specifically expressed at high levels to facilitate the establishment of bivalent domains [139] in genes associated with termination of the GC reaction (IRF4, PRDM1) or with other GC relevant functions (CDKN1A) [140, 141]. In line with the essential function of its histone methyltransferase activity in the GC, conditional deletion of *Ezh2* in the mouse was sufficient to prevent GC formation and affinity maturation [140, 141].

Somatic heterozygous mutations of the *EZH2* gene have been identified in 22 % of GCB-DLBCL [142]. Most of these events replace a single evolutionarily conserved residue (Y641) within the protein SET domain, leading to increased levels of H3K27me3 through altered catalytic specificity of EZH2 for its substrates [143–145]. In mice, expression of the mutant $EZH2^{Y641F}$ allele induces GC hyperplasia and, when combined with BCL2 deregulation, accelerates the development of mature B-cell lymphomas [140, 141]. Notably, selective small molecule inhibitors of EZH2 have shown specific activity in preclinical studies independently of the presence of somatic mutations, particularly in GCB-DLBCL [140, 146, 147], supporting their evaluation in early clinical trials [148].

8.5.2.4 Mutations of the Gα13 pathway

The positioning and confinement of GC B cells within the B-cell follicle is regulated by the activity of two G protein-coupled receptors that, among B-cell types, are expressed specifically in the GC: sphingosine 1 phosphate receptor 2 (S1PR2) [149] and the orphan purinergic receptor P2Y, G-protein coupled 8 (P2RY8) [150]. These receptors recruit two closely related G proteins (Gα12 and Gα13) in response to lipid ligands, and stimulate Rho activity via specific guanine nucleotide exchange factors (Rho-GEFs), ultimately suppressing pAKT signaling and cell migration. The importance of the Gα13 inhibitory circuit in the pathogenesis of GCB-DLBCLs is underscored by the presence of recurrent inactivating mutations in several of its components, including *S1PR2*, *GNA13*, and, more rarely, *ARHGEF1* and *P2RY8* (~ 20 % of cases) (Fig. 8.5) [86, 88, 150]. Consistently, deletion of these genes in the mouse is associated with increased GC B-cell survival and disruption of the GC architecture, followed by dissemination of GC B cells to the lymph and bone marrow [150]; most importantly, *GNA13*, *ARHGEF1*, and, as shown in earlier studies, *S1PR2* [151] knock-out mice develop lymphomas exhibiting features of GCB-DLBCL [150]. These data suggest that disruption of this pathway exerts a dual function during lymphomagenesis by abrogating Gα13-mediated inhibitory signals on both cell migration and AKT signaling.

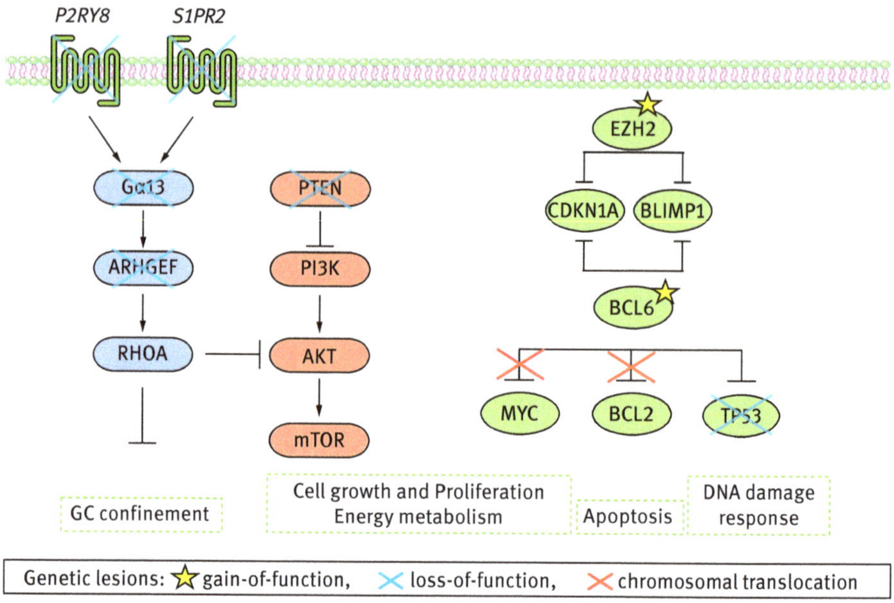

Fig. 8.5: **Disrupted signaling pathways in GCB-DLBCL.** Genetic lesions preferentially associated with GCB-DLBCL include chromosomal translocations of BCL2 (up to 35 % of cases) and/or MYC (10–15 % of cases), which lead to their ectopic expression in part by allowing them to bypass BCL6-mediated transcriptional repression; mutually exclusive chromosomal translocations (15 % of cases) and point mutations (10 % of cases) of BCL6, causing its deregulated expression; and gain-of-function mutations of EZH2 (~ 20 % of cases), which induce epigenetic remodeling in various lymphoma-relevant genes, such as CDKN1A, in part cooperating with BCL6. Additionally, loss of PTEN expression is observed in 55 % of cases, as a consequence of genetic deletions (15 %) and amplifications of miR-17–92, resulting in activation of the PI3K/Akt/mTOR signaling pathway.

8.5.2.5 Other Lesions

Approximately 55 % of GCB-DLBCL, but only 14 % of ABC-DLBCL have lost the expression of the tumor suppressor PTEN [152]. This finding is ascribed in part to the presence of mutually exclusive chromosomal deletions of 10q and amplifications of the oncogenic miR-17–92 micro-RNA cluster (collectively, ~ 20 % of patients) [153], and was inversely correlated with activation of the phosphatidylinositol 3 kinase (PI3K)/AKT pathway and sensitivity to pharmacologic PI3K inhibition, revealing potential therapeutic opportunities [152].

Also more common in GCB-DLBCL (~ 8 % of cases) are mutations of **FOXO1**, a transcription factor that is expressed in the GC DZ and is required for DZ formation, as documented by the phenotype of a conditional *Foxo1* knock-out mouse model, which lacks the DZ gene expression program and forms LZ-only GCs upon immunization [154, 155]. *FOXO1* mutations cluster around a T24 phosphorylation site that is required for its AKT-mediated nuclear-cytoplasmic translocation and inactivation, and were suggested to enhance its activity by preventing nuclear export following PI3K signaling [156]. Nonetheless, a systematic examination of the consequences of these mutations in the context of B cells is still lacking. The preferential enrichment of *FOXO1* mutations in patients with aggressive disease suggests a role for prognostication and risk stratification [156].

While the large catalogue of additional genetic alterations found in DLBCL and not yet functionally tested precludes their discussion, two potentially relevant ones include a recurrent amplicon on chromosome 2p encompassing the *REL* gene, which encodes for a NF-κB subunit, and (ASHM-dependent) mutations of the *SGK1* gene, which encodes a PI3K-regulated kinase implicated in the regulation of FOXO transcription factors, NF-κB, and NOTCH signaling [157, 158].

8.5.3 Genetic lesions associated with ABC-DLBCL

A distinctive feature of the ABC-DLBCL genome is the presence of multiple genetic aberrations that converge on two main molecular circuits: the NF-κB signaling pathway, leading to its constitutive activation, and the BCL6-BLIMP1 axis, leading to the blockade of terminal differentiation (Fig. 8.6). Additionally, ABC-DLBCL is characterized by recurrent amplifications of the *BCL2* locus, observed in more than one third of patients [153], homozygous deletions or lack of expression of the *CDKN2A/B* tumor suppressor genes (30 % of cases) [153], and gains or amplifications of a genomic region encompassing the *SPIB* locus on chromosome 19q, which is generally highly expressed in this malignancy (27 % of cases) [153].

8.5.3.1 Genetic lesions leading to constitutive activation of the NF-κB signaling pathway

Initial evidence for the constitutive activation of NF-κB in ABC-DLBCL came from GEP studies that revealed the enrichment of numerous NF-κB target genes specifically in this subgroup [36], and was later reinforced by the detection of NF-κB subunits in the nuclei of the tumor cells [159]. The demonstration that pharmacologic or genetic inhibition of NF-κB was toxic to ABC-, but not GCB-DLBCL cell lines provided further support to the central role of this signaling cascade in ABC-type lymphoma growth and survival [36]. Indeed, subsequent investigations offered a genetic explanation for this phenotypic feature by uncovering the presence of structural alterations in multiple molecules whose common denominator is the ability to induce activation of NF-κB; these include components of the BCR signaling cascade, downstream adaptors of the Toll-like receptor (TLR)/IL1R pathway, and direct positive or negative regulators of the NF-κB transcription complex.

Mutations activating the BCR signaling pathway. Mature B cells require “tonic”, antigen-independent signaling from the BCR for their survival [160, 161]; however,

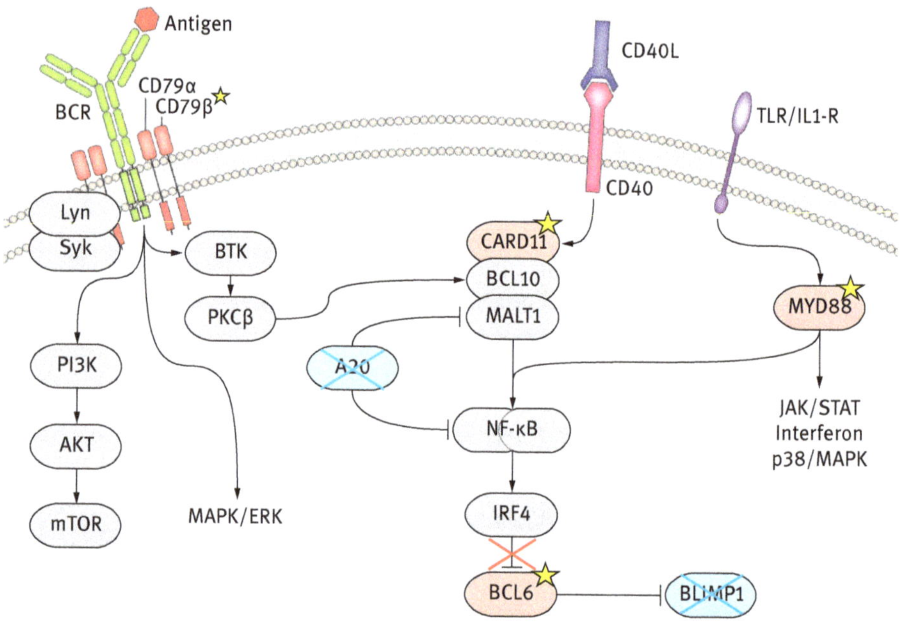

Fig. 8.6: **Disrupted signaling pathways in ABC-DLBCL.** ABC-DLBCL is defined by multiple genetic alterations that cluster around two main biological programs: constitutive activation of the NF-κB transcription complex and block of plasma cell differentiation. In normal LZ GC B cells, NF-κB activity can be induced by a variety of signals, including engagement of the BCR by the antigen, interaction of the CD40 receptor with the CD40L presented by T cells, and Toll-like receptors, leading to the upregulated expression of numerous target genes (e.g., IRF4 and A20). IRF4, in turn, extinguishes the GC program by terminating the expression of BCL6 and releasing the expression of BLIMP1, another master regulator of plasma cell differentiation. The A20 negative regulator terminates NF-κB responses via a negative feedback loop. Subversion of the BCR and NF-κB signaling pathways may fuel malignant transformation by hijacking the anti-apoptotic and pro-proliferative functions of NF-κB while mutually exclusive alterations deregulating BCL6 or inactivating BLIMP1 lead to block in terminal B cell differentiation (modified with permission from [182]).

ABC-DLBCL cells were found to display a "chronic, active" form of BCR signaling that is associated with the presence on the plasma membrane of clustered BCRs akin to those formed in antigen-stimulated B cells, requires CARD11, and is sustained by genetic alterations in proximal members of the BCR signal transduction cascade [162]. This can be attributed in 20 % of patients to gain-of-function somatic mutations affecting the immunoreceptor tyrosine-based activation motifs (ITAMs) of the IG superfamily member ***CD79B*** (or, rarely, *CD79A*) [162]. Most of these events replace the first tyrosine residue (Y196) in the cytoplasmic tail of the two BCR subunits, and are thought to maintain BCR signaling chronically active by attenuating the phosphorylation and activation of the Lyn kinase, which is necessary for internalization of the surface BCR and serves as a negative feedback regulator. Consistent with this model, ABC-DLBCL cases with *CD79B* mutations show greater abundance of BCR on their surface [162].

In another ~9 % of ABC-DLBCL (and a smaller subset of GCB-DLBCL), activation of BCR and NF-κB can be attributed to oncogenic mutations of ***CARD11*** [163], a component of the "signalosome" complex whose coordinated recruitment is needed for proper transduction of BCR signaling [164]. DLBCL-associated mutations are confined to the protein coiled-coil domain and were shown to enhance the ability of CARD11 to transactivate NF-κB target genes, in part by favoring the formation of cytosolic aggregates and the recruitment of downstream effector molecules [163].

Mutations of the *MYD88* gene. *MYD88* encodes an adaptor molecule in the TLR signaling pathway, which mediates activation of NF-κB as well as type I interferon responses [165], and is affected in ~30 % of ABC-DLBCL cases by a prototypical L265P substitution that targets an evolutionarily invariant residue within the protein TIR (Toll/IL1 receptor) domain [166]. A functional attribute of this mutation is the ability to induce the spontaneous assembly and activation of a protein complex containing the kinases IRAK1 and IRAK4, leading to engagement of the NF-κB signaling pathway. Additionally, MYD88 mutated cases show autocrine transcriptional activation of the JAK/STAT3 signaling cascade, which is also a phenotypic trait of ABC-DLBCL and appears to be required for their survival [166, 167]. Other mutations in the MYD88 TIR domain are observed in both ABC- and GCB-DLBCL, but their significance remains to be established.

Mutations of TNFAIP3. Almost one third of ABC-DLBCLs and fewer GCB-DLBCLs harbor biallelic truncating mutations and/or focal deletions inactivating the ***TNFAIP3*** gene [159, 168]. This gene (also called *A20*) encodes a dual function ubiquitin-modification enzyme that is normally involved in the termination of NF-κB responses triggered by TLR and BCR stimulation, as demonstrated by the overactive phenotype observed in A20 knock-out mice [169]. A20 removes K63-linked regulatory ubiquitines from its substrates via its OTU domain, and subsequently conjugates K48-linked ubiquitines via its zinc finger domains, targeting them for proteasome-mediated degradation. The loss of these domains in A20-mutated tumors may thus contribute to lymphomagenesis by inducing inappropriately prolonged NF-κB responses. A tumor suppressor role for A20 in ABC-DLBCL is supported by the observation that reconstitution of A20-null cell lines with vectors expressing wild-type A20 results in cytoplasmic re-localization of the NF-κB complex and suppression of its activity, leading to apoptosis [159, 168].

At a lower frequency, mutations or aberrant expression were found in a variety of other genes encoding for NF-κB positive and negative regulators, overall accounting for > 50 % of all ABC-DLBCL [159] and suggesting that yet unidentified lesions may be responsible for BCR and NF-κB activity in the remaining fraction of cases.

The lesions described above do not fully explain the reliance of ABC-DLBCL cases on chronic BCR signaling and constitutively active NF-κB, as knockdown of several BCR proximal and distal subunits is toxic to wild-type ABC-DLBCL. Thus, additional (genetic or extrinsic) mechanisms of addiction must exist in these tumors, which may also include stimulation of the BCR by foreign or self-antigens. The dependence of ABC-DLBCL from the BCR signaling pathway is underscored by recent clinical studies reporting successful results with the use of agents that inhibit Bruton's tyrosine kinase (BTK), the molecule linking BCR to NF-κB, particularly in patients with this molecular subtype and even in the absence of mutations [170].

Of note, BCR emanates signals to multiple downstream effectors besides canonical NF-κB, including PI3K, ERK and NF-AT, all of which may play important roles in the neoplastic transformation of ABC-DLBCL cells. These vulnerabilities offer a unique opportunity for the development of combinatorial therapeutic strategies, as suggested by the cooperative toxicity observed upon combined inhibition of NF-κB and PI3K [163].

8.5.3.2 Genetic lesions preventing terminal differentiation

The differentiation of GC B cells into plasma cells requires BLIMP1 [171], a transcriptional repressor that is selectively upregulated in a subset of LZ B cells committed to plasmacytic differentiation and in all plasma cells [172, 173]. BLIMP1 orchestrates this transition by extinguishing the expression of several mature B-cell genes, including its targets PAX5 and BCL6 [174]. In contrast, a variety of genetic and epigenetic mechanisms abrogate its function in nearly all ABC-DLBCL cases, suggesting that blockade of post-GC differentiation is a relevant pathogenic mechanism in this lymphoma type (Fig. 8.6). In 25 % of patients, both copies of the *PRDM1* gene are lost owing to truncating mutations, loss-of-function missense mutations, and/or genomic deletions [26–28]. Another 25 % of patients lack BLIMP1 expression as the result of direct transcriptional repression by constitutively active, translocated *BCL6* alleles [26], while alternative modes of inactivation could explain the absence of BLIMP1 in the remaining fraction of cases, including epigenetic silencing or high expression of the ETS-family factor SPIB. Consistent with this model, conditional ablation of *Prdm1* in mouse GC B cells, either alone or in combination with a constitutively active IκB kinase, leads to the development of DLBCL with ABC-like phenotype [26, 175]. *BCL6* rearrangements and *PRDM1* inactivation are mutually exclusive [26, 87], supporting a complementary role for these lesions in promoting lymphomagenesis.

8.5.4 Genetic lesions associated with histologic transformation of FL and CLL

Insights into the mechanisms underlying the histologic transformation of low-grade lymphomas into DLBCL were offered by recent genomic analyses that interrogated the hierarchy of somatic mutations in sequential pre- and post-transformation biopsies from CLL and FL patients. These studies revealed that the transformation of CLL into DLBCL (known as Richter Syndrome) derives largely from the dominant CLL clone through a linear pattern of evolution that involves the maintenance of the CLL-associated lesions and the acquisition of new genetic alterations affecting *NOTCH1* (mutated in 35 % of cases), *CDKN2A/B* (homozygously deleted in 30 % of cases), *TP53* (genetically inactivated in as many as 60 % of cases) and *MYC* (amplified or translocated in ~30 % of cases) [176]. Conversely, FL and tFL arise by divergent evolution from a common mutated precursor clone that independently acquired distinct genetic aberrations, among which *CDKN2A/B* loss, *TP53* loss, *MYC* translocations, *B2M* inactivation and ASHM are significantly enriched in tFL [93, 94]. Genomic analysis of these tumors also showed that, while being morphologically undistinguishable from *de novo* DLBCL, Richter Syndrome and tFL are each characterized by a largely unique mutational landscape featuring the combination of genetic aberrations otherwise not commonly observed [94, 176]. This distinction has critical implications for the development of rationally designed combinatorial therapies.

8.6 Double-hit DLBCL

Approximately 5 to 10 % of DLBCL have concurrent chromosomal translocations of *MYC* and *BCL2* or, less likely, *BCL6* [134, 177]. This group of tumors, referred to as "double-hit lymphomas" (DHL), is gaining increasing attention due to its aggressive clinical course and extremely poor prognosis, the overall survival being less than 2 years from diagnosis when treated with R-CHOP induction regimens [177, 178]. Because of the rarity of the disease, a comprehensive characterization of the genetic make-up of DHL has not been done yet; however, multiple mechanisms can explain the particularly aggressive behavior of these tumors in relationship to the potent oncogenic activity of the involved oncogenes. These are exemplified by increased cell growth/proliferation, energy metabolism, protein biosynthesis, and genomic instability associated with DNA replication stress (via MYC), or protection from apoptosis (via BCL2). Consistent with the preferential distribution of *MYC* and *BCL2* translocations in GCB-like DLBCL [179], DHLs tend to belong to this molecular subtype, although the majority of cases are histologically classified as "B-cell lymphoma with features intermediate between Burkitt lymphoma and diffuse large B-cell lymphoma" [2].

More recently, retrospective analyses have found that high co-expression of MYC and BCL2, independent of the presence of chromosomal translocations, is also a strong predictor of poor clinical outcome [180, 181], suggesting alternative underlying mechanisms and extending the concept of "double-hit lymphoma" to protein expression ("double-expressor" lymphomas) [178]. Different from DHL, "double-expressor" DLBCL appear to be primarily of the ABC subgroup [180]. In the future, larger prospective cohorts of patients will need to be studied to validate these findings and translate them into definitive recommendations for general use, such as the definition of precise inclusion criteria and the establishment of optimal cut-offs for positivity of MYC and BCL2 expression. The involvement of distinct underlying events in "double-hit" and "double-expressor" DLBCL questions whether therapeutic approaches that may turn successful in the former category can be translated into the latter.

8.7 Concluding remarks

During the past decade, our understanding of the pathogenesis of DLBCL has improved substantially, owing in part to the implementation of powerful genomic technologies. We have now identified a number of pathways that are recurrently dysregulated by genetic alterations in this disease, and are emerging as central players in tumor development and maintenance. These lesions reveal vulnerabilities in the lymphoma cells, which are often uniquely associated with distinct lymphoma subtypes and could thus be exploited for patient stratification and for the design of more effective, possibly less toxic targeted therapeutic approaches. Indeed, the information gained from these studies is already being translated into the development and clinical testing (or repositioning) of novel drugs/drug combinations directed against specifically disrupted programs in DLBCL. This is the case for small molecule inhibitors of EZH2 and BCL2, as well as inhibitors of BCR and NF-κB activity. Among these drugs, the BTK inhibitor ibrutinib is rapidly emerging as a novel paradigm for the treatment of ABC-DLBCL [5]. While these strategies are expected to impact the standard of care for this malignancy, the complexity of the involved pathways and the overall heterogeneity of the disease, which may also implicate non-genetic mechanisms, underscores the need of precise patient stratification in order to identify sensitive and resistant cases.

References

[1] Siegel RL, Miller KD, Jemal A. Cancer statistics, 2015. CA Cancer J Clin 2015;65:5–29.
[2] Swerdlow SH, Campo E, Harris NL, et al. WHO Classification of Tumors of Haematopoietic and Lymphoid Tissues. Lyon: World Health Organization of Tumor (Bosman, F. T., Jaffe, E. S. Lakhani, S. R. & Ohgaki, H, Eds.), International Agency for Research on Cancer (IARC); 2008.
[3] Montoto S, Fitzgibbon J. Transformation of indolent B-cell lymphomas. Journal of Clinical Oncology: Official Journal of the American Society of Clinical Oncology 2011;29:1827–1834.
[4] Rossi D, Gaidano G. Richter syndrome. Adv Exp Med Biol 2013;792:173–191.
[5] Roschewski M, Staudt LM, Wilson WH. Diffuse large B-cell lymphoma-treatment approaches in the molecular era. Nat Rev Clin Oncol 2014;11:12–23.
[6] Shaffer AL, 3rd, Young RM, Staudt LM. Pathogenesis of human B cell lymphomas. Annual Review of Immunology 2012;30:565–610.
[7] Jacob J, Kassir R, Kelsoe G. In situ studies of the primary immune response to (4-hydroxy-3-nitrophenyl)acetyl. I. The architecture and dynamics of responding cell populations. The Journal of Experimental Medicine 1991;173:1165–1175.
[8] De Silva NS, Klein U. Dynamics of B cells in germinal centres. Nature Reviews Immunology 2015;15:137–148.

[9] Victora GD, Nussenzweig MC. Germinal centers. Annual Review of Immunology 2012;30:429–457.
[10] Allen CD, Okada T, Tang HL, Cyster JG. Imaging of germinal center selection events during affinity maturation. Science 2007;315:528–531.
[11] Schwickert TA, Lindquist RL, Shakhar G, et al. In vivo imaging of germinal centres reveals a dynamic open structure. Nature 2007;446:83–87.
[12] Hauser AE, Junt T, Mempel TR, et al. Definition of germinal-center B cell migration in vivo reveals predominant intrazonal circulation patterns. Immunity 2007;26:655–667.
[13] Allen CD, Okada T, Cyster JG. Germinal-center organization and cellular dynamics. Immunity 2007;27:190–202.
[14] Victora GD, Schwickert TA, Fooksman DR, et al. Germinal center dynamics revealed by multiphoton microscopy with a photoactivatable fluorescent reporter. Cell 2010;143:592–605.
[15] Basso K, Dalla-Favera R. Roles of BCL6 in normal and transformed germinal center B cells. Immunological Reviews 2012;247:172–183.
[16] Kuppers R, Klein U, Hansmann ML, Rajewsky K. Cellular origin of human B-cell lymphomas. The New England Journal of Medicine 1999;341:1520–1529.
[17] Kuppers R, Dalla-Favera R. Mechanisms of chromosomal translocations in B cell lymphomas. Oncogene 2001;20:5580–5594.
[18] Pasqualucci L, Neumeister P, Goossens T, et al. Hypermutation of multiple proto-oncogenes in B-cell diffuse large-cell lymphomas. Nature 2001;412:341–346.
[19] Alizadeh AA, Eisen MB, Davis RE, et al. Distinct types of diffuse large B-cell lymphoma identified by gene expression profiling. Nature 2000;403:503–511.
[20] Lim KH, Yang Y, Staudt LM. Pathogenetic importance and therapeutic implications of NF-kappaB in lymphoid malignancies. Immunological Reviews 2012;246:359–378.
[21] Young RM, Staudt LM. Targeting pathological B cell receptor signalling in lymphoid malignancies. Nature Reviews Drug Discovery 2013;12:229–243.
[22] Baron BW, Nucifora G, McCabe N, Espinosa R, 3rd, Le Beau MM, McKeithan TW. Identification of the gene associated with the recurring chromosomal translocations t(3;14)(q27;q32) and t(3;22)(q27;q11) in B-cell lymphomas. Proceedings of the National Academy of Sciences USA 1993;90:5262–5266.
[23] Ye BH, Lista F, Lo Coco F, et al. Alterations of a zinc finger-encoding gene, BCL-6, in diffuse large-cell lymphoma. Science 1993;262:747–750.
[24] Ye BH, Rao PH, Chaganti RS, Dalla-Favera R. Cloning of bcl-6, the locus involved in chromosome translocations affecting band 3q27 in B-cell lymphoma. Cancer Research 1993;53:2732–2735.
[25] Kerckaert JP, Deweindt C, Tilly H, Quief S, Lecocq G, Bastard C. LAZ3, a novel zinc-finger encoding gene, is disrupted by recurring chromosome 3q27 translocations in human lymphomas. Nature Genetics 1993;5:66–70.
[26] Mandelbaum J, Bhagat G, Tang H, et al. BLIMP1 is a tumor suppressor gene frequently disrupted in activated B cell-like diffuse large B cell lymphoma. Cancer Cell 2010;18:568–579.
[27] Pasqualucci L, Compagno M, Houldsworth J, et al. Inactivation of the PRDM1/BLIMP1 gene in diffuse large B cell lymphoma. The Journal of Experimental Medicine 2006;203:311–317.
[28] Tam W, Gomez M, Chadburn A, Lee JW, Chan WC, Knowles DM. Mutational analysis of PRDM1 indicates a tumor-suppressor role in diffuse large B-cell lymphomas. Blood 2006;107:4090–4100.
[29] Rosenwald A, Wright G, Leroy K, et al. Molecular diagnosis of primary mediastinal B cell lymphoma identifies a clinically favorable subgroup of diffuse large B cell lymphoma related to Hodgkin lymphoma. The Journal of Experimental Medicine 2003;198:851–862.

[30] Savage KJ, Monti S, Kutok JL, et al. The molecular signature of mediastinal large B-cell lymphoma differs from that of other diffuse large B-cell lymphomas and shares features with classical Hodgkin lymphoma. Blood 2003;102:3871–3879.
[31] Wright G, Tan B, Rosenwald A, Hurt EH, Wiestner A, Staudt LM. A gene expression-based method to diagnose clinically distinct subgroups of diffuse large B cell lymphoma. Proceedings of the National Academy of Sciences USA 2003;100:9991–9996.
[32] Victora GD, Dominguez-Sola D, Holmes AB, Deroubaix S, Dalla-Favera R, Nussenzweig MC. Identification of human germinal center light and dark zone cells and their relationship to human B-cell lymphomas. Blood 2012;120:2240–2248.
[33] Lossos IS, Alizadeh AA, Eisen MB, et al. Ongoing immunoglobulin somatic mutation in germinal center B cell-like but not in activated B cell-like diffuse large cell lymphomas. Proceedings of the National Academy of Sciences USA 2000;97:10209–10213.
[34] Klein U, Casola S, Cattoretti G, et al. Transcription factor IRF4 controls plasma cell differentiation and class-switch recombination. Nature Immunology 2006;7:773–782.
[35] Sciammas R, Shaffer AL, Schatz JH, Zhao H, Staudt LM, Singh H. Graded expression of interferon regulatory factor-4 coordinates isotype switching with plasma cell differentiation. Immunity 2006;25:225–236.
[36] Davis RE, Brown KD, Siebenlist U, Staudt LM. Constitutive nuclear factor kappaB activity is required for survival of activated B cell-like diffuse large B cell lymphoma cells. The Journal of Experimental Medicine 2001;194:1861–1874.
[37] Lenz G, Nagel I, Siebert R, et al. Aberrant immunoglobulin class switch recombination and switch translocations in activated B cell-like diffuse large B cell lymphoma. The Journal of Experimental Medicine 2007;204:633–643.
[38] Steidl C, Gascoyne RD. The molecular pathogenesis of primary mediastinal large B-cell lymphoma. Blood 2011;118:2659–2669.
[39] Rosenwald A, Wright G, Chan WC, et al. The use of molecular profiling to predict survival after chemotherapy for diffuse large-B-cell lymphoma. The New England Journal of Medicine 2002;346:1937–1947.
[40] Lenz G, Wright G, Dave SS, et al. Stromal gene signatures in large-B-cell lymphomas. The New England Journal of Medicine 2008;359:2313–2323.
[41] Thieblemont C, Briere J, Mounier N, et al. The germinal center/activated B-cell subclassification has a prognostic impact for response to salvage therapy in relapsed/refractory diffuse large B-cell lymphoma: a bio-CORAL study. Journal of Clinical Oncology: Official Journal of the American Society of Clinical Oncology 2011;29:4079–4087.
[42] Salles G, de Jong D, Xie W, et al. Prognostic significance of immunohistochemical biomarkers in diffuse large B-cell lymphoma: a study from the Lunenburg Lymphoma Biomarker Consortium. Blood 2011;117:7070–7078.
[43] Nyman H, Adde M, Karjalainen-Lindsberg ML, et al. Prognostic impact of immunohistochemically defined germinal center phenotype in diffuse large B-cell lymphoma patients treated with immunochemotherapy. Blood 2007;109:4930–4935.
[44] Choi WW, Weisenburger DD, Greiner TC, et al. A new immunostain algorithm classifies diffuse large B-cell lymphoma into molecular subtypes with high accuracy. Clinical Cancer Research: An Official Journal of the American Association for Cancer Research 2009;15:5494–5502.
[45] Hans CP, Weisenburger DD, Greiner TC, et al. Confirmation of the molecular classification of diffuse large B-cell lymphoma by immunohistochemistry using a tissue microarray. Blood 2004;103:275–282.
[46] de Jong D, Xie W, Rosenwald A, et al. Immunohistochemical prognostic markers in diffuse large B-cell lymphoma: validation of tissue microarray as a prerequisite for broad clinical applications (a study from the Lunenburg Lymphoma Biomarker Consortium). J Clin Pathol 2009;62:128–138.

[47] Scott DW, Mottok A, Ennishi D, et al. Prognostic Significance of Diffuse Large B-Cell Lymphoma Cell of Origin Determined by Digital Gene Expression in Formalin-Fixed Paraffin-Embedded Tissue Biopsies. Journal of Clinical Oncology 2015;33(26):2848–2856.
[48] Monti S, Savage KJ, Kutok JL, et al. Molecular profiling of diffuse large B-cell lymphoma identifies robust subtypes including one characterized by host inflammatory response. Blood 2005;105:1851–1861.
[49] Chen L, Monti S, Juszczynski P, et al. SYK inhibition modulates distinct PI3K/AKT- dependent survival pathways and cholesterol biosynthesis in diffuse large B cell lymphomas. Cancer Cell 2013;23:826–838.
[50] Caro P, Kishan AU, Norberg E, et al. Metabolic signatures uncover distinct targets in molecular subsets of diffuse large B cell lymphoma. Cancer Cell 2012;22:547–560.
[51] Polo JM, Juszczynski P, Monti S, et al. Transcriptional signature with differential expression of BCL6 target genes accurately identifies BCL6-dependent diffuse large B cell lymphomas. Proceedings of the National Academy of Sciences USA 2007;104:3207–3212.
[52] Cerchietti LC, Ghetu AF, Zhu X, et al. A small-molecule inhibitor of BCL6 kills DLBCL cells in vitro and in vivo. Cancer Cell 2010;17:400–411.
[53] Revy P, Muto T, Levy Y, et al. Activation-induced cytidine deaminase (AID) deficiency causes the autosomal recessive form of the Hyper-IgM syndrome (HIGM2). Cell 2000;102:565–575.
[54] de Villartay JP, Fischer A, Durandy A. The mechanisms of immune diversification and their disorders. Nature Reviews Immunology 2003;3:962–972.
[55] Zhang J, MacLennan IC, Liu YJ, Lane PJ. Is rapid proliferation in B centroblasts linked to somatic mutation in memory B cell clones? Immunol Lett 1988;18:297–299.
[56] Phan RT, Dalla-Favera R. The BCL6 proto-oncogene suppresses p53 expression in germinal-centre B cells. Nature 2004;432:635–639.
[57] Phan RT, Saito M, Basso K, Niu H, Dalla-Favera R. BCL6 interacts with the transcription factor Miz-1 to suppress the cyclin-dependent kinase inhibitor p21 and cell cycle arrest in germinal center B cells. Nature Immunology 2005;6:1054–1060.
[58] Ranuncolo SM, Polo JM, Dierov J, et al. Bcl-6 mediates the germinal center B cell phenotype and lymphomagenesis through transcriptional repression of the DNA-damage sensor ATR. Nature Immunology 2007;8:705–714.
[59] Ranuncolo SM, Polo JM, Melnick A. BCL6 represses CHEK1 and suppresses DNA damage pathways in normal and malignant B-cells. Blood Cells, Molecules & Diseases 2008;41:95–99.
[60] Muramatsu M, Kinoshita K, Fagarasan S, Yamada S, Shinkai Y, Honjo T. Class switch recombination and hypermutation require activation-induced cytidine deaminase (AID), a potential RNA editing enzyme. Cell 2000;102:553–563.
[61] Pasqualucci L, Bhagat G, Jankovic M, et al. AID is required for germinal center-derived lymphomagenesis. Nature Genetics 2008;40:108–112.
[62] Ramiro AR, Jankovic M, Eisenreich T, et al. AID is required for c-myc/IgH chromosome translocations in vivo. Cell 2004;118:431–438.
[63] Robbiani DF, Bothmer A, Callen E, et al. AID is required for the chromosomal breaks in c-myc that lead to c-myc/IgH translocations. Cell 2008;135:1028–1038.
[64] Dalla-Favera R, Pasqualucci L. Molecular Genetics of Lymphomas. In: Mauch PM, Armitage, J.O., Coiffier, B., Dalla-Favera, R., Harris, N.L., ed. Non-Hodgkin's Lymphoma. Philadelphia: Lippincott Williams & Wilkins; 2010:825–843.
[65] Akasaka H, Akasaka T, Kurata M, et al. Molecular anatomy of BCL6 translocations revealed by long-distance polymerase chain reaction-based assays. Cancer Research 2000;60:2335–2341.
[66] Chen W, Iida S, Louie DC, Dalla-Favera R, Chaganti RS. Heterologous promoters fused to BCL6 by chromosomal translocations affecting band 3q27 cause its deregulated expression during B-cell differentiation. Blood 1998;91:603–607.

[67] Yoshida S, Kaneita Y, Aoki Y, Seto M, Mori S, Moriyama M. Identification of heterologous translocation partner genes fused to the BCL6 gene in diffuse large B-cell lymphomas: 5′-RACE and LA – PCR analyses of biopsy samples. Oncogene 1999;18:7994–7999.
[68] Dominguez-Sola D, Victora GD, Ying CY, et al. The proto-oncogene MYC is required for selection in the germinal center and cyclic reentry. Nature Immunology 2012;13:1083–1091.
[69] Saito M, Novak U, Piovan E, et al. BCL6 suppression of BCL2 via Miz1 and its disruption in diffuse large B cell lymphoma. Proceedings of the National Academy of Sciences USA 2009;106:11294–11299.
[70] Neuberger MS. Antibody diversification by somatic mutation: from Burnet onwards. Immunology and Cell Biology 2008;86:124–132.
[71] Pasqualucci L, Migliazza A, Fracchiolla N, et al. BCL-6 mutations in normal germinal center B cells: evidence of somatic hypermutation acting outside Ig loci. Proceedings of the National Academy of Sciences USA 1998;95:11816–11821.
[72] Shen HM, Peters A, Baron B, Zhu X, Storb U. Mutation of BCL-6 gene in normal B cells by the process of somatic hypermutation of Ig genes. Science 1998;280:1750–1752.
[73] Gordon MS, Kanegai CM, Doerr JR, Wall R. Somatic hypermutation of the B cell receptor genes B29 (Igbeta, CD79b) and mb1 (Igalpha, CD79a). Proceedings of the National Academy of Sciences USA 2003;100:4126–4131.
[74] Liu M, Duke JL, Richter DJ, et al. Two levels of protection for the B cell genome during somatic hypermutation. Nature 2008;451:841–845.
[75] Pavri R, Gazumyan A, Jankovic M, et al. Activation-induced cytidine deaminase targets DNA at sites of RNA polymerase II stalling by interaction with Spt5. Cell 2010;143:122–133.
[76] Yamane A, Resch W, Casellas R. Deep-sequencing identification of the genomic targets of the cytidine deaminase AID and its cofactor RPA in B lymphocytes. Nature Immunology 2011;12:62–69.
[77] Cerri M, Capello D, Muti G, et al. Aberrant somatic hypermutation in post-transplant lymphoproliferative disorders. British Journal of Haematology 2004;127:362–364.
[78] Deutsch AJ, Aigelsreiter A, Staber PB, et al. MALT lymphoma and extranodal diffuse large B-cell lymphoma are targeted by aberrant somatic hypermutation. Blood 2007;109:3500–3504.
[79] Gaidano G, Pasqualucci L, Capello D, et al. Aberrant somatic hypermutation in multiple subtypes of AIDS-associated non-Hodgkin lymphoma. Blood 2003;102:1833–1841.
[80] Montesinos-Rongen M, Van Roost D, Schaller C, Wiestler OD, Deckert M. Primary diffuse large B-cell lymphomas of the central nervous system are targeted by aberrant somatic hypermutation. Blood 2004;103:1869–1875.
[81] Vakiani E, Basso K, Klein U, et al. Genetic and phenotypic analysis of B-cell post-transplant lymphoproliferative disorders provides insights into disease biology. Hematological Oncology 2008;26:199–211.
[82] Longerich S, Basu U, Alt F, Storb U. AID in somatic hypermutation and class switch recombination. Current Opinion in Immunology 2006;18:164–174.
[83] Khodabakhshi AH, Morin RD, Fejes AP, et al. Recurrent targets of aberrant somatic hypermutation in lymphoma. Oncotarget 2012;3:1308–1319.
[84] Meng FL, Du Z, Federation A, et al. Convergent transcription at intragenic super-enhancers targets AID-initiated genomic instability. Cell 2014;159:1538–1548.
[85] Qian J, Wang Q, Dose M, et al. B cell super-enhancers and regulatory clusters recruit AID tumorigenic activity. Cell 2014;159:1524–1537.
[86] Morin RD, Mendez-Lago M, Mungall AJ, et al. Frequent mutation of histone-modifying genes in non-Hodgkin lymphoma. Nature 2011;476:298–303.
[87] Pasqualucci L, Trifonov V, Fabbri G, et al. Analysis of the coding genome of diffuse large B-cell lymphoma. Nature Genetics 2011;43:830–837.

[88] Lohr JG, Stojanov P, Lawrence MS, et al. Discovery and prioritization of somatic mutations in diffuse large B-cell lymphoma (DLBCL) by whole-exome sequencing. Proceedings of the National Academy of Sciences USA 2012;109:3879–3884.
[89] Zhang J, Grubor V, Love CL, et al. Genetic heterogeneity of diffuse large B-cell lymphoma. Proceedings of the National Academy of Sciences USA 2013;110:1398–13403.
[90] Jiang Y, Soong TD, Wang L, Melnick AM, Elemento O. Genome-wide detection of genes targeted by non-Ig somatic hypermutation in lymphoma. PLoS One 2012;7:e40332.
[91] Green MR, Gentles AJ, Nair RV, et al. Hierarchy in somatic mutations arising during genomic evolution and progression of follicular lymphoma. Blood 2013;121:1604–1611.
[92] Green MR, Kihira S, Liu CL, et al. Mutations in early follicular lymphoma progenitors are associated with suppressed antigen presentation. Proceedings of the National Academy of Sciences USA 2015;112:E1116–25.
[93] Okosun J, Bodor C, Wang J, et al. Integrated genomic analysis identifies recurrent mutations and evolution patterns driving the initiation and progression of follicular lymphoma. Nature Genetics 2014;46:176–181.
[94] Pasqualucci L, Khiabanian H, Fangazio M, et al. Genetics of follicular lymphoma transformation. Cell Reports 2014;6:130–140.
[95] Pasqualucci L, Dominguez-Sola D, Chiarenza A, et al. Inactivating mutations of acetyltransferase genes in B-cell lymphoma. Nature 2011;471:189–195.
[96] Goodman RH, Smolik S. CBP/p300 in cell growth, transformation, and development. Genes & Development 2000;14:1553–1577.
[97] Petrij F, Giles RH, Dauwerse HG, et al. Rubinstein-Taybi syndrome caused by mutations in the transcriptional co-activator CBP. Nature 1995;376:348–351.
[98] Bereshchenko OR, Gu W, Dalla-Favera R. Acetylation inactivates the transcriptional repressor BCL6. Nature Genetics 2002;32:606–613.
[99] Gu W, Roeder RG. Activation of p53 sequence-specific DNA binding by acetylation of the p53 C-terminal domain. Cell 1997;90:595–606.
[100] Tang Y, Zhao W, Chen Y, Zhao Y, Gu W. Acetylation is indispensable for p53 activation. Cell 2008;133:612–626.
[101] Shilatifard A. The COMPASS family of histone H3K4 methylases: mechanisms of regulation in development and disease pathogenesis. Annu Rev Biochem 2012;81:65–95.
[102] Wendel H, Melnick A. Nature Medicine 2015.
[103] Zhang J. Disruption of KMT2D perturbs germinal center B cell development and promotes lymphomagenesis. Nature Medicine 2015.
[104] Hatzi K, Melnick A. Breaking bad in the germinal center: how deregulation of BCL6 contributes to lymphomagenesis. Trends in Molecular Medicine 2014;20:343–352.
[105] Basso K, Saito M, Sumazin P, et al. Integrated biochemical and computational approach identifies BCL6 direct target genes controlling multiple pathways in normal germinal center B cells. Blood 2010;115:975–984.
[106] Ci W, Polo JM, Cerchietti L, et al. The BCL6 transcriptional program features repression of multiple oncogenes in primary B cells and is deregulated in DLBCL. Blood 2009;113:5536–5548.
[107] Tunyaplin C, Shaffer AL, Angelin-Duclos CD, Yu X, Staudt LM, Calame KL. Direct repression of prdm1 by Bcl-6 inhibits plasmacytic differentiation. Journal of Immunology 2004;173:1158–1165.
[108] Iqbal J, Greiner TC, Patel K, et al. Distinctive patterns of BCL6 molecular alterations and their functional consequences in different subgroups of diffuse large B-cell lymphoma. Leukemia 2007;21:2332–2343.

[109] Ye BH, Chaganti S, Chang CC, et al. Chromosomal translocations cause deregulated BCL6 expression by promoter substitution in B cell lymphoma. The EMBO Journal 1995;14:6209–6217.
[110] Butler MP, Iida S, Capello D, et al. Alternative translocation breakpoint cluster region 5' to BCL-6 in B-cell non-Hodgkin's lymphoma. Cancer Research 2002;62:4089–4094.
[111] Pasqualucci L, Migliazza A, Basso K, Houldsworth J, Chaganti RS, Dalla-Favera R. Mutations of the BCL6 proto-oncogene disrupt its negative autoregulation in diffuse large B-cell lymphoma. Blood 2003;101:2914–223.
[112] Wang X, Li Z, Naganuma A, Ye BH. Negative autoregulation of BCL-6 is bypassed by genetic alterations in diffuse large B cell lymphomas. Proceedings of the National Academy of Sciences USA 2002;99:15018–15023.
[113] Saito M, Gao J, Basso K, et al. A signaling pathway mediating downregulation of BCL6 in germinal center B cells is blocked by BCL6 gene alterations in B cell lymphoma. Cancer Cell 2007;12:280–292.
[114] Ying CY, Dominguez-Sola D, Fabi M, et al. MEF2B mutations lead to deregulated expression of the oncogene BCL6 in diffuse large B cell lymphoma. Nature Immunology 2013;14:1084–1092.
[115] Duan S, Cermak L, Pagan JK, et al. FBXO11 targets BCL6 for degradation and is inactivated in diffuse large B-cell lymphomas. Nature 2012;481:90–93.
[116] Cattoretti G, Pasqualucci L, Ballon G, et al. Deregulated BCL6 expression recapitulates the pathogenesis of human diffuse large B cell lymphomas in mice. Cancer Cell 2005;7:445–455.
[117] Cresswell P, Ackerman AL, Giodini A, Peaper DR, Wearsch PA. Mechanisms of MHC class I-restricted antigen processing and cross-presentation. Immunological Reviews 2005;207:145–157.
[118] Challa-Malladi M, Lieu YK, Califano O, et al. Combined genetic inactivation of beta2-Microglobulin and CD58 reveals frequent escape from immune recognition in diffuse large B cell lymphoma. Cancer Cell 2011;20:728–740.
[119] Fangazio M, Dominguez-Sola D, Tabbo F, et al. Genetic mechanisms of immune escape in diffuse large B cell lymphoma. Blood 2014;124.
[120] Moingeon P, Chang HC, Wallner BP, Stebbins C, Frey AZ, Reinherz EL. CD2-mediated adhesion facilitates T lymphocyte antigen recognition function. Nature 1989;339:312–314.
[121] de Miranda NF, Georgiou K, Chen L, et al. Exome sequencing reveals novel mutation targets in diffuse large B-cell lymphomas derived from Chinese patients. Blood 2014;124:2544–2553.
[122] Lo Coco F, Gaidano G, Louie DC, Offit K, Chaganti RS, Dalla-Favera R. p53 mutations are associated with histologic transformation of follicular lymphoma. Blood 1993;82:2289–2295.
[123] Monti S, Chapuy B, Takeyama K, et al. Integrative analysis reveals an outcome-associated and targetable pattern of p53 and cell cycle deregulation in diffuse large B cell lymphoma. Cancer Cell 2012;22:359–372.
[124] Hata AN, Engelman JA, Faber AC. The BCL2 Family: Key Mediators of the Apoptotic Response to Targeted Anticancer Therapeutics. Cancer Discov 2015;5:475–487.
[125] Iqbal J, Sanger WG, Horsman DE, et al. BCL2 translocation defines a unique tumor subset within the germinal center B-cell-like diffuse large B-cell lymphoma. The American Journal of Pathology 2004;165:159–166.
[126] Schuetz JM, Johnson NA, Morin RD, et al. BCL2 mutations in diffuse large B-cell lymphoma. Leukemia 2012;26:1383–1390.
[127] Barrans SL, Evans PA, O'Connor SJ, et al. The t(14;18) is associated with germinal center-derived diffuse large B-cell lymphoma and is a strong predictor of outcome. Clinical Cancer Research: An Official Journal of the American Association for Cancer Research 2003;9:2133–2139.

[128] Conacci-Sorrell M, McFerrin L, Eisenman RN. An overview of MYC and its interactome. Cold Spring Harbor Perspectives in Medicine 2014;4:a014357.
[129] Dominguez-Sola D, Ying CY, Grandori C, et al. Non-transcriptional control of DNA replication by c-Myc. Nature 2007;448:445–451.
[130] Dalla-Favera R, Bregni M, Erikson J, Patterson D, Gallo RC, Croce CM. Human c-myc onc gene is located on the region of chromosome 8 that is translocated in Burkitt lymphoma cells. Proceedings of the National Academy of Sciences USA 1982;79:7824–7827.
[131] Dalla-Favera R, Martinotti S, Gallo RC, Erikson J, Croce CM. Translocation and rearrangements of the c-myc oncogene locus in human undifferentiated B-cell lymphomas. Science 1983;219:963–967.
[132] Ladanyi M, Offit K, Jhanwar SC, Filippa DA, Chaganti RS. MYC rearrangement and translocations involving band 8q24 in diffuse large cell lymphomas. Blood 1991;77:1057–1063.
[133] Karube K, Campo E. MYC alterations in diffuse large B-cell lymphomas. Semin Hematol 2015;52:97–106.
[134] Barrans S, Crouch S, Smith A, et al. Rearrangement of MYC is associated with poor prognosis in patients with diffuse large B-cell lymphoma treated in the era of rituximab. Journal of Clinical Oncology: Official Journal of the American Society of Clinical Oncology 2010;28:3360–3365.
[135] Mahmoudi T, Verrijzer CP. Chromatin silencing and activation by Polycomb and trithorax group proteins. Oncogene 2001;20:3055–3066.
[136] Czermin B, Melfi R, McCabe D, Seitz V, Imhof A, Pirrotta V. Drosophila enhancer of Zeste/ESC complexes have a histone H3 methyltransferase activity that marks chromosomal Polycomb sites. Cell 2002;111:185–196.
[137] Cao R, Wang L, Wang H, et al. Role of histone H3 lysine 27 methylation in Polycomb-group silencing. Science 2002;298:1039–1043.
[138] Ezhkova E, Pasolli HA, Parker JS, et al. Ezh2 orchestrates gene expression for the stepwise differentiation of tissue-specific stem cells. Cell 2009;136:1122–1135.
[139] Bernstein BE, Mikkelsen TS, Xie X, et al. A bivalent chromatin structure marks key developmental genes in embryonic stem cells. Cell 2006;125:315–326.
[140] Beguelin W, Popovic R, Teater M, et al. EZH2 is required for germinal center formation and somatic EZH2 mutations promote lymphoid transformation. Cancer Cell 2013;23:677–692.
[141] Caganova M, Carrisi C, Varano G, et al. Germinal center dysregulation by histone methyltransferase EZH2 promotes lymphomagenesis. The Journal of Clinical Investigation 2013;123:5009–5022.
[142] Morin RD, Johnson NA, Severson TM, et al. Somatic mutations altering EZH2 (Tyr641) in follicular and diffuse large B-cell lymphomas of germinal-center origin. Nature Genetics 2010;42:181–185.
[143] Sneeringer CJ, Scott MP, Kuntz KW, et al. Coordinated activities of wild-type plus mutant EZH2 drive tumor-associated hypertrimethylation of lysine 27 on histone H3 (H3K27) in human B-cell lymphomas. Proceedings of the National Academy of Sciences USA 2010;107:20980–20985.
[144] Yap DB, Chu J, Berg T, et al. Somatic mutations at EZH2 Y641 act dominantly through a mechanism of selectively altered PRC2 catalytic activity, to increase H3K27 trimethylation. Blood 2011;117:2451–2459.
[145] McCabe MT, Graves AP, Ganji G, et al. Mutation of A677 in histone methyltransferase EZH2 in human B-cell lymphoma promotes hypertrimethylation of histone H3 on lysine 27 (H3K27). Proceedings of the National Academy of Sciences USA 2012;109:2989–2994.
[146] Knutson SK, Wigle TJ, Warholic NM, et al. A selective inhibitor of EZH2 blocks H3K27 methylation and kills mutant lymphoma cells. Nature Chemical Biology 2012;8:890–896.

[147] McCabe MT, Ott HM, Ganji G, et al. EZH2 inhibition as a therapeutic strategy for lymphoma with EZH2-activating mutations. Nature 2012;492:108–112.
[148] A Study Evaluating CPI-1205 in Patients With B-Cell Lymphomas. Accessed December 1, 2015. https://clinicaltrials.gov/ct2/show/NCT02395601?term=EZH2inhibitors&rank=3.
[149] Green JA, Cyster JG. S1PR2 links germinal center confinement and growth regulation. Immunological Reviews 2012;247:36–51.
[150] Muppidi JR, Schmitz R, Green JA, et al. Loss of signalling via Galpha13 in germinal centre B-cell-derived lymphoma. Nature 2014; 516:254–258.
[151] Cattoretti G, Mandelbaum J, Lee N, et al. Targeted disruption of the S1P2 sphingosine 1-phosphate receptor gene leads to diffuse large B-cell lymphoma formation. Cancer Research 2009;69:8686–8692.
[152] Pfeifer M, Grau M, Lenze D, et al. PTEN loss defines a PI3K/AKT pathway-dependent germinal center subtype of diffuse large B-cell lymphoma. Proceedings of the National Academy of Sciences USA 2013;110:12420–12425.
[153] Lenz G, Wright GW, Emre NC, et al. Molecular subtypes of diffuse large B-cell lymphoma arise by distinct genetic pathways. Proceedings of the National Academy of Sciences USA 2008;105:13520–13525.
[154] Dominguez-Sola D, Kung J, Holmes AB, et al. The FOXO1 transcription factor instructs the germinal center dark zone program. Immunity 2015; doi:10.106.
[155] Sander et al. PI3 Kinase and Foxo1 Transcription Factor Activity Differentially Control B Cells in the Germinal Center Light and Dark Zones. Immunity 2015; doi: 10.1016.
[156] Trinh DL, Scott DW, Morin RD, et al. Analysis of FOXO1 mutations in diffuse large B-cell lymphoma. Blood 2013;121:3666–3674.
[157] Brunet A, Park J, Tran H, Hu LS, Hemmings BA, Greenberg ME. Protein kinase SGK mediates survival signals by phosphorylating the forkhead transcription factor FKHRL1 (FOXO3a). Molecular and Cellular Biology 2001;21:952–965.
[158] Tai DJ, Su CC, Ma YL, Lee EH. SGK1 phosphorylation of IkappaB Kinase alpha and p300 Up-regulates NF-kappaB activity and increases N-Methyl-D-aspartate receptor NR2A and NR2B expression. The Journal of Biological Chemistry 2009;284:4073–4089.
[159] Compagno M, Lim WK, Grunn A, et al. Mutations of multiple genes cause deregulation of NF-kappaB in diffuse large B-cell lymphoma. Nature 2009;459:717–721.
[160] Kraus M, Alimzhanov MB, Rajewsky N, Rajewsky K. Survival of resting mature B lymphocytes depends on BCR signaling via the Igalpha/beta heterodimer. Cell 2004;117:787–800.
[161] Lam KP, Kuhn R, Rajewsky K. In vivo ablation of surface immunoglobulin on mature B cells by inducible gene targeting results in rapid cell death. Cell 1997;90:1073–1083.
[162] Davis RE, Ngo VN, Lenz G, et al. Chronic active B-cell-receptor signalling in diffuse large B-cell lymphoma. Nature 2010;463:88–92.
[163] Lenz G, Davis RE, Ngo VN, et al. Oncogenic CARD11 mutations in human diffuse large B cell lymphoma. Science 2008;319:1676–1679.
[164] Thome M. CARMA1, BCL-10 and MALT1 in lymphocyte development and activation. Nature Reviews Immunology 2004;4:348–359.
[165] Lin SC, Lo YC, Wu H. Helical assembly in the MyD88-IRAK4-IRAK2 complex in TLR/IL-1R signalling. Nature 2010;465:885–890.
[166] Ngo VN, Young RM, Schmitz R, et al. Oncogenically active MYD88 mutations in human lymphoma. Nature 2011;470:115–119.
[167] Lam LT, Wright G, Davis RE, et al. Cooperative signaling through the signal transducer and activator of transcription 3 and nuclear factor-{kappa}B pathways in subtypes of diffuse large B-cell lymphoma. Blood 2008;111:3701–3713.
[168] Kato M, Sanada M, Kato I, et al. Frequent inactivation of A20 in B-cell lymphomas. Nature 2009;459:712–716.

[169] Boone DL, Turer EE, Lee EG, et al. The ubiquitin-modifying enzyme A20 is required for termination of Toll-like receptor responses. Nature Immunology 2004;5:1052–1060.

[170] Wilson WH, Young RM, Schmitz R, et al. Targeting B cell receptor signaling with ibrutinib in diffuse large B cell lymphoma. Nature Medicine 2015;21:922–926.

[171] Shapiro-Shelef M, Lin KI, McHeyzer-Williams LJ, Liao J, McHeyzer-Williams MG, Calame K. Blimp-1 is required for the formation of immunoglobulin secreting plasma cells and pre-plasma memory B cells. Immunity 2003;19:607–620.

[172] Angelin-Duclos C, Cattoretti G, Lin KI, Calame K. Commitment of B lymphocytes to a plasma cell fate is associated with Blimp-1 expression in vivo. Journal of Immunology 2000;165:5462–5471.

[173] Cattoretti G, Angelin-Duclos C, Shaknovich R, Zhou H, Wang D, Alobeid B. PRDM1/Blimp-1 is expressed in human B-lymphocytes committed to the plasma cell lineage. J Pathol 2005;206:76–86.

[174] Shaffer AL, Lin KI, Kuo TC, et al. Blimp-1 orchestrates plasma cell differentiation by extinguishing the mature B cell gene expression program. Immunity 2002;17:51–62.

[175] Calado DP, Zhang B, Srinivasan L, et al. Constitutive canonical NF-kappaB activation cooperates with disruption of BLIMP1 in the pathogenesis of activated B cell-like diffuse large cell lymphoma. Cancer Cell 2010;18:580–589.

[176] Fabbri G, Khiabanian H, Holmes AB, et al. Genetic lesions associated with chronic lymphocytic leukemia transformation to Richter syndrome. The Journal of Experimental Medicine 2013;210:2273–2288.

[177] Aukema SM, Siebert R, Schuuring E, et al. Double-hit B-cell lymphomas. Blood 2011;117:2319–2331.

[178] Dunleavy K. Double-hit lymphomas: current paradigms and novel treatment approaches. Hematology Am Soc Hematol Educ Program 2014;2014:107–112.

[179] Green TM, Young KH, Visco C, et al. Immunohistochemical double-hit score is a strong predictor of outcome in patients with diffuse large B-cell lymphoma treated with rituximab plus cyclophosphamide, doxorubicin, vincristine, and prednisone. Journal of Clinical Oncology: Official Journal of the American Society of Clinical Oncology 2012;30:3460–3467.

[180] Hu S, Xu-Monette ZY, Tzankov A, et al. MYC/BCL2 protein coexpression contributes to the inferior survival of activated B-cell subtype of diffuse large B-cell lymphoma and demonstrates high-risk gene expression signatures: a report from The International DLBCL Rituximab-CHOP Consortium Program. Blood 2013;121:4021–4031; quiz 250.

[181] Perry AM, Alvarado-Bernal Y, Laurini JA, et al. MYC and BCL2 protein expression predicts survival in patients with diffuse large B-cell lymphoma treated with rituximab. British Journal of Haematology 2014;165:382–391.

[182] Pasqualucci L, Dalla-Favera R. The genetic landscape of diffuse large B-cell lymphoma. Semin Hematol 2015;52:67–76.

Christian Steidl and Randy D. Gascoyne

9 The biology of primary mediastinal large B-cell lymphoma

9.1 Introduction

Primary mediastinal large B-cell lymphoma (PMBCL) is a mature lymphoid neoplasm representing ~10–15 % of all large cell lymphomas. PMBCL was traditionally classified as a variant of DLBCL, the most common aggressive lymphoma, characterized by its primary clinical manifestation as an anterior mediastinal mass. However, pivotal molecular studies based on gene expression profiling (GEP) and gene mutational analyses identified a distinct pathobiology of PMBCL and molecular overlap with nodular sclerosis classical Hodgkin lymphoma (CHL) [1]. From a historical point of view, distinguishing PMBCL from other forms of DLBCL can be viewed as a paradigm of molecularly informed reclassification in evolving classification systems. Validated by many published clinical, pathology and molecular studies, PMBCL is now recognized as a unique entity by the World Health Organization classification of lymphoid cancers [2].

In the 4th edition of the WHO classification a new borderline category was created to capture the cases with transitional features intermediate between PMBCL and nodular sclerosis Hodgkin lymphoma (NSHL) [2]. These unclassifiable cases typically occur in the mediastinum and are commonly referred to as mediastinal grey zone lymphomas (MGZL). Clinically, morphologically, immunophenotypically and biologically PMBCL overlaps with NSHL. This observation together with reports of both histologies presenting as composite lymphoid malignancies or as synchronous/metachronous lymphomas in the same patient, provided further evidence in support of this new grey zone category. In contrast to PMBCL and NSHL where there is a female predominance, MGZL is more common in men. In a proportion of cases, MGZL can occur in patients without clinical or radiographic evidence of mediastinal involvement. These latter cases tend to occur in older patients in contrast to those characterized by mediastinal involvement [3]. MGZL can be broadly categorized as two types based on morphology and immunophenotype: one where the histology is more reminiscent of CHL but the cells strongly express a number of B-cell antigens in addition to CHL antigens, and one that includes lymphomas resembling DLBCL or PMBCL sharing overlapping features with typical NSHL. There is often an asynchrony between the morphology and the immunophenotype. The genetic aspects of MGZL will also be briefly discussed in this chapter.

Treatment of large B-cell lymphomas remains a challenge, although major improvements have been made in the past decade. Specifically, in DLBCL, clinical outcomes have significantly improved by the addition of targeted immunotherapy (rituximab, R) to the previous standard of care consisting of polychemotherapy (cyclophosphamide, doxorubicin, vincristine, prednisone; CHOP) [4]. However, major clinical challenges remain, including the management of patients with relapsed or progressive disease and reducing treatment-related side effects. Toward the goal of a better understanding of the pathogenesis of large cell lymphoma and developing biomarkers, recent molecular studies using gene expression profiling and next generation sequencing have identified molecular subgroups of large cell lymphoma that are characterized by changes in gene expression and specific gene mutations. The current lack of a biological understanding of treatment failure and associated biomarkers is even more prominent in PMBCL, as this disease is only recently recognized as a separate entity in most clinical studies, and biological studies of PMBCL are still relatively uncommon in comparison to DLBCL. Overall, the lack of biological insight has hampered the design of clinical trials and the development of targeted treatment approaches capitalizing on the increasing knowledge of the pathogenesis of PMBCL. However, based on very recent advances, the field anticipates the emergence of tailored therapies for patients on the strength of more accurate diagnostic tools and the identification of biological characteristics that can be molecularly targeted.

In this chapter we will focus on the molecular hallmarks of PMBCL, the underlying somatic mutational landscape, and outline opportunities for clinical translation.

9.2 Cell of origin and gene expression profiling

To elucidate lymphomagenesis, inferences about the putative cell of origin giving rise to the malignant clone have been made in PMBCL based on gene expression profiling, immunohistochemistry and mutational analysis of various loci that are typical targets of somatic hypermutation (SHM). Cumulatively, these data suggest that PMBCL is of germinal center or post-germinal center origin [5, 6]. The partly preserved B-cell immunophenotype typically shows positivity for surface CD19, CD20, CD22, CD79a and frequent nuclear positivity of the transcriptional regulators BOB1, PU.1, OCT2, PAX5, BCL6 and IRF4 [5, 7]. However, the mature B-cell phenotype is incomplete [5–10]. Most PMBCL cases also stain positively for CD30, but the staining intensity is often weaker as compared to HL or anaplastic large cell lymphoma [11, 12]. In contrast to HL, the malignant cells typically lack expression of CD15 and do not harbor markers of latent EBV infection. SHM of the immunoglobulin (*IG*) genes and *BCL6* are considered reliable markers indicating transit of a B cell through the germinal center (GC) and, in agreement with the proposed (post-) germinal center origin of PMBCL, isotype-switched *IG* genes and *BCL6* gene mutations unequivocally show the footprint of SHM [6, 13, 14]. Of note, the spatial distribution and targeted motifs of hypermutation in the *BCL6* noncoding region were found to be distinct from DLBCL and FL suggesting specificity of the SHM process and a distinct clonal evolution in PMBCL [15]. Interestingly, activation-induced cytidine deaminase (*AID*), a gene that plays a key role in SHM and class switch recombination (CSR), has been shown to be constitutively expressed in PMBCL and the PMBCL-derived cell line KARPAS-1106P [16] shows ongoing SHM, but not CSR, involving *IG* genes and *BCL6* [17].

Although these data are consistent with a germinal center origin of PMBCL and migration into the thymus, another potential explanation is that PMBCL is derived from AID-expressing nongerminal center B lymphocytes residing in the medullary thymus [18]. Anatomical localization, presence of thymic remnants (Hassall's corpuscles and epithelial cysts) suggests that the cell of origin is a lymphocyte resident in the thymus [2]. This is further supported by an immunophenotype that corresponds to thymic medullary asteroid B cells [5, 9, 19–29]. However, the PMBCL phenotype corresponding to this putative cell of origin may result from additive features derived from sequential exposure to multiple environments. Specifically, a lack of CD21 expression on malignant cells suggests that PMBCL is not internally derived from thymic GC reactions, but rather from a subset of B cells that may have migrated to the thymic medulla [30–33]. Taken together, it remains possible that PMBCL is derived from a transitory, mature B cell that has passed through a peripheral GC reaction and has migrated to the thymic medulla to become a resident asteroid B cell [8].

Gene expression profiling confirmed previous immunophenotypic findings validating that PMBCL expresses a mature B-cell phenotype with comparable expression levels of CD19, CD20, CD22, CD79A, CD79B and OCT2 as compared to DLBCL [34, 35]. However, differences in transcript expression for proteins that distinguish between PMBCL and DLBCL were also observed (*CD23*, *MAL*, *REL* and *TRAF1*) [34–36]. The PMBCL gene expression profile was also found to share features with the previously analyzed HL cell lines, providing additional evidence that PMBCL was a distinct disease from DLBCL [34, 37]. Moreover, further studies revealed hypomethylation and elevated transcript levels of genes involved in NF-κB and JAK-STAT signaling [35, 36, 38].

The PMBCL-specific gene expression profile might primarily reflect the specific cell-of-origin context, however, little is known about the additional genetic changes during PMBCL pathogenesis that establish the full malignant phenotype. While mutational events in NF-κB and JAK-STAT signaling leading to constitutive pathway activation and altered inflammatory expression profiles are widely recognized (see Section 9.3 "Molecular hallmarks of PMBCL"), a recent study has implicated the transcription factor FOXO1 in the pathogenesis of PMBCL establishing a link to the JAK-STAT pathway. Specifically, the authors found that, similar to CHL, FOXO1 expression was reduced or absent in almost all cases, strongly suggesting a tumor suppressor gene function [39]. Interestingly, FOXO1 mRNA expression was inversely correlated with JAK2 expression and *in vitro* inhibition of JAK2 partially restored FOXO1 expression in two PMBCL-derived cell lines. Additional studies are needed to more comprehensively describe perturbations of transcription factor networks and associated phenotypes.

9.3 Molecular hallmarks of PMBCL

PMBCL is recognized as a distinct entity in the 2008 WHO Classification of Tumors of Haematopoietic and Lymphoid Tissues, owing to the distinct gene expression profile, immunophenotype and clinical presentation that is markedly different from both activated B-cell (ABC)-type and germinal center B-cell (GCB)-type DLBCL [2]. Although a pathological and clinical spectrum including ABC/GCB-type DLBCL, PMBCL, MGZL, NSHL has been recognized for a long time, pivotal gene expression profiling data and more recent genomic data define PMBCL as a unique lymphoid cancer characterized by molecular and clinical features that are shared with the related entities of the spectrum. (Fig. 9.1) [40]. Summarizing the knowledge about genetic alterations, pathway perturbations and expression phenotypes, the most prominent PMBCL molecu-

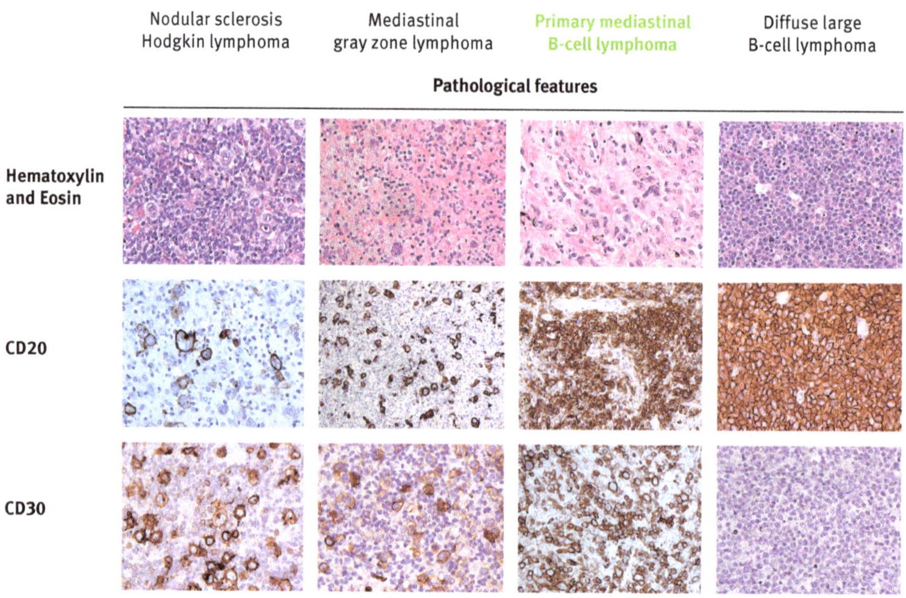

Clinical features

	Nodular sclerosis Hodgkin lymphoma	Mediastinal gray zone lymphoma	Primary mediastinal B-cell lymphoma	Diffuse large B-cell lymphoma
Approximate median age	30 years	30 years	35 years	65 years
Gender predominance	female	male > female	female	male ≥ female
Typical manifestation	supraclavicular LN/ mediastinal	mediastinal	mediastinal/ supraclavicular LN	nodal
Bone marrow involvement	uncommon	rare	rare	16 %

Fig. 9.1: **Pathological spectrum of related diseases with clinical manifestations in the mediastinum.** Representative hematoxylin/eosin (H&E) and immunohistochemical stains are shown, and the typical clinical features are contrasted between the related entities.

lar hallmarks include activation of the JAK-STAT and NF-κB pathways, and avoiding immune destruction [1]. Recently, high-resolution genomics approaches have added texture to our knowledge about these pathways and facilitated the discovery of a number of genetic events that underlie these phenotypic changes including point mutations, genomic rearrangements and copy number changes [41–43]. Interestingly, most of these genetic events are present across the aforementioned pathobiological spectrum providing additional evidence for the molecular relatedness of these diseases. As novel treatments emerge that target JAK-STAT signaling and the crosstalk of malignant cells with cells in the tumor microenvironment via immune checkpoint blockade, a better understanding of these molecular pathways in the context of PMBCL is needed to further improve targeted therapeutic approaches and develop biomarkers as potential companion diagnostics (see also Section 9.3.6 "Biomarkers and therapeutic considerations").

9.3.1 The mutational landscape of primary mediastinal large B-cell lymphoma

The mutational landscape of PMBCL is based on the collection of individual studies interrogating specific candidate genes and pathways [44–51] and one very recently published study using whole genome and transcriptome sequencing [41]. Many of the observed recurrent mutations are also encountered in the related entities of the pathological spectrum (CHL, DLBCL, and MGZL) with similar mutational patterns [1]. Related next-generation sequencing-based studies encompass whole genome and transcriptome sequencing of HL-derived cell lines [42, 52, 53], a whole-exome sequencing study of enriched Hodgkin Reed Sternberg cells [54] and a targeted capture sequencing study of primary testicular DLBCL [55]. Many of the recognized somatic gene mutations fall into pathways that have been previously identified as phenotypic hallmarks of the disease and provide evidence for genetic dependence on constitutive pathway perturbations.

Table 9.1 gives an overview of the recurrent gene mutations in PMBCL that can be broadly categorized into genomic rearrangements, copy number changes and coding sequence mutations. Most strikingly, the mutations are involved in deregulation of JAK-STAT and NF-κB signaling pathways, as well as acquired immune privilege as discussed in the following paragraphs.

9.3.2 JAK-STAT pathway signaling

Signaling through the JAK-STAT pathway in PMBCL is the result of paracrine interleukin receptor-mediated signaling and constitutive activation by various somatic gene mutations in pathway genes (Tab. 9.1). This pathway activation is one of the shared features between PMBCL and CHL, supported by gene expression profiling and

Tab. 9.1: Somatic gene mutations in primary mediastinal large B-cell lymphoma.

Gene	Pattern	Proposed function	Frequency	Reference
I. Genomic Rearrangements				
CIITA	Balanced/unbalanced rearrangements with a cluster breakpoint region in Intron 1	– CIITA loss of function – Overexpression of fusion partners (e.g. PDL1/PDL2)	38 %	Steidl et al. [52]
PDL1/ PDL2	Balanced/unbalanced rearrangements. Two main breakpoint regions (5′ and 3′ in both genes) are described	– Overexpression of PDL1/PDL2 domains – Increased fusion transcript stability	20 %	Twa et al. [42]
II. Copy number changes				
JAK2	Amplification of chromosome 9p24.1 that typically includes multiple genes with a minimally amplified region containing JAK2, PDL1, PDL2 and JMJD2C	JAK2 overexpression leading to constitutive JAK-STAT signaling	63 %	Rui et al. [56]
PDL1/PDL2		Overexpression of PDL1/PDL2 resulting in T-cell anergy in the microenvironment		
JMJD2C		Altered histone H3 modifications promoting proliferation and survival.		
REL	Amplification of chromosome 2p16 with a minimally amplified region containing REL and BCL11A	Overexpression of the REL oncogene (NF-κB complex component) resulting in constitutive pathway activation	75 %	Joos et al. [45] Weniger et al. [50]
BCL11A		Deregulation of the BCL11A oncogene as a transcriptional repressor leading to lymphomagenic expression changes		Weniger et al. [51]

Tab. 9.1: (continued)

Gene	Pattern	Proposed function	Frequency	Reference
III. Coding sequence mutations				
SOCS1	Mostly small deletions, frameshift and nonsense mutations throughout the gene	Deletion of SOCS1 functional domains leading to constitutive JAK-STAT pathway activation	45–100 %	Melzner et al. [46] Gunawardana et al. [41]
PTPN1	Scattered nonsense and missense mutations with mutation hotspot in exon 1	Complete or partial loss of PTP1B phosphatase activity leading to constitutive JAK-STAT pathway activation	20 %	Gunawardana et al. [41]
TNFAIP3	Scattered, mostly nonsense mutations	Inactivation of A20 leading to constitutive NF-κB pathway activation	36 %	Schmitz et al. [49]
MYC	Rearrangements of the second/third exon; single nucleotide variations in P2 promoter region	Potentially altered transcriptional regulation / Chromatin remodeling leading to lymphomagenic expression changes	25 %	Scarpa et al. [48]
TP53	Missense mutations targeting exons 6 to 8	Functional abrogation of the p53 pathway	13 %	Scarpa et al. [48]
STAT6	Missense mutations with predominance in DNA binding domain	The exact consequences are still unknown, but DNA binding properties are likely altered	36 %	Guiter et al. [44]
KIT	Missense mutations with predominance in kinase domain	Potentially damaging mutations of unclear functional significance	12.5 %	Nagel et al. [47]

functional studies of specific somatic gene mutations found in PMBCL clinical samples and derived cell lines.

In normal B cells the pathway is activated through binding of cytokines to their cognate receptor on the cell surface causing tyrosine phosphorylation of JAK that is constitutively bound to the receptor [57]. Heterodimerization of interleukin receptor subunits interacting with the ligand is critically needed in this process. JAK molecules are then transphosphorylated, as are tyrosine residues on the receptor, which act as docking sites for STAT proteins. The STAT molecules are also phosphorylated as a result of JAK kinase activity to form homo- or heterodimers before translocating to the nucleus to activate expression of target genes. Intact paracrine signaling via IL13 stimulation has been reported to lead to increased but reversible levels of phosphorylated STAT5 in the PMBCL-derived cell line Med-B1 [58], while IL4 or IL13 stimulation induces an inflammatory phenotype in the PMBCL-derived cell line KARPAS-1106P similar to CHL [59]. The constitutive activation of the pathway is evidenced by recurrent somatic gene alterations on multiple levels including amplification of JAK2, deletions or inactivating mutations of the negative regulators *SOCS1* and *PTPN1*, and mutations in *STAT6*. Thus, as a result of both paracrine and constitutive pathway activation, STAT6 phosphorylation has been suggested as a reliable marker to distinguish PMBCL from other large cell lymphomas [44].

Based on target gene and more recently genome-wide mutation studies, somatic point mutations of *SOCS1* are amongst the most frequently found alteration in PMBCL. Melzner et al. report point mutations including biallelic mutations in 45 % of cases [46] and whole transcriptome analysis (RNA-seq) identified single nucleotide variants and indels in 7 of 7 clinical PMBCL cases [41]. The latter study also confirmed previously identified *SOCS1* mutations in the cell lines Med-B1 and KARPAS-1106P. These mutations are shared across the spectrum of CHL, nodular lymphocyte predominant Hodgkin lymphoma, PMBCL and DLBCL [60, 61]. Functionally, *SOCS1* mutations are shown to abrogate SOCS box function leading to hyperphosphorylation of JAK2 and STAT5 [46].

Moreover, RNA-seq was used to identify mutations in an additional negative regulator of JAK-STAT signaling, known to be critically involved in the pathogenesis of PMCBL. Gunawardana and colleagues found *PTPN1* coding sequence mutations in 22 % of PMBCL and 20 % of CHL cases, consisting of nonsense, missense and frameshift mutations [41]. *PTPN1* encodes PTP1B, a nonreceptor member of the superfamily of protein tyrosine phosphatases that acts as a negative regulator of IL-4-induced STAT6 signaling [62]. These nonsense and missense mutations were characterized as deleterious leading to reduction or loss of *PTPN1* phosphatase activity and subsequent hyperphosphorylation of STAT molecules. Immunohistochemistry revealed a close correlation with undetectable protein.

In addition, structural genomic changes have been discovered involving JAK-STAT deregulation. Specifically, gains and amplifications of chromosome 9p have been described as one of the most common somatic gene alterations occurring in more than half of all PMBCL patients [43, 45]. The minimally amplified region contains multiple genes, including *JAK2*, *CD274 (PDL1)*, *PDCD1LG2 (PDL2)* and *JMJD2C* [56], and it is very likely that the various effects on JAK-STAT signaling (JAK2), and the later described acquired immune privilege (PDL1/2) (see below) and chromatin remodeling (JMJD2C) act synergistically in the pathogenesis of PMBCL [1].

Recurrent point mutations in *STAT6* were found in 36 % of PMBCL cases, completing the spectrum of pathway hits that might contribute to altered transcriptional programs. Interestingly, knockdown of STAT6 by RNA interference enhanced the sensitivity of PMBCL-derived cell lines to R-CHOP components providing proof of concept that JAK-STAT pathway perturbation might influence treatment outcomes [63]. However, the exact functional consequences of STAT6 interference and mutations in the DNA binding domain still need to be determined.

There are no mutational data reported for MGZL; however, gains and amplifications targeting 9p24 involving the *JAK2/PD-L1/2* loci are found in 55 % of cases and were more common in those cases resembling PMBCL [64].

9.3.3 NF-κB signaling

Early GEP studies revealed that several genes of the nuclear factor kappa B (NF-κB) signaling pathway were overexpressed in PMBCL as compared to DLBCL and that this gene expression pattern more closely resembled cHL [34, 35]. The NF-κB signaling pathway is referenced as one of the most important deregulated pathways in B-cell lymphomas, and specifically in CHL and PMBCL [1, 65]. Studies in PMBCL-derived cell lines confirmed overexpression of an NF-κB activation signature and nuclear localization of NF-κB transcription factor complexes in contrast to the activated B-cell (ABC)- and germinal center B (GCB)-type DLBCL [36], and small molecule inhibition of the IκB kinase (IKK) complex was toxic to KARPAS-1106P cells demonstrating pathway addiction [66]. However, the underlying genetic causes for pathway activation are not yet fully elucidated.

More recently, specific somatic gene mutations and structural genomic changes in pathway genes have added to the evidence that NF-κB pathway activation is clonally selected for during PMBCL pathogenesis. Most prominently, copy number gains and amplifications of the *REL* gene locus (chromosome 2p16.1) have been found in 50 % or more cases using various cytogenetic techniques including array comparative genomic hybridization and fluorescence *in situ* hybridization [43, 50, 67], a finding that was correlated with nuclear localization of REL protein indicating increased pathway activity. Moreover, chromosomal amplification of the NF-κB regulators *BCL10* (1p22) and *MALT1* (18q21) are described in a significant minority of cases [43].

Beyond copy number changes, recurrent somatic point mutations have been described that influence NF-κB signaling. One of these, point mutations of *TNFAIP3* (encoding A20), has been reported as a novel driver mutation in PMBCL and CHL. The

A20 protein acts as a ubiquitin-modifying enzyme that interacts with RIP1, TRAF1 and TRAF2, thereby inhibiting NF-κB signaling downstream of TNFR receptor engagement [68]. Inactivating gene mutations were found in 36 % of PMBCL establishing *TNFAIP3* as a tumor suppressor gene [49]. Specifically in PMBCL, a biallelic inactivation mechanism was described in the A20-null cell line KARPAS-1106P and, earlier array CGH studies of primary cases had already shown chromosomal 6q23.3 deletions harboring the *TNFAIP3* locus in 30 % of patients [43, 69]. These findings complement observations across a spectrum of B-cell lymphomas in which somatic *TNFAIP3* mutations and/or deletion have been described, including mantle cell lymphoma [70], ABC-type DLBCL [70], marginal zone B-cell lymphoma [71], mucosa-associated lymphoid tissue (MALT) lymphomas [73, 73] and AIDS-related B-cell lymphomas [74]. Gains including amplifications of 2p16.1 involving the *REL/BCL11A* locus are found in 33 % of MGZL cases [64]. MGZL cases do not harbor *MYC* translocations, but gains involving this region are seen in 27 % of cases.

9.3.4 Acquired immune privilege

The crosstalk between the malignant lymphoma cells and the cells in the tumor microenvironment is prominently referenced in the literature and the specific composition of the microenvironment represents an important feature for subtyping of many lymphomas [75]. Emerging studies have started to elucidate specific genetic events harbored by the malignant cells that alter the interplay with the immune microenvironment in PMBCL and CHL. The largely overlapping mechanisms that lead to an immune privilege phenotype can be broadly categorized into (i) overexpression of surface molecules, (ii) downregulation of surface molecules and, (iii) induction of a favorable regulatory immunological niche [76]. Although not directly shown in PMBCL, support for the latter mechanism can be drawn from the paradigm in CHL where elevated surface expression of galectin-1 (encoded by *LGALS1*) has been shown to skew the composition of the microenvironment towards increased T helper 2 and T regulatory cells [77]. Importantly, from a biology point of view, various immunological selection pressures influenced by host-specific factors are likely to result in the diversity of lymphoma-associated microenvironments and variable immune cell crosstalk between malignant and nonmalignant cells that is selected for during disease progression.

9.3.4.1 Loss of MHC molecules

Loss of major histocompatibility (MHC) molecules, in particular MHC class II is frequent in PMBCL [9, 10, 78] and has been linked to inferior outcome [10], similar to observations in DLBCL [79, 80] and CHL [81]. Moreover, MHC class II loss was found to be correlated with decreased infiltrating cytotoxic T cells [82] suggesting escape from immune surveillance as a potential mechanism [83]. Missing a direct link between HLA class II loss and reduced numbers of CD8+ cytotoxic T cells, it is likely that T helper cells are involved in modulating the observed cytotoxic T-cell response. Searching for genomic correlates of this finding, microsatellite analysis and copy number analysis identifying allelic imbalances on chromosome 6p harboring the MHC class I and II genes, although the number of affected cases was low [43, 78]. Interestingly, genomic imbalances of 6p are also found in cases of immune-privilege site diffuse large B-cell lymphoma suggesting immune privilege as a shared feature between both entities [84]. Whole transcriptome sequencing analysis of the HL cell line KMH2 and subsequent screening of primary CHL and PMBCL samples by fluorescence *in situ* hybridization (FISH) revealed recurrent chromosomal rearrangement involving the *CIITA* gene in 15 % and 38 %, respectively [52]. *CIITA* is localized on chromosome 16p encoding for the master transcriptional regulator of MHC class II expression and is tightly correlated with MHC class II expression in PMBCL [10]. *CIITA* rearrangements involve promiscuous partners with heterogeneous consequences, one of which is the reduction of MHC class II expression resulting from the dominant-negative function of CIITA-FLJ27352 observed in KMH2 cells [52]. In MGZL, breaks disrupting *CIITA* were found in 27 % of cases, with others showing a signal pattern consistent with amplification of the locus [64]. The cases with apparent disruption of the *CIITA* locus did not correlate with HLA-DR expression by tumor cells.

9.3.4.2 Overexpression of programmed death ligands

Early gene expression profiling studies revealed high expression of programmed death ligand 2 (*PDCD1LG2/PDL2*) as part of a gene signature distinguishing PMBCL from ABC- and GCB-type DLBCL [34, 35]. Integrative analysis of copy number and gene expression data in subsequent studies revealed that both *PDL2* and the neighboring *PDL1* gene with similar function are harbored in a commonly amplified region on chromosome 9p24, and presence of this amplification was correlated with increased gene expression in the majority of PMBCL cases [85]. Moreover, recurrent genomic rearrangements involving the 9p locus were discovered by next generation sequencing, and FISH analysis in a PMBCL extension cohort revealed a prevalence of these rearrangements in 20 % of cases, resulting in PDL1 and PDL2 gene fusions with various gene fusion partners [42, 52]. The cumulative incidence of both copy number changes and genomic rearrangements speaks to the importance of these events in the pathogenesis of PMBCL. The resulting upregulation of PDL1/2 and/or expression of PDL1/2-containing fusion genes establishes these events as major hallmarks of the disease. Functionally, *in vitro* studies showed that ectopic expression of *PDL*-containing gene fusions inactivated admixed T cells in co-culture systems, and that this inactivation is reversible by PD-1 receptor or ligand blockade. These preclinical data lend further support to the idea of testing checkpoint inhibitors that can block the PD-1 axis in PMBCL and related diseases expressing programmed death ligands. In this context, rearrangements of PDL1/2 with related strong surface expression of the ligands were specifically seen in IP-DLBCL of the testis, providing a biologic rationale for checkpoint inhibitor therapy in this subset of large cell lymphoma [55]. Moreover, in the cases with co-amplification of *JAK2* and programmed death ligands on chromosome 9p24, it was shown that PDL1/2 expression can be enhanced through a STAT-mediated transcriptional mechanism [85]. These data suggest that combined inhibition of JAK-STAT and immune checkpoints might be therapeutically synergistic in these patients.

9.3.5 Epigenetic changes

Molecular studies in PMBCL focusing on epigenetic changes are uncommon in comparison to the wealth of data in nodal DLBCL [86, 87]. Recently, a few studies have established the importance of individual key genes as targets of methylation or at least as loci involved in epigenetic regulation in PMBCL. This published work covers a spectrum of approaches ranging from methylation studies, characterization of key enzyme regulating histone marks and miRNA profiling. Scarpa and colleagues studied p16 promoter methylation and found hypermethylation with or without concomitant genomic deletion in a limited number of cases [48]. In another study, the methylation status of a number of target genes affecting gene expression was studied across a spectrum of lymphomas including six PMBCL samples. Hypermethylation of LRP12, CDH1, BMPER, DUSP4, and BMP7 was found to reliably distinguish these lymphomas from nonlymphoma control samples. However, PMBCL-specific patterns were not studied. In a methylation profiling study of PMBCL, CHL and MGZL interrogating 1,505 CpG sites harboring over 800 genes, a specific methylation pattern of PMBCL was described that was also used to place MGZL in between CHL and PMBCL, based on unsupervised clustering of methylation patterns [38]. Specific genes of interest encompassed *IFNG, EPHA7, CCL3, MMP9, DAPK1* as well as multiple genes in the NF-κB pathway.

Specifically, JAK2 and JMJD2C have been drawn into focus based on studies of histone H3 modifications essential to tumor cell proliferation and survival [56]. JMJD2C functions as a demethylase removing histone H3 lysine 9 trimethylation marks, and thereby diminishing the recruitment of HP1α and heterochromatin formation [88]. Furthermore, phosphorylated JAK2 cooperates in chromatin remodeling through phosphorylation of histone H3 tyrosine 41 and HP1α leading to formation of active chromatin [89]. These findings were unique to PMBCL and CHL cell lines with the 9p24 amplicon harboring both the *JAK2* and *JMJD2C* genes, in contrast to DLBCL cell lines without 9p amplification, highlighting the specificity of this observation linked to overexpression of both genes. In an RNA interference screen with JAK2 and JMJD2C, *MYC* was identified as one of the most regulated genes, targeted through a mechanism that altered the chromatin structure of the *MYC* locus [56].

Iqbal and colleagues used miRNA expression profiling to distinguish subsets of B-cell lymphoma including PMBCL, and found a number of miRNAs differentially expressed between PMBCL and ABC/GCB-type DLBCL (5 miRNA classifier) [90]. These miRNAs include miR-365, -193b, -629, -423-5p and -15a, of which miR-365 and -193b were specifically expressed in PMBCL. Previous studies also established miR-155 as highly expressed in PMBCL and CHL [91]. Further studies are needed to determine the functional consequences of these miRNAs in the pathogenesis of PMBCL.

9.3.6 Biomarkers and therapeutic considerations

PMBCL is a distinct clinicopathological entity and should be approached as such therapeutically. However, due to the rarity of PMBCL and its recent recognition as a distinct entity, the published experience in PMBCL is limited to retrospective studies or subgroup analyses in trials that encompass both PMBCL and the parent entity of DLBCL. With these historical and incidence-related obstacles in mind, a better understanding of the pathobiology and identification of PMBCL-specific drug targets might be pivotal for treatment advances in this disease. In the following paragraphs we will briefly discuss the current treatment standards, novel opportunities for targeted treatments and emerging biomarkers.

9.3.6.1 Treatment standard

While there is no clear consensus on a defined treatment standard, combined modality treatments including rituximab (anti-CD20) immunotherapy have overall yielded quite favorable outcomes. Consolidative mediastinal radiation remains part of standard therapy for a significant proportion of patients [92, 93]. However, given the young age and female gender of most patients with this disease, radiation leads to long-term sequelae such as increased risk of breast cancer and secondary malignancies, as well as ischemic heart disease [94]. Therefore, it will be critical in future studies to assess the need for consolidating radiation therapy, and to develop strategies to remove routine mediastinal radiation, while maintaining high cure rates.

Despite the overall relatively favorable outcomes, the 3-year progression-free/event-free survival has been variably reported between 68 % and 91 % (for R-CHOP studies) and between 78 % and 95 % (for R + dose intense studies), likely due to selection bias of patients in retrospective series [95–101]. In aggregate, the optimal combination chemotherapy and the role that consolidative radiotherapy plays in treating PMBCL is yet to be determined [102]. Recent phase II data suggests that clinical benefit can be achieved using the dose-adjusted EPOCH + rituximab regimen (DA-EPOCHR), so-called infusional therapy. This more dose-intense strategy maybe preferred, as its use appears to obviate the need for radiotherapy [95]. Moreover, ongoing studies in PMBCL are investigating if end-of-therapy FDG-PET can identify patients who might not require radiation. However, a British Columbia study demonstrated that a sizeable number of FDG-PET-negative patients who have received R-CHOP subsequently relapsed [101]. A number of studies suggest that end of treatment FDG-PET specifically in PMBCL has a low positive predictive value, in contrast to the high clinical accuracy of the technique in other aggressive lymphomas [93, 95]. Optimal treatment strategies in MGZL are less established, as outcomes following DA-EPOCH-R are inferior as compared with PMBCL [103].

9.3.6.2 Novel therapies

As described in the molecular hallmarks section, a number of new therapeutic targets have recently emerged (Fig. 9.2). Due to the clinical, pathological, and biological relatedness of CHL and PMBCL, additional therapeutic rationale might also be carried over into PMBCL based on advances in CHL. Although only limited specific data are currently available, the most promising approaches are projected to encompass immune checkpoint inhibitors, JAK-STAT pathway inhibitors, antibody-drug conjugates such as brentuximab vedotin, and cell-based immunotherapy using chimeric antigen receptor (CAR) T cells.

Brentuximab vedotin is an antibody drug conjugate consisting of a chimeric anti-CD30 monoclonal antibody linked to the microtubule inhibitor monomethyl auristatin E (MMAE) [104]. It has been shown to induce responses in a high percentage of patients with relapsed and refractory CHL and is currently investigated in frontline therapy [105, 106]. Due to the variable expression of CD30 in PMBCL, brentuximab vedotin is also under investigation in combination with R-CHOP in newly diagnosed PMBCL and MGZL.

Since programmed death ligand expression and underlying structural genomic changes have been shown to be hallmarks of PMBCL occurring in the vast majority of cases [42, 85, 107], the PD-1 checkpoint pathway emerges as a very promising target to therapeutically reverse immune privilege cultivated by the tumor cells. Immune checkpoint inhibitors have demonstrated exciting activity in a wide range of solid cancers [108–110]. and also in relapsed aggressive lymphoma [111]. Very recently, nivolumab, a fully humanized IgG-4 blocking antibody, has been tested in relapsed and refractory HL, and an early report demonstrates promising activity in this disease population, with objective response rates of 87 % and complete response rates of 17 % [112]. The availability of PDL1/2 mutation- or expression-related biomarkers (see above) in PMBCL supports a potential companion diagnostic approach for future immune checkpoint inhibitor therapy.

The previously described activation of the JAK-STAT signaling pathway and the presence of multiple somatic gene mutations in both PMBCL and CHL provide a strong rationale for therapeutic pathway inhibition in both diseases, and selective JAK2 inhibition has been shown to decrease PMBCL and HL growth *in vitro* and *in vivo* [113]. Moreover, the multikinase inhibitor, SB1518 (pacritinib), preferentially inhibiting FLT3 and JAK2, demonstrated encouraging clinical activity in relapsed or refractory HL and NHL. However, conclusions about therapeutic effects on PMBCL are lacking due to the

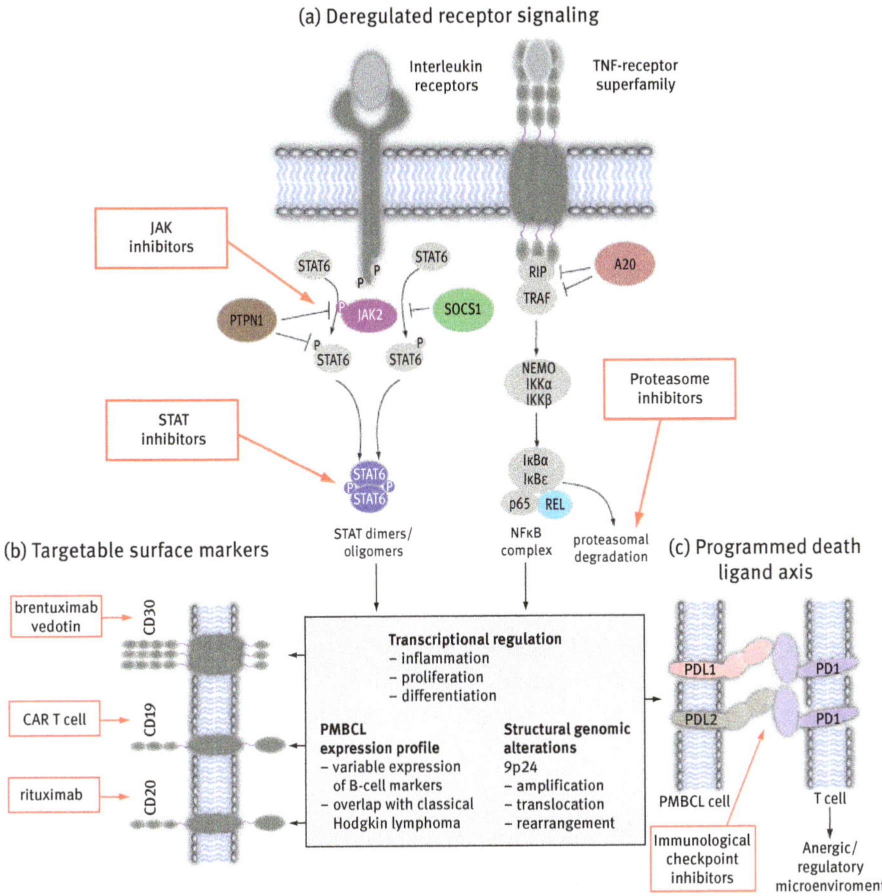

Fig. 9.2: **Targetable molecular features of primary mediastinal large B-cell lymphoma.** A. JAK-STAT and NF-κB signaling are the target of multiple genetic hits (highlighted in color). Pathway inhibitors are shown in the red boxes. B. Specific surface markers, as part of a PMBCL-specific expression profile, are shown with corresponding therapeutic approaches. C. Overexpression of programmed death ligands, as a result of structural genomic alterations, represents a promising therapeutic target. Immune checkpoint inhibition has the potential to restore and accentuate anti-tumor immune responses.

low case number [114]. Another JAK inhibitor (ruxolitinib), an oral JAK1/2 inhibitor, is FDA-approved for myelofibrosis and might be an interesting drug that could be repurposed for use in B-cell lymphomas with active JAK-STAT signaling. From a biomarker perspective, it will be important to study if specific pathway mutations and overall pathway activity might influence sensitivity to conventional and novel therapeutics targeting the JAK-STAT pathway ("predictive biomarkers"). Given the inferior outcomes of young men with MGZL and the involvement of many of the same oncogenic pathways, testing novel biological agents in this disease setting is recommended.

Finally, a recent report on the use of genetically modified T cells, expressing an anti-CD19 chimeric antigen receptor (CAR), has raised hope for an alternative treatment approach using cell-based immunotherapy in lymphoma [115]. In this report, the authors found high complete remission rates in relapsed and refractory DLBCL and indolent lymphomas. As most PMBCL cases show expression of CD19, inclusion of PMBCL in future CAR T cell-based therapies should be tested for their ability to induce durable remissions in the relapsed/refractory setting.

9.4 Concluding remarks

Significant progress has been achieved in the last decade in our fundamental understanding of the biology and pathogenesis of PMBCL. Importantly, defining and unique oncogenic pathway perturbations set PMBCL aside from nodal DLBCL and provide justification for its recognition as a unique disease entity in the current WHO classification of lymphoid neoplasms. It will be imperative that future clinical trials testing therapies in PMBCL distinguish these cases from more typical nodal or extranodal DLBCL. MGZL continues to be recognized with increasing frequency, including cases without clinically apparent mediastinal disease. Large multicenter studies highlight the inferior survival of MGZL and thus studies testing novel agents alone or in combination are needed to improve overall survival of this more aggressive grey zone entity [3].

References

[1] Steidl C, Gascoyne RD. The molecular pathogenesis of primary mediastinal large B-cell lymphoma. Blood 2011;118:2659–2669.

[2] Swerdlow SH, Campo E, Harris NL, et al. eds. WHO Classification of Tumours of Haematopoietic and Lymphoid Tissues. Lyon, France: IARC 2008.

[3] Evens AM, Kanakry JA, Sehn LH, et al. Gray zone lymphoma with features intermediate between classical hodgkin lymphoma and diffuse large B-cell lymphoma: Characteristics, outcomes, and prognostication among a large multicenter cohort. Am J Hematol 2015.

[4] Sehn LH, Donaldson J, Chhanabhai M, et al. Introduction of combined CHOP plus rituximab therapy dramatically improved outcome of diffuse large B-cell lymphoma in British Columbia. J Clin Oncol 2005;23:5027–5033.

[5] Pileri SA, Gaidano G, Zinzani PL, et al. Primary mediastinal B-cell lymphoma: high frequency of BCL-6 mutations and consistent expression of the transcription factors OCT-2, BOB.1, and PU.1 in the absence of immunoglobulins. Am J Pathol 2003;162:243–253.

[6] Csernus B, Timar B, Fulop Z, et al. Mutational analysis of IgVH and BCL-6 genes suggests thymic B-cells origin of mediastinal (thymic) B-cell lymphoma. Leuk Lymphoma 2004;45:2105–2110.

[7] de Leval L, Ferry JA, Falini B, Shipp M, Harris NL. Expression of bcl-6 and CD10 in primary mediastinal large B-cell lymphoma: evidence for derivation from germinal center B cells? Am J Surg Pathol 2001;25:1277–1282.

[8] Moller P, Moldenhauer G, Momburg F, et al. Mediastinal lymphoma of clear cell type is a tumor corresponding to terminal steps of B cell differentiation. Blood 1987;69:1087–1095.
[9] Moller P, Lammler B, Herrmann B, Otto HF, Moldenhauer G, Momburg F. The primary mediastinal clear cell lymphoma of B-cell type has variable defects in MHC antigen expression. Immunology 1986;59:411–417.
[10] Roberts RA, Wright G, Rosenwald AR, et al. Loss of major histocompatibility class II gene and protein expression in primary mediastinal large B-cell lymphoma is highly coordinated and related to poor patient survival. Blood 2006;108:311–318.
[11] Falini B, Venturi S, Martelli M, et al. Mediastinal large B-cell lymphoma: clinical and immunohistological findings in 18 patients treated with different third-generation regimens. Br J Haematol 1995;89:780–789.
[12] Higgins JP, Warnke RA. CD30 expression is common in mediastinal large B-cell lymphoma. Am J Clin Pathol 1999;112:241–247.
[13] Capello D, Vitolo U, Pasqualucci L, et al. Distribution and pattern of BCL-6 mutations throughout the spectrum of B-cell neoplasia. Blood 2000;95:651–659.
[14] Leithauser F, Bauerle M, Huynh MQ, Moller P. Isotype-switched immunoglobulin genes with a high load of somatic hypermutation and lack of ongoing mutational activity are prevalent in mediastinal B-cell lymphoma. Blood 2001;98:2762–2770.
[15] Malpeli G, Barbi S, Moore PS, et al. Primary mediastinal B-cell lymphoma: hypermutation of the BCL6 gene targets motifs different from those in diffuse large B-cell and follicular lymphomas. Haematologica 2004;89:1091–1099.
[16] Nacheva E, Dyer MJ, Metivier C, et al. B-cell non-Hodgkin's lymphoma cell line (Karpas 1106) with complex translocation involving 18q21.3 but lacking BCL2 rearrangement and expression. Blood 1994;84:3422–3428.
[17] Popov SW, Moldenhauer G, Wotschke B, et al. Target sequence accessibility limits activation-induced cytidine deaminase activity in primary mediastinal B-cell lymphoma. Cancer Res 2007;67:6555–6564.
[18] Barth TF, Leithauser F, Joos S, Bentz M, Moller P. Mediastinal (thymic) large B-cell lymphoma: where do we stand? Lancet Oncol 2002;3:229–234.
[19] al-Sharabati M, Chittal S, Duga-Neulat I, et al. Primary anterior mediastinal B-cell lymphoma. A clinicopathologic and immunohistochemical study of 16 cases. Cancer 1991;67:2579–2587.
[20] Calaminici M, Piper K, Lee AM, Norton AJ. CD23 expression in mediastinal large B-cell lymphomas. Histopathology 2004;45:619–624.
[21] Copie-Bergman C, Plonquet A, Alonso MA, et al. MAL expression in lymphoid cells: further evidence for MAL as a distinct molecular marker of primary mediastinal large B-cell lymphomas. Modern Pathology: An Official Journal of the United States and Canadian Academy of Pathology, Inc 2002;15:1172–1180.
[22] Dorfman DM, Shahsafaei A, Alonso MA. Utility of CD200 immunostaining in the diagnosis of primary mediastinal large B cell lymphoma: comparison with MAL, CD23, and other markers. Modern Pathology: An Official Journal of the United States and Canadian Academy of Pathology, Inc 2012;25:1637–1643.
[23] Hofmann WJ, Momburg F, Moller P. Thymic medullary cells expressing B lymphocyte antigens. Human Pathology 1988;19:1280–1287.
[24] Isaacson PG, Norton AJ, Addis BJ. The human thymus contains a novel population of B lymphocytes. Lancet 1987;2:1488–1491.
[25] Lazzarino M, Orlandi E, Paulli M, et al. Primary mediastinal B-cell lymphoma with sclerosis: an aggressive tumor with distinctive clinical and pathologic features. Journal of Clinical Oncology: Official Journal of the American Society of Clinical Oncology 1993;11:2306–2313.

[26] Oschlies I, Burkhardt B, Salaverria I, et al. Clinical, pathological and genetic features of primary mediastinal large B-cell lymphomas and mediastinal gray zone lymphomas in children. Haematologica 2011;96:262-268.
[27] Salama ME, Rajan Mariappan M, Inamdar K, Tripp SR, Perkins SL. The value of CD23 expression as an additional marker in distinguishing mediastinal (thymic) large B-cell lymphoma from Hodgkin lymphoma. International Journal of Surgical Pathology 2010;18:121–128.
[28] Spencer J, Choy M, Hussell T, Papadaki L, Kington JP, Isaacson PG. Properties of human thymic B cells. Immunology 1992;75:596–600.
[29] Yamamoto W, Nakamura N, Tomita N, et al. Clinicopathological analysis of mediastinal large B-cell lymphoma and classical Hodgkin lymphoma of the mediastinum. Leukemia & Lymphoma 2013;54:967–972.
[30] Akashi K, Richie LI, Miyamoto T, Carr WH, Weissman IL. B lymphopoiesis in the thymus. J Immunol 2000;164:5221–5226.
[31] Moldenhauer G, Popov SW, Wotschke B, et al. AID expression identifies interfollicular large B cells as putative precursors of mature B-cell malignancies. Blood 2006;107:2470–2473.
[32] Mori S, Inaba M, Sugihara A, et al. Presence of B cell progenitors in the thymus. J Immunol 1997;158:4193–4199.
[33] Perera J, Meng L, Meng F, Huang H. Autoreactive thymic B cells are efficient antigen-presenting cells of cognate self-antigens for T cell negative selection. Proceedings of the National Academy of Sciences of the United States of America 2013;110:17011–17016.
[34] Rosenwald A, Wright G, Leroy K, et al. Molecular diagnosis of primary mediastinal B cell lymphoma identifies a clinically favorable subgroup of diffuse large B cell lymphoma related to Hodgkin lymphoma. J Exp Med 2003;198:851–862.
[35] Savage KJ, Monti S, Kutok JL, et al. The molecular signature of mediastinal large B-cell lymphoma differs from that of other diffuse large B-cell lymphomas and shares features with classical Hodgkin lymphoma. Blood 2003;102:3871–3879.
[36] Feuerhake F, Kutok JL, Monti S, et al. NFkappaB activity, function, and target-gene signatures in primary mediastinal large B-cell lymphoma and diffuse large B-cell lymphoma subtypes. Blood 2005;106:1392–1399.
[37] Calvo KR, Traverse-Glehen A, Pittaluga S, Jaffe ES. Molecular profiling provides evidence of primary mediastinal large B-cell lymphoma as a distinct entity related to classic Hodgkin lymphoma: implications for mediastinal gray zone lymphomas as an intermediate form of B-cell lymphoma. Advances in Anatomic Pathology 2004;11:227–238.
[38] Eberle FC, Rodriguez-Canales J, Wei L, et al. Methylation profiling of mediastinal gray zone lymphoma reveals a distinctive signature with elements shared by classical Hodgkin's lymphoma and primary mediastinal large B-cell lymphoma. Haematologica 2011;96:558–566.
[39] Xie L, Ritz O, Leithauser F, et al. FOXO1 downregulation contributes to the oncogenic program of primary mediastinal B-cell lymphoma. Oncotarget 2014;5:5392–5402.
[40] Grant C, Dunleavy K, Eberle FC, Pittaluga S, Wilson WH, Jaffe ES. Primary mediastinal large B-cell lymphoma, classic Hodgkin lymphoma presenting in the mediastinum, and mediastinal gray zone lymphoma: what is the oncologist to do? Curr Hematol Malig Rep 2011;6:157–163.
[41] Gunawardana J, Chan FC, Telenius A, et al. Recurrent somatic mutations of PTPN1 in primary mediastinal B cell lymphoma and Hodgkin lymphoma. Nat Genet 2014;46:329–335.
[42] Twa DD, Chan FC, Ben-Neriah S, et al. Genomic rearrangements involving programmed death ligands are recurrent in primary mediastinal large B-cell lymphoma. Blood 2014;123:2062–2065.
[43] Wessendorf S, Barth TF, Viardot A, et al. Further delineation of chromosomal consensus regions in primary mediastinal B-cell lymphomas: an analysis of 37 tumor samples using high-resolution genomic profiling (array-CGH). Leukemia 2007;21:2463–2469.

[44] Guiter C, Dusanter-Fourt I, Copie-Bergman C, et al. Constitutive STAT6 activation in primary mediastinal large B-cell lymphoma. Blood 2004;104:543–549.

[45] Joos S, Otano-Joos MI, Ziegler S, et al. Primary mediastinal (thymic) B-cell lymphoma is characterized by gains of chromosomal material including 9p and amplification of the REL gene. Blood 1996;87:1571–1578.

[46] Melzner I, Bucur AJ, Bruderlein S, et al. Biallelic mutation of SOCS-1 impairs JAK2 degradation and sustains phospho-JAK2 action in the MedB-1 mediastinal lymphoma line. Blood 2005;105:2535–2542.

[47] Nagel PD, Stenzinger A, Feld FM, et al. KIT mutations in primary mediastinal B-cell lymphoma. Blood Cancer J 2014;4:e241.

[48] Scarpa A, Moore PS, Rigaud G, et al. Molecular features of primary mediastinal B-cell lymphoma: involvement of p16INK4A, p53 and c-myc. Br J Haematol 1999;107:106–113.

[49] Schmitz R, Hansmann ML, Bohle V, et al. TNFAIP3 (A20) is a tumor suppressor gene in Hodgkin lymphoma and primary mediastinal B cell lymphoma. J Exp Med 2009;206:981–989.

[50] Weniger MA, Gesk S, Ehrlich S, et al. Gains of REL in primary mediastinal B-cell lymphoma coincide with nuclear accumulation of REL protein. Genes Chromosomes Cancer 2007;46:406–415.

[51] Weniger MA, Pulford K, Gesk S, et al. Gains of the proto-oncogene BCL11A and nuclear accumulation of BCL11A(XL) protein are frequent in primary mediastinal B-cell lymphoma. Leukemia 2006;20:1880–1882.

[52] Steidl C, Shah SP, Woolcock BW, et al. MHC class II transactivator CIITA is a recurrent gene fusion partner in lymphoid cancers. Nature 2011;471:377–381.

[53] Liu Y, Abdul Razak FR, Terpstra M, et al. The mutational landscape of Hodgkin lymphoma cell lines determined by whole-exome sequencing. Leukemia 2014;28:2248–2251.

[54] Reichel J, Chadburn A, Rubinstein PG, et al. Flow sorting and exome sequencing reveal the oncogenome of primary Hodgkin and Reed-Sternberg cells. Blood 2015;125:1061–1072.

[55] Twa DD, Mottok A, Chan FC, et al. Recurrent genomic rearrangements in primary testicular lymphoma. J Pathol 2015;236:136–141.

[56] Rui L, Emre NC, Kruhlak MJ, et al. Cooperative epigenetic modulation by cancer amplicon genes. Cancer Cell 2010;18:590–605.

[57] Darnell JE, Jr., Kerr IM, Stark GR. Jak-STAT pathways and transcriptional activation in response to IFNs and other extracellular signaling proteins. Science 1994;264:1415–1421.

[58] Raia V, Schilling M, Bohm M, et al. Dynamic mathematical modeling of IL13-induced signaling in Hodgkin and primary mediastinal B-cell lymphoma allows prediction of therapeutic targets. Cancer Res 2011;71:693–704.

[59] Andersson E, Schain F, Sjoberg J, Bjorkholm M, Claesson HE. Interleukin-13 stimulation of the mediastinal B-cell lymphoma cell line Karpas-1106P induces a phenotype resembling the Hodgkin lymphoma cell line L1236. Exp Hematol 2010;38:116–123.

[60] Mottok A, Renne C, Willenbrock K, Hansmann ML, Brauninger A. Somatic hypermutation of SOCS1 in lymphocyte-predominant Hodgkin lymphoma is accompanied by high JAK2 expression and activation of STAT6. Blood 2007;110:3387–3390.

[61] Weniger MA, Melzner I, Menz CK, et al. Mutations of the tumor suppressor gene SOCS-1 in classical Hodgkin lymphoma are frequent and associated with nuclear phospho-STAT5 accumulation. Oncogene 2006;25:2679–2684.

[62] Lu X, Malumbres R, Shields B, et al. PTP1B is a negative regulator of interleukin 4-induced STAT6 signaling. Blood 2008;112:4098–4108.

[63] Haberle MT, Kelsch E, Dorsch K, Moller P, Ritz O. Knock-down of BCL6 / STAT6 sensitizes primary B cell lymphoma cells for treatment with current therapeutic agents. Oncoscience 2014;1:283–286.

[64] Eberle FC, Salaverria I, Steidl C, et al. Gray zone lymphoma: chromosomal aberrations with immunophenotypic and clinical correlations. Mod Pathol 2011;24:1586–1597.

[65] Kuppers R. The biology of Hodgkin's lymphoma. Nat Rev Cancer 2009;9:15–27.

[66] Lam LT, Davis RE, Pierce J, et al. Small molecule inhibitors of IkappaB kinase are selectively toxic for subgroups of diffuse large B-cell lymphoma defined by gene expression profiling. Clin Cancer Res 2005;11:28–40.

[67] Bea S, Zettl A, Wright G, et al. Diffuse large B-cell lymphoma subgroups have distinct genetic profiles that influence tumor biology and improve gene-expression-based survival prediction. Blood 2005;106:3183–3190.

[68] Song HY, Rothe M, Goeddel DV. The tumor necrosis factor-inducible zinc finger protein A20 interacts with TRAF1/TRAF2 and inhibits NF-kappaB activation. Proc Natl Acad Sci USA 1996;93:6721–6725.

[69] Kimm LR, deLeeuw RJ, Savage KJ, et al. Frequent occurrence of deletions in primary mediastinal B-cell lymphoma. Genes Chromosomes Cancer 2007;46:1090–1097.

[70] Honma K, Tsuzuki S, Nakagawa M, et al. TNFAIP3/A20 functions as a novel tumor suppressor gene in several subtypes of non-Hodgkin lymphomas. Blood 2009;114:2467–2475.

[71] Novak U, Rinaldi A, Kwee I, et al. The NF-{kappa}B negative regulator TNFAIP3 (A20) is inactivated by somatic mutations and genomic deletions in marginal zone lymphomas. Blood 2009;113:4918–4921.

[72] Chanudet E, Huang Y, Ichimura K, et al. A20 is targeted by promoter methylation, deletion and inactivating mutation in MALT lymphoma. Leukemia 2010;24:483–487.

[73] Kato M, Sanada M, Kato I, et al. Frequent inactivation of A20 in B-cell lymphomas. Nature 2009;459:712–716.

[74] Giulino L, Mathew S, Ballon G, et al. A20 (TNFAIP3) genetic alterations in EBV-associated AIDS-related lymphoma. Blood 2011;117:4852–4854.

[75] Scott DW, Gascoyne RD. The tumour microenvironment in B cell lymphomas. Nat Rev Cancer 2014;14:517–534.

[76] Mottok A, Steidl C. Genomic alterations underlying immune privilege in malignant lymphomas. Curr Opin Hematol 2015;22:343–354.

[77] Juszczynski P, Ouyang J, Monti S, et al. The AP1-dependent secretion of galectin-1 by Reed Sternberg cells fosters immune privilege in classical Hodgkin lymphoma. Proc Natl Acad Sci USA 2007;104:13134–13139.

[78] Rigaud G, Moore PS, Taruscio D, et al. Alteration of chromosome arm 6p is characteristic of primary mediastinal B-cell lymphoma, as identified by genome-wide allelotyping. Genes Chromosomes Cancer 2001;31:191–195.

[79] Rosenwald A, Wright G, Chan WC, et al. The use of molecular profiling to predict survival after chemotherapy for diffuse large-B-cell lymphoma. N Engl J Med 2002;346:1937–1947.

[80] Rimsza LM, Leblanc ML, Unger JM, et al. Gene expression predicts overall survival in paraffin-embedded tissues of diffuse large B-cell lymphoma treated with R-CHOP. Blood 2008;112:3425–3433.

[81] Diepstra A, van Imhoff GW, Karim-Kos HE, et al. HLA class II expression by Hodgkin Reed-Sternberg cells is an independent prognostic factor in classical Hodgkin's lymphoma. J Clin Oncol 2007;25:3101–3108.

[82] Farinha P, Steidl C, Rimsza LM, Savage KJ, Connors JM, Gascoyne RD. HLA-DR Protein Expression Correlates with Non-Neoplastic T-Cell Infiltration and Predicts Survival in Patients with Primary Mediastinal Large B Cell Lymphoma (PMBCL) Treated with CHOP Chemotherapy. ASH Annual Meeting Abstracts 2009;114:133.

[83] Rimsza LM, Roberts RA, Miller TP, et al. Loss of MHC class II gene and protein expression in diffuse large B-cell lymphoma is related to decreased tumor immunosurveillance and poor

patient survival regardless of other prognostic factors: a follow-up study from the Leukemia and Lymphoma Molecular Profiling Project. Blood 2004;103:4251–4258.

[84] Riemersma SA, Jordanova ES, Schop RF, et al. Extensive genetic alterations of the HLA region, including homozygous deletions of HLA class II genes in B-cell lymphomas arising in immune-privileged sites. Blood 2000;96:3569–3577.

[85] Green MR, Monti S, Rodig SJ, et al. Integrative analysis reveals selective 9p24.1 amplification, increased PD-1 ligand expression, and further induction via JAK2 in nodular sclerosing Hodgkin lymphoma and primary mediastinal large B-cell lymphoma. Blood 2010;116:3268–3277.

[86] Morin RD, Johnson NA, Severson TM, et al. Somatic mutations altering EZH2 (Tyr641) in follicular and diffuse large B-cell lymphomas of germinal-center origin. Nat Genet 2010;42:181–185.

[87] Pasqualucci L, Dominguez-Sola D, Chiarenza A, et al. Inactivating mutations of acetyltransferase genes in B-cell lymphoma. Nature 2011;471:189–195.

[88] Cloos PA, Christensen J, Agger K, et al. The putative oncogene GASC1 demethylates tri- and dimethylated lysine 9 on histone H3. Nature 2006;442:307–311.

[89] Dawson MA, Bannister AJ, Gottgens B, et al. JAK2 phosphorylates histone H3Y41 and excludes HP1alpha from chromatin. Nature 2009;461:819–822.

[90] Iqbal J, Shen Y, Huang X, et al. Global microRNA expression profiling uncovers molecular markers for classification and prognosis in aggressive B-cell lymphoma. Blood 2015;125:1137–1145.

[91] Kluiver J, Poppema S, de Jong D, et al. BIC and miR-155 are highly expressed in Hodgkin, primary mediastinal and diffuse large B cell lymphomas. J Pathol 2005;207:243–249.

[92] Zinzani PL, Martelli M, Bertini M, et al. Induction chemotherapy strategies for primary mediastinal large B-cell lymphoma with sclerosis: a retrospective multinational study on 426 previously untreated patients. Haematologica 2002;87:1258–1264.

[93] Martelli M, Ceriani L, Zucca E, et al. [18F]fluorodeoxyglucose positron emission tomography predicts survival after chemoimmunotherapy for primary mediastinal large B-cell lymphoma: results of the International Extranodal Lymphoma Study Group IELSG-26 Study. J Clin Oncol 2014;32:1769–1775.

[94] Castellino SM, Geiger AM, Mertens AC, et al. Morbidity and mortality in long-term survivors of Hodgkin lymphoma: a report from the Childhood Cancer Survivor Study. Blood 2011;117:1806–1816.

[95] Dunleavy K, Pittaluga S, Maeda LS, et al. Dose-adjusted EPOCH-rituximab therapy in primary mediastinal B-cell lymphoma. N Engl J Med 2013;368:1408–1416.

[96] Soumerai JD, Hellmann MD, Feng Y, et al. Treatment of primary mediastinal B-cell lymphoma with rituximab, cyclophosphamide, doxorubicin, vincristine and prednisone is associated with a high rate of primary refractory disease. Leuk Lymphoma 2014;55:538–543.

[97] Rieger M, Osterborg A, Pettengell R, et al. Primary mediastinal B-cell lymphoma treated with CHOP-like chemotherapy with or without rituximab: results of the Mabthera International Trial Group study. Ann Oncol 2011;22:664–670.

[98] Xu LM, Fang H, Wang WH, et al. Prognostic significance of rituximab and radiotherapy for patients with primary mediastinal large B-cell lymphoma receiving doxorubicin-containing chemotherapy. Leuk Lymphoma 2013;54:1684–1690.

[99] Vassilakopoulos TP, Pangalis GA, Katsigiannis A, et al. Rituximab, cyclophosphamide, doxorubicin, vincristine, and prednisone with or without radiotherapy in primary mediastinal large B-cell lymphoma: the emerging standard of care. Oncologist 2012;17:239–249.

[100] Moskowitz C, Hamlin PA, Jr., Maragulia J, Meikle J, Zelenetz AD. Sequential Dose-Dense RCHOP Followed by ICE Consolidation (MSKCC protocol 01-142) without Radiotherapy for

Patients with Primary Mediastinal Large B Cell Lymphoma. ASH Annual Meeting Abstracts 2010;116:420.
[101] Savage KJ, Yenson PR, Shenkier T, et al. The Outcome of Primary Mediastinal Large B-Cell Lymphoma (PMBCL) in the R-CHOP Treatment Era. ASH Annual Meeting Abstracts 2012;120:303.
[102] Savage KJ. Primary mediastinal large B-cell lymphoma. Oncologist 2006;11:488–495.
[103] Wilson WH, Pittaluga S, Nicolae A, et al. A prospective study of mediastinal gray-zone lymphoma. Blood 2014;124:1563–1569.
[104] Provencio M, Sanchez A, Sanchez-Beato M. New drugs and targeted treatments in Hodgkin's lymphoma. Cancer Treatment Reviews 2014;40:457–464.
[105] de Claro RA, McGinn K, Kwitkowski V, et al. U.S. Food and Drug Administration approval summary: brentuximab vedotin for the treatment of relapsed Hodgkin lymphoma or relapsed systemic anaplastic large-cell lymphoma. Clin Cancer Res 2012;18:5845–5849.
[106] Monjanel H, Deville L, Ram-Wolff C, et al. Brentuximab vedotin in heavily treated Hodgkin and anaplastic large-cell lymphoma, a single centre study on 45 patients. British Journal of Haematology 2014;166:306–308.
[107] Shi M, Roemer MG, Chapuy B, et al. Expression of programmed cell death 1 ligand 2 (PD-L2) is a distinguishing feature of primary mediastinal (thymic) large B-cell lymphoma and associated with PDCD1LG2 copy gain. Am J Surg Pathol 2014;38:1715–1723.
[108] Naidoo J, Page DB, Wolchok JD. Immune modulation for cancer therapy. Br J Cancer 2014;111:2214–2219.
[109] Topalian SL, Hodi FS, Brahmer JR, et al. Safety, activity, and immune correlates of anti-PD-1 antibody in cancer. N Engl J Med 2012;366:2443–2454.
[110] Brahmer JR, Tykodi SS, Chow LQ, et al. Safety and activity of anti-PD-L1 antibody in patients with advanced cancer. N Engl J Med 2012;366:2455–2465.
[111] Armand P, Nagler A, Weller EA, et al. Disabling immune tolerance by programmed death-1 blockade with pidilizumab after autologous hematopoietic stem-cell transplantation for diffuse large B-cell lymphoma: results of an international phase II trial. J Clin Oncol 2013;31:4199–4206.
[112] Ansell SM, Lesokhin AM, Borrello I, et al. PD-1 blockade with nivolumab in relapsed or refractory Hodgkin's lymphoma. N Engl J Med 2015;372:311–319.
[113] Hao Y, Chapuy B, Monti S, Sun HH, Rodig SJ, Shipp MA. Selective JAK2 inhibition specifically decreases Hodgkin lymphoma and mediastinal large B-cell lymphoma growth in vitro and in vivo. Clin Cancer Res 2014;20:2674–2683.
[114] Younes A, Romaguera J, Fanale M, et al. Phase I study of a novel oral Janus kinase 2 inhibitor, SB1518, in patients with relapsed lymphoma: evidence of clinical and biologic activity in multiple lymphoma subtypes. J Clin Oncol 2012;30:4161-4167.
[115] Kochenderfer JN, Dudley ME, Kassim SH, et al. Chemotherapy-refractory diffuse large B-cell lymphoma and indolent B-cell malignancies can be effectively treated with autologous T cells expressing an anti-CD19 chimeric antigen receptor. J Clin Oncol 2015;33:540–549.

James L. Rubenstein

10 Molecular pathogenesis of primary central nervous system lymphoma

10.1 Introduction

A significant body of evidence obtained via multiple investigators using different approaches demonstrates that the molecular pathogenesis of primary central nervous system lymphoma (PCNSL) is distinctive compared to systemic lymphomas of the large B-cell type. First, at least 90 % of non-Hodgkin's lymphoma (NHL) cases that present in the central nervous system (CNS) will, upon routine clinical staging, prove to be confined to the brain, leptomeninges, and intraocular structures and thus be classified as PCNSL. Over the natural history of the disease it is rare for PCNSL to manifest outside of the brain, highlighting this unique CNS tropism. In addition, PCNSL tumors are associated with an inferior prognosis compared to other localized extranodal NHL tumors confined to a single extranodal site, such as bone. Finally, CNS lymphomas typically exhibit a heightened responsiveness to methotrexate-based therapy compared to systemic lymphomas of the same histology [1]. Given the reproducible evidence for distinct transcriptional features in PCNSL [2–5] as well as the fact that the disease requires therapeutic regimens that are distinct from its systemic counterpart, PCNSL is recognized as a unique subtype of large B-cell lymphoma by the WHO working group [6]. The goal of this chapter is to review the molecular properties of PCNSL and to highlight those that are distinct from systemic lymphomas of the large B-cell type.

10.2 Etiology

The best established risk factors for PCNSL include acquired and/or congenital immunodeficiency states. PCNSL is an AIDS-defining illness that is associated with a low CD4 cell count (< 50 cells/ml) and Epstein–Barr Virus (EBV) infection. In systemic AIDS-related lymphomas, EBV infection of the lymphoma clone has been shown to be predictive of secondary CNS involvement [7]. Congential immunodeficiency states such as severe-combined or common-variable immunodeficiency, Wiskott–Aldrich syndrome or ataxia-telangiectasia carry a ~ 4 % risk of PCNSL. Post-transplant lymphoproliferative disorder (PTLD) involving the brain develops in 1–2 % of renal transplant recipients and in 2–7 % recipients of cardiac, lung and liver transplants. CNS PTLD is associated with EBV with iatrogenic T-cell immunodeficiency induced by agents such as mycophenolate mofetil (Cell Cept) [8]. Among PCNSL patients without evidence of overt immunesuppression, EBV infection of the lymphoma is rare [9].

10.3 Clinical features and pathogenesis

Like glioblastoma, PCNSL is a highly infiltrative neoplasm that has been characterized as a "whole brain disease" (Fig. 10.1) [10]. For this reason, the radiographic appearance of the tumor often markedly underestimates disease burden and extent, and like malignant astrocytomas, PCNSL is not amenable to curative resection [10]. One of the archetypical histologic features of PCNSL is that of angiotropism or angiocentricity, i.e. the accumulation of lymphoma cells around blood vessels, a phenotype that likely contributes to disruption of the blood-brain barrier and enables radiographic detection of lesions via pathologic contrast enhancement. PCNSL is commonly diagnosed as a solitary mass, typically with vasogenic edema and mass effect. The frequency of multiple lesions is increased among immunesuppressed patients [11].

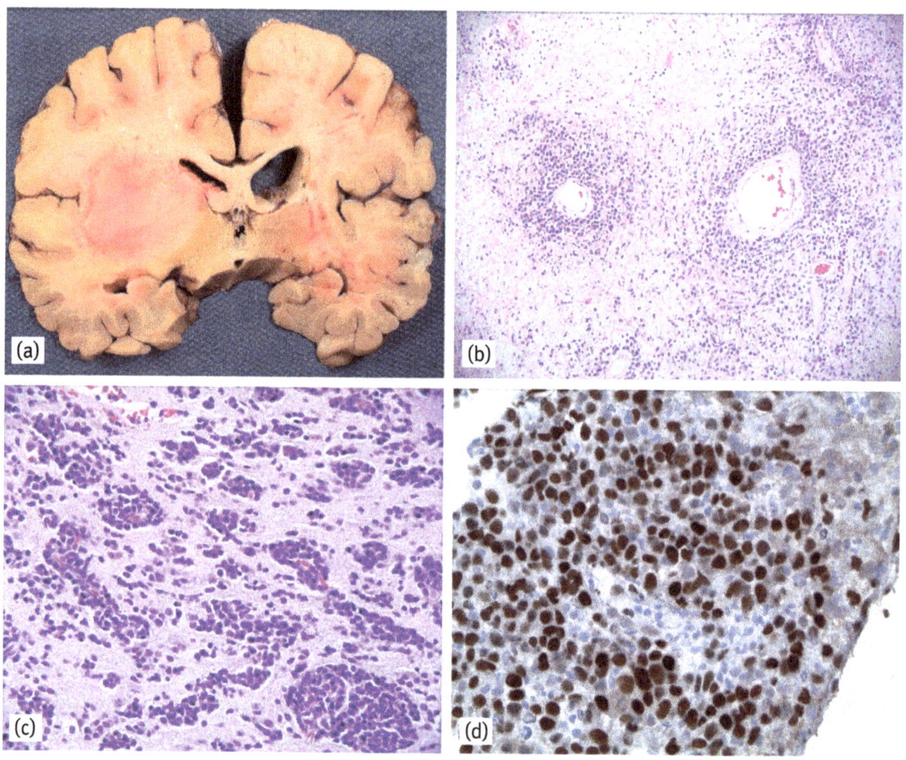

Fig. 10.1: **Pathologic features of PCNSL.** (a) Diffuse large B-cell lymphoma (DLBCL) involving the left parietal lobe and basal ganglia exhibits marked mass effect, subependymal spread and invasion of the lateral ventricle at relapse, upon progression with high-dose methotrexate and rituximab-based chemotherapy (courtesy of Ray Sobel, M.D., Stanford University School of Medicine). (b) DLBCL cells exhibiting angiotropic growth pattern in a diagnostic specimen of PCNSL (100 ×, H&E). (c) Invasive growth of DLBCL cells along cerebral vasculature in PCNSL (200 ×, H&E). (d) High expression of MYC by DLBCL cells in diagnostic specimen of PCNSL, as demonstrated by immunohistochemistry (400 ×) (courtesy of Eric Hsi, M.D., Cleveland Clinic). (Blood Journal permission pending.)

While CNS lymphomas that present in the brain are typically classified as PCNSL, there is reproducible evidence that subclinical tumor-related clones are detectable in the blood and bone marrow of PCNSL patients, suggesting that the CNS microenvironment might promote a more aggressive phenotype, mandating immediate clinical intervention [12, 13].

Intraocular disease is a common manifestation of PCNSL: approximately twenty percent of PCNSL patients present with involvement of the retina, vitreous, and uvea. An important clinical principle is that apparently localized intraocular lymphoma (IOL) will ultimately disseminate within the brain in greater than 80 % of cases, and therefore, detection of IOL mandates staging of the neuroaxis including cerebrospinal fluid (CSF) evaluation and brain MRI as well as therapies that address this risk for brain involvement [14].

Approximately 95 % of PCNSL are large B-cell lymphoma; other histologies that present as CNS lymphomas without systemic involvement include T-cell (2 %) [15], lymphoblastic, Burkitt, and marginal zone lymphomas. PCNSL is distinguished from dural-based marginal zone lymphomas as these have a distinct pathogenesis, typically do not disseminate within the brain parenchyma, and share overlapping radiographic features with meningioma on MRI [16].

10.4 Molecular pathogenesis of PCNSL

Determination of the unique genetic features of PCNSL poses a greater challenge than for systemic DLBCL, both because of the rarity of this neoplasm and because of the paucity of material available for investigational studies. Most specimens are obtained by stereotactic needle biopsy or via cytologic analysis of CSF. Despite these challenges, several investigations have focused on the elucidation of the features of PCNSL, large cell type. Between 50 % to 80 % of PCNSL tumors express BCL6 by immunohistochemistry [17] and at least 95 % stain positive for MUM-1; therefore the majority of PCNSL cases are of a late germinal center B-cell or an activated B-cell immunophenotype [9]. Immunohistochemical characterization of tumors from patients that participated in the recent CALGB 50202 trial demonstrated that high BCL6 expression by PCNSL tumors correlated with refractory disease and shorter progression-free and overall survival [20], thus representing a potentially useful molecular prognostic biomarker (Fig. 10.1). The adverse prognostic significance of high BCL6 in PCNSL was recently confirmed in another prospective trial, by Korfel and colleagues [18]. Between 56–93 % of PCNSL express BCL2 [9, 17].

Transcriptional profile studies of PCNSL identified several potential mediators of disease pathogenesis, including high Pim 1 and 2 expression as well as high MYC [3, 19]. Increased MYC in PCNSL was later confirmed in the recent CALGB 50202 study (Fig. 10.1) [20]. Selective upregulation of microRNAs (miRNAs) associated with the MYC pathway (miR-17-5p, miR-20a, miR-9) was also demonstrated in an analysis comparing miRNA expression profiles of PCNSL vs. nodal diffuse large B -ell lymphoma (DLBCL) [21]. Circulating extracellular microRNAs in the CSF such as miR-21, miR-19, and miR-92a have also been described in PCNSL, raising the possibility of their use as clinical biomarkers [22].

Because an abnormal functioning of the immunoglobulin variable (IgV) region genes-associated somatic hypermutation mechanism (also known as aberrant somatic hypermutation) has an established role in the pathogenesis of DLBCL, Montesinos-Rongen et al. pursued the potential contribution of this mechanism in the development of PCNSL. These investigators detected a high relative mutational frequency for Ig heavy chain V region (*IGHV)* genes in PCNSL. Moreover, it was determined that PCNSL tumors commonly displayed aberrant somatic hypermutation involving four proto-oncogenes – *PAX5*, *PIM1*, *c-MYC*, and *RhoH/TTF*, genes that have an established role in B-cell development and differentiation as well as in the regulation of proliferation and apoptosis [23]. The mean mutation frequencies were 2- to 5-fold higher in PCNSL compared to extraneural DLBCL [23]. Such high mutation frequencies may arise during a prolonged interaction of the tumor cell (or its precursor) in the GC microenvironment [24]. A recent whole exome sequencing effort on PCNSL identified mutations in several protein-coding genes, including *CARD11*, *CD79B* and *CD79A*, *TLR2*, *TLR6*, and *TLR10*, as well as genes involved in cell cycle regulation (*CCND3*, *CDK18* and *CDK20*) and in chromatin structure/histone modifications, including *CREBBP*, *KMT2D*, *ARID1A*, *ARID1B*, *SMARCA4*, *SMARCA5* and *SMARCC1* [25].

The $p16^{\mathrm{INK4a}}$ gene is frequently inactivated by either homozygous deletion (40–50 %) or 5'-CpG hypermethylation (15–30 %) in PCNSL tumors [26]. Inactivation of the *CDKN2A/B* genes, encoding for $\mathrm{p14^{ARF}}$ and $\mathrm{p16^{INK4a}}$, by either homozygous deletion or promoter hypermethylation is likely an important step in the molecular pathogenesis of PCNSL. The $\mathrm{p14^{ARF}}$ protein normally induces growth arrest as well as stabilizes p53 protein in the cell nucleus. Its deletion has been reported in glioblastomas and systemic NHL. The *CDKN2A* and *CDKN2B* genes are often co-deleted in human neoplasms; moreover, mice lacking the murine homologue of $p14^{\mathrm{ARF}}$ develop a variety of tumors, including lymphomas, gliomas and sarcomas [14, 27–29]. In contrast, mutations in the *TP53* gene have been observed in only a small proportion of PCNSL specimens.

Comparative genomic hybridization has also been used to identify other potentially significant genetic lesions in PCNSL. Gains have been detected on chromosome 12 as well as on the long arms of chromosomes 1, 7, and 18. Gain on chromosome 12 appears to be the most common chromosomal alteration, specifically in the 12q region harboring STAT6, MDM2, CDK4 and GLI1 [14, 28, 30].

Reduced DNA copy number was detected on chromosome 6 and the short arms of chromosomes 17 and 18. Common genomic aberrations in PCNSL include losses on chromosome 6p21 that harbor the *HLA* loci [31–33], as well as broad deletions involving chromosome 6q. Chromosome 6q deletions, in particular 6q21-23, a locus containing the putative tumor suppressor gene *PTPRK*, may be most frequent and occur in between 40 % and 60 % of PCNSL tumors [34]. Candidate tumor suppressor genes linked to deleted loci on chromosome 6q include *PRDM1*, a tumor suppressor and regulator of B-cell differentiation [35], *PTPRK*, a protein tyrosine phosphatase that participates in cell adhesion signaling events [36], and *TNFAIP3* (*A20*), a regulator of NF-κB signaling [37].

Cady and colleagues from the Mayo Clinic used interphase fluorescence in situ hybridization (FISH) to evaluate four candidate genomic loci that they hypothesized to have prognostic significance in PCNSL tumors: 6q22-23, as well as translocations involving BCL6, MYC and IgH. PCNSL tumors isolated from 75 immunocompetent patients treated at a single institution were analyzed. Del 6q22 was noted to be the most prevalent genomic aberration, occurring in 45 % of tumors; BCL6 translocations were the next most common, detected in 17 % of tumors. The presence of del(6)(q22) and/or a BCL6 translocation was associated with inferior overall survival (p=0.0097). Notably, del (6)(q22) may be more common in PCNSL (45 %) than in systemic DLBCL (25 %) and IGH translocations are less common in PCNSL (13 %) than in systemic DLBCL (45–51 %) [31].

Further evidence for the aberrant activation of the NF-κB pathway is supported by recurrent gains in DNA copy number for *MALT* [38] as well as activating mutations of *CARD11* [39] and *MYD88*. The activating exchange of leucine to proline at position 265 of *MYD88* may be enriched in PCNSL and has been demonstrated to occur in 38 % to 50 % of cases [40, 41].

Several lines of investigation support a role for the JAK/STAT signaling pathway as a mediator of pro-survival signals in PCNSL. Interleukin-4, a B-cell growth factor that mediates intracellular signals via JAK/STAT, is upregulated at the transcript and protein level within the vascular microenvironment in PCNSL tumors [3]. Increased concentration of IL-10 (another first messenger in the JAK/STAT signaling) is upregulated in the vitreous and CSF in PCNSL and, in independent studies, correlated with adverse prognosis [42, 43]. Finally, intratumoural JAK1 transcripts are upregulated and JAK1 activation in PCNSL has been demonstrated [3, 44]. Elevated IL-10 expression plus activation of JAK/STAT signaling are consistent with aberrant activation of the MyD88 pathway in PCNSL [45] (Fig. 10.2).

In addition, *CD79B*, a component of the B-cell receptor signaling pathway, is mutated in approximately 20 % of cases, providing evidence that dysregulation of the B-cell receptor and NF-κB contribute to the pathogenesis of PCNSL [46]. Silencing of the cell cycle regulator *CDKN2A* occurs in 50 % of CNS lymphomas and is linked to inferior outcome [38, 40].

A variety of evidence demonstrates upregulation of translational regulatory proteins EIF4A, EIF4E and EIF5 in PCNSL [3]. Notably, there is recent preclinical evidence that pharmacologic antagonism of translational initiation targeting EIF4E may be effective, supporting further analysis of this potential strategy in PCNSL [3, 47].

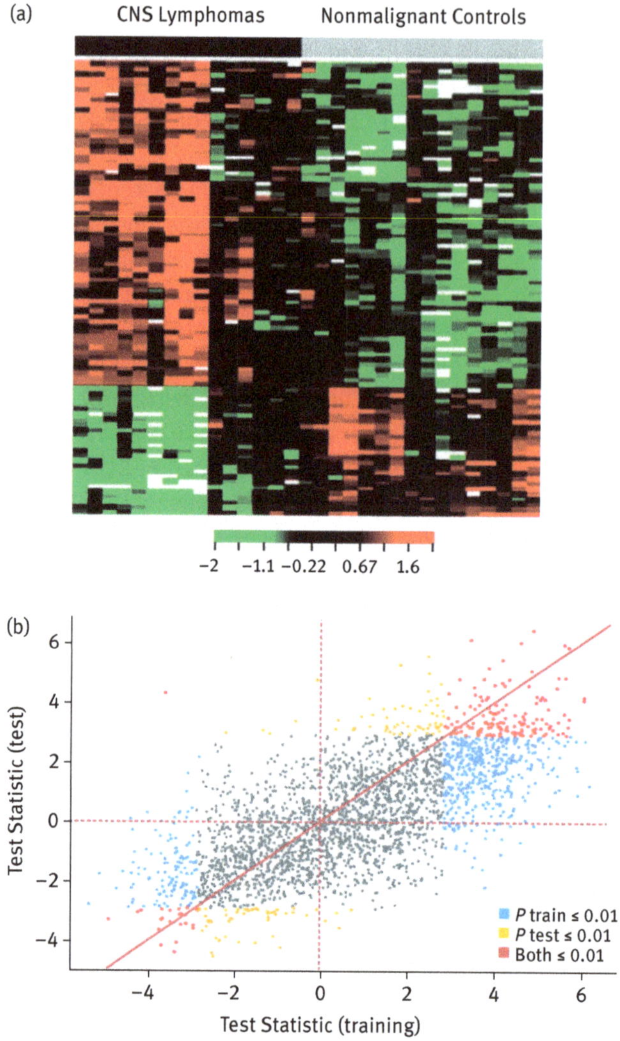

Fig. 10.2: (a) Differential expression of CSF proteins that distinguish CNS lymphoma from patients without CNS malignancy as defined by two-dimensional liquid chromatography combined with mass spectrometry, using training and validation sets of patients. Values are shown as a color scale with positive values represented by red, negative represented by green, and black corresponding to values close to zero. Missing values are shown in white. (b) The concordance between training and test set was exceptional, as shown by comparing the test statistics from the two data sets. (J. Clin Oncol. permission pending.)

10.5 Role of the tumor microenvironment in PCNSL

The molecular basis for tropism and selective dissemination of lymphoma within the brain are biological questions that are fundamental to the pathogenesis of PCNSL. Expression of chemokines CXCL12 (SDF-1) and CXCL-13 (B-lymphocyte chemoattractant) within PCNSL tumors has been confirmed by several investigators [48–50], and chemotactic responsiveness to these molecules by CNS lymphoma tumor cells was recently demonstrated, providing evidence for their role as neurotropic factors. Moreover, high CXCL-13 concentration in the CSF of affected patients correlates with adverse prognosis, supporting its role as a pro-survival factor in PCNSL. In addition, the determination of the CSF concentration of CXCL-13, as well as of IL-10, may be useful in facilitating diagnosis of CNS lymphoma; bivariate upregulated expression of each molecule in CSF has a diagnostic sensitivity at least two-fold greater than cytology or flow-cytometry. In a multicenter investigation that involved discovery and validation cohorts, the positive predictive value of bivariate elevation of IL-10 plus CXCL-13 in the CSF was 95 % in the identification of newly diagnosed PCNSL [51].

Notably, the expression of the B-cell chemokines CXCL13 and SDF-1 by the retinal pigment epithelium has also been demonstrated in primary intraocular lymphomas [52]. Ectopic expression of these chemokines within the intraocular compartment likely contributes to lymphoma homing to the retinal pigment epithelium from the choroidal circulation [53].

While under physiologic conditions, the brain is assumed to be an immunologically quiescent or "privileged site," histopathologic studies of diagnostic specimens of PCNSL often reveals a robust inflammatory response, with infiltrating reactive T cells and activated macrophages. Of note, reactive, perivascular T-cell infiltrates in PCNSL may be predictive of favorable outcome, supporting the development of immunotherapies that potentiate T-cell-mediated immune surveillance [54]. A significant proportion of infiltrating myeloid cells exhibit markers of alternative activation, consistent with Th2 responses via IL-4 and IL-10 signaling, thus potentially promoting tumor progression and immune suppression. On the other hand, it has been demonstrated that infiltrating myeloid cells in CNS lymphoma express components of the complement cascade, thus providing key components of the innate immune response that may contribute to the efficacy of the therapeutic anti-CD20 antibody rituximab [55].

A major proportion of genes that reproducibly distinguished PCNSL and systemic DLBCL in transcriptional studies are components of the extracellular matrix. For example, osteopontin and chitinase-3-like 1 were reproducibly noted to be expressed at higher levels in PCNSL [56]. High protein expression of osteopontin and chitinase 3-like 1 in PCNSL were confirmed by proteomic analysis using 2-D- LC/MS (Fig. 10.2) [42]. By contrast, collagen type VI, laminin alpha-4 and lumican, a keratin sulfate proteoglycan, were noted to be consistently expressed at higher levels in systemic DLBCL. The differential expression of collagen subtypes between CNS and systemic DLBCL may impact the apoptotic threshold to specific types of genotoxic stress. For example, the expression of extracellular matrix proteins in the "lymph node signature" has been associated with outcome in systemic large B-cell lymphoma [57]. Also, increased expression of collagen type VI may modulate chemotherapy drug resistance [58]. It is plausible that the differential components of the extracellular matrix in CNS vs. non-CNS lymphoma may contribute to the differential sensitivity of these tumors to genotoxic agents such as high-dose methotrexae.

10.6 Concluding remarks

A variety of evidence demonstrates that PCNSL has properties of activated B-cell lymphomas; however, immunophenotyping, transcriptional profiling and genetic studies have identified significant differences that may contribute to the unique clinical behavior of this tumor. The majority of PCNSL are BCL6+MUM1+, consistent with an activated germinal center phenotype; moreover, there is reproducible evidence based upon prospective clinical trial data from two recent studies that, unlike systemic DLBCL, high BCL6 expression in PCNSL is predictive of adverse outcome. On the other hand, PCNSLs harbor multiple mutations that lead to aberrant NF-κB activation, via activating mutations of *MYD88* as well as deletions on chromosome 6q including the *TNFAIP3* gene, a negative regulator of NF-κB.

While there has been significant improvement in survival for patients with PCNSL, survival rates will certainly plateau with the current armamentarium of therapeutic agents. Given that PCNSL impacts an older population, a significant proportion of patients are not eligible for intensive therapies such as whole brain radiation or high-dose chemotherapy. There is therefore a need for the development of novel agents that target key survival pathways in order to continue to make progress in this disease [59, 60]. Potential targets include the B-cell receptor signaling pathway, Toll-like receptor signaling pathway, survival signals mediated by MYC, Pim kinases and NF-kB, as well as immune suppressive signals from the CNS microenvironment. Key molecular and physiologic differences that contribute to the challenge of treating CNS lymphoma include the blood-brain barrier as well as distinct components of the tumor extracellular matrix. It is hoped that continued progress in the understanding of the molecular pathogenesis of PCNSL will provide insights toward the development of new therapeutic strategies that are necessary to make further progress in this disease.

Acknowledgements

Supported by the National Institutes of Health, R01CA139-83-01A1, and by the Leukemia & Lymphoma Society (JLR).

References

[1] Bokstein F, Lossos A, Lossos IS, Siegal T. Central nervous system relapse of systemic non-Hodgkin's lymphoma: results of treatment based on high-dose methotrexate combination chemotherapy. Leuk Lymphoma 2002;43:587–593.

[2] Jordanova ES, Riemersma SA, Philippo K, Giphart-Gassler M, Schuuring E, Kluin PM. Hemizygous deletions in the HLA region account for loss of heterozygosity in the majority of diffuse large B-cell lymphomas of the testis and the central nervous system. Genes Chromosomes Cancer 2002;35:38–48.

[3] Rubenstein JL, Fridlyand J, Shen A, et al. Gene expression and angiotropism in primary CNS lymphoma. Blood 2006;107:3716–3723.

[4] Tun HW, Personett D, Baskerville KA, et al. Pathway analysis of primary central nervous system lymphoma. Blood 2008;111:3200–3210.

[5] Booman M, Szuhai K, Rosenwald A, et al. Genomic alterations and gene expression in primary diffuse large B-cell lymphomas of immune-privileged sites: the importance of apoptosis and immunomodulatory pathways. J Pathol 2008;216:209–217.

[6] Campo E, Swerdlow SH, Harris NL, Pileri S, Stein H, Jaffe ES. The 2008 WHO classification of lymphoid neoplasms and beyond: evolving concepts and practical applications. Blood 2011;117:5019–5032.

[7] Cingolani A, Gastaldi R, Fassone L, et al. Epstein–Barr virus infection is predictive of CNS involvement in systemic AIDS-related non-Hodgkin's lymphomas. J Clin Oncol 2000;18:3325–3330.

[8] Schabet M. Epidemiology of primary CNS lymphoma. J Neurooncol 1999;43:199–201.

[9] Camilleri-Broet S, Criniere E, Broet P, et al. A uniform activated B-cell-like immunophenotype might explain the poor prognosis of primary central nervous system lymphomas: analysis of 83 cases. Blood 2006;107:190–196.

[10] Lai R, Rosenblum MK, DeAngelis LM. Primary CNS lymphoma: a whole-brain disease? Neurology 2002;59:1557–1562.

[11] Fine HA, Mayer RJ. Primary central nervous system lymphoma. Ann Intern Med 1993;119:1093–1104.

[12] McCann KJ, Ashton-Key M, Smith K, Stevenson FK, Ottensmeier CH. Primary central nervous system lymphoma: tumor-related clones exist in the blood and bone marrow with evidence for separate development. Blood 2009;113:4677–4680.

[13] Jahnke K, Hummel M, Korfel A, et al. Detection of subclinical systemic disease in primary CNS lymphoma by polymerase chain reaction of the rearranged immunoglobulin heavy-chain genes. J Clin Oncol 2006;24:4754–4757.

[14] Rubenstein JL, Treseler P, O'Brien JM. Pathology and genetics of primary central nervous system and intraocular lymphoma. Hematol Oncol Clin North Am 2005;19:705–717, vii.

[15] Shenkier TN, Blay JY, O'Neill BP, et al. Primary CNS lymphoma of T-cell origin: a descriptive analysis from the international primary CNS lymphoma collaborative group. J Clin Oncol 2005;23:2233–2239.

[16] Tu PH, Giannini C, Judkins AR, et al. Clinicopathologic and genetic profile of intracranial marginal zone lymphoma: a primary low-grade CNS lymphoma that mimics meningioma. J Clin Oncol 2005;23:5718–5727.

[17] Braaten KM, Betensky RA, de Leval L, et al. BCL-6 expression predicts improved survival in patients with primary central nervous system lymphoma. Clin Cancer Res 2003;9:1063–1069.

[18] Kreher S, Johrens K, Strehlow F, et al. Prognostic impact of B-cell lymphoma 6 in primary CNS lymphoma. Neuro Oncol 2015.

[19] Rubenstein JL, Fridlyand J, Abrey L, et al. Phase I study of intraventricular administration of rituximab in patients with recurrent CNS and intraocular lymphoma. J Clin Oncol 2007;25:1350–1356.

[20] Rubenstein JL, Hsi ED, Johnson JL, et al. Intensive Chemotherapy and Immunotherapy in Patients With Newly Diagnosed Primary CNS Lymphoma: CALGB 50202 (Alliance 50202). J Clin Oncol 2013;31:3061–3068.

[21] Fischer L, Hummel M, Korfel A, Lenze D, Joehrens K, Thiel E. Differential micro-RNA expression in primary CNS and nodal diffuse large B-cell lymphomas. Neuro Oncol 2011;13:1090–1098.

[22] Baraniskin A, Kuhnhenn J, Schlegel U, et al. Identification of microRNAs in the cerebrospinal fluid as marker for primary diffuse large B-cell lymphoma of the central nervous system. Blood 2011;117:3140–3146.

[23] Montesinos-Rongen M, Roost DV, Schaller C, Wiestler OD, Deckert M. Primary diffuse large B-cell lymphomas of the central nervous system are targeted by aberrant somatic hypermutation. Blood 2004;103:1869–1875.

[24] Montesinos-Rongen M, Kuppers R, Schluter D, et al. Primary central nervous system lymphomas are derived from germinal-center B cells and show a preferential usage of the V4-34 gene segment. Am J Pathol 1999;155:2077–2086.

[25] Vater I, Montesinos-Rongen M, Schlesner M, et al. The mutational pattern of primary lymphoma of the central nervous system determined by whole-exome sequencing. Leukemia 2015;29:677–685.

[26] Nakamura M, Sakaki T, Hashimoto H, et al. Frequent alterations of the p14(ARF) and p16(INK4a) genes in primary central nervous system lymphomas. Cancer Res 2001;61:6335–6339.

[27] Kamijo T, Bodner S, van de Kamp E, Randle DH, Sherr CJ. Tumor spectrum in ARF-deficient mice. Cancer Res 1999;59:2217–2222.

[28] Nakamura M, Shimada K, Ishida E, Konishi N. Histopathology, pathogenesis and molecular genetics in primary central nervous system lymphomas. Histol Histopathol 2004;19:211–219.

[29] Mrugala MM, Rubenstein JL, Ponzoni M, Batchelor TT. Insights into the biology of primary central nervous system lymphoma. Curr Oncol Rep 2009;11:73–80.

[30] Harada K, Nishizaki T, Kubota H, Harada K, Suzuki M, Sasaki K. Distinct primary central nervous system lymphoma defined by comparative genomic hybridization and laser scanning cytometry. Cancer Genet Cytogenet 2001;125:147–150.

[31] Cady FM, O'Neill BP, Law ME, et al. Del(6)(q22) and BCL6 rearrangements in primary CNS lymphoma are indicators of an aggressive clinical course. J Clin Oncol 2008;26:4814–4819.

[32] Harada K, Nishizaki T, Kubota H, Suzuki M, Sasaki K. Distinct primary central nervous system lymphoma defined by comparative genomic hybridization and laser scanning cytometry. Cancer Genet Cytogenet 2001;125:147–150.

[33] Boonstra R, Koning A, Mastik M, van den Berg A, Poppema S. Analysis of chromosomal copy number changes and oncoprotein expression in primary central nervous system lymphomas: frequent loss of chromosome arm 6q. Virchows Arch 2003;443:164–169.

[34] Weber T, Weber RG, Kaulich K, et al. Characteristic chromosomal imbalances in primary central nervous system lymphomas of the diffuse large B-cell type. Brain Pathol 2000;10:73–84.

[35] Courts C, Montesinos-Rongen M, Brunn A, et al. Recurrent inactivation of the PRDM1 gene in primary central nervous system lymphoma. J Neuropathol Exp Neurol 2008;67:720–727.

[36] Nakamura M, Kishi M, Sakaki T, et al. Novel tumor suppressor loci on 6q22-23 in primary central nervous system lymphomas. Cancer Res 2003;63:737–741.

[37] Braggio E, McPhail ER, Macon W, et al. Primary central nervous system lymphomas: a validation study of array-based comparative genomic hybridization in formalin-fixed paraffin-embedded tumor specimens. Clin Cancer Res 2011;17:4245–4253.

[38] Schwindt H, Vater I, Kreuz M, et al. Chromosomal imbalances and partial uniparental disomies in primary central nervous system lymphoma. Leukemia 2009;23:1875–1884.

[39] Montesinos-Rongen M, Schmitz R, Brunn A, et al. Mutations of CARD11 but not TNFAIP3 may activate the NF-kappaB pathway in primary CNS lymphoma. Acta Neuropathol 2010;120:529–535.

[40] Gonzalez-Aguilar A, Idbaih A, Boisselier B, et al. Recurrent mutations of MYD88 and TBL1XR1 in primary central nervous system lymphomas. Clin Cancer Res 2012;18:5203–5211.

[41] Montesinos-Rongen M, Godlewska E, Brunn A, Wiestler OD, Siebert R, Deckert M. Activating L265P mutations of the MYD88 gene are common in primary central nervous system lymphoma. Acta Neuropathol 2011;122:791–792.

[42] Roy S, Josephson SA, Fridlyand J, et al. Protein biomarker identification in the CSF of patients with CNS lymphoma. J Clin Oncol 2008;26:96–105.

[43] Sasayama T, Nakamizo S, Nishihara M, et al. Cerebrospinal fluid interleukin-10 is a potentially useful biomarker in immunocompetent primary central nervous system lymphoma (PCNSL). Neuro Oncol 2012;14:368–380.

[44] Sung CO, Kim SC, Karnan S, et al. Genomic profiling combined with gene expression profiling in primary central nervous system lymphoma. Blood 2011;117:1291–1300.

[45] Ngo VN, Young RM, Schmitz R, et al. Oncogenically active MYD88 mutations in human lymphoma. Nature 2011;470:115–119.

[46] Montesinos-Rongen M, Schafer E, Siebert R, Deckert M. Genes regulating the B cell receptor pathway are recurrently mutated in primary central nervous system lymphoma. Acta Neuropathol 2012;124:905–906.

[47] Muta D, Makino K, Nakamura H, Yano S, Kudo M, Kuratsu J. Inhibition of eIF4E phosphorylation reduces cell growth and proliferation in primary central nervous system lymphoma cells. J Neurooncol 2011;101:33–39.

[48] Fischer L, Korfel A, Pfeiffer S, et al. CXCL13 and CXCL12 in central nervous system lymphoma patients. Clin Cancer Res 2009;15:5968–5973.

[49] Smith JR, Braziel RM, Paoletti S, Lipp M, Uguccioni M, Rosenbaum JT. Expression of B-cell-attracting chemokine 1 (CXCL13) by malignant lymphocytes and vascular endothelium in primary central nervous system lymphoma. Blood 2003;101:815–821.

[50] Smith JR, Falkenhagen KM, Coupland SE, Chipps TJ, Rosenbaum JT, Braziel RM. Malignant B cells from patients with primary central nervous system lymphoma express stromal cell-derived factor-1. Am J Clin Pathol 2007;127:633–641.

[51] Rubenstein JL, Wong VS, Kadoch C, et al. CXCL13 plus interleukin 10 is highly specific for the diagnosis of CNS lymphoma. Blood 2013;121:4740–4748.

[52] Chan CC, Shen D, Hackett JJ, Buggage RR, Tuaillon N. Expression of chemokine receptors, CXCR4 and CXCR5, and chemokines, BLC and SDF-1, in the eyes of patients with primary intraocular lymphoma. Ophthalmology 2003;110:421–426.

[53] Falkenhagen KM, Braziel RM, Fraunfelder FW, Smith JR. B-Cells in ocular adnexal lymphoproliferative lesions express B-cell attracting chemokine 1 (CXCL13). Am J Ophthalmol 2005;140:335–337.

[54] Ponzoni M, Berger F, Chassagne-Clement C, et al. Reactive perivascular T-cell infiltrate predicts survival in primary central nervous system B-cell lymphomas. Br J Haematol 2007;138:316–323.

[55] Kadoch C, Li J, Wong VS, et al. Complement activation and intraventricular rituximab distribution in recurrent central nervous system lymphoma. Clin Cancer Res 2014;20:1029–1041.

[56] Rubenstein JL, Shen A, Batchelor TT, Kadoch C, Treseler P, Shuman MA. Differential gene expression in central nervous system lymphoma. Blood 2009;113:266-7; author reply 7–8.
[57] Rosenwald A, Wright G, Chan WC, et al. The use of molecular profiling to predict survival after chemotherapy for diffuse large-B-cell lymphoma. N Engl J Med 2002;346:1937–1947.
[58] Sherman-Baust CA, Weeraratna AT, Rangel LB, et al. Remodeling of the extracellular matrix through overexpression of collagen VI contributes to cisplatin resistance in ovarian cancer cells. Cancer Cell 2003;3:377–386.
[59] Ponzoni M, Issa S, Batchelor TT, Rubenstein JL. Beyond high-dose methotrexate and brain radiotherapy: novel targets and agents for primary CNS lymphoma. Ann Oncol 2014;25:316–322.
[60] Rubenstein JL, Gupta NK, Mannis GN, Lamarre AK, Treseler P. How I treat CNS lymphomas. Blood 2013;122:2318–2330.

Davide Rossi and Gianluca Gaidano

11 Chronic lymphocytic leukemia

11.1 Introduction

Chronic lymphocytic leukemia (CLL) is an indolent B-cell malignancy characterized by the accumulation of mature CD5+/CD19+/CD23+ lymphocytes with weak surface expression of a monoclonal immunoglobulin (Ig) [1]. The clinical course of CLL patients ranges from a very indolent condition, with a nearly normal life expectancy, to rapidly progressive leading to early death. Asymptomatic patients are managed with watch and wait until the development of symptoms that result from cytopenia, adenopathy, or splenomegaly, as outlined by the 2008 International Workshop on CLL [2]. Although modern treatment strategies are highly effective in most cases of CLL, a challenging subgroup of patients show poor response to standard chemo-immunotherapy regimens. CLL is part of a spectrum of pathological conditions. At one end of the spectrum, CLL may be preceded by monoclonal B-cell lymphocytosis (MBL), a state in which a smaller size B-cell clone is present, typically in the absence of symptoms. At the other end of the spectrum, CLL may transform into an aggressive lymphoma, a complication known as Richter syndrome (RS) [3, 4].

11.2 Clinical background

An extensive array of effective therapeutic options is available when treatment is required, though the disease still remains incurable. Combination chemo-immunotherapy with fludarabine, cyclophosphamide and rituximab is the most effective regimen for young and fit patients, since it yields excellent long-term results. Immunotherapy-based approaches such as alemtuzumab and allogeneic stem cell transplantation, provide effective disease control in treatment-refractory patients. As CLL often affects elderly individuals, more tolerable therapeutic approaches have been successfully applied, such as chemo-immunotherapy combining chlorambucil with an anti-CD20 antibody (rituximab, obinutuzumab, ofatumumab), and bendamustine-based regimens [5]. Most recently, therapies targeting the B-cell receptor signaling pathway, such as ibrutinib and idelalisib, have shown high response rates and tolerability [6].

CLL pathogenesis might be viewed as cooperation between tumor and microenvironmental mechanisms, sometimes, though rarely, acting in the context of a constitutive genetic background that is predisposed to CLL development (Fig. 11.1). A decade of molecular acquisitions, including the significant contribution of recent genomic studies, has disclosed the genetics of CLL to a certain extent pointing to the deregulation of a series of molecular cellular programs in leukemia initiation, progression and

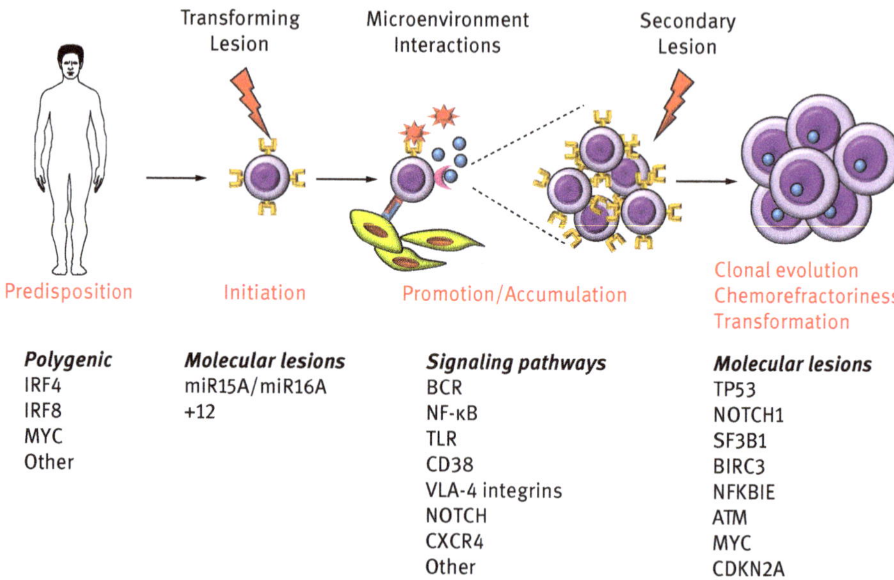

Fig. 11.1: **General overview of the molecular pathogenesis of CLL.** Molecular and biological mechanisms involved in CLL predisposition, initiation, promotion, and clonal evolution.

treatment resistance (Fig. 11.2). In addition to intrinsic genetic abnormalities involved in tumor clone initiation and promotion, CLL pathogenesis is also promoted by external factors, including contacts between CLL cells and the microenvironment through a series of surface receptors. Among these contacts, B-cell receptor (BCR) stimulation by antigens essentially acts as a spark that switches on tumor B cells translating extrinsic stimuli into signaling cascades that impact on tumor cell survival and proliferation [7].

Fig. 11.2: **Mutated genes in CLL.** Word clouds representing genes that are affected by mutations in CLL (COSMIC v71). The size of the font is proportional to the prevalence of gene lesions.

11.3 B-cell receptor signaling pathway

At variance with other B-cell tumors, genes encoding for components of the BCR signaling machinery are usually not targeted by somatic mutations in CLL [8–10]. Therefore, BCR pathway activation in CLL cells can be viewed as antigen-dependent, resulting from BCR ligation of auto-antigens and/or microbial antigens that are present in the microenvironment, or autonomous (antigen-independent). Several observations support the importance of the BCR in disease development and evolution of CLL [11].

The load of somatic hypermutation (SHM) in the rearranged immunoglobulin heavy variable (*IGHV*) genes, which is associated with antigen binding capacity and functions of the BCR, is also one of the most important prognostic biomarkers in CLL [12, 13]. Cases carrying unmutated *IGHV* genes (U-CLL) experience an aggressive clinical course with clonal evolution and resistance to therapy, which translates into a shorter survival compared with patients carrying mutated *IGHV* genes (M-CLL) [12, 13]. At the functional level, U-CLL cells display augmented BCR signaling compared with M-CLL cells, which may reflect functional differences related to the type of antigenic interaction through the BCR [14]. The BCR of U-CLL displays polyreactivity against auto-antigens present on apoptotic cells and against various bacterial and viral epitopes [15]. Such BCR polyreactivity of U-CLL significantly correlates with clinical aggressiveness, suggesting that a BCR capable of binding multiple epitopes may be more prone to sustained signaling. This observation provides a biologic interpretation for the poor prognosis that is typical of U-CLL. In CLL carrying mutated *IGHV* genes (M-CLL), the accumulation of SHM within the VH complementarity determining region (CDR) and the relative paucity of such mutations in the VH framework regions provide evidence of antigen selection. Although SHM across the VH domain can refine antigen-binding specificity and confers affinity maturation, in the case of autoreactive BCR, it may also serve as an editing mechanism by altering critical positions that might change antigen specificity. In line with this concept, nonpolyreactive M-CLL may acquire polyreactivity when reverted to the germline configuration [16]. At the functional level, M-CLL cases usually fail to respond to BCR stimulation, supporting the notion that they have undergone receptor desensitization as a consequence of chronic antigenic stimulation [15]. Failure to respond to BCR stimulation, because of lack of epitope binding capacity due to BCR editing or because of BCR signaling desensitization, provides a biologic interpretation for the good prognosis typical of M-CLL.

In ~ 30 % of CLL, specific *IGHV* genes partner with specific *IGHD-J* genes, and specific immunoglobulin light chain variable (*IGLV*) genes with specific *IGLJ* genes, leading to remarkably similar, "stereotyped" VH CDR3 and BCR. These data suggest that stereotyped BCR have been selected to bind a restricted set of antigenic epitopes and support the concept of antigen-driven expansion of CLL [17]. CLL subsets defined by specific stereotyped configurations of the BCR share common clinic-biologic characteristics. For instance, genetic aberrations are differentially distributed among stereotyped subsets, with *SF3B1* mutations and 11q22-23 deletion predominating in subset #2, and *NOTCH1* mutations and trisomy 12 in subset #8 [18, 19]. These findings imply that common genetic aberrations can be acquired and/or selected in the context of shared BCR signaling of a given subset. The unique immunogenetic profile of major stereotyped subsets may underlie a distinct clinical outcome. Indeed, subset #4 patients experience a particularly indolent disease course, whereas subsets #1, #2, and #8 are aggressive and enriched in cases requiring treatment, with the latter also displaying the highest risk for RS transformation among all CLL [17, 20].

Increasing evidence implicates also antigen-independent self-association of BCR as an additional key feature in CLL. The BCR of CLL may induce antigen-independent cell-autonomous signaling, which is dependent on the VH CDR3 structure, and, in particular, on the presence of a specific internal epitope produced by the configuration of this region. Indeed, in ~ 50 % of CLL patients, such self-recognizing epitopes located in the VH CDR3 region causes a tonic, constitutive and autonomous signal of the BCR [21].

The importance of the BCR in CLL prompted the development of therapeutic strategies based on the inhibition of key actionable targets in this signaling pathway. BTK and the PI3Kδ isoform are essential for BCR signaling (Fig. 11.3). BTK is a non-receptor tyrosine kinase that is rapidly activated upon BCR engagement. BTK activation results in PLCγ2 phosphorylation, which in turn triggers Ca^{2+}-mediated signaling, NF-κB upregulation, proliferation and tumor cell retention within the protecting and supporting microenvironment. PI3Kδ activation upon BCR signaling generates PIP3, which acts as a pivotal scaffold in the plasma membrane that recruits key components of the functional signaling complex downstream of the BCR including BTK, PLCγ2, and AKT. Consistent with the addiction of CLL cells to BCR, inhibition of BTK or PI3Kδ leads to reduced activation and proliferation of tumor cells [6]. Ibrutinib blocks BTK activity by irreversible binding of the C481 residue of the protein, while idelalisib is a potent and selective inhibitor of ATP binding at the PI3Kδ catalytic domain. Ibrutinib and idelalisib induce high rates of durable remissions in CLL, including patients with high-risk genetic lesions [22].

Among CLLs developing acquired resistance to ibrutinib, genomic studies have disclosed mutations of *BTK* that affect the C481 residue and interfere with the binding of the drug to the target, or gain of function mutations that activate the downstream PLCγ2 molecule to bypass BTK inhibition. *BTK* mutations occur in a minority (~ 8 %) of CLL treated with ibrutinib and confirm that BTK is a relevant pharmacologic target of the drug from a genetic perspective [23, 24]. Ibrutinib resistance mutations are biologically irrelevant in the absence of the selective pressure of the drug [23, 24]. Accordingly, subclones harboring drug resistance mutations are not detectable in CLL before ibrutinib treatment starts, conceivably because they are scattered within the tumor [25]. Indeed, according to a computational evolutionary model based on kinetic parameters of CLL patients treated with ibrutinib, targeted drug resistant mutants are predicted to exist in patients before initiation of targeted therapy, although they only

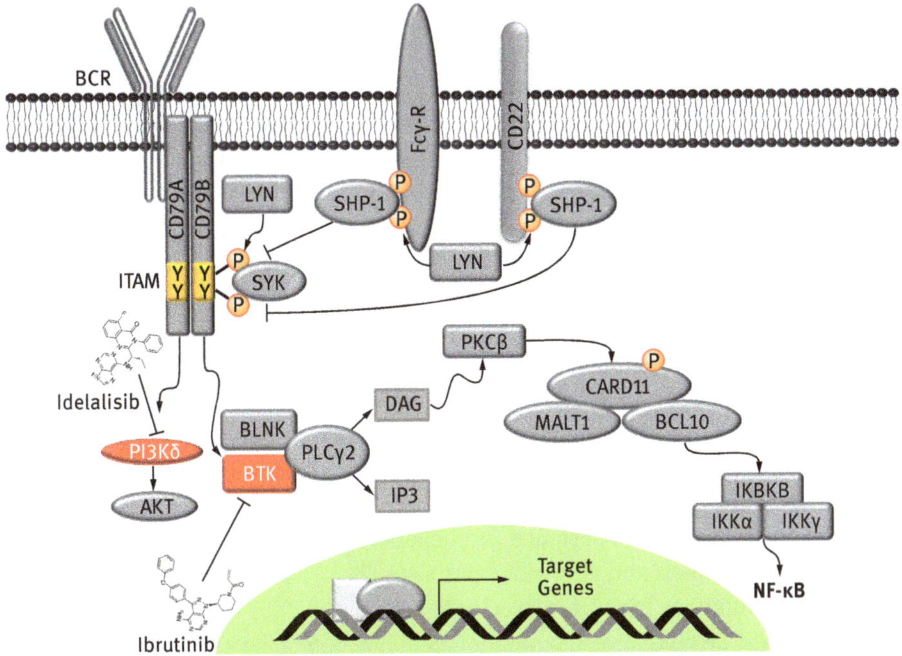

Fig. 11.3: **B-cell receptor (BCR) signaling pathway.** The BCR is composed of a membrane immunoglobulin bound to CD79A and CD79B. Antigen binding induces signaling initiation by the recruitment of SYK and LYN to the ITAM domains of CD79A/CD79B, and the subsequent activation of BTK, PI3Kδ and PLCγ2. BTK leads to phosphorylation of PLCγ2, which, in turns, promotes the production of DAG and IP3. IP3 activates PKCβ, leading to CARD11 phosphorylation, recruitment of MALT1 and BCL10 into the CBM multiprotein complex, activation of the IKBKB kinase, and NF-κB signaling initiation. Ibrutinib blocks BTK activity by irreversible binding of the C481 residue of the protein, while idelalisib is a potent and selective inhibitor of the PI3Kδ catalytic domain.

comprise a minority of the overall tumor burden, namely ~ 1 mutant cell out of 10,000–100,000 tumor cells [26].

11.4 Notch signaling pathway

The NOTCH receptor genes encode a family of heterodimeric transmembrane proteins (NOTCH1 to NOTCH4) that function as ligand-activated transcription factors. When the NOTCH receptors interact with their ligands through the extracellular subunit, two consecutive proteolytic cleavages of the NOTCH proteins are initiated and lead to pathway activation. Upon activation, the cleaved intracellular portion of the NOTCH receptors (ICN) translocates into the nucleus, where it forms a complex with the RBPJ transcription factor. In the absence of NOTCH signaling, RBPJ binds DNA in a sequence-specific manner and acts as a repressor of transcription. In the presence of

NOTCH signaling, displacement of co-repressors bound to RBPJ by the active ICN allows the recruitment of co-activators, such as MAML1, and histone acetyltransferases, such as EP300, to create an activation complex and to modify the expression of a number of target genes, including *MYC* and NF-κB signaling components. Throughout its transcriptional program, NOTCH plays an important role in a number of cellular functions, including cell fate specification, proliferation, and apoptosis [27]. The most prominent mechanism of NOTCH signal suppression is operated through its PEST [proline (P), glutamic acid (E), serine (S), and threonine (T) rich] domain. The PEST domain of ICN is recognized by the FBXW7 ubiquitin protein ligase that terminates NOTCH signaling by directing ICN towards proteasomal degradation [28].

In CLL, two genes of the NOTCH pathway are recurrently mutated, namely *NOTCH1* and *FBXW7*. *NOTCH1* mutations characterize ~ 10 % of unselected CLL and are mainly represented by frameshift or nonsense events clustering within exon 34, including the highly recurrent c.7544_7545delCT deletion (~ 80 % of all mutations) [8, 29] (Fig. 11.4). By taking advantage of this mutational spectrum, PCR based strategies not requiring DNA sequencing have been designed for the rapid detection of the c.7544_7545delCT mutation for diagnostic and prognostic purposes [29].

NOTCH1 mutations in CLL are selected to disrupt the PEST domain of the protein, resulting in NOTCH1 impaired degradation, stabilization of the active ICN, and deregulated NOTCH signaling. Consistent with this notion, a number of cellular pathways, including those controlling cell metabolism and cell cycle progression, are deregulated in CLL harboring *NOTCH1* mutations [30, 31]. *NOTCH1* is preferentially targeted in specific biological groups of CLL. In fact, *NOTCH1* mutations are significantly more common in CLL with unmutated *IGHV* gens, and are enriched in CLL harboring +12 [31, 32]. Mutations of *FBXW7* have been detected in ~ 2 % of unselected CLL and are predicted to cause loss of function by disrupting the WD40 domain of the protein. *FBXW7* mutations are likely to functionally mimic the PEST domain mutations of *NOTCH1*, thus resulting into constitutively active NOTCH signaling. Consistently, *FBXW7* mutations in CLL distribute in a mutually exclusive manner with *NOTCH1* mutations and, similar to *NOTCH1* mutations, cluster within cases harboring unmutated *IGHV* genes and +12 [10].

From a clinical perspective, the presence of *NOTCH1* mutations at the time of CLL presentation identifies a group of patients with intermediate-risk disease (~ 40 % of cases are alive at 10 years, accounting for a ~ 50 % reduction of the expected survival compared to the general population) and higher likelihood to transform into RS (cumulative incidence of transformation at 10 years of ~ 50 %) [29, 33, 34]. Among CLL requiring treatment, cases harboring *NOTCH1* mutations apparently do not benefit from the addition of an anti-CD20 monoclonal antibody to chemotherapy. Indeed, among CLL harboring *NOTCH1* mutations, treatment with fludarabine, cyclophosphamide, rituximab (FCR) does not result in the expected increase in minimal residual disease (MRD) response (46 % vs 50 %) nor in an improvement in progression free survival (PFS; median 34 months vs 33 months) or overall survival (OS; median 79 months vs

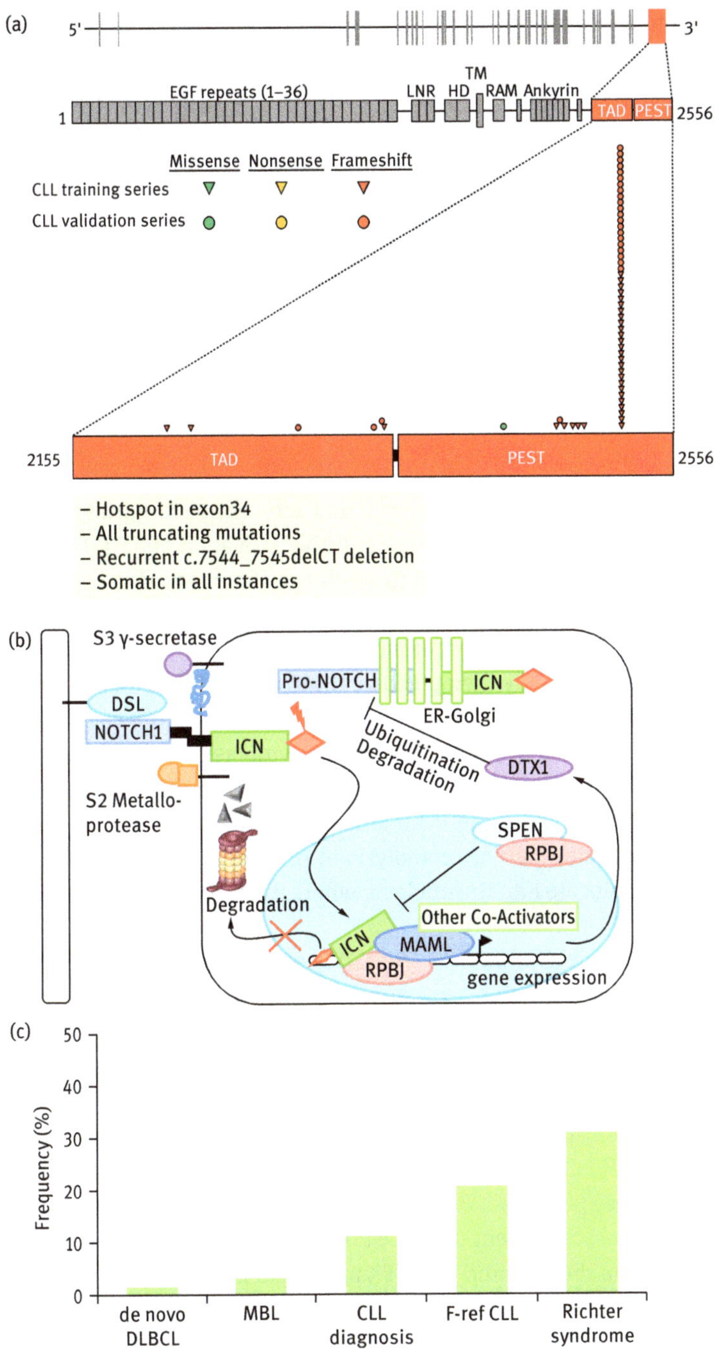

Fig. 11.4: ***NOTCH1* mutations in CLL.** Panel (a): Molecular spectrum of *NOTCH1* mutations in CLL. Panel (b): The NOTCH1 pathway. Panel (c): Prevalence of *NOTCH1* mutations in different clinical phases of the disease.

85 months) compared to treatment with the sole fludarabine, cyclophosphamide (FC) [35]. Similarly, among *NOTCH1* mutated CLLs, treatment with chlorambucil plus ofatumumab does not result in an improvement in PFS (median 17 months vs. 10 months) compared to treatment with the sole chlorambucil [36]. Although lower efficacy of therapeutic anti-CD20 monoclonal antibodies in *NOTCH1* mutated CLL seems to be consistent across two independent trials, the precise biological mechanisms underlying this clinical observation still remain to be clarified. Elucidation of the mechanistic basis would strengthen the rationale for guiding treatment based on *NOTCH1* mutation status. Furthermore, it will be interesting to see if a similar observation can be made with other anti-CD20 monoclonal antibodies.

11.5 Other signaling pathways

Malignant B cells manipulate signaling pathways that are central to normal B-cell homeostasis, many of which contribute to the constitutive activation of NF-κB [37]. In CLL, NF-κB signaling is generally upregulated through specific interactions between protective microenvironmental niches and CLL cells. At least in a fraction of cases, CLL cells gain active NF-κB signaling by mutating NF-κB genes.

The noncanonical NF-κB pathway is engaged by CD40 and BAFF receptors. Upon receptor binding, the TRAF3/MAP3K14-TRAF2/BIRC3 negative regulatory complex of noncanonical NF-κB signaling is disrupted, allowing the cytoplasmic release and stabilization of MAP3K14, the central activating kinase of noncanonical NF-κB signaling. The stabilized MAP3K14 activates the IKKα kinase, which in turn directly phosphorylates NF-κB/p100, inducing partial proteolysis of p100 to p52. The p52 protein dimerizes with RelB to translocate into the nucleus, where it regulates gene transcription [38].

The Baculoviral IAP repeat containing 3 (*BIRC3*) gene, which cooperates in the TRAF3/MAP3K14-TRAF2/BIRC3 negative regulatory complex of noncanonical NF-κB signaling, is mutated in ~ 3 % of unselected CLL [39]. At the biochemical level, *BIRC3* mutations cause the truncation of the C-terminal RING domain of the BIRC3 protein, whose E3 ubiquitin ligase activity is essential for switching off MAP3K14 through proteasomal degradation, thus leading to constitutive noncanonical NF-κB activation. From a clinical standpoint, *BIRC3* mutations identify a genetic subgroup of cases characterized by poor risk disease [39]. Also, *BIRC3* mutations confer complete resistance to ibrutinib in *in vitro* studies, consistent with the ability of mutant BIRC3 to activate NF-κB signaling downstream the BCR and in a BTK independent fashion [40].

NF-κB comprises a small family of transcription factors, including the NF-κB/Rel members RelA, RelB, c-Rel, NF-κB1, and NF-κB2. These proteins are kept inactive by cytoplasmic association with the IκB inhibitory proteins. The NF-κB inhibitor epsilon (NFKBIE) belongs to the IκB inhibitory protein family and counteracts NF-κB activation via cytoplasmic retention of the Rel proteins [38]. *NFKBIE* is affected by a recurrent 4bp deletion in ~5 % unselected CLL [41, 42]. *NFKBIE* mutation results in protein truncation, reduced inhibitory interaction with the Rel transcription factor, and enhanced NF-κB activation. Though the precise clinical implication of *NFKBIE* mutations remains to be clarified, their enrichment among CLLs presenting in advanced stage suggests that they might be involved in disease progression.

In B cells, Toll-like receptors (TLR) are central to the BCR-independent response to antigens by sensing a variety of pathogen-associated molecular patterns derived from bacteria, viruses, and fungi. Adaptor proteins, including the myeloid differentiation factor 88 (MYD88), are essential for initiating the TLR signaling. MYD88 has a modular structure with a death domain (DD) at the N terminus, and a Toll-IL-1 receptor (TIR) domain at the C terminus. The TIR domain of MYD88 is crucial for signal transduction since it mediates contacts with the intracellular TIR domains of the TLRs upon signaling activation. The DD domain allows oligomerization of the active MYD88 and its interaction with the respective DD of the serine-threonine kinases IRAK1-4, thus resulting in a multimeric complex. This complex propagates the signal and leads to activation of a series of cascades and transcription factors, such as NF-κB, AP-1 and STAT3 [43]. *MYD88* is mutated in ~3 % of unselected CLL, while they are enriched in a specific clinical subgroup of CLL characterized by young age at presentation, mutated *IGHV* genes and expected survival similar to that of the age-sex matched normal population [44]. Most *MYD88* mutations in CLL are represented by the L265P missense substitution, which affects the evolutionary conserved beta-beta loop of the TIR domain of MYD88, suggesting that it has been selected to change the structure of MYD88 to allow spontaneous homodimerization and recruitment of IRAK1 and IRAK4. Consistently, in B-cell tumors, mutant *MYD88* results in uncontrolled formation of the MYD88/IRAK complex, which translates into the recruitment of TRAF6, constitutive phosphorylation of TAK1 and, ultimately, the elevation of NF-κB activity and cytokine secretion [43].

11.6 Splicing

Splicing of precursor messenger RNA (pre-mRNA) and formation of mature mRNA through the removal of introns in protein-encoding genes is carried out in the nucleus by the spliceosome, a complex of five small nuclear ribonucleoproteins (snRNPs). These spliceosome components are required for normal constitutive and alternative splicing. Alternative splicing can generate numerous transcript variants for each gene, thus adding to genomic complexity and potentially contributing to tumorigenesis. *SF3B1* is a core component of the U2 snRNP, that recognizes the 3′ splice site at the intron-exon junctions and orchestrates the excision of introns from pre-mRNA to form mature mRNA. Structurally, the SF3B1 protein has two well-defined regions: (i) the N-terminal amino acid region, that contains several protein-binding motifs and functions as a scaffold to facilitate its interaction with other splicing factors such as U2AF65 and SF3B14; (ii) the C-terminal region, that contains 22 nonidentical tandem repeats of the HEAT motif that meander around the SF3b complex, enclosing SF3B14 [45].

SF3B1 mutations occur with a prevalence that ranges from 7 % to 10 % of unselected CLL, are enriched in cases harboring unmutated *IGHV* genes, and tend to co-occur with *ATM* deletion or mutation. *SF3B1* mutations in CLL are generally represented by missense nucleotide changes that recurrently target hotspots (codons 662, 666, 700, 704, 742), with a single amino-acid substitution (K700E) accounting for ~ 50 % of all *SF3B1* mutations [9, 10, 46] (Fig. 11.5).

The pathogenic role of *SF3B1* mutations is supported by their clustering in evolutionarily conserved hotspots within the inner surfaces of the HEAT structure at the supposed sites of interaction with RNA and cofactors. This observation suggests that mutation in *SF3B1* could possibly mediate alteration in its normal function through change in the physical interactions of the SF3B1 protein with its binding partners, thus leading to a defective spliceosome complex that is incapable of performing the correct splicing steps. Accordingly, transcriptome profiling shows increased usage of alternative 3' splice sites generating known and new alternative splice junctions in tumors harboring mutant *SF3B1*. These events induce lineage specific expression of alternative mRNAs that are translated into novel proteins or aberrant mRNAs that are decayed by cells. In particular, *SF3B1* mutant CLL carry multiple aberrantly spliced genes in pathways that are important for the malignant transformation of B cells, including the NF-κB, BCR signaling, and FOS/JUN transcriptional network [9, 10].

From a clinical perspective, the presence of *SF3B1* mutations at the time of CLL presentation identifies a group of patients with intermediate-risk disease (40 % of cases are alive at 10 years, accounting for a 48 % reduction of the expected survival compared to the general population) [33]. Among CLL patients requiring treatment, *SF3B1* mutations can potentially help refine prognostication of treatment relapse, though they do not represent a predictive biomarker for treatment tailoring. Indeed, in the UK Leukemia Research Found CLL4 trial [47], the *SF3B1* status did not impact on the chance of achieving response to chemotherapy, though patients harboring *SF3B1* mutations showed a shorter PFS (29 months vs 39 months) than *SF3B1* wild-type cases. Consistently, in the CLL8 trial of the German CLL Study Group [35], the *SF3B1* status did not impact on the chance of achieving clinical or MRD responses, but strongly affected PFS, that was shorter in patients harboring *SF3B1* mutations than in wild-type cases, independent of whether they received FC (28 months vs 34 months) or FCR (42 months vs 59 months).

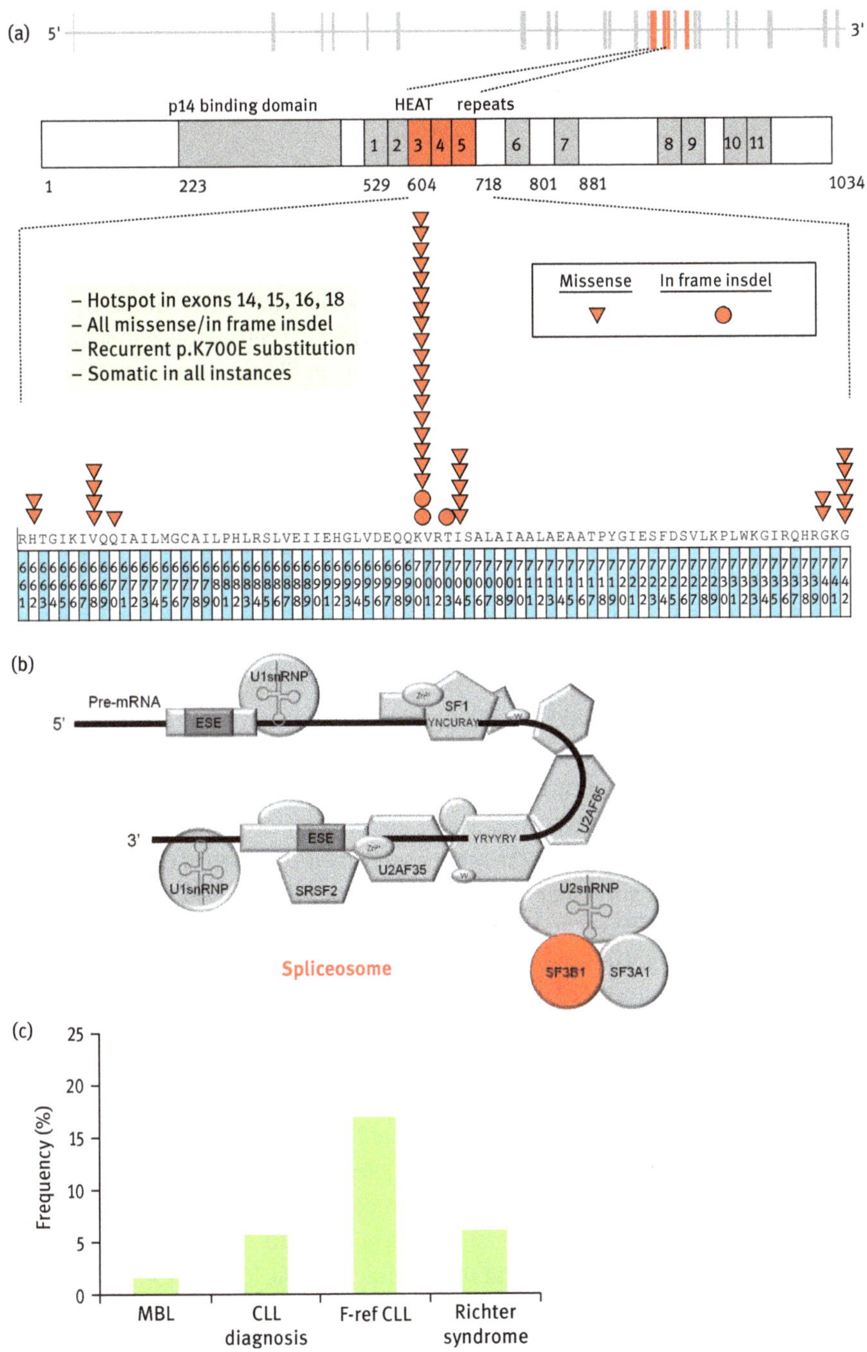

Fig. 11.5: ***SF3B1* mutations in CLL.** Panel (a): Molecular spectrum of *SF3B1* mutations in CLL. Panel (b): SF3B1 in the spliceosome machinery. Panel (c): Prevalence of *SF3B1* mutations in different clinical phases of the disease.

11.7 Apoptotic and cell cycle checkpoints

A number of cell cycle and apoptosis regulators are recurrently affected in CLL, including *miR15A* and mi*R16A*, *TP53*, ATM *and CDKN2A*.

Deletion of 13q14 is the most frequent genetic lesion of CLL, occurring in 50–60 % of cases [48]. The minimal deleted region on 13q14 contains the *miR15A* and *miR16A* microRNAs [49]. In normal cells, these two microRNAs appear to inhibit expression of multiple genes, including *BCL2*, the cyclins *CCND1* and *CCND3*, and cyclin-dependent kinase 6 (*CDK6*) [50, 51]. Deletion of *miR15A*/*miR16A* abrogates this inhibitory effect and favors the constitutive survival and cycling of tumor B cells. Mouse models have formally proven the pathogenetic role of 13q14 deletion in CLL development. However, lymphoproliferations observed in mice harboring 13q14 deletion are characterized by low penetrance and long latency, thus suggesting that the development of a full-blown phenotype in mice might require accumulation of additional genetic alterations [51].

In a relevant fraction (~ 25 %) of CLL patients, deletion of 13q14 occurs in the absence of any concomitant driver genetic lesion. Patients harboring solely 13q14 deletion have an excellent clinical outcome with a progression rate of less than 1 % per year and an expected survival only slightly lower than that of the general population [33]. *BCL2* is one of the genes that are upregulated in CLL as a consequence of *miR15A*/mi*R16A deletion.* Consistent with the central contribution of *BCL2* activation in the pathogenesis of CLL, selective inhibition of BCL2 through the BH3 mimetic ABT-199 molecule results in high response rates in relapsed or refractory patients, including those harboring high-risk genetic abnormalities [52].

The tumor suppressor gene *TP53* codes for a central regulator of the DNA-damage-response pathway, and its activation leads to cell-cycle arrest and DNA repair, apoptosis, or senescence through both transcription-dependent and transcriptional-independent activities. *TP53* may be disrupted in CLL by deletions, mutations or a combination of both (Fig. 11.6). The deletion of 17p13 always contains the *TP53* locus, and is found in 3–8 % unselected CLL at diagnosis and in 8–12 % CLL at the time of first line treatment [48, 53, 54]. On the other hand, specific subgroups of patients affected by fludarabine-refractory CLL or RS show a higher incidence of 17p13 deletion that may be detected in up to 30 %–40 % of cases [55, 56]. Mutations represent the most frequent form of *TP53* inactivation in CLL and are frequently (80–90 % of the cases) accompanied by the loss of the second allele through 17p13 deletion. At diagnosis, the incidence of *TP53* mutation has been reported to be 4–8 % [57–59]. As disease progresses, the incidence of *TP53* mutations raises to 10–12 % at the time of first line treatment [35, 60, 61], 40 % in fludarabine-refractory CLL [62], and 50–60 % in RS [56]. Overall, 95 % of mutations are localized within the central DNA-binding domain of TP53, impairing DNA binding and transactivation of target genes [63].

The clinical importance of *TP53* abnormalities in CLL is tightly linked to the poor prognosis marked by this genetic lesion and its close association with chemorefractoriness, as documented by a number of observational studies and prospective

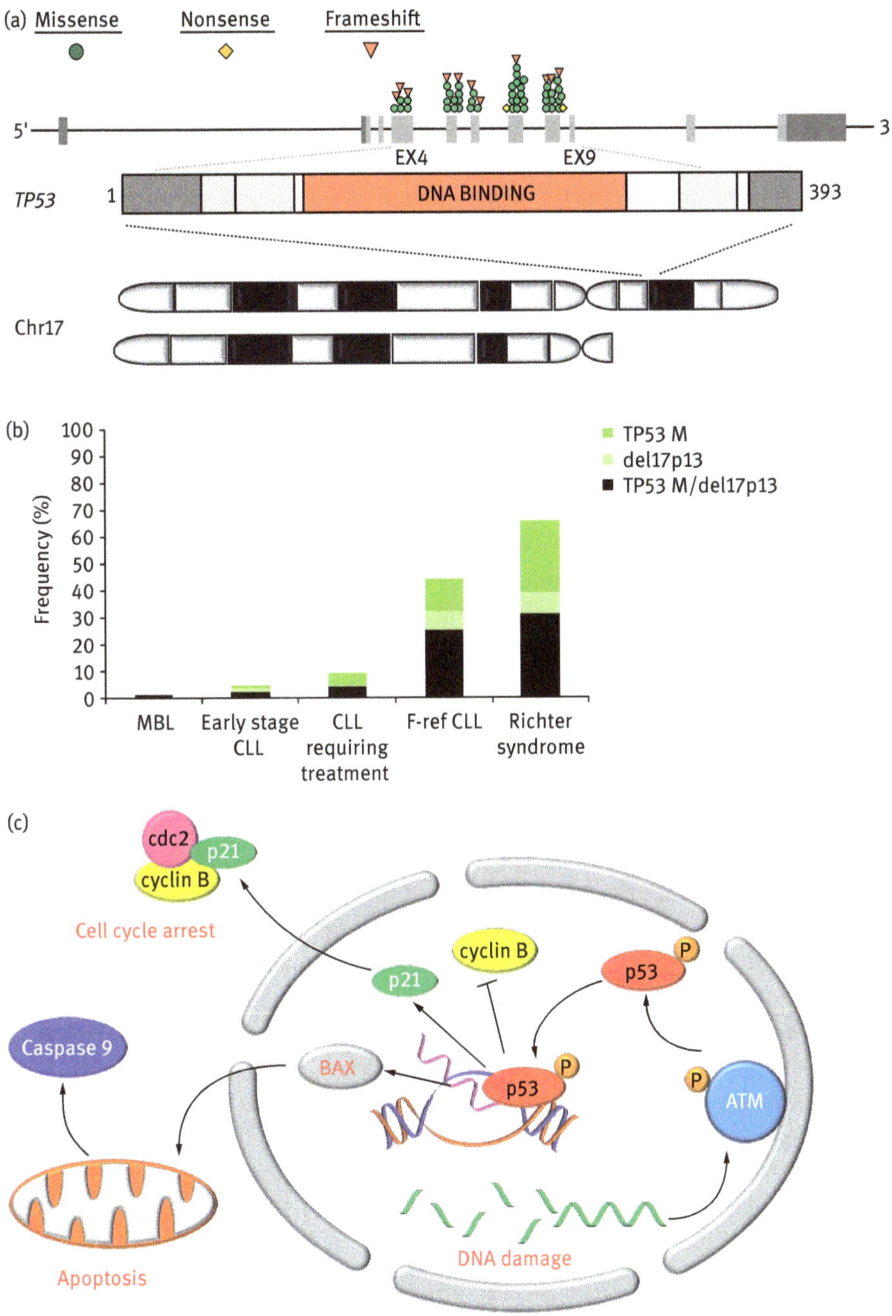

Fig. 11.6: **TP53 lesions in CLL.** Panel (a): TP53 maps on chromosome 17p and is affected by both point mutations and deletions of the locus in CLL. Panel (b): Prevalence of TP53 genetic lesions in different clinical phases of the disease. Panel (c): TP53 is activated in the apoptotic checkpoint pathway upon DNA damage.

trials led in both the chemotherapy and immuno-chemotherapy era. Among unselected newly diagnosed CLLs, patients harboring 17p13 deletion have the worst outcome, with an estimated median overall survival (OS) of 3–5 years [48]. This is consistent with the notion that newly diagnosed CLL with 17p13 deletion frequently (~ 50 % of cases) present in advanced stage or show a rapidly progressive disease that requires treatment shortly after diagnosis (median time to first treatment of 9 months) [48].

The outcome of patients with 17p13 deletion and need for treatment is very poor. Patients with 17p13 deletion very rarely achieve complete response after chemo/chemo-immunotherapy, as demonstrated by: (i) the German CLL Study Group CLL8 trial [54], in which CLL harboring 17p13 deletion had a complete response rate of only 2 % compared to 21 % and 44 % among CLL lacking 17p13 deletion and treated with FC or FCR, respectively; and (ii) the UK Leukemia Research Foundation CLL4 trial [53], in which the group of patients with 17p13 deletion had a very low rate (5 %) of complete/nodular partial responses. Such poor response to chemo+/- immunotherapy translates into an estimated OS in the range of 2 to 3 years from the time of frontline treatment for 17p13 deleted patients.

A number of retrospective studies suggest that, in addition to 17p13 deletion, also *TP53* mutations, even in the absence of 17p13 deletion, predict poor outcome in CLL. These observations have been confirmed within the frame of two large prospective studies led in the setting of previously untreated cases. In the UK Leukemia Research Foundation CLL4 trial [61], *TP53* mutations were significantly associated with poorer overall response rates (27 % v 83 %) and shorter PFS (5 year PFS: 5 % v 17 %) and OS (5 year OS: 20 % v 59 %) compared to *TP53* wild-type patients. In the CLL8 trial of the German CLL study Group [35], *TP53* mutated patients were less likely to achieve response (FC arm: 51 % vs. 92 %; FCR arm: 98 % vs 75 %) and MRD negativity (FC arm: 7 % vs. 37 %; FCR arm: 18 % vs. 68 %), and showed a significantly shorter PFS (FC arm: 12 months vs. 35 months; FCR arm: 15 months vs. 59 months) and OS (FC arm: 30 months vs. 89 months; FCR arm: median not reached vs. 42 months) than *TP53* wild-type cases. The prognostic impact of *TP53* mutations was similarly evident in both treatment groups (FC and FCR), suggesting that the addition of rituximab to chemotherapy does not alleviate the negative impact of *TP53* lesions.

Based on these data, the sole genetic studies recommended by guidelines in CLL are those aimed at detecting *TP53* defects. Deletion of 17p13 is the sole cytogenetic abnormality that is recommended to be tested by FISH in CLL patients requiring treatment [2, 65, 66]. Because CLL with *TP53* mutations carry a poor prognosis, current guidelines by the European Research Initiative on CLL (ERIC) warrant integration of *TP53* mutation analysis into the evaluation of CLL patients before treatment initiation. Sanger sequencing is the currently recommended approach for *TP53* mutation analysis [64]. However, due to its limited sensitivity, conventional Sanger sequencing misclassifies as wild-type those CLL cases harboring *TP53* mutations at low clonal abundance (< 10 % of the alleles). Such small *TP53* mutated subclones occur in a significant fraction of CLL, have the same unfavorable prognostic impact as clonal *TP53* defects, and anticipate the development of a chemo-refractory phenotype among CLL patients requiring treatment. Thanks to its high sensitivity (down to 1–0.1 %), deep next generation sequencing is capable of detecting these minor, but clinically relevant, *TP53* mutated subclones. Therefore, deep next generation sequencing should be considered as a useful tool for a comprehensive assessment of *TP53* disruption in CLL [67].

Combinations of alemtuzumab with steroids are amongst the most potent therapies for patients harboring *TP53* abnormalities, yielding response rates of 88 % in previously untreated cases, with 65 % of cases achieving a complete response [68]. The BTK inhibitor ibrutinib as single agent or combined to rituximab induces a response rate of 68–97 % in patients with *TP53* abnormalities [69–72]. A similar proportion of responses is observed with the PI3Kδ inhibitor idelalisib combined to rituximab [73]. Although these results appear significantly better than every previous historical control in CLL with *TP53* abnormalities, alemtuzumab, ibrutinib and idelalisib *per se* do not promise long-lasting remissions. Therefore, these patients should be considered for alternative treatments within clinical trials whenever possible. Use of FCR, alemtuzumab-based regimens or new drugs may be considered as a debulking strategy, but allogeneic stem cell transplantation should still be offered and discussed in patients with sufficient physical fitness and an available donor [74].

The *ATM* gene encodes a nuclear serine/threonine kinase whose activity is induced by chromosomal double-strand breaks that arise endogenously or after exposure to DNA-damaging agents, including chemotherapeutic drugs. ATM protects the integrity of the genome by regulating the cell-cycle arrest at G1/S and G2/M to prevent processing of damaged DNA, and by activating DNA-repair pathways or, alternatively, inducing apoptosis if the DNA damage cannot be repaired. As for *TP53*, the *ATM* gene is inactivated in CLL by both deletion and/or somatic mutations, which result in impaired DNA damage responses. Deletion of 11q22-23 always includes *ATM* and occurs in less than 10 % newly diagnosed CLL, while its prevalence raises to ~ 20 % at the time of first treatment [48, 53, 54]. Deletion 11q22-23 co-occurs with *ATM* mutations in 30–40 % of patients. *ATM* mutations generally consist in a mixture of missense substitutions distributed across the *ATM* coding sequence, with no clear hotspots, and have been observed in ~ 10–15 % of newly diagnosed patients and in ~ 15 % of progressive CLL requiring first treatment. By combining mutations and deletions, genetic lesions of *ATM* occur in ~ 20 % of diagnostic samples of CLL and in ~ 35 % of cases requiring first treatment [75–77].

From a clinical perspective, the presence of *ATM* deletions at the time of CLL presentation identifies a group of patients with intermediate-risk disease (40 % of cases are alive at 10 years, accounting for a 48 % reduction of the expected survival compared to the general population) [33]. Among CLLs requiring treatment, the presence of 11q22-23 deletions alone or combined to *ATM* mutations associates with poor response to chemotherapy. In the UK Leukemia Research Found CLL4 trial [77], where CLL patients were treated with various chemotherapy approaches not including an anti-CD20 monoclonal antibody, cases with both *ATM* mutation and 11q22-q23 deletion showed a significantly reduced response rate and PFS (46 % and 7 months, respectively) compared to those with wild-type *ATM* (84 % and 28 months, respectively), 11q22-q23 deletion alone (72 % and 17 months, respectively), or *ATM* mutation alone (87 % and 30 months, respectively). Consistently, OS for patients with biallelic *ATM* alterations was significantly reduced compared to those with wild-type *ATM* or *ATM* mutations alone (42 vs 85 vs 77 months, respectively). The addition of rituximab to chemotherapy significantly improved the outcome of CLL patients harboring *ATM* lesions. In the CLL8 trial of the German CLL Study Group [54], treatment with FCR significantly increased the complete response rate (51 % vs 15 %) and PFS (64 % vs 32 % at 3 years) in CLL patients with 11q22-23 deletions compared to FC alone. However, even among CLLs treated with FCR, 11q22-23 deletion still remains an adverse factor that, similar to unmutated *IGHV* genes, identifies a group of patients with intermediate-risk disease that are projected to progress in a relatively short timeframe.

CDKN2A (also known as p16) is a negative regulator of cell cycle progression from G1 phase to S phase that acts through the inhibition of cyclin dependent kinases. *CDKN2A* is recurrently affected by focal and homozygous losses in ~ 30 % of CLL that have transformed to RS, thus representing one of the most frequent genetic abnormalities occurring in this fatal complication of CLL. The notions that *CDKN2A* deletions are rare in unselected CLL and may be acquired at the time of transformation, points to a direct implication of this lesion in the development of RS [78, 79].

11.8 MYC network

MYC is a transcription factor that lies at the crossroads of cell cycle and cell growth by acting downstream of many signal transduction pathways. Functional studies have documented that *MYC* deregulation in CLL is part of a program of downstream events mainly driven by microenvironmental interactions. However, in a fraction of CLL, the *MYC* gene may be activated in a microenvironment-independent fashion by somatic structural lesions, including translocations with immunoglobulin and nonimmunoglobulin loci (~ 1 % of unselected CLL), gain/amplification of 8q24 (~ 5 % of unselected CLL), and point mutations (~ 1 % of unselected CLL) [80–82]. *MYC* lesions associate with progressive CLL and accumulate in the advanced phases of the disease. In addition, consistent with a more general role of *MYC* deregulation in driving the transformation from indolent to aggressive B-cell lymphoproliferations, *MYC* abnormalities are particularly enriched in ~ 30–40 % of RS cases [56, 79].

MYC belongs to a transcriptional regulating network that also includes *MAX*, a cofactor required for DNA binding by the various members of this network, as well as a group of putative *MYC* antagonists, namely *MNT*, *MXD1-4* and *MGA*. In physiological conditions, the MYC protein activates gene transcription through heterodimerization with MAX and then by binding to the E-box DNA recognition sequences in its target gene promoters. Conversely, heterodimers of MAX with MNT, MXD1-4 and MGA antagonize MYC-dependent gene expression regulation by transcriptional repression of the same E-box sequences. On these bases, the MYC oncoprotein transforming ability depends also on the balance between the MYC-MAX and MNT/MXD1-4/MGA-MAX complexes. In CLL, this balance may be disrupted not only by *MYC* activating lesions, but also by alterations targeting *MYC* antagonists. Indeed, the *MYC* antagonist *MNT* is known to be recurrently affected by focal losses in a small proportion of CLL, as documented by high resolution SNP array [82]. In addition, *MGA*, another MYC oncoprotein antagonist, is targeted by focal and recurrent gene deletions or truncating point mutations in a fraction of CLL patients [83]. Though the clinical impact of these lesions needs to be formally defined in CLL, *MGA* abnormalities seem to be enriched in cases displaying high-risk clinical features, namely fludarabine-refractoriness or transformation into RS.

Overall, these findings suggest multiple genetic mechanisms of *MYC* activation in CLL that, albeit occurring at low frequency as single events, may account for a significant fraction of cases when integrated into a common pathway.

11.9 Telomere function

Telomeres are specialized DNA-protein structures that cap the ends of all eukaryotic chromosomes and have important roles in genome stability. They are maintained by the enzyme telomerase and protected by the shelterin complex, composed of the duplex telomere-binding proteins TRF1 and TRF2 and the single-stranded telomere-binding protein POT1. *POT1* mutations occur in 1–5 % of unselected CLL, while they are enriched in *IGHV* unmutated cases and in cases presenting in advanced stage, thus suggesting an association between *POT1* mutations and disease progression. POT1 contains two highly conserved oligonucleotide/oligosaccharide-binding (OB) folds through which it binds to the 3′ terminus of the single-stranded overhang of the telomere. The majority of *POT1* mutations in CLL cluster within both OB folds, abrogating the binding of POT1 to telomeric single-stranded DNA. As a consequence, *POT1* mutations induce a dysfunctional telomeric phenotype resulting in the accumulation of chromosomal alterations. On these bases, *POT1* mutations are postulated to be one of the mechanisms promoting the genomic instability observed in a fraction of CLL [84].

11.10 Clonal architecture and evolution of CLL

Genomic studies have disclosed the complexity of clonal architecture of CLL and provided the proof of principle that genetically diverse subclones may be admixed with a dominant tumor clone (intratumoral heterogeneity). Also, sequential genomic approaches have documented that the molecular lesions of CLL are temporally ordered in a specific fashion rather than being randomly accumulated. Some genetic lesions (i.e. 13q14 deletion and +12) are clonally represented in all tumor cells and present since the early disease phases. On these bases, 13q14 deletion and +12 are interpreted as early events responsible for driving the founder clone of the tumor [85, 86]. This is consistent with the occurrence of 13q14 deletion and +12 in MBL, a condition that often anticipates overt CLL, and with the experimental evidence that 13q14 deletion causes CLL in mice [3, 51]. Conversely, other mutations are considered secondary events because they are initially restricted to a subclone of the tumor and become predominant in the later phases of the disease (i.e. *TP53* mutations). The occurrence of such subclones is a common source of drug resistance [67].

Longitudinal genetic data disclose two distinct and mutually exclusive evolutionary paths of CLL. The first evolutionary path involves CLL initially harboring +12 and *NOTCH1* mutations. In this path, clonal evolution proceeds toward the development of *TP53* abnormalities and *BIRC3* mutations. The second evolutionary path involves CLL initially harboring 13q14 deletion and proceeds toward the development of *SF3B1* mutations, *BIRC3* mutations and *TP53* abnormalities. This notion, along with mutually exclusive distribution of +12 and 13q14 deletion and the atypical presentation of +12 CLL compared to 13q14 deleted CLL, support the hypothesis that at least two distinct major subtypes of CLL exist [86].

11.11 Inherited susceptibility of CLL

A number of epidemiological evidences are in support of the concept that, along with somatic genetic lesions, also the constitutive genetic background contributes to CLL development. First, the incidence of CLL varies across geographic areas, being higher in Europe and European descent and lower in South and East Asia [87]. Second, the incidence of CLL remains low in individuals of Asian descent born in the United States, suggesting a predominant contribution of the genetic background, rather than environmental factors, to the differential distribution of CLL risk across geographic areas [88]. Third, relatives of CLL patients have an increased risk of developing CLL [89]. While not exclusively a consequence of genetic predisposition, familial aggregation provides strong evidence for inherited genetic factors playing a role in CLL development. Failure to identify a disease-causing gene through family-based linkage studies led to the assumption that CLL follows a polygenic model of inherited cancer susceptibility. Under this assumption, the overall risk of CLL in a given subject stems from the co-inheritance of multiple low-risk variants that have been progressively disclosed by unbiased genome-wide association (GWAS) studies. Previously unexpected novel CLL risk loci and genes have been identified, including functional polymorphisms affecting genes that have well-known implications in CLL, such as *IRF4*, *MYC* and *POT1*. Consistent with the hypothesis that CLL susceptibility follows a polygenic model, the risk of CLL associated with these polymorphisms is modest, when considered individually, while it increases with the increase in the numbers of risk alleles that are co-inherited. Collectively, the variants identified thus far (n=30) account for 19 % of the familial risk of CLL [90–93]. It is therefore likely that a substantial number of additional common variants remain to be discovered by further GWAS-based initiatives.

11.12 Concluding remarks

In the era of personalized medicine, the challenge for the treatment of patients with CLL involves correctly matching targeted therapies to the unique genetic and clonal composition of each individual tumor. Genomic studies have produced an unprecedented body of knowledge regarding the cellular programs involved in CLL pathogenesis and their implications as both prognostic/predictive factors and actionable targets. Next generation sequencing approaches have the potential to make testing of these genetic markers cheaper, faster and more widely available.

Clonal evolution studies have revealed those pathways that are altered by early genetic events in CLL and conceivably responsible for driving the founder clone of the tumor, thus representing ideal therapeutic targets for treatments aiming at disease eradication. Technical advances including ultra-deep next generation sequencing have allowed to reach deep levels of resolution of the genetic portrait of CLL, providing a precise definition of its subclonal genetic architecture. This approach has shown that small subclones harboring drug resistant mutations may be admixed with a large number of wild-type cells, thus being far below the levels of detection of conventional mutation assays. Treatments that do not take into account such clonal diversity and target only the dominant clone leave the possibility that one of these minor and resistant clones will then replicate and become dominant, leading to recurrence of the tumor. Thus, targeting both minor and dominant CLL clones through drug combinations incorporating novel agents might represent a pivotal strategy for treating CLL in the most efficient way.

Finally, beside the identification of actionable targets, genomic studies are also revealing new and previously unexpected mechanisms of targeted drug resistance in CLL, as exemplified by the discovery of mutations associated with ibrutinib-refractoriness.

References

[1] Müller-Hermelink HK, Montserrat E, Catovsky D, Campo E, Harris NL, Stein H. Chronic lymphocytic leukemia/small lymphocytic lymphoma. In: Swerdlow SH, Campo E, Harris NL, (eds): World Health Organization Classification of Tumours, Pathology and Genetics of Tumours of Haematopoietic and Lymphoid Tissues. Lyon, France: IARC; 2008:180–182.

[2] Hallek M, Cheson BD, Catovsky D, et al. Guidelines for the diagnosis and treatment of chronic lymphocytic leukemia: a report from the International Workshop on Chronic Lymphocytic Leukemia updating the National Cancer Institute-Working Group 1996 guidelines. Blood 2008;111:5446–5456.
[3] Ghia P, Caligaris-Cappio F. Monoclonal B-cell lymphocytosis: right track or red herring? Blood 2012;119:4358–4362.
[4] Rossi D, Gaidano G. Richter syndrome. Adv Exp Med Biol 2013a;792:173–191.
[5] Ghia P, Hallek M. Management of chronic lymphocytic leukemia. Haematologica 2014;99:965–972.
[6] Byrd JC, Jones JJ, Woyach JA, Johnson AJ, Flynn JM. Entering the era of targeted therapy for chronic lymphocytic leukemia: impact on the practicing clinician. J Clin Oncol 2014a;32:3039–3047.
[7] Gaidano G, Foà R, Dalla-Favera R. Molecular pathogenesis of chronic lymphocytic leukemia. J Clin Invest 2012;122:3432–3438.
[8] Fabbri G, Rasi S, Rossi D, et al. Analysis of the chronic lymphocytic leukemia coding genome: role of NOTCH1 mutational activation. J Exp Med 2011;208:1389–1401.
[9] Quesada V, Conde L, Villamor N, et al. Exome sequencing identifies recurrent mutations of the splicing factor SF3B1 gene in chronic lymphocytic leukemia. Nat Genet 2011;44:47–52.
[10] Wang L, Lawrence MS, Wan Y, et al. SF3B1 and other novel cancer genes in chronic lymphocytic leukemia. N Engl J Med 2011;365:2497–2506.
[11] Vardi A, Agathangelidis A, Sutton LA, Ghia P, Rosenquist R, Stamatopoulos K. Immunogenetic studies of chronic lymphocytic leukemia: revelations and speculations about ontogeny and clinical evolution. Cancer Res 2014;74:4211–4216.
[12] Hamblin TJ, Davis Z, Gardiner A, Oscier DG, Stevenson FK. Unmutated Ig V(H) genes are associated with a more aggressive form of chronic lymphocytic leukemia. Blood 1999;94:1848–1854
[13] Damle RN, Wasil T, Fais F, et al. Ig V gene mutation status and CD38 expression as novel prognostic indicators in chronic lymphocytic leukemia. Blood 1999;94:1840–1847.
[14] Mockridge CI, Potter KN, Wheatley I, Neville LA, Packham G, Stevenson FK. Reversible anergy of sIgM-mediated signaling in the two subsets of CLL defined by VH-gene mutational status. Blood 2007;109:4424–4431.
[15] Agathangelidis A, Ntoufa S, Stamatopoulos K. B-cell receptor and antigens in CLL. Adv Exp Med Biol 2013;792:1–24.
[16] Herve M, Xu K, Ng YS, et al. Unmutated and mutated chronic lymphocytic leukemias derive from self-reactive B cell precursors despite expressing different antibody reactivity. J Clin Invest 2005;115:1636–1643.
[17] Agathangelidis A, Darzentas N, Hadzidimitriou A, et al. Stereotyped B-cell receptors in one-third of chronic lymphocytic leukemia: a molecular classification with implications for targeted therapies. Blood 2012;119:4467–4475.
[18] Rossi D, Spina V, Bomben R, et al. Association between molecular lesions and specific B-cell receptor subsets in chronic lymphocytic leukemia. Blood 2013b;121:4902–4905.
[19] Strefford JC, Sutton LA, Baliakas P, et al. Distinct patterns of novel gene mutations in poor-prognostic stereotyped subsets of chronic lymphocytic leukemia: the case of SF3B1 and subset #2. Leukemia 2013;27:2196–2199.
[20] Rossi D, Spina V, Cerri M, et al. Stereotyped B-cell receptor is an independent risk factor of chronic lymphocytic leukemia transformation to Richter syndrome. Clin Cancer Res. 2009a;15:4415–4422.
[21] Dühren-von Minden M, Übelhart R, Schneider D, et al. Chronic lymphocytic leukaemia is driven by antigen-independent cell-autonomous signalling. Nature 2012;489:309–312.

[22] Wiestner A. BCR pathway inhibition as therapy for chronic lymphocytic leukemia and lymphoplasmacytic lymphoma. Hematology Am Soc Hematol Educ Program 2014;2014:125–134.
[23] Furman RR, Cheng S, Lu P, et al. Ibrutinib resistance in chronic lymphocytic leukemia. N Engl J Med 2014;370:2352–2354.
[24] Woyach JA, Furman RR, Liu TM, et al. Resistance mechanisms for the Bruton's tyrosine kinase inhibitor ibrutinib. N Engl J Med 2014;370:2286–2294.
[25] Famà R, Bomben R, Rasi S, et al. Ibrutinib-naïve chronic lymphocytic leukemialacks Bruton tyrosine kinase mutations associated with treatment resistance. Blood 2014;124:3831–3833.
[26] Komarova NL, Burger JA, Wodarz D. Evolution of ibrutinib resistance in chronic lymphocytic leukemia (CLL). Proc Natl Acad Sci USA 2014;111:13906–13911.
[27] Lobry C, Oh P, Aifantis I. Oncogenic and tumor suppressor functions of Notch in cancer: it's NOTCH what you think. J Exp Med 2011;208:1931–1935.
[28] Welcker M, Clurman BE. FBW7 ubiquitin ligase: a tumour suppressor at the crossroads of cell division, growth and differentiation. Nat Rev Cancer 2008;8:83–93.
[29] Rossi D, Rasi S, Fabbri G, et al. Mutations of NOTCH1 are an independent predictor of survival in chronic lymphocytic leukemia. Blood 2012a;119:521–529.
[30] Puente XS, Pinyol M, Quesada V, et al. Whole-genome sequencing identifies recurrent mutations in chronic lymphocytic leukaemia. Nature 2011;475:101–105.
[31] Del Giudice I, Rossi D, Chiaretti S, et al. NOTCH1 mutations in +12 chronic lymphocytic leukemia(CLL) confer an unfavorable prognosis, induce a distinctive transcriptional profiling and refine the intermediate prognosis of +12 CLL. Haematologica 2012;97:437–441.
[32] Balatti V, Bottoni A, Palamarchuk A, et al. NOTCH1 mutations in CLL associated with trisomy 12. Blood 2012;119:329–331.
[33] Rossi D, Rasi S, Spina V, et al. Integrated mutational and cytogenetic analysis identifies new prognostic subgroups in chronic lymphocytic leukemia. Blood 2013c;121:1403–1412.
[34] Villamor N, Conde L, Martínez-Trillos A, et al. NOTCH1 mutations identify a genetic subgroup of chronic lymphocytic leukemia patients with high risk of transformation and poor outcome. Leukemia 2013;27:1100–1106.
[35] Stilgenbauer S, Schnaiter A, Paschka P, et al. Gene mutations and treatment outcome in chronic lymphocytic leukemia: results from the CLL8 trial. Blood 2014;123:3247–3254.
[36] Tausch E, Beck P, Schlenk RF, et al. NOTCH1 Mutation and Treatment Outcome In CLL Patients Treated With Chlorambucil (Chl) Or Ofatumumab-Chl (O-Chl): Results From The Phase III Study Complement 1 (OMB110911). Blood 2013;122:527.
[37] Spina V, Martuscelli L, Rossi D. Molecular deregulation of signaling in lymphoid tumors. Eur J Haematol E-pub on Apr 16, 2015;doi: 10.1111/ejh.12567.
[38] Staudt LM. Oncogenic activation of NF-kappaB. Cold Spring Harb Perspect Biol 2010;2:a000109.
[39] Rossi D, Fangazio M, Rasi S, et al. Disruption of BIRC3 associates with fludarabine chemorefractoriness in TP53 wild-type chronic lymphocytic leukemia. Blood 2012b;119:2854–2862.
[40] Rahal R, Frick M, Romero R, et al. Pharmacological and genomic profiling identifies NF-κB-targeted treatment strategies for mantle cell lymphoma. Nat Med 2014;20:87–92.
[41] Damm F, Mylonas E, Cosson A, et al. Acquired initiating mutations in early hematopoietic cells of CLL patients. Cancer Discov 2014;4:1088–1101.
[42] Mansouri L, Sutton LA, Ljungstrom V, et al. Recurrent Mutations within the Nfkbie gene: A Novel Mechanism for NF-κB Deregulation in Aggressive Chronic Lymphocytic Leukemia. Blood 2014;124:297.
[43] Rossi D. Role of MYD88 in lymphoplasmacytic lymphoma diagnosis and pathogenesis. Hematology Am Soc Hematol Educ Program 2014a;2014:113–118.

[44] Martínez-Trillos A, Pinyol M, Navarro A, et al. Mutations in TLR/MYD88 pathway identify a subset of young chronic lymphocytic leukemia patients with favorable outcome. Blood 2014;123:3790–3796.
[45] Bonnal S, Vigevani L, Valcárcel J. The spliceosome as a target of novel antitumour drugs. Nat Rev Drug Discov 2012;11:847–859.
[46] Rossi D, Bruscaggin A, Spina V, et al. Mutations of the SF3B1 splicing factor in chronic lymphocytic leukemia: association with progression and fludarabine-refractoriness. Blood 2011a;118:6904–6908.
[47] Oscier DG, Rose-Zerilli MJ, Winkelmann N, et al. The clinical significance of NOTCH1 and SF3B1 mutations in the UK LRF CLL4 trial. Blood 2013;121:468–475.
[48] Döhner H, StilgenbauerS, Benner A, et al. Genomic aberrations and survival in chronic lymphocytic leukemia. N Engl J Med 2000;343:1910–1916.
[49] Calin GA, Dumitru CD, Shimizu M, et al. Frequent deletions and down-regulation of micro-RNA genes miR15 and miR16 at 13q14 in chronic lymphocytic leukemia. Proc Natl Acad Sci USA 2002;99:15524–1529.
[50] Cimmino A, Calin GA, Fabbri M, et al. miR-15 and miR-16 induce apoptosis by targeting BCL2. Proc Natl Acad Sci USA 2005;102:13944–13949.
[51] Klein U, Lia M, Crespo M, et al. The DLEU2/miR-15a/16-1 cluster controls B cell proliferation and its deletion leads to chronic lymphocytic leukemia. Cancer Cell 2010;17:28–40.
[52] Seymour J. ABT-199 for Chronic Lymphocytic Leukemia. Clin Adv Hematol Oncol 2014;12:698–700.
[53] Catovsky D, Richards S, Matutes E, et al. Assessment of fludarabine plus cyclophosphamide for patients with chronic lymphocytic leukaemia (the LRF CLL4 Trial): a randomised controlled trial. Lancet 2007;370:230–239.
[54] Hallek M, Fischer K, Fingerle-Rowson G, et al. Addition of rituximab to fludarabine and cyclophosphamide in patients with chronic lymphocytic leukaemia: a randomised, open-label, phase 3 trial. Lancet 2010;376:1164–1174.
[55] Stilgenbauer S, Zenz T, Winkler D, et al. Subcutaneous alemtuzumab in fludarabine-refractory chronic lymphocytic leukemia: clinical results and prognostic marker analyses from the CLL2H study of the German Chronic Lymphocytic Leukemia Study Group. J Clin Oncol 2009;27:3994–4001.
[56] Rossi D, Spina V, Deambrogi C, et al. The genetics of Richter syndrome reveals disease heterogeneity and predicts survival after transformation. Blood 2011b;117:3391–3401.
[57] Zenz T, Kröber A, Scherer K, et al. Monoallelic TP53 inactivation is associated with poor prognosis in chronic lymphocytic leukemia: results from a detailed genetic characterization with long-term follow-up. Blood 2008;112:3322–3329.
[58] Rossi D, Cerri M, Deambrogi C, et al. The prognostic value of TP53 mutations in chronic lymphocytic leukemia is independent of Del17p13: implications for overall survival and chemorefractoriness. Clin Cancer Res 2009b;15:995–1004.
[59] Panagiotis, B, Hadzidimitriou A, Sutton LA, et al. Recurrent mutations refine prognosis in chronic lymphocytic leukemia. Leukemia 2015;29:329–336.
[60] Zenz T, Eichhorst B, Busch R, et al. TP53 mutation and survival in chronic lymphocytic leukemia. J Clin Oncol 2010a;28:4473–4479.
[61] Gonzalez D, Martinez P, Wade R, et al. Mutational status of the TP53 gene as a predictor of response and survival in patients with chronic lymphocytic leukemia: results from the LRF CLL4 trial. J Clin Oncol 2011;29:2223–2229.
[62] Schnaiter A, Paschka P, Rossi M, et al. NOTCH1, SF3B1, and TP53 mutations in fludarabine-refractory CLL patients treated with alemtuzumab: results from the CLL2H trial of the GCLLSG. Blood 2013;122:1266–1270.

[63] Zenz T, Vollmer D, Trbusek M, et al. TP53 mutation profile in chronic lymphocytic leukemia: evidence for a disease specific profile from a comprehensive analysis of 268 mutations. Leukemia 2010b;24:2072–2079.
[64] Pospisilova S, Gonzalez D, Malcikova J, et al. ERICrecommendations on TP53 mutation analysis in chronic lymphocytic leukemia. Leukemia 2012;26:1458–1461.
[65] Ghielmini M, Vitolo U, Kimby E, et al. ESMO Guidelines consensus conference on malignant lymphoma 2011 part 1: diffuse large B-cell lymphoma (DLBCL), follicular lymphoma (FL) and chronic lymphocytic leukemia (CLL). Ann Oncol 2013;24:561–576.
[66] Zelenetz AD, Gordon LI, Wierda WG, et al. Chronic lymphocytic leukemia/small lymphocytic lymphoma, version 1.2015. J Natl Compr Canc Netw 2015;13:326–362.
[67] Rossi D, Khiabanian H, Spina V, et al. Clinical impact of small TP53 mutated subclones in chronic lymphocytic leukemia. Blood 2014b;123:2139–2147.
[68] Pettitt AR, Jackson R, Carruthers S, et al. Alemtuzumab in combination with methylprednisolone is a highly effective induction regimen for patients with chronic lymphocytic leukemia and deletion of TP53: final results of the national cancer research institute CLL206 trial. J Clin Oncol 2012;30:1647–1655.
[69] Byrd JC, Furman RR, Coutre SE, et al. Targeting BTK with ibrutinibin relapsed chronic lymphocytic leukemia. N Engl J Med 2013;369:32–42.
[70] Burger JA, Keating MJ, Wierda WG, et al. Safety and activity of ibrutinib plus rituximab for patients with high-risk chronic lymphocytic leukaemia: a single-arm, phase 2 study. Lancet Oncol 2014;15:1090–1099.
[71] Byrd JC, Brown JR, O'Brien S, et al. Ibrutinib versus ofatumumab in previously treated chronic lymphoid leukemia. N Engl J Med 2014;371:213–223.
[72] Farooqui MZ, Valdez J, Martyr S, et al. Ibrutinib for previously untreated and relapsed or refractory chronic lymphocytic leukaemiawith TP53 aberrations: a phase 2, single-arm trial. Lancet Oncol 2015;16:169–176.
[73] Furman RR, Sharman JP, Coutre SE, et al. Idelalisib and rituximab in relapsed chronic lymphocytic leukemia. N Engl J Med 2014;370:997–1007.
[74] Dreger P, Schetelig J, Andersen N, et al. Managing high-risk CLL during transition to a new treatment era: stem cell transplantation or novel agents? Blood 2014;124:3841–3849.
[75] Guarini A, Marinelli M, Tavolaro S, et al. ATM gene alterations in chronic lymphocytic leukemia patients induce a distinct gene expression profile and predict disease progression. Haematologica 2012;97:47–55.
[76] Ouillette P, Li J, Shaknovich R, et al. Incidence and clinical implications of ATM aberrations in chronic lymphocytic leukemia. Genes Chromosomes Cancer 2012;51:1125–1132.
[77] Skowronska A, Parker A, Ahmed G, et al. Biallelic ATM inactivation significantly reduces survival in patients treated on the United Kingdom LeukemiaResearch Fund Chronic Lymphocytic Leukemia 4 trial. J Clin Oncol 2012;30:4524–4532.
[78] Chigrinova E, Rinaldi A, Kwee I, et al. Two main genetic pathways lead to the transformation of chronic lymphocytic leukemia to Richter syndrome. Blood 2013;122:2673–2682.
[79] Fabbri G, Khiabanian H, Holmes AB, et al. Genetic lesions associated with chronic lymphocytic leukemia transformation to Richter syndrome. J Exp Med 2013;210:2273–2288.
[80] Huh YO, Lin KI, Vega F, et al. MYC translocation in chronic lymphocytic leukaemia is associated with increased prolymphocytes and a poor prognosis. Br J Haematol 2008;142:36–44.
[81] Brown JR, Hanna M, Tesar B, et al. Integrative genomic analysis implicates gain of PIK3CA at 3q26 and MYC at 8q24 in chronic lymphocytic leukemia. Clin Cancer Res 2012;18:3791–3802
[82] Edelmann J, Holzmann K, Miller F, et al. High-resolution genomic profiling of chronic lymphocytic leukemia reveals new recurrent genomic alterations. Blood 2012;120:4783–4794.

[83] De Paoli L, Cerri M, Monti S, et al. MGA, a suppressor of MYC, is recurrently inactivated in high risk chronic lymphocytic leukemia. Leuk Lymphoma 2013;54:1087–1090.

[84] Ramsay AJ, Quesada V, Foronda M, et al. POT1 mutations cause telomere dysfunction in chronic lymphocytic leukemia. Nat Genet 2013;45:526–530.

[85] Landau DA, Carter SL, Stojanov P, et al. Evolution and impact of subclonal mutations in chronic lymphocytic leukemia. Cell 2013;152:714–726.

[86] Wang J, Khiabanian H, Rossi D, et al. Tumor evolutionary directed graphs and the history of chronic lymphocytic leukemia. Elife, e-pub on Dec 11, 2014; doi: 10.7554/eLife.02869.

[87] Goldin LR, Slager SL. Familial CLL: genes and environment. Hematology Am Soc Hematol Educ Program 2007:339–345.

[88] Gale RP, Cozen W, Goodman MT, Wang FF, Bernstein L. Decreased chronic lymphocytic leukemia incidence in Asians in Los Angeles County. Leuk Res 2000;24:665–669.

[89] Yuille MR, Matutes E, Marossy A, Hilditch B, Catovsky D, Houlston RS. Familial chronic lymphocytic leukaemia: a survey and review of published studies. Br J Haematol 2000;109:794–799.

[90] Sellick GS, Goldin LR, Wild RW, et al. A high-density SNP genome-wide linkage search of 206 families identifies susceptibility loci for chronic lymphocytic leukemia. Blood 2007;110:3326–3333.

[91] Di Bernardo MC, Crowther-Swanepoel D, Broderick P, et al. A genome-wide association study identifies six susceptibility loci for chronic lymphocytic leukemia. Nat Genet 2008;40:1204–1210.

[92] Crowther-Swanepoel D, Broderick P, Di Bernardo MC, et al. Common variants at 2q37.3, 8q24.21, 15q21.3 and 16q24.1 influence chronic lymphocytic leukemia risk. Nat Genet 2010;42:132–136.

[93] Speedy HE, Di Bernardo MC, Sava GP, et al. A genome-wide association study identifies multiple susceptibility loci for chronic lymphocytic leukemia. Nat Genet 2014;46:56–60.

Zachary R. Hunter, Jorge J. Castillo, and Steven P. Treon

12 Waldenström's macroglobulinemia

12.1 Introduction

Waldenström's macroglobulinemia (WM) is an uncommon B-cell lymphoma, that corresponds to the histopathological classification of IgM secreting lymphomplasmacytic lymphoma (LPL) by the World Health Organization [1, 2]. Most cases of LPL are WM, with IgA, IgG, and nonsecreting LPL making up less than 5 % of cases. Since the first clinical description of WM by Jan Waldenström in 1944, considerable progress has been made in the diagnosis and care of patients, but it is only recently that the molecular pathology for WM has begun to be understood [3, 4].

12.2 Epidemiology and patient characteristics

WM has a relatively low age adjusted incidence rate of 3.4 per million among males and 1.7 per million among females in the United States of America [5, 6]. The median age of diagnosis for WM is 62 years. Patients typically present with significant bone marrow (BM) infiltration by LPL cells, which contributes to cytopenias. Patients can also exhibit adenopathy and/or splenomegaly at the time of diagnosis in 15 % and 10 % of cases, respectively [4]. The median overall survival of WM is 5–12 years, and is impacted by risk stratification based on the WM International Prognostic Scoring System [7]. However, there are indications that the median survival for WM patients is improving based on recent SEER data [8]. Many of the clinical disease manifestations of WM are related to paraprotein production by the tumor cells, including symptomatic hyperviscosity, peripheral neuropathy, cryoglobulinemia, cold-agglutinin disease, and light chain amyloidosis [4, 9, 10]. Most WM patients also present with hypogammaglobulinemia of the uninvolved immunoglobulins, with IgA (63 %), IgG (58 %), or both (49 %) being subnormal at the time of diagnosis [11]. Subnormal IgA and IgG immunoglobulin levels do not typically improve with therapy, and may contribute to recurring sino-bronchial infections in WM patients [4].

There are a number of hematological malignancies that have similar clinical presentations with WM including IgM secreting multiple myeloma (MM), chronic lymphocytic leukemia (CLL), and marginal zone lymphoma (MZL) [2, 4, 12]. All of these entities, including WM, can evolve from monoclonal gammopathy of undetermined significance (MGUS), suggesting the possibility of shared genetic events or predispositions between these B-cell lineage neoplasms [13–15]. There are multiple lines of evidence suggesting the existence of genetic predispositions to WM. These studies include genetic linkage studies and epidemiological analysis that show up to a quarter of WM patients as having a first or second degree relative with either WM or a related lymphoid malignancy [16–18]. A number of these studies have also shown an increased risk for other indolent non-Hodgkin's lymphomas (iNHL), diffuse large B-cell lymphoma (DLBCL), and CLL in family members of WM patients, as well as secondary cancers in WM patients themselves [19–21]. Similar familial associations also exist with other immunological disorders in healthy relatives, including hypogammaglobulinemia or hypergammaglobulinemia (typically polyclonal IgM), autoantibody (particularly thyroid) production, and presence of hyper-responsive B cells [22, 23]. Increased expression of the BCL2 gene has been suggested to underlie familial predisposition to WM in one study [23]. The presence of either a personal or familial history of autoimmune disorders, including autoimmune hemolytic anemia and Sjögren syndrome, have also been associated with an increased risk of WM [24]. While the role of environmental factors in WM remains to be clarified, several infections including hepatitis C virus (HCV), human immunodeficiency virus (HIV), and rickettsiosis have also been suggested as etiological [25, 26]. In the case of HCV, one large follow-up study failed to confirm an association with WM using both serological and molecular diagnostic studies for HCV infection [26, 27].

12.3 Nature of the WM clone

Examination of the B-cell clone(s) found in the bone marrow of WM patients reveals a range of differentiation from small lymphocytes with large focal deposits of surface immunoglobulins, to lymphoplasmacytic cells, to mature plasma cells that contain intracytoplasmic IgM (Fig. 12.1) [28]. Circulating clonal B cells are often detectable in patients with WM, though these patients typically do not manifest a frank lymphocytosis [28, 29]. WM cells express the monoclonal IgM, and some clonal cells also express surface IgD. The characteristic immunophenotypic profile of WM lymphoplasmacytic cells includes the expression of the pan B-cell markers CD19, CD20 (including FMC7), CD22, and CD79 [30, 31]. Expression of CD5, CD10 and CD23 can be present in 10–20 % of cases, and their presence does not exclude the diagnosis of WM [32]. In addition, multiparameter flow cytometric analysis has also identified CD25 and CD27 as being characteristic of the WM clone, and that a $CD22^{dim}/CD25^{+}/CD27^{+}/IgM^{+}$ population can be observed among clonal B lymphocytes in IgM MGUS patients who ultimately progressed to WM [33].

The nature of the clonal cell in WM as being derived from a post-germinal center B cell is supported by the distribution of somatic mutations in the immunoglobulin genes, which demonstrate increased nonsynonymous versus silent mutation rates in complement determining regions, consistent along with somatic hypermutation [34, 35]. A strong preferential usage of VH3/JH4 gene families without intraclonal variation, and without evidence for any isotype-switched transcripts has also been shown [36, 37]. Taken together, these data indicate that WM may originate from an

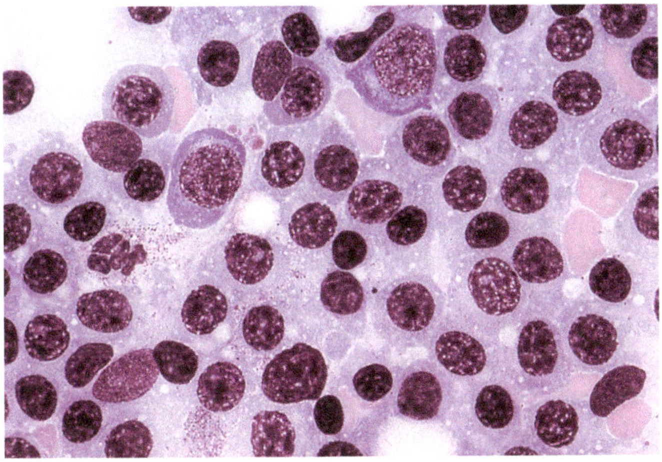

Fig. 12.1: Aspirate from a patient with Waldenström's macroglobulinemia demonstrating excess mature lymphocytes, lymphoplasmacytic cells and plasma cells (courtesy of Marvin Stone M.D.).

IgM^+ and/or IgM^+ IgD^+ memory B cell. Normal IgM^+ memory B cells localize in bone marrow, where they mature to IgM-secreting cells [38].

12.4 Chromosomal abnormalities in WM

No recurrent translocations have been described in WM, further distinguishing it from IgM myeloma and other B-cell malignancies [12, 39, 40]. However, karyotyping, FISH, and array-based studies have identified recurrent chromosomal abnormalities. These include deletions in chromosome 6q21-23 in 40 %–60 % of WM patients, with concordant gains in 6p in 41 % of 6q deleted patients [17, 41, 42]. Gains in chromosomes 3q, 4, 18, 8q and Xq as well as losses of 11q23, 13q14, and 17p have also been described in up to 20 % of WM cases [41, 43]. As 6q deletions represent the most recurrent cytogenetic finding, there has been a great deal of interest in identifying the minimally deleted regions and target genes within this locus. Two such candidates are TNFAIP3, a negative regulator of nuclear factor kappa B signaling (NF-κB) and PRDM1, a master regulator of B-cell development [17, 44]. The removal of an NF-κB negative regulator is of particular interest as the phosphorylation and translocation of NF-κB into the nucleus is a crucial event for WM cell survival [45]. This has led to speculation that the success of proteasome inhibitor therapy in WM is due in part to inhibition of the degradation of negative regulators of NF-κB such as inhibitor of kappa B (IκB) [45–47].

12.5 MYD88 somatic mutations

The advent of next generation sequencing led to the discovery of a somatic mutation in MYD88 resulting in a leucine to proline substitution at position 265 ($MYD88^{L265P}$) in over 90 % of WM patients [48]. MYD88 serves as an adaptor molecule for the Toll-like receptor (TLR) and IL1R families by binding directly to the receptor or, as in the case of TLR4, through the adaptor molecule TIRAP. Upon receptor activation, MYD88 undergoes a conformational change promoting homodimerization resulting in the recruitment and phosphorylation of IRAK4 and IRAK1 [49–51]. The $MYD88^{L265P}$ somatic mutation induces constitutive homodimerization and downstream MYD88 signaling independent of receptor activation (Fig. 12.2) [48, 52, 53]. This leads to the activation of TRAF6 and subsequent NF-κB activation via IκBα phosphorylation [54]. This constitutive pathway activation is significant given the established importance of NF-κB signaling in WM cell growth and survival [45]. $MYD88^{L265P}$ can also trigger NF-κB signaling through an IRAK-independent pathway by direct interaction with phosphorylated BTK in WM cells, suggesting parallel pathways for NF-κB activation [53]. MYD88 knockdown and dimerization inhibitor studies have confirmed increased WM LPC apoptosis, and decreases in the phosphorylation and nuclear translocation of NF-κB in WM patient's BM LPC and WM cell lines [53, 55]. IRAK phosphorylation and association of phosphorylated BTK with the MYD88 complex was also reduced in these experiments. The knockdown or kinase inhibition of BTK directly was also sufficient to reduce NF-κB signaling and promote apoptosis [53].

The high prevalence of *MYD88* L265P has been confirmed in large series which have used either Sanger or PCR based sequencing [56–61]. Activating mutations in MYD88, including $MYD88^{L265P}$ are not unique to WM, and were first described in activated B-cell (ABC) subtyped DLBCL, where they were observed in 29 % of patients [52]. Multiple MYD88 mutation types have been reported for ABC DLBCL, including L265P, V217F, S219C, M232T, S243N, and T294P. In contrast, MYD88 mutations in WM patients are predominately L265P. The mutation is typically heterozygous, though reports of copy number amplifications and acquired uniparental disomies (aUPD) affecting the MYD88 locus resulting in increased/homozygous expression have been reported [40, 59]. The high prevalence of the $MYD88^{L265P}$ mutation distinguishes WM from a number of related B-cell disorders such as MZL (6–8 %)and CLL (3 %), and has not been observed in IgM MM [61–63]. Likewise, central nervous system (CNS) involvement by WM, known as Bing Neel syndrome, has been shown to harbor $MYD88^{L265P}$ that is detectible by highly sensitive allele specific polymerase chain reaction (AS-PCR) assays of the spinal fluid [64]. In a similar manner, it is IgM, but not IgG or IgA MGUS, that typically precedes the diagnosis of WM, and $MYD88^{L265P}$ can be detected in approximately 50–80 % of IgM MGUS patients using AS-PCR assays, but not in IgG or IgA MGUS [57, 58, 61, 65]. This data suggest that $MYD88^{L265P}$ occurs very early in WM oncogenesis.

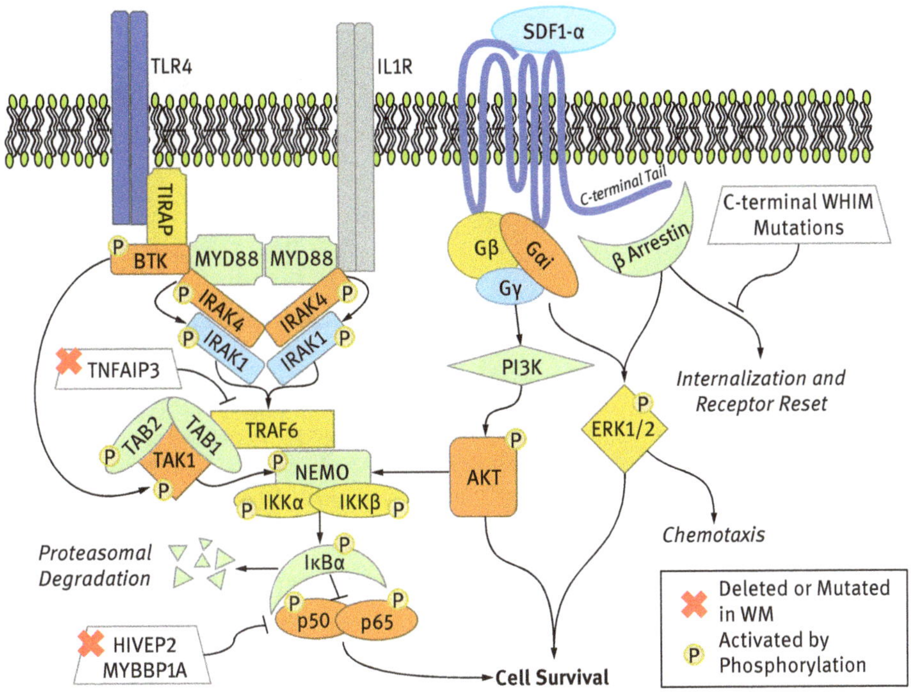

Fig. 12.2: **The signaling pathways of Waldenström's macroglobulinemia.** MYD88^{L265P} constitutively dimerizes and triggers auto-phosphorylation of IRAK4, with subsequent recruitment and phosphorylation of IRAK1. This results in the cytosolic release of membrane-bound TRAF6 from IRAK1. The TAK1 binding proteins, TAB1 and TAB2, promote binding of TRAF6 to TAK1, triggering phosphorylation of the IKK complex (NEMO/IKKγ, IKKα, and IKKβ). Phosphorylated BTK associates with MYD88^{L265P} and activates TAK1 independent of IRAK1/4. IKK activation leads to IκBα phosphorylation and its proteasomal degradation. The release of NF-κB p65 and p50 subunits and their subsequent phosphorylation allow them to translocate to the nucleus and drive NF-κB-dependent pro-survival signaling. CXCR4 is activated by its ligand, SDF1-α that results in the exchange of guanosine diphosphate for guanosine triphosphate in $G_{\alpha i}$. This exchange releases the activated heterotrimeric g-proteins resulting in the activation of PI3K and ERK1/2. Phosphorylation of the phosphatidylinositol by PI3K recruits AKT to the membrane, where it is phosphorylated. Activated AKT promotes cell survival and signals through the IKK complex. Activation of CXCR4 results in the phosphorylation of its C-terminal tail serines that recruit β-arrestins. β-arrestins signal through the ERK1/2 and promote the termination of the signal through receptor internalization. Deletion of the C-terminal tail in CXCR4^{WHIM} patients impairs the internalization process resulting in constitutive signaling.

The clinical presentation of WM patients at diagnosis is partly determined by MYD88 mutation status [58, 61, 66]. These studies demonstrated that patients harboring the MYD88^{L265P} mutation have significantly increased bone marrow involvement, higher serum IgM levels, and lower serum IgA levels. Despite the lower level of disease burden associated with MYD88 wild-type status, a ten-fold higher risk of death was associated with this genotype in one study [66] (Fig. 12.3). While the molecular pathol-

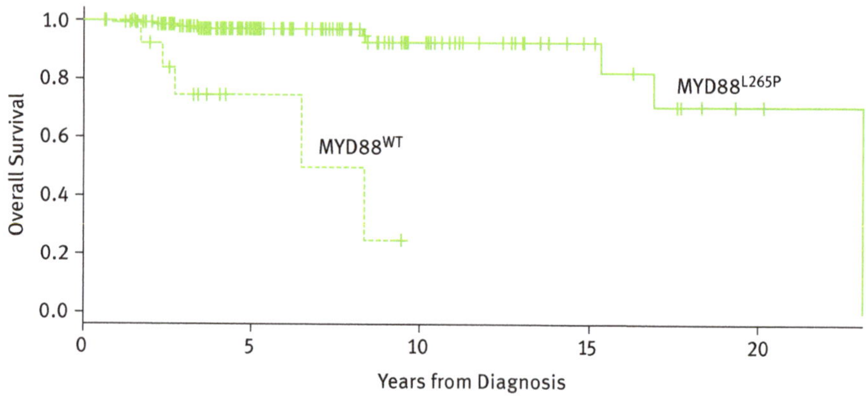

Fig. 12.3: Kaplan–Meier overall survival curves from diagnosis for 175 WM patients stratified by *MYD88* mutation status (16 *MYD88*^{WT}, 159 *MYD88*^{L265P}). Median follow up was 4.8 years and found increased mortality within the *MYD88*^{WT} population (P < 0.0001 by log-rank analysis).

ogy of MYD88^{WT} WM disease remains to be clarified, one study reported that two of three MYD88^{WT} patients had nonsynonymous somatic mutations within the MLL2 gene [40]. These two patients were further noted to have high levels of circulating WM LPC and the LPC were CD23^{+}, which are both uncommon findings in WM patients [2, 32]. This is noteworthy as MLL2 mutations are found with a high frequency in other lymphomas including 89 % of follicular lymphoma and 32 % of DLBCL [67].

Small studies have indicated that changes in MYD88^{L265P} allele burden, measured by AS-PCR in WM patient BM samples, correlate with BM response to therapy, and studies are ongoing to determine the clinical utility of this approach [61]. Efforts to detect MYD88^{L265P} in the peripheral blood have demonstrated a sensitivity of 0.983 (95 % CI 0.895–0.999), compared to matching BM-based assays in untreated WM patients [62]. However, AS-PCR accuracy is significantly reduced in patients who have received prior therapy with a sensitivity of 0.738 (95 % CI 0.607–0.838) complicating the use of blood based assays in this population. Specificity for the mutation was 100 % in these studies regardless of treatment status.

12.6 CXCR4 somatic mutations

After MYD88^{L265P}, the next most commonly mutated gene in WM is CXCR4, which can be observed in 29 % of patient BM LPC and nearly exclusively in those harboring the MYD88^{L265P} mutation [40, 66]. CXCR4 is a G protein coupled receptor (GPCR) that, in WM, induces cell survival and migration to the BM mediated by its ligand, SDF-1α (CXCL12 at the gene level) [68]. This chemokine receptor consists of seven transmembrane helices, the loops of which form the extracellular ligand-binding, pocket for SDF-1α and the intracellular signaling region that interacts with the heterotrimeric G proteins (Fig. 12.2). The intracellular carboxyl terminal tail (C-terminal) of CXCR4 does not participate in G protein signaling or ligand binding but plays an important regulatory role [69]. Upon SDF-1α binding, CXCR4 activates $G_{\alpha i}$ by catalyzing the exchange of its guanosine diphosphate (GDP) for guanosine triphosphate (GTP), releasing the heterotrimer from the receptor and dissociating the $G_\beta G_\gamma$ heterodimer from $G_{\alpha i}$ [70, 71]. The $G_\beta G_\gamma$ subunits signal through the phospholipase C (PLC) and phosphoinositide 3-kinase (PI3K) pathways until inhibited by another GDP bound $G_{\alpha i}$. The $G_{\alpha i}$ subunit remains active, inhibiting adenylate cyclase and signaling through PI3K until the GTP is hydrolyzed to GDP by a regulator of G protein signaling family member (RGS). This G protein based signaling will continue until the C-terminal tail of CXCR4 is activated and the receptor internalized. Upon ligand bindings, several kinases, including G protein receptor kinase (GRK) and protein kinase C (PKC) family members, phosphorylate serines in the C-terminal tail which recruit β-arrestins that transport the receptor to clathrin coated pits for internalization where the receptor can be either reset or degraded. During this process, β-arrestins signal through the mitogen activated protein kinase pathway (MAPK) [72].

The WM somatic mutations manifest as nonsense or frameshift mutations exclusively within the C-terminal domain [40, 66]. The mutations are typically heterozygous, though homozygous and compound heterozygous cases have been observed. This mutational pattern mirrors the germline variants found in familial warts, hypogammaglobulinemia, infection, and myelokathexis (WHIM) syndrome. In both diseases, these mutations occur after the last of the seven transmembrane helices, destroying the C-terminal tail but leaving the regions responsible for ligand binding and downstream signaling via G proteins intact [40, 73]. The loss of regulatory serines in the C-terminal tail of CXCR4 impairs receptor internalization via β-arrestins, thereby prolonging G protein and β-arrestin signaling [69, 72, 74]. In WHIM syndrome, this results in impaired B-cell class switching and constitutive homing to the bone marrow leading to peripheral neutropenia [75]. While this method of GPCR somatic activation was first described in CXCR4 in WM patients, it affects a regulatory mechanism common to many GPCR chemokine receptors and a similar somatic mutation pattern can be found in the CCR4 c-terminal domain in up to 26 % of patients with adult T-cell leukemia/lymphoma [40, 76, 77].

WM cell lines transduced to express these WHIM-like CXCR4 mutations ($CXCR4^{WHIM}$) likewise exhibited impaired receptor internalization and increased signaling through AKT and MAPK1 compared to CXCR4 wild-type ($CXCR4^{WT}$) or empty vector lines when stimulated with SDF-1α [77, 78]. Incubating with a CXCR4 inhibitor prior to SDF-1α stimulation can mitigate these effects. In addition, more robust signaling through MAPK1 in the S338X nonsense transduced cell lines was observed when compared with two separate CXCR4 frameshift lines, indicating that some differences may exist within the WHIM-like spectrum of mutations [79]. Differences between nonsense ($CXCR4^{WHIM\text{-}NS}$) and frameshift ($CXCR4^{WHIM\text{-}FS}$) mutations can be seen in patient characteristics at diagnosis. While patients with any $CXCR4^{WHIM}$ mutation have decreased rates of adenopathy, patients with CXCR4$^{WHIM\text{-}NS}$ mutations have significantly increased BM involvement and serum IgM levels compared with CXCR4$^{WHIM\text{-}FS}$ and CXCR4^{WT} patients (Fig. 12.4) [66]. Likewise, CXCR4$^{WHIM\text{-}NS}$ patients presented with significantly higher rates of symptomatic disease and serum hyperviscosity syndrome (Fig. 12.5). No differences in overall survival based on CXCR4 mutation status have been observed.

The CXCR4^{WHIM} mutations may also offer a targeted approach to the therapy of WM given the successful employment of the CXCR4 agonist plexifor in the treatment of WHIM syndrome patients [80]. This is particularly pertinent as the WM cell lines BCWM.1 and MWCL-1 transduced with the S338X CXCR4$^{WHIM\text{-}NS}$ mutation demonstrate resistance to a variety of anti-neoplastic agents including the alkylator bendamustine,

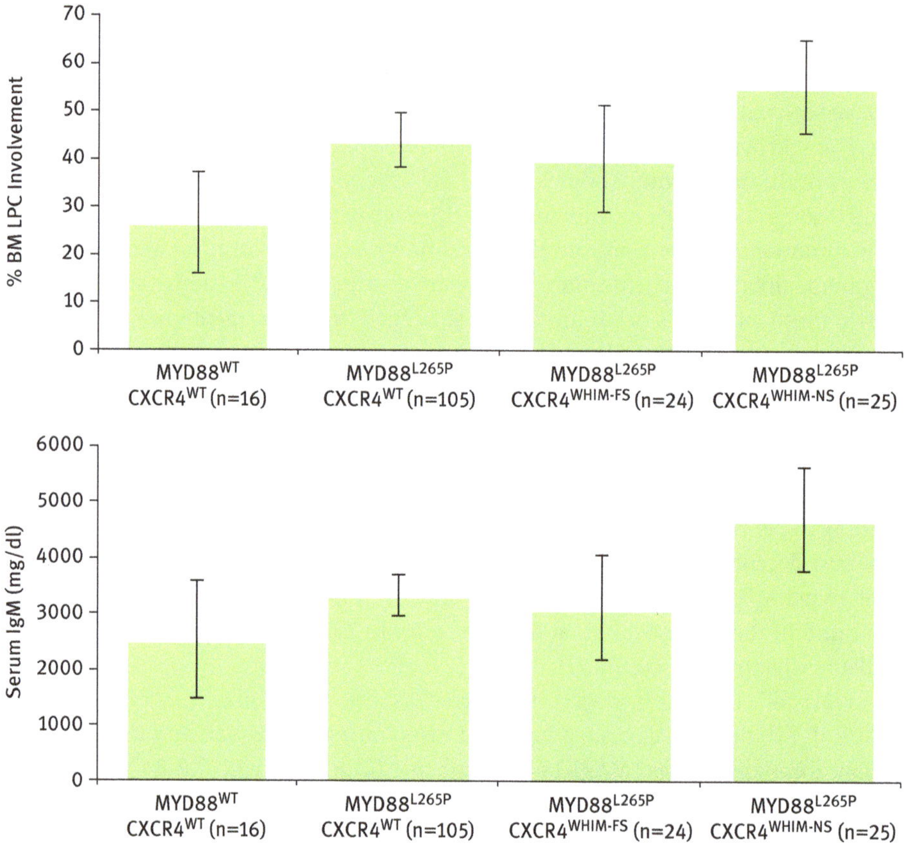

Fig. 12.4: Differences in BM involvement and serum IgM levels by CXCR4 and *MYD88* mutation status. Data represents mean values with two times the standard error of the mean. P = 0.05, 0.01, and 0.01 for serum IgM in CXCR4$^{WHIM\text{-}NS}$ versus CXCR4$^{WHIM\text{-}FS}$ MYD888^{L265P}CXCR4^{WT}, and MYD88^{WT} respectively. P=0.05 and 0.005 for WM BM involvement in CXCR$^{WHIM\text{-}NS}$ versus CXCR4$^{WHIM\text{-}FS}$ and MYD88^{WT} respectively.

MYD88	**Wild Type**	**L265P**		
CXCR4	**Wild Type**		**WHIM**	**WHIM-NS**
Genotype *Prevalence*	MYD88^{WT}CXCR4^{WT} ~10 % of patients Poor prognosis/Overall Survival Low serum IgM Low BM involvement	MYD88^{L265P}CXCR4^{WT} ~60 % of patients	MYD88^{L265P}CXCR4^{WHIM} ~30 % of patients Low adenopathy Therapeutic resistance	MYD88^{L265P}CXCR4$^{WHIM-NS}$ ~50 % of CXCR4^{WHIM} High BM High IgM Hyperviscocity syndrome More symptomatic

Fig. 12.5: Chart of WM patient characteristics at diagnosis by *MYD88* and *CXCR4* mutation status.

the nucleoside analog fludarabine, the proteasome inhibitor bortezomib, and the PI3K inhibitor idelalisib in the presence of SDF-1α [79]. Patients with any type of CXCR4^{WHIM} mutation also have lower overall and major responses to the BTK inhibitor ibrutinib in relapsed or refractory WM disease [81]. However, in one clinical trial using a combination of rituximab, dexamethasone, and carflizomib, no differences in response based on CXCR4 genotype were observed suggesting that the protective effects of CXCR4 signaling due to WHIM mutations may apply to certain therapeutics [47].

12.7 Other recurrent mutations

Several additional somatically altered genes are associated with WM including the chromatin remodeling proteins, ARID1A and ARID1B [40, 82]. In one WM whole genome sequencing study, ARID1A mutations were found in 5/30 (17 %) cases, one of which was opposite a somatic deletion and another made homozygous by an aUPD. The alternate family member ARID1B is contained within the region of chromosome 6q that is commonly deleted in WM patients [17, 42, 44, 82]. Both ARID1A and ARID1B are members of the SWI/SNF family of proteins, which are known to be mutated in other neoplastic malignancies wherein they are thought to exert their effects via TP53 and CDKN1A regulation [83–85]. Moreover, TP53 itself is either somatically deleted or mutated in 5–10 % of WM patients [40, 43, 44].

Two studies identified somatic mutations in CD79B in 7–9 % of WM cases, which included one report of a CD79B mutation that was homozygous by an aUPD event [40, 86]. CD79B is mutated in 12–18 % of ABC-DLBCL, where it is thought to activate B-cell receptor signaling [87, 88]. CD79B mutations were also reported in 23 % of immune privileged DLBCL cases, with one study reporting CD79B mutations in 11/18 (61 %) of cases of primary DLBCL of the CNS [88, 89]. These represent the DLBCL populations with the highest rate of MYD88^{L265P} mutations and indicate possible synergistic crosstalk between CD79B and MYD88 mutations. Many of the other mutations associated with WM impact NF-κB signaling and the TLR4/MYD88 pathway. MYBBP1A was mutated in 2/30 (7 %) patients and is thought to inhibit NF-κB activity by repressing RELA [40, 90]. Somatic copy loss of HIVEP2 is present in 23/30 (77 %) patients and competes with NF-κB for DNA binding sites [40, 91]. Biallelic loss of TRAF3, a negative regulator of NF-κB, has been reported in two studies at a frequency of 2–3 % of WM patients [40, 41]. Finally, partial methylation of DLEU7 was reported in a study in all 12/12 patients [86]. DLEU7 inhibits transmembrane activator and CAML interactor (TACI; TNFRSF13B at the gene level) and B-cell maturation antigen (BCMA; TNFRSF17 at the gene level), leading to suppression of NF-κB signaling [92].

12.8 Microenvironmental interactions in WM

Increased numbers of mast cells are found in the bone marrow of WM patients, wherein they are usually admixed with tumor aggregates [1, 30, 93]. The role of mast cells in WM has been investigated in one study wherein co-culture of primary autologous or mast cell lines with WM LPC resulted in dose-dependent WM cell proliferation and/or tumor colony formation, primarily through CD40 ligand (CD40L; CD40LG at gene level) signaling [93]. Furthermore, WM cells release soluble CD27 (sCD27), which may be mediated by cleavage of membrane bound CD27 by matrix metalloproteinase 8 (MMP8) [94]. sCD27 levels are elevated in the serum of WM patients, and follow disease burden in mice engrafted with WM cells, as well as in WM patients [95]. Importantly, sCD27 triggers the upregulation of CD40L as well as a proliferation inducing ligand (APRIL; TNFSF13 at gene level) on both mast cells derived from WM patients and mast cell lines through its receptor CD70. Modeling in mice engrafted with a CD70 blocking antibody shows inhibition of tumor cell growth, suggesting that WM cells require a microenvironmental support system for their growth and survival (Fig. 12.6) [95]. High levels of CXCR4 and very late antigen-4 (VLA-4; an ITGA4 and ITGB1 dimer) have also been observed in WM cells [68]. In blocking experiments studies, CXCR4 was shown to support migration of WM cells, while VLA-4 contributed to adhesion of WM cells to bone marrow stromal cells.

12.9 Concluding remarks

Waldenström's macroglobulinemia is a distinct B-cell lymphoproliferative disorder characterized by production of an IgM paraprotein, bone marrow involvement with LPC, and highly prevalent somatic mutations in MYD88 and CXCR4. Somatic mutations in MYD88 and CXCR4 impact disease presentation, while MYD88 mutation status is a determinant of overall survival. MYD88 and CXCR4 inhibitors are under investigation for the treatment of WM. Soluble CD27 release by WM cells triggers growth and survival signaling by neighboring mast cells. Inhibitors targeting microenvironmental signaling are under development.

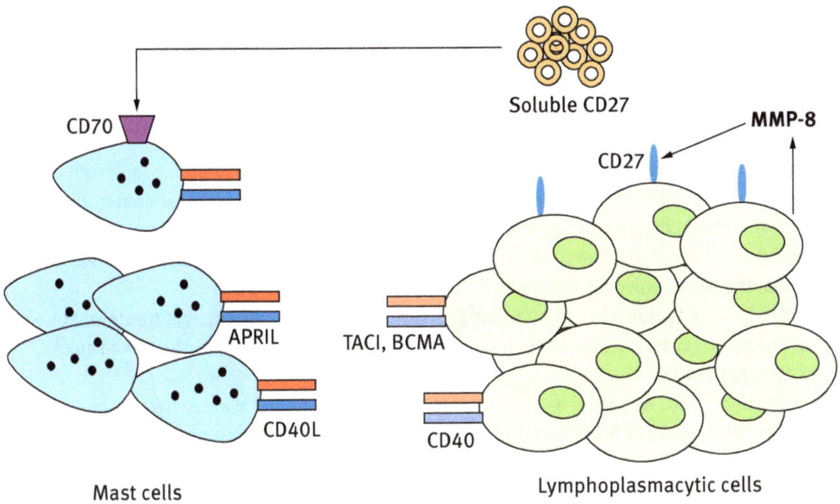

Fig. 12.6: **Microenvironmental signaling in WM.** Soluble CD27 is cleaved from the WM LPC by MMP8 and activates CD70 on mast cells admixed with the tumor. CD70 promotes the expression of APRIL and CD40L on the mast cells. APRIL binds to TACI and BCMA, while CD40L binds to CD40. These receptors are expressed on the WM LPC and promote cell growth and survival.

References

[1] Swerdlow SH, Campo E, Harris NL, et al. WHO Classification of Tumours of Haematopoietic and Lymphoid Tissues. 2008.

[2] Owen RG, Treon SP, Al-Katib A, et al. Clinicopathological definition of Waldenstrom's macroglobulinemia: consensus panel recommendations from the Second International Workshop on Waldenstrom's Macroglobulinemia. Semin Oncol 2003;30(2):110–115.

[3] Waldenström J. Incipient myelomatosis or "essential" hyperglobulinemia with fibrinogenopenia – A new syndrome? Acta Med Scand 1944;216(117).

[4] Treon SP. How I treat Waldenström macroglobulinemia. Blood 2009;114(12):2375–2385.

[5] Groves FD, Travis LB, Devesa SS, Ries LA, Fraumeni JF. Waldenström's macroglobulinemia: incidence patterns in the United States, 1988-1994. Cancer 1998;82(6):1078–1081.

[6] Herrinton LJ, Weiss NS. Incidence of Waldenström's macroglobulinemia. Blood 1993;82(10):3148–3150.

[7] Morel P, Duhamel A, Gobbi P, et al. International prognostic scoring system for Waldenstrom macroglobulinemia. Blood 2009;113(18):4163–4170.

[8] Castillo JJ, Olszewski AJ, Cronin AM, Hunter ZR, Treon SP. Survival trends in Waldenström macroglobulinemia: an analysis of the Surveillance, Epidemiology and End Results database. Blood 2014;123(25):3999–4000.

[9] Levine T, Pestronk A, Florence J, et al. Peripheral neuropathies in Waldenström's macroglobulinaemia. J Neurol Neurosurg Psychiatry 2006;77(2):224–248.

[10] Menke MN, Feke GT, McMeel JW, Branagan A, Hunter Z, Treon SP. Hyperviscosity-related retinopathy in waldenstrom macroglobulinemia. Arch Ophthalmol 2006;124(11):1601–1606.

[11] Hunter ZR, Manning RJ, Hanzis C, et al. IgA and IgG hypogammaglobulinemia in Waldenström's macroglobulinemia. Haematologica 2010;95(3):470–475.

[12] Schuster SR, Rajkumar SV, Dispenzieri A, et al. IgM multiple myeloma: disease definition, prognosis, and differentiation from Waldenstrom's macroglobulinemia. Am J Hematol 2010;85(11):853–855.
[13] Kyle RA, Therneau TM, Rajkumar SV, et al. Long-term follow-up of IgM monoclonal gammopathy of undetermined significance. Blood 2003;102(10):3759–3764.
[14] Landgren O, Kristinsson SY, Goldin LR, et al. Risk of plasma cell and lymphoproliferative disorders among 14621 first-degree relatives of 4458 patients with monoclonal gammopathy of undetermined significance in Sweden. Blood 2009;114(4):791–795.
[15] Weinhold N, Johnson DC, Rawstron AC, et al. Inherited genetic susceptibility to monoclonal gammopathy of unknown significance. Blood 2014;123(16):2513–2517.
[16] Treon SP, Tripsas C, Hanzis C, et al. Familial disease predisposition impacts treatment outcome in patients with Waldenström macroglobulinemia. Clin Lymphoma Myeloma Leuk 2012;12(6):433–437.
[17] Treon SP, Hunter ZR, Aggarwal a, et al. Characterization of familial Waldenstrom's macroglobulinemia. Ann Oncol 2006;17(3):488–494.
[18] McMaster ML, Goldin LR, Bai Y, et al. Genomewide linkage screen for Waldenstrom macroglobulinemia susceptibility loci in high-risk families. Am J Hum Genet 2006;79(4):695–701.
[19] Kristinsson SY, Björkholm M, Goldin LR, McMaster ML, Turesson I, Landgren O. Risk of lymphoproliferative disorders among first-degree relatives of lymphoplasmacytic lymphoma/Waldenstrom macroglobulinemia patients: a population-based study in Sweden. Blood 2008;112(8):3052–3056.
[20] Morra E, Varettoni M, Tedeschi A, et al. Associated cancers in Waldenström macroglobulinemia: clues for common genetic predisposition. Clin Lymphoma Myeloma Leuk 2013;13(6):700–703.
[21] Hanzis C, Ojha RP, Hunter Z, et al. Associated malignancies in patients with Waldenström's macroglobulinemia and their kin. Clin Lymphoma Myeloma Leuk 2011;11(1):88–92.
[22] Ögmundsdóttir HM, Steingrímsdóttir H, Haraldsdóttir V. Familial paraproteinemia: hyperresponsive B-cells as endophenotype. Clin Lymphoma Myeloma Leuk 2011;11(1):82–84.
[23] Ogmundsdóttir HM, Sveinsdóttir S, Sigfússon A, Skaftadóttir I, Jónasson JG, Agnarsson BA. Enhanced B cell survival in familial macroglobulinaemia is associated with increased expression of Bcl-2. Clin Exp Immunol 1999;117(2):252–260.
[24] Kristinsson SY, Koshiol J, Björkholm M, et al. Immune-related and inflammatory conditions and risk of lymphoplasmacytic lymphoma or Waldenstrom macroglobulinemia. J Natl Cancer Inst 2010;102(8):557–567.
[25] Koshiol J, Gridley G, Engels EA, McMaster ML, Landgren O. Chronic immune stimulation and subsequent Waldenström macroglobulinemia. Arch Intern Med 2008;168(17):1903–1909.
[26] Silvestri F, Barillari G, Fanin R, et al. Risk of hepatitis C virus infection, Waldenström's macroglobulinemia, and monoclonal gammopathies. Blood 1996;88(3):1125–1126.
[27] Leleu X, Connor KO, Ho AW, et al. Hepatitis C viral infection is not associated with Waldenström's macroglobulinemia. Am J Hematol 2007;82(1):83–84.
[28] Preud'homme JL, Seligmann M. Immunoglobulins on the surface of lymphoid cells in Waldenström's macroglobulinemia. J Clin Invest 1972;51(3):701–705.
[29] Smith BR, Robert NJ, Ault KA. In Waldenstrom's macroglobulinemia the quantity of detectable circulating monoclonal B lymphocytes correlates with clinical course. Blood 1983;61(5):911–904.
[30] San Miguel JF, Vidriales MB, Ocio E, et al. Immunophenotypic analysis of Waldenstrom's macroglobulinemia. Semin Oncol 2003;30(2):187–195.

[31] Owen RG, Barrans SL, Richards SJ, et al. Waldenström macroglobulinemia. Development of diagnostic criteria and identification of prognostic factors. Am J Clin Pathol 2001;116(3):420–428.

[32] Hunter ZR, Branagan AR, Manning R, et al. CD5, CD10, and CD23 expression in Waldenstrom's macroglobulinemia. Clin Lymphoma 2005;5(4):246–249.

[33] Paiva B, Montes MC, García-Sanz R, et al. Multiparameter flow cytometry for the identification of the Waldenström's clone in IgM MGUS and Waldenström's Macroglobulinemia: New criteria for differential diagnosis and risk stratification. Leukemia 2013.

[34] Wagner SD, Martinelli V, Luzzatto L. Similar patterns of V kappa gene usage but different degrees of somatic mutation in hairy cell leukemia, prolymphocytic leukemia, Waldenstrom's macroglobulinemia, and myeloma. Blood 1994;83(12):3647–3653.

[35] Aoki H, Takishita M, Kosaka M, Saito S. Frequent somatic mutations in D and/or JH segments of Ig gene in Waldenström's macroglobulinemia and chronic lymphocytic leukemia (CLL) with Richter's syndrome but not in common CLL. Blood 1995;85(7):1913–1919.

[36] Shiokawa S, Suehiro Y, Uike N, Muta K, Nishimura J. Sequence and expression analyses of mu and delta transcripts in patients with Waldenström's macroglobulinemia. Am J Hematol 2001;68(3):139–143.

[37] Sahota SS, Forconi F, Ottensmeier CH, et al. Typical Waldenstrom macroglobulinemia is derived from a B-cell arrested after cessation of somatic mutation but prior to isotype switch events. Blood 2002;100(4):1505–1507.

[38] Paramithiotis E, Cooper MD. Memory B lymphocytes migrate to bone marrow in humans. Proc Natl Acad Sci USA 1997;94(1):208–212.

[39] Ackroyd S, O'Connor SJM, Owen RG. Rarity of IgH translocations in Waldenström macroglobulinemia. Cancer Genet Cytogenet 2005;163(1):77–80.

[40] Hunter ZR, Xu L, Yang G, et al. The genomic landscape of Waldenstom's Macroglobulinemia is characterized by highly recurring MYD88 and WHIM-like CXCR4 mutations, and small somatic deletions associated with B-cell lymphomagenesis. Blood 2013;123(11):1637–1646.

[41] Braggio E, Keats JJ, Leleu X, et al. High-resolution genomic analysis in Waldenström's macroglobulinemia identifies disease-specific and common abnormalities with marginal zone lymphomas. Clin Lymphoma Myeloma 2009;9(1):39–42.

[42] Schop RFJ, Van Wier SA, Xu R, et al. 6q deletion discriminates Waldenström macroglobulinemia from IgM monoclonal gammopathy of undetermined significance. Cancer Genet Cytogenet 2006;169(2):150–153.

[43] Nguyen-Khac F, Lambert J, Chapiro E, et al. Chromosomal aberrations and their prognostic value in a series of 174 untreated patients with Waldenstrom's macroglobulinemia. Haematologica 2013;98(4):649–654.

[44] Braggio E, Keats JJ, Leleu X, et al. Identification of copy number abnormalities and inactivating mutations in two negative regulators of nuclear factor-kappaB signaling pathways in Waldenstrom's macroglobulinemia. Cancer Res 2009;69(8):3579–3588.

[45] Leleu X, Eeckhoute J, Jia X, et al. Targeting NF-kappaB in Waldenstrom macroglobulinemia. Blood 2008;111(10):5068–5077.

[46] Treon SP, Hunter ZR, Matous J, et al. Multicenter clinical trial of bortezomib in relapsed/refractory Waldenstrom's macroglobulinemia: results of WMCTG Trial 03-248. Clin Cancer Res 2007;13(11):3320–3325.

[47] Treon SP, Tripsas CK, Meid K, et al. Carfilzomib, rituximab and dexamethasone (CaRD) is active and offers a neuropathy-sparing approach for proteasome-inhibitor based therapy in Waldenstrom's macroglobulinemia. Blood 2014;124(4):503–510.

[48] Treon SP, Xu L, Yang G, et al. MYD88 L265P somatic mutation in Waldenström's macroglobulinemia. N Engl J Med 2012;367(9):826–833.

[49] Cohen L, Henzel WJ, Baeuerle PA. IKAP is a scaffold protein of the IkappaB kinase complex. Nature 1998;395(6699):292–296.

[50] Loiarro M, Gallo G, Fantò N, et al. Identification of critical residues of the MyD88 death domain involved in the recruitment of downstream kinases. J Biol Chem 2009;284(41):28093–28103.

[51] Lin S-C, Lo Y-C, Wu H. Helical assembly in the MyD88-IRAK4-IRAK2 complex in TLR/IL-1R signalling. Nature 2010;465(7300):885–890.

[52] Ngo VN, Young RM, Schmitz R, et al. Oncogenically active MYD88 mutations in human lymphoma. Nature 2011;470(7332):115–119.

[53] Yang G, Zhou Y, Liu X, et al. A mutation in MYD88 (L265P) supports the survival of lymphoplasmacytic cells by activation of Bruton tyrosine kinase in Waldenström macroglobulinemia. Blood 2013;88(7):1222–1232.

[54] Kawagoe T, Sato S, Matsushita K, et al. Sequential control of Toll-like receptor-dependent responses by IRAK1 and IRAK2. Nat Immunol 2008;9(6):684–691.

[55] Loiarro M, Sette C, Gallo G, et al. Peptide-mediated interference of TIR domain dimerization in MyD88 inhibits interleukin-1-dependent activation of NF-{kappa}B. J Biol Chem 2005;280(16):15809–15814.

[56] Treon SP, Hunter ZR. A new era for Waldenstrom macroglobulinemia: MYD88 L265P. Blood 2013;121(22):4434–4436.

[57] Jiménez C, Sebastián E, Del Carmen Chillón M, et al. MYD88 L265P is a marker highly characteristic of, but not restricted to, Waldenström's macroglobulinemia. Leukemia 2013;27(8):1722–1728.

[58] Varettoni M, Arcaini L, Zibellini S, et al. Prevalence and clinical significance of the MYD88 (L265P) somatic mutation in Waldenstrom's macroglobulinemia and related lymphoid neoplasms. Blood 2013;121(13):2522–2528.

[59] Poulain S, Roumier C, Decambron A, et al. MYD88 L265P mutation in Waldenstrom's macroglogulinemia. Blood 2013;121(22):4504–4511.

[60] Ansell SM, Hodge LS, Secreto FJ, et al. Activation of TAK1 by MYD88 L265P drives malignant B-cell Growth in non-Hodgkin lymphoma. Blood Cancer J 2014;4:e183.

[61] Xu L, Hunter ZR, Yang G, et al. MYD88 L265P in Waldenstrom's Macroglobulinemia, IgM Monoclonal Gammopathy, and other B-cell Lymphoproliferative Disorders using Conventional and Quantitative Allele-Specific PCR. Blood 2013;121(11):2051–2058.

[62] Xu L, Hunter ZR, Yang G, et al. Detection of MYD88 L265P in peripheral blood of patients with waldenström's macroglobulinemia and IgM monoclonal gammopathy of undetermined significance. Leukemia 2014;28(8):1698–1707.

[63] Puente XS, Pinyol M, Quesada V, et al. Whole-genome sequencing identifies recurrent mutations in chronic lymphocytic leukaemia. Nature 2011;475(7354):3–7.

[64] Poulain S, Boyle EM, Roumier C, et al. MYD88 L265P mutation contributes to the diagnosis of Bing Neel syndrome. Br J Haematol 2014;167(4):506–513.

[65] Baldini L, Goldaniga M, Guffanti A, et al. Immunoglobulin M monoclonal gammopathies of undetermined significance and indolent Waldenstrom's macroglobulinemia recognize the same determinants of evolution into symptomatic lymphoid disorders: proposal for a common prognostic scoring system. J Clin Oncol 2005;23(21):4662–4668.

[66] Treon SP, Cao Y, Xu L, Yang G, Liu X, Hunter ZR. Somatic mutations in MYD88 and CXCR4 are determinants of clinical presentation and overall survival in Waldenstrom's macroglobulinemia. Blood 2014;123(18):2791–2796.

[67] Morin RD, Mendez-Lago M, Mungall AJ, et al. Frequent mutation of histone-modifying genes in non-Hodgkin lymphoma. Nature 2011;476(7360):298–303.

[68] Ngo HT, Leleu X, Lee J, et al. SDF-1/CXCR4 and VLA-4 interaction regulates homing in Waldenstrom macroglobulinemia. Blood 2008;112(1):150–158.
[69] Haribabu B, Richardson RM, Fisher I, et al. Regulation of human chemokine receptors CXCR4. Role of phosphorylation in desensitization and internalization. J Biol Chem 1997;272(45):28726–28731.
[70] Teicher BA, Fricker SP. CXCL12 (SDF-1)/CXCR4 pathway in cancer. Clin Cancer Res 2010;16(11):2927–2931.
[71] Cojoc M, Peitzsch C, Trautmann F, Polishchuk L, Telegeev GD, Dubrovska A. Emerging targets in cancer management: role of the CXCL12/CXCR4 axis. Onco Targets Ther 2013;6:1347–1361.
[72] Lagane B, Chow KYC, Balabanian K, et al. CXCR4 dimerization and beta-arrestin-mediated signaling account for the enhanced chemotaxis to CXCL12 in WHIM syndrome. Blood 2008;112(1):34–44.
[73] Wu B, Chien EYT, Mol CD, et al. Structures of the CXCR4 chemokine GPCR with small-molecule and cyclic peptide antagonists. Science (80-) 2010;330(6007):1066–1071.
[74] Orsini MJ. Trafficking of the HIV Coreceptor CXCR4. Role of arrestins and identification of residues in the c-terminal tail that mediate receptor internalization. J Biol Chem 1999;274(43):31076–31086.
[75] Mc Guire PJ, Cunningham-Rundles C, Ochs H, Diaz GA. Oligoclonality, impaired class switch and B-cell memory responses in WHIM syndrome. Clin Immunol 2010;135(3):412–421.
[76] Nakagawa M, Schmitz R, Xiao W, et al. Gain-of-function CCR4 mutations in adult T cell leukemia/lymphoma. J Exp Med 2014;211(13):2497–2505.
[77] Cao Y, Hunter ZR, Liu X, et al. The WHIM-like CXCR4(S338X) somatic mutation activates AKT and ERK, and promotes resistance to ibrutinib and other agents used in the treatment of Waldenstrom's Macroglobulinemia. Leukemia 2014.
[78] Roccaro AM, Sacco A, Jimenez C, et al. C1013G/CXCR4 acts as a driver mutation of tumor progression and modulator of drug resistance in lymphoplasmacytic lymphoma. Blood 2014;123(26):4120–4131.
[79] Cao Y, Hunter ZR, Liu X, et al. CXCR4 WHIM-like frameshift and nonsense mutations promote ibrutinib resistance but do not supplant MYD88(L265P) -directed survival signalling in Waldenström macroglobulinaemia cells. Br J Haematol 2014.
[80] Dale DC, Bolyard AA, Kelley ML, et al. The CXCR4 antagonist plerixafor is a potential therapy for myelokathexis, WHIM syndrome. Blood 2011;118(18):4963–4966.
[81] Tripsas CK, Yang G, Cao Y, et al. A Prospective Multicenter Study Of The Bruton's Tyrosine Kinase Inhibitor Ibrutinib In Patients With Relapsed Or Refractory Waldenstrom's Macroglobulinemia. Blood 2013;122(21):251.
[82] Nagl NG, Wang X, Patsialou A, Van Scoy M, Moran E. Distinct mammalian SWI/SNF chromatin remodeling complexes with opposing roles in cell-cycle control. EMBO J 2007;26(3):752–763.
[83] Jones S, Li M, Parsons DW, et al. Somatic mutations in the chromatin remodeling gene ARID1A occur in several tumor types. Hum Mutat 2012;33(1):100–103.
[84] Guan B, Wang T-L, Shih I-M. ARID1A, a factor that promotes formation of SWI/SNF-mediated chromatin remodeling, is a tumor suppressor in gynecologic cancers. Cancer Res 2011;71(21):6718–6727.
[85] Inoue H, Giannakopoulos S, Parkhurst CN, et al. Target genes of the largest human SWI/SNF complex subunit control cell growth. Biochem J 2011;434(1):83–92.
[86] Poulain S, Roumier C, Galiègue-Zouitina S, et al. Genome wide SNP array identified multiple mechanisms of genetic changes in Waldenstrom macroglobulinemia. Am J Hematol. 2013 Nov;88(11):948–54.
[87] Davis RE, Ngo VN, Lenz G, et al. Chronic active B-cell-receptor signalling in diffuse large B-cell lymphoma. Nature 2010;463(7277):88–92.

[88] Kraan W, Horlings HM, van Keimpema M, et al. High prevalence of oncogenic MYD88 and CD79B mutations in diffuse large B-cell lymphomas presenting at immune-privileged sites. Blood Cancer J 2013;28(3):719–720.

[89] Yamada S, Ishida Y, Matsuno A, Yamazaki K. Primary diffuse large B-cell lymphomas of central nervous system exhibit remarkably high prevalence of oncogenic MYD88 and CD79B mutations. Leuk Lymphoma 2014;1–18.

[90] Owen HR, Elser M, Cheung E, Gersbach M, Kraus WL, Hottiger MO. MYBBP1a is a novel repressor of NF-kappaB. J Mol Biol 2007;366(3):725–736.

[91] Iwashita Y, Fukuchi N, Waki M, Hayashi K, Tahira T. Genome-wide repression of NF-κB target genes by transcription factor MIBP1 and its modulation by O-linked β-N-acetylglucosamine (O-GlcNAc) transferase. J Biol Chem 2012;287(13):9887–9900.

[92] Palamarchuk A, Efanov A, Nazaryan N, et al. 13q14 deletions in CLL involve cooperating tumor suppressors. Blood 2010;115(19):3916–3922.

[93] Tournilhac O, Santos DD, Xu L, et al. Mast cells in Waldenstrom's macroglobulinemia support lymphoplasmacytic cell growth through CD154/CD40 signaling. Ann Oncol 2006;17(8):1275–1282.

[94] Zhou Y, Liu X, Xu L, et al. Matrix metalloproteinase-8 is overexpressed in waldenström's macroglobulinemia cells, and specific inhibition of this metalloproteinase blocks release of soluble CD27. Clin Lymphoma Myeloma Leuk 2011;11(1):172–175.

[95] Ho AW, Hatjiharissi E, Ciccarelli BT, et al. CD27-CD70 interactions in the pathogenesis of Waldenstrom macroglobulinemia. Blood 2008;112(12):4683–4689.

Marta Chesi and P. Leif Bergsagel

13 Molecular pathogenesis of multiple myeloma

13.1 Introduction

Multiple myeloma (MM) is a post-germinal center plasma cell malignancy that forms multiple bone-destroying bone marrow (myel-) tumors (-oma), aptly called plasma cell myeloma by the World Health Organization. It is estimated that in 2015 in the USA 27,000 people will be diagnosed with MM, and 11,000 will die with MM [1]. In contrast to stable or declining trends for the incidence of most cancers, the delay-adjusted incidence of MM appears to have increased at a rate of 0.7 % per year from 1975–2007, and 3.3 % per year from 2007–2012 [2]. In the past, apparent increases in the incidence of MM have frequently been associated with increased case ascertainment due to improving access to and quality of health care. In contrast, the mortality rate has been declining for much of this time (1994–2009) because of improved treatments that have increased the five-year survival rate reported in the SEER database from 27 % in 1988 to 47 % in 2008 [2].

13.2 Clinical stages of plasma cell neoplasms: a new definition of multiple myeloma

MM mostly – perhaps always – is preceded by a pre-malignant MGUS (monoclonal gammopathy of undetermined significance) in the absence of end-organ damage [3], which is present in 4 % of individuals over the age of 50 [4]. Presently there are no molecular or phenotypic markers that unequivocally discriminate MGUS from MM tumor cells, however MGUS can sporadically progress to symptomatic MM expressing the same monoclonal immunoglobulin at an average rate of ~ 1 % per year. Asymptomatic/smoldering MM has a larger tumor mass than MGUS but sporadically progresses to symptomatic MM at an average rate of ~ 10 % per year for the first five years, and then decreasing to a rate of ~ 3 % per year, suggesting that SMM is not a uniform disease [5]. Recently, investigators have been able to identify subsets of patients with SMM that progress very quickly to MM (~ 80 % at 2 years). This has led to a revised definition of MM that includes these high risk SMM patients, so that their treatment can be started before they develop end-organ damage (Fig. 13.1). Genetic biomarkers in SMM associated with increased risk of progression to MM are t(4;14) and del17p (50 % at 2 years) [6], and to a lesser extent add1q and hyperdiploidy (HRD) [7]. Extramedullary MM, often manifested as primary or secondary plasma cell leukemia (PCL), is a more aggressive stage of disease [8, 9]. MM cell lines (MMCLs) are almost always generated from extramedullary MM tumors indicating their independence from the BM microenvironment [10].

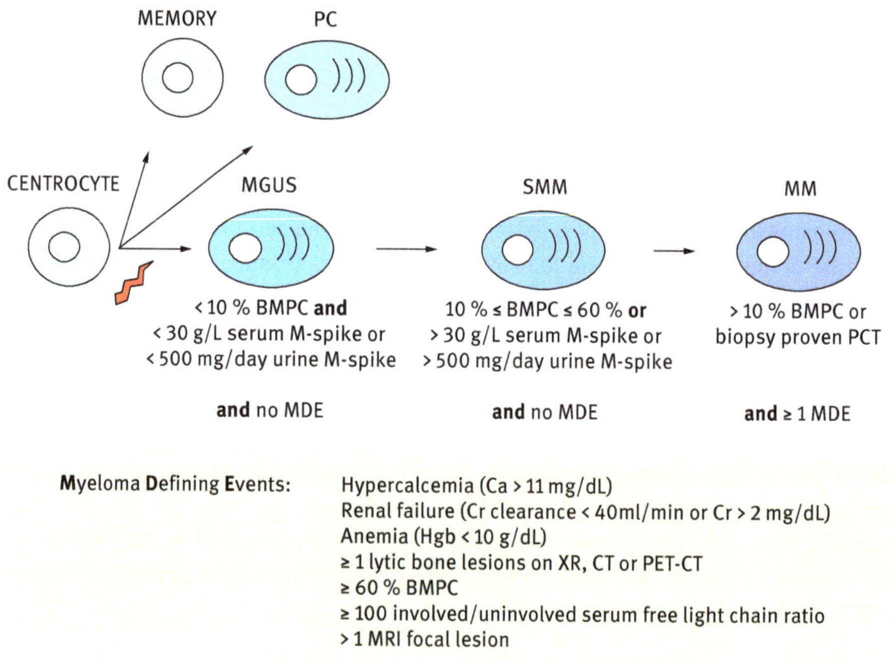

Fig. 13.1: **Origin and multistage clinical progression of multiple myeloma.** MGUS is a premalignant stage of MM and results from a transformation event occurring in the germinal center. Diagnostic criteria distinguishing MGUS from smoldering MM and MM are indicated, as well as MM defining events.

The immunoglobulin genes in MGUS and MM cells have a high degree of somatic mutations of the variable regions, indicative of extensive exposure to somatic hypermutation and antigen selection in germinal centers [11, 12]. Most MGUS and MM tumors have undergone class switch recombination of the heavy chain, similar to long-lived bone marrow plasma cells, resulting in expression of IgG or IgA, and uncommonly IgE or IgD. Nonetheless, ~ 1 % have not undergone switch recombination and express IgM. Furthermore, no heavy chain, and only a light chain is expressed in ~ 15 % of patients, probably as a result of post-germinal center inactivation of heavy chain expression by a variety of mechanisms, including IGH translocations [13, 14].

13.3 Chromosome translocations of the immunoglobulin genes are present in half of patients with MGUS and MM

Errors during the B-cell specific DNA modification processes of class switch recombination and somatic hypermutation are thought to be responsible for IgH chromosomal translocations, since the breakpoints are usually adjacent to IgH switch regions, although sometimes to VDJ sequences [15]. Although there is frequent subclonal heterogeneity in MM [16], the primary chromosome translocations are identified at all stages of disease progression, consistent with representing primary oncogenic events in MGUS and MM. Results from several studies indicate that the prevalence of IGH translocations is ~ 45 % for MGUS tumors, ~ 55 % for intramedullary MM tumors, ~ 80 % in primary PCL, and ~ 80 % in MMCL [9, 17–22]. There is very little information about IGK or IGL rearrangements. However, IGL rearrangements were identified in about 15–20 % of MMCL or advanced MM tumors, whereas IGK rearrangements are much less frequent, i.e., ~ 2 % of MMCL or advanced MM tumors [22].

13.4 Three groups of primary IgH translocations in ~ 40 % of MM

Primary IGH translocations are defined as recurrent translocations that are present in both MGUS and MM tumors, and although very likely, it is unknown if they represent the initial oncogenic event [23] (Fig. 13.2). Seven recurrent IGH translocations comprise three translocation groups, 18 % Cyclin D (15 % CCND1 at 11q13, 1 % CCND2 at 12p13, 2 % CCND3 at 6p21), 8 % MAF (5 % MAF at 16q23, 2 % MAFB at 20q12, < 1 % MAFA at

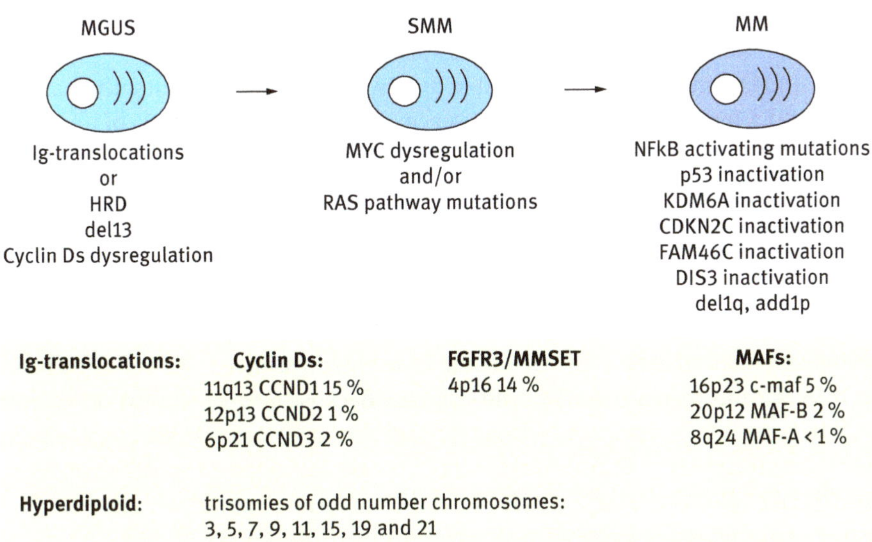

Fig. 13.2: **Primary and progression genetic events in multiple myeloma.** Clonal chromosome translocations involving the immunoglobulin loci and Cyclin Ds, FGFR3/MMSET or MAFs genes are mediated by errors during B-cell specific rearrangements events and are detected already at the MGUS stage. Like hyperdiploidy, they are thought to represent the primary transforming events leading to immortalization of the MGUS clone. They all converge on the *cis* or *trans* dysregulation of a D type cyclin. MYC dysregulation or (K)RAS activating mutations are likely to cause malignant progression. A variety of secondary, subclonal mutations further drive MM growth.

8q24.3) and 14 % MMSET/FGFR3 at 4p16 [15, 18, 20, 24–26]. The prevalence of t(4;14) (4 %) and t(14;16) is decreased, and t(11;14) increased in MGUS compared to MM. It is tempting to speculate that this different prevalence results from either a more or less rapid progression from MGUS to MM in these diverse cytogenetic groups. It is interesting that the t(11;14) is detected in 40 % of monoclonal gammopathies associated with primary amyloidosis, in 7/8 patients with IgM MM, 2/2 with IgE MM, and 11/14 with nonsecretory MM, but presently there is no biological explanation for these observations [27–29]. Clonal mutations in the first exon of CCND1 have been identified in ~ 10 % of t(11;14) patients, and were associated with decreased overall survival in the MRCXI but not the MRCIX clinical trial [30].

While cyclin D translocations only cause dysregulated expression of a single cyclin D gene, MAF translocations dysregulate a transcription factor that directly dysregulates CCND2 as well as numerous other genes, including adhesion proteins that increase the interaction of the MM cells with the bone marrow microenvironment [31]. This is evident in the MAF type of MMCL cells, which *in vitro* all adhere to plastic petri dishes. Subclonal MAF and MAFB mutations were detected in 12.5 % of t(14;16) and 25 % of t(14;20) patients and cluster in the basic-leucine zipper motif, with no effects on survival. Interestingly, a unique APOBEC mutational signature has been identified in the MAF groups, characterized by C > T, C > G and C > A mutations in a TpC context, and distinct from a generic C > T transition at CpG dinucleotides. The APOBEC3A and -3B genes are transcriptionally regulated by MAFs [30].

The t(4;14) translocation is unusual in that two genes are concomitantly dysregulated: MMSET by the IgH intronic enhancer on der4, and in about 80 % of t(4;14) tumors, FGFR3 by the IgH 3'enhancer on der14. Because activating mutations of FGFR3 are identified in ~ 17 % of t(4;14) MM, it is likely that activated FGFR3 plays an important role in tumorigenesis [30, 32], something that has also been shown using mouse models for wild-type FGFR3 [33]. On the other hand, the lack of expression of FGFR3 in 20 % of t(4;14) MM patients, often due to loss of der14, indicates that FGFR3 may also be dispensable and its oncogenic function replaced by RAS mutations [32]. MMSET is a histone methyltransferase for H3K36, and its overexpression causes global increase in H3K36me2 and decrease in H3K27me3 [34–36]. Because MMSET expression is never lost, it is thought to be the critical oncogenic target in t(4;14) MM. MMSET has also been shown to be recruited to sites of double-strand breaks following its Ser102 phosphorylation by ATM [37]. Importantly, MMSET appears to be an appropriate target for drug therapy, as inhibition of expression suppresses growth of MM [35].

13.5 Hyperdiploidy with multiple trisomies is another primary genetic event

About half of MM cases are characterized by HRD, typically containing 48–60 chromosomes, frequently with multiple trisomies of eight chromosomes (3, 5, 7, 9, 11, 15, 19, 21) [38, 39]. Interphase FISH studies revealed that HRD is less frequent in MGUS compared to MM [18, 40]. Primary IGH translocations are present in ~ 15 % of HRD MM but ~ 70 % of non-HRD (NHRD) MM [18]. Although not completely distinct, it appears that primary IgH translocations and HRD are associated with the two major transformation pathways of MGUS/MM pathogenesis representing early initiating events. The causes and molecular consequences of HRD have not been determined, however, it is notable that the fraction of MM tumors with HRD increases and that with IGH translocations decreases with age [41]. There is a remarkable parallel between MM and acute lymphocytic leukemia, which is also characterized by tumors having either translocations or HRD, the latter group, as in MM, having a better prognosis.

13.6 Universal dysregulation of a cyclin D/RB axis in MM

As noted above, a cyclin D gene is directly dysregulated by chromosome translocation in the cyclin D group, and CCND2 is a direct target of MAF in the MAF group. The MMSET group has elevated expression of CCND2, although the mechanism for this is not known. The HRD group with trisomy 11 has biallelic expression of CCND1, and most of the remaining patients express CCND2. About 1 % of MM does not express a CCND gene, most of which lack expression and have biallelic inactivation of RB [34]. Therefore, it appears that dysregulation of the cyclin D/RB axis is an early and unifying event in the pathogenesis of MGUS/MM.

13.7 Secondary genetic events associated with progression of MGUS to MM

A variety of genetic events have been cataloged in MM samples that in general occur with increasing frequency in MGUS > SMM > MM > MMCL, and probably drive disease progression (Fig. 13.2).

13.7.1 MYC

Rearrangements of the oncogene MYC were recently identified in half of patients with MM [24]. The expression of MYC increases progressively from normal PC > MGUS > MM without MYC rearrangements > MM with MYC rearrangements, with the expression of

MYC almost always monoallelic in the presence of rearrangements (i.e. *cis* dysregulation), and biallelic in their absence (i.e. *trans* dysregulation), suggesting that the progression of MGUS to MM is associated with genetic dysregulation of MYC. Importantly, the sporadic activation of MYC in the post-germinal center B cells of a strain of mouse (C57BL/6) that spontaneously develop monoclonal gammopathy, but not in one (BALB/c) that does not, universally resulted in the progression of the MGUS to MM, indicating that the genetic dysregulation of MYC can also drive the progression of MGUS to MM in some cases [24]. Genetic rearrangements, rarely primary translocation, juxtapose MYC to a promiscuous array of enhancers (mostly superenhancers) from elsewhere in the genomes, although in some cases only duplication of a CTCF site located ~ 500kb telomeric of MYC is identified (Fig. 13.3). Of note, in one study the prevalence of MYC rearrangements in the t(11;14) is 25 %, but in HRD is 66 % [24], while another study found a negative correlation with t(4;14), and positive correlation with HRD [30]. The frequency of involvement of immunoglobulin enhancers is similar across the genetic subgroups of MM, while nonimmunoglobulin enhancers are much more frequent in HRD. These findings provide support for the hypothesis that different genetic subgroups of MM have distinct mechanisms of disease progression. Although it has been difficult to target dysregulated MYC protein in cancer, it is now possible to specifically target superenhancer mediated dysregulated expression of MYC RNA us-

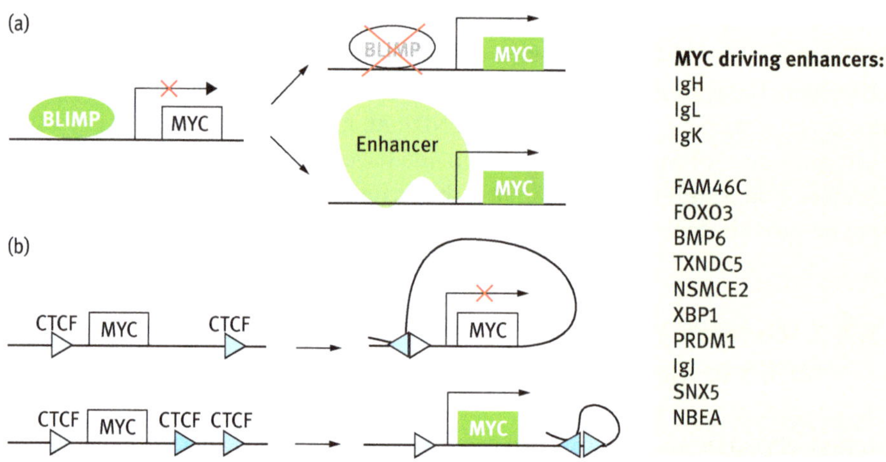

Fig. 13.3: **Three alternative ways of dysregulating MYC expression in multiple myeloma.** (a) The binding of the transcriptional repressor BLIMP on the MYC promoter inhibits MYC expression in terminal differentiated plasma cells. BLIMP mutations or juxtaposition of enhancers or superenhancers to the MYC locus dysregulate its expression in MM. The identity of recurrent (found more than once) enhancers driving MYC expression is indicated in the gray box. (b) Chromosome looping occurring between two CTCF sites, one just upstream the MYC promoter and the other one ~ 500kb downstream contributes to insulate the MYC locus and repress MYC transcription. Frequent duplication of the downstream CTCF site alters the 3D chromatin conformation and releases the insulation of the MYC locus.

ing drugs that compete with bromodomain and end terminal repeat family recognition of acetylated histones [42].

13.7.2 Deletion 13

Monosomy 13 is reported in MGUS (e.g. t(4;14) or t(14;16) MGUS) but also associated with progression in MM (e.g. in t(11;14)) [15]. It is more frequent in HRD than NHRD MM. Simple haploinsuffiency of RB1 has been postulated to be oncogenic [20]. Whole exome sequencing studies have identified that 3 % of HRD, but 21 % of NHRD MM have missense mutations of DIS3, a ribonuclease subunit of the exosome complex [43].

13.7.3 RAS pathway

The prevalence of activating mutations of *NRAS* or *KRAS* is about 20 % each in MM [20]. Whereas mutations of *NRAS* are present in 7 % of MGUS, no mutations of *KRAS* have been reported [15]. This supports the hypothesis that activating mutations of KRAS may represent a critical transition point between MGUS and MM. MM cell survival depends on maintaining expression of activated, but not wild-type, RAS, identifying the former as a good target for drug therapy [44]. In ~6 % of MM BRAF mutations have been identified, usually, but not always, in patients lacking NRAS or KRAS mutations. The presence of MM patients with different combination of subclones harboring mutations of NRAS, KRAS, BRAF or FGFR3, all causing constitutive MAPK activation, is an example of convergent evolution, indicating these mutations play similar, overlapping roles in tumorigenesis [32, 43].

NF-κB pathway mutations. The BM microenvironment is rich in ligands (BAFF, APRIL) that stimulate receptors (TACI, BCMA, APRIL) on normal and malignant PC, eventually leading to nuclear localization of NFKB1 and NFKB2 and transcription of their target genes, explaining the dependency of MM cells on the BM microenvironment [45]. A promiscuous array of mutations that inactivate negative regulators (cIAP1/2, TRAF2, TRAF3, CYLD, TNFAIP3), and activate positive regulators (NIK, NFKB1, NKFB2, TACI, LTBR, CD40) of the NFKB pathway have been described in about 20 % of untreated MM, and about half of MMCL. These mutations are predicted to decrease the MM dependence on BM microenvironment, expanding the PC niche and contributing to growth outside the BM.

Del17p. Loss of signal for a FISH probe including TP53 is seen in ~10 % of untreated MM, more frequently with disease progression, and usually identifies deletion of the p arm of chromosome 17 [9, 20]. A third of untreated MM with del17p have mutations of TP53, something rarely seen in patients without del17p [46]. Del17p, even in the absence of TP53 mutations, is an independent poor prognostic factor for overall survival. Del17p has been shown to be associated with decreased TP53 expression, and an abrogated TP53 response to genotoxic and nongenotoxic stress, implicating the TP53 pathway even in tumors lacking mutations [47].

Del1p and Add1q. These copy number abnormalities frequently co-occur and both are associated with a poor prognosis in MM [20, 48]. The gain of 1q21, seen in ~ 40 % of patients, is associated with jumping translocations of 1q12. Similar genetic aberrations could be induced in the lymphocytes of individuals with constitutional translocations of 1q12 by exposure to hypomethylating agents. These observations support the hypothesis that hypomethylation of the highly repetitive regions of 1q12 can result in amplification of any region juxtaposed to it [49]. It is not clear which gene(s) is(are) responsible for driving the selection for amplification of 1q. On the short arm, in contrast, monoallelic deletions of two different regions are associated with a poor outcome: *CDKN2C* at 1p32.3 in 11 % of cases and *FAM46C* at 1p12 in 19 % of cases [50]. Homozygous deletion of *CDKN2C* is identified in ~ 5 % of untreated MM and in ~ 35 % of MMCL, and is correlated with a higher proliferative index, whereas the heterozygous deletion is not. FAM46C is an interferon stimulated gene mutated in ~ 10 % of untreated MM, often associated with HRD [43].

13.8 High-risk MM is associated with subclonal tumor heterogeneity

Within the MM tumor population there are distinct subclones with different mutations that independently play a role in tumor response and disease progression [16, 51]. Subclonal tumor heterogeneity is common in untreated MM. At relapse following treatment the degree of heterogeneity is reduced, presumably as a result of elimination of the more sensitive subclones by treatment [43]. There appears to be more subclonal heterogeneity in patients with high-risk MM [16, 43, 51] with alternating clonal "tides" under therapeutic selective pressure. These results are consistent with the hypothesis that the subclones compete for scarce resources and that low-dose monotherapy may simply eliminate the more sensitive subclone, allowing space for the more aggressive drug-resistant subclone to expand. They bolster a therapeutic approach using multidrug combinations in high-risk patients, that have greater subclonal heterogeneity, and for patients with stable genomes and low-risk disease therapy with one or two drugs used in sequence.

13.9 Concluding remarks and future directions

Although we have made significant progress in dissecting the molecular pathogenesis of plasma cell neoplasms, a number of important questions remain unanswered. What are the events (genetic, epigenetic, microenvironment) that drive the progression of MGUS to SMM to MM? Is there a specific molecular defect underlying hyperdiploidy? Why do malignant PC have a different immunophenotype from normal PC? What are the critical determinants of the interaction of MM cells with the BM microenvironment, and how can they be targeted? To what extent are MM cells dependent on expression of a PC phenotype, and can we target this PC phenotype dependency (e.g. IRF4) [52]? Critical to answering these and other questions will be careful studies of serial patient samples, and pre-clinical models that reproduce the genetic, orthotopic, and immunocompetent aspects of MM cells [53, 54].

References

[1] Siegel RL, Miller KD, Jemal A. Cancer statistics, 2015. CA Cancer J Clin 2015;65:5–29.

[2] Howlader N, Noone A, Krapcho M, et al., eds. SEER Cancer Statistics Review, 1975-2012. Bethesda, MD: National Cancer Institute; 2015.

[3] Weiss BM, Abadie J, Verma P, Howard RS, Kuehl WM. A monoclonal gammopathy precedes multiple myeloma in most patients. Blood 2009;113:5418–5422.

[4] Kyle RA, Therneau TM, Rajkumar SV, et al. Prevalence of monoclonal gammopathy of undetermined significance. N Engl J Med 2006;354:1362–1369.

[5] Kyle RA, Durie BG, Rajkumar SV, et al. Monoclonal gammopathy of undetermined significance (MGUS) and smoldering (asymptomatic) multiple myeloma: IMWG consensus perspectives risk factors for progression and guidelines for monitoring and management. Leukemia 2010;24:1121–1127.

[6] Rajkumar SV, Dimopoulos MA, Palumbo A, et al. International Myeloma Working Group updated criteria for the diagnosis of multiple myeloma. Lancet Oncol 2014;15:e538–48.

[7] Neben K, Jauch A, Hielscher T, et al. Progression in smoldering myeloma is independently determined by the chromosomal abnormalities del(17p), t(4;14), gain 1q, hyperdiploidy, and tumor load. J Clin Oncol 2013;31:4325–4332.

[8] Malpas JS, Bergsagel DE, Kyle R, Anderson K. Multiple Myeloma: Biology and Management. Oxford: Oxford University Press; 2004.

[9] Tiedemann RE, Gonzalez-Paz N, Kyle RA, et al. Genetic aberrations and survival in plasma cell leukemia. Leukemia 2008;22:1044–1052.

[10] Drexler HG, Matsuo Y. Malignant hematopoietic cell lines: in vitro models for the study of multiple myeloma and plasma cell leukemia. Leuk Res 2000;24:681–703.

[11] Bakkus MH, Heirman C, Van Riet I, Van Camp B, Thielemans K. Evidence that multiple myeloma Ig heavy chain VDJ genes contain somatic mutations but show no intraclonal variation. Blood 1992;80:2326–2335.

[12] Kuehl WM, Bergsagel PL. Multiple myeloma: evolving genetic events and host interactions. Nat Rev Cancer 2002;2:175–187.

[13] Dispenzieri A, Katzmann JA, Kyle RA, et al. Prevalence and risk of progression of light-chain monoclonal gammopathy of undetermined significance: a retrospective population-based cohort study. Lancet 2010;375:1721–1728.

[14] Zingone A, Kuehl WM. Pathogenesis of monoclonal gammopathy of undetermined significance and progression to multiple myeloma. Semin Hematol 2011;48:4–12.

[15] Kuehl WM, Bergsagel PL. Molecular pathogenesis of multiple myeloma and its premalignant precursor. J Clin Invest 2012;122:3456–463.

[16] Keats JJ, Chesi M, Egan JB, et al. Clonal competition with alternating dominance in multiple myeloma. Blood 2012;120:1067–1076.

[17] Avet-Loiseau H, Facon T, Grosbois B, et al. Oncogenesis of multiple myeloma: 14q32 and 13q chromosomal abnormalities are not randomly distributed, but correlate with natural history, immunological features, and clinical presentation. Blood 2002;99:2185–2191.

[18] Chiecchio L, Dagrada GP, Ibrahim AH, et al. Timing of acquisition of deletion 13 in plasma cell dyscrasias is dependent on genetic context. Haematologica 2009;94:1708–1713.

[19] Fonseca R, Bailey RJ, Ahmann GJ, et al. Genomic abnormalities in monoclonal gammopathy of undetermined significance. Blood 2002;100:1417–1424.

[20] Fonseca R, Bergsagel PL, Drach J, et al. International Myeloma Working Group molecular classification of multiple myeloma: spotlight review. Leukemia 2009;23:2210–2221.

[21] Fonseca R, Blood E, Rue M, et al. Clinical and biologic implications of recurrent genomic aberrations in myeloma. Blood 2003;101:4569–4575.

[22] Gabrea A, Martelli ML, Qi Y, et al. Secondary genomic rearrangements involving immunoglobulin or MYC loci show similar prevalences in hyperdiploid and nonhyperdiploid myeloma tumors. Genes Chromosomes Cancer 2008;47:573–590.

[23] Gabrea A, Bergsagel PL, Michael Kuehl W. Distinguishing primary and secondary translocations in multiple myeloma. DNA Repair 2006;5:1225–1233.

[24] Affer M, Chesi M, Chen WD, et al. Promiscuous MYC locus rearrangements hijack enhancers but mostly super-enhancers to dysregulate MYC expression in multiple myeloma. Leukemia 2014;28:1725–1735.

[25] Bergsagel PL, Kuehl WM. Molecular pathogenesis and a consequent classification of multiple myeloma. J Clin Oncol 2005;23:6333–6338.

[26] Zhan F, Huang Y, Colla S, et al. The molecular classification of multiple myeloma. Blood 2006;108:2020–2028.

[27] Bochtler T, Hegenbart U, Cremer FW, et al. Evaluation of the cytogenetic aberration pattern in amyloid light chain amyloidosis as compared with monoclonal gammopathy of undetermined significance reveals common pathways of karyotypic instability. Blood 2008;111:4700–4705.

[28] Hayman SR, Bailey RJ, Jalal SM, et al. Translocations involving the immunoglobulin heavy-chain locus are possible early genetic events in patients with primary systemic amyloidosis. Blood 2001;98:2266–2268.

[29] Avet-Loiseau H, Garand R, Lode L, Harousseau JL, Bataille R. Translocation t(11;14)(q13;q32) is the hallmark of IgM, IgE, and nonsecretory multiple myeloma variants. Blood 2003;101:1570–1571.

[30] Walker BA, Wardell CP, Murison A, et al. APOBEC family mutational signatures are associated with poor prognosis translocations in multiple myeloma. Nat Commun 2015;6:6997.

[31] Hurt EM, Wiestner A, Rosenwald A, et al. Overexpression of c-maf is a frequent oncogenic event in multiple myeloma that promotes proliferation and pathological interactions with bone marrow stroma. Cancer Cell 2004;5:191–199.

[32] Chesi M, Brents LA, Fly SA, et al. Activated fibroblast growth factor receptor 3 is an oncogene that contributes to tumor progression in multiple myeloma. Blood 2001;97:729–736.

[33] Zingone A, Cultraro CM, Shin D-M, et al. Ectopic expression of wild-type FGFR3 cooperates with MYC to accelerate development of B-cell lineage neoplasms. Leukemia 2010;24:1171–1178.
[34] Bergsagel PL, Kuehl WM, Zhan F, Sawyer J, Barlogie B, Shaughnessy J, Jr. Cyclin D dysregulation: an early and unifying pathogenic event in multiple myeloma. Blood 2005;106:296–303.
[35] Martinez-Garcia E, Popovic R, Min D-J, et al. The MMSET histone methyl transferase switches global histone methylation and alters gene expression in t(4;14) multiple myeloma cells. Blood 2011;117:211–220.
[36] Popovic R, Martinez-Garcia E, Giannopoulou EG, et al. Histone methyltransferase MMSET/NSD2 alters EZH2 binding and reprograms the myeloma epigenome through global and focal changes in H3K36 and H3K27 methylation. PLoS Genet 2014;10:e1004566.
[37] Pei H, Zhang L, Luo K, et al. MMSET regulates histone H4K20 methylation and 53BP1 accumulation at DNA damage sites. Nature 2011;470:124–128.
[38] Fonseca R, Debes-Marun CS, Picken EB, et al. The recurrent IgH translocations are highly associated with nonhyperdiploid variant multiple myeloma. Blood 2003;102:2562–2567.
[39] Smadja NV, Leroux D, Soulier J, et al. Further cytogenetic characterization of multiple myeloma confirms that 14q32 translocations are a very rare event in hyperdiploid cases. Genes Chromosomes Cancer 2003;38:234–239.
[40] Chng WJ, Van Wier SA, Ahmann GJ, et al. A validated FISH trisomy index demonstrates the hyperdiploid and nonhyperdiploid dichotomy in MGUS. Blood 2005;106:2156–2161.
[41] Ross FM, Ibrahim AH, Vilain-Holmes A, et al. Age has a profound effect on the incidence and significance of chromosome abnormalities in myeloma. Leukemia 2005;19:1634–642.
[42] Delmore JE, Issa GC, Lemieux ME, et al. BET bromodomain inhibition as a therapeutic strategy to target c-Myc. Cell 2011;146:904–917.
[43] Lohr JG, Stojanov P, Carter SL, et al. Widespread Genetic Heterogeneity in Multiple Myeloma: Implications for Targeted Therapy. Cancer Cell 2014;25:91–101.
[44] Steinbrunn T, Stühmer T, Gattenlöhner S, et al. Mutated RAS and constitutively activated Akt delineate distinct oncogenic pathways, which independently contribute to multiple myeloma cell survival. Blood 2011;117:1998–2004.
[45] Keats JJ, Fonseca R, Chesi M, et al. Promiscuous mutations activate the noncanonical NF-kappaB pathway in multiple myeloma. Cancer Cell 2007;12:131–144.
[46] Lode L, Eveillard M, Trichet V, et al. Mutations in TP53 are exclusively associated with del(17p) in multiple myeloma. Haematologica 2010;95:1973–1976.
[47] Teoh PJ, Chung TH, Sebastian S, et al. p53 haploinsufficiency and functional abnormalities in multiple myeloma. Leukemia 2014;28:2066–2074.
[48] Walker BA, Leone PE, Chiecchio L, et al. A compendium of myeloma-associated chromosomal copy number abnormalities and their prognostic value. Blood 2010;116:e56–65.
[49] Sawyer JR, Tian E, Heuck CJ, et al. Evidence of an epigenetic origin for high-risk 1q21 copy number aberrations in multiple myeloma. Blood 2015;125:3756–379.
[50] Boyd KD, Ross FM, Walker BA, et al. Mapping of chromosome 1p deletions in myeloma identifies FAM46C at 1p12 and CDKN2C at 1p32.3 as being genes in regions associated with adverse survival. Clin Cancer Res 2011;17:7776–7784.
[51] Walker BA, Wardell CP, Melchor L, et al. Intraclonal heterogeneity and distinct molecular mechanisms characterize the development of t(4;14) and t(11;14) myeloma. Blood 2012;120:1077–1086.
[52] Shaffer AL, Emre NC, Lamy L, et al. IRF4 addiction in multiple myeloma. Nature 2008;454:226–231.

[53] Chesi M, Robbiani DF, Sebag M, et al. AID-dependent activation of a MYC transgene induces multiple myeloma in a conditional mouse model of post-germinal center malignancies. Cancer Cell 2008;13:167–180.
[54] Chesi M, Matthews GM, Garbitt VM, et al. Drug response in a genetically engineered mouse model of multiple myeloma is predictive of clinical efficacy. Blood 2012;120:376–385.

Ralf Küppers

14 Biology of Hodgkin's lymphoma

14.1 Introduction

Hodgkin's lymphoma (HL) is named after Thomas Hodgkin, who first described this disease in 1832 [1]. A peculiar type of large cell always seen in this lymphoma was later characterized by Dorothy Reed and Carl Sternberg [2, 3]. We now call the mononuclear tumor cells Hodgkin cells and their bi- or multinucleated large variants Reed–Sternberg cells [4]. HL was the first human lymphoma to be identified and named, although its lymphoid origin was proven only in the mid-1990s, resulting in its renaming from Hodgkin's disease to Hodgkin's lymphoma in the 2001 WHO lymphoma classification [5].

HL is subdivided into a classical form, accounting for about 95 % of cases, and a subgroup called nodular lymphocyte predominant HL (NLPHL) [4]. Classical HL is further subdivided into nodular sclerosis, mixed cellularity, lymphocyte-rich classical and lymphocyte-depleted HL. This subdivision is based on differences in the morphology and immunophenotype of the Hodgkin and Reed–Sternberg (HRS) cells and differences in the histological picture [4]. In NLPHL, the tumor cells were originally called L&H cells, for lymphocytic and histiocytic, but they were renamed into lymphocyte predominant (LP) cells [4]. In about 30–40 % of cases, HRS cells in classical HL are infected by Epstein–Barr virus (EBV), but such an infection is not seen in NLPHL.

HL is one of the most frequent lymphomas in the Western world, with about 3–4 new cases per year per 100,000 persons. With multiagent chemotherapy and/or radiation therapy, about 80–90 % of patients can nowadays be cured [6].

For a number of reasons, HL has been and still is the most enigmatic lymphoma. First, the pathognomonic HRS cells usually account for only about 1 % of cells in the tissue, the vast majority of the cellular infiltrate being composed of a mixture of a large variety of cells of the hematopoietic system, including T cells, B cells, macrophages, eosinophils, plasma cells, and others [4, 7]. It was for a long time unresolved whether the HRS cells indeed represent a monoclonal population of malignant cells, or whether Hodgkin's disease represents an inflammatory condition. Second, it was unclear how this diverse cellular infiltrate is established and what its role in the pathophysiology of the disease is. Third, HRS cells in classical HL have a very unusual morphology and immunophenotype that does not resemble any normal cell of the immune system, so that the cellular origin of HRS cells was unknown for a long time. Fourth, the presence of the bi- or multinucleated Reed–Sternberg cells is essential for the diagnosis of a classical HL [4], but their generation and their relationship to the mononuclear Hodgkin cells was enigmatic. Finally, due to the technical difficulties of isolating and molecularly analyzing the rare HRS cells, the identification of genetic lesions involved in the pathogenesis of HL lagged much behind comparative studies for B-cell non-Hodgkin's lymphomas (B-NHL). However, for each of the aspects mentioned, we have made major progress, so that many aspects of the biology of HL have become much clearer, even though important issues still remain to be resolved.

14.2 The cellular origin of the tumor cells in HL

As the phenotype of HRS cells does not resemble any normal cell in the immune system (Tab. 14.1) [8], the origin of these cells was debated for a long time. Genetic studies finally clarified the cellular origin of these cells. B lymphocytes undergo specific rearrangements of their immunoglobulin (Ig) genes during their early development in the bone marrow, so that each mature B cell is equipped with a unique B-cell receptor. Because such rearrangements of the Ig heavy and light chain genes are highly specific for B lineage cells, the detection of rearranged Ig genes represents a genetic trait of a B-cell derivation, even if the cells do not show a clear B-cell phenotype and more. By molecular analysis of microdissected HRS cells, it was shown that HRS cells are B lineage cells, because they harbor rearranged Ig heavy and light chain genes [9–12]. The sequence analysis of these rearrangements provided further important insights into the histogenetic origin of HRS cells. In nearly all cases of classical HL, the HRS cells showed heavily mutated Ig variable (IgV) region genes [9, 10, 12–15]. Such mutations happen specifically in antigen-activated B cells in the course of T-cell dependent immune responses in the histological structure of the germinal center (GC) through the

Tab. 14.1: Comparison of HRS and LP cells.

Feature	HRS cells	LP cells
Diagnostic immunophenotype	$CD30^+$, $CD15^{+/-}$, $CD45^-$, $CD75^-$, EMA^-	$CD30^-$, $CD15^-$, $CD45^+$, $CD75^+$, EMA^+
B-cell marker expression (selected)	$BOB.1^-$, $CD20^{-/+}$, $CD40^+$, $CD79^-$, $MUM1^+$, $OCT2^-$, $PAX5^{low}$, sIg^-	$BOB.1^+$, $CD20^+$, $CD40^+$, $CD79^+$, $MUM1^{low}$, $OCT2^+$, $PAX5^+$, sIg^+
GC B-cell factor expression	$AICD^-$, $BCL6^{-/+}$, $HGAL^-$, $SERPINA9^-$	$AICD^+$, $BCL6^+$, $HGAL^+$, $SERPINA9^+$
Non-B-cell marker expression (selected)	$ID2^+$, $NOTCH1^+$, $CSFR1^+$, $GATA3^+$, $CD3^{-/+}$, granzyme $B^{-/+}$, $CCL17^+$	$ID2^-$, $NOTCH1^-$, $CSFR1^-$, $GATA3^-$, $CD3^-$, granzyme B^-, $CCL17^-$
Multinuclear cells	Always present	Not present
Presumed cellular origin	Crippled GC B cell	Selected GC B cell
EBV infection	Yes (30–40 % of cases)	No

sIg: surface immunoglobulin

process of somatic hypermutation [16]. Therefore, the regular detection of mutated IgV genes argues for a GC or post-GC origin of HRS cells (Tab. 14.2). Unexpectedly, in about a quarter of cases, destructive mutations were found in originally functional heavy or light chain V gene rearrangements [9, 10, 13, 15]. Such destructive mutations physiologically happen in mutating GC B cells, but normally very quickly and efficiently cause the death of the respective B cell through apoptosis. Indeed, apoptosis is the default program of GC B cells [22], and only if the cells manage to acquire affinity-increasing mutations they are positively selected and may leave the GC to become long-lived memory B cells or plasma cells. The detection of crippling mutations in the HRS cells of 25 % of cases of cHL showed that at least this fraction of cases is derived from a peculiar population of pre-apoptotic GC B cells [23]. However, as most disadvantageous mutations that cause apoptosis of GC B cells cannot be easily identified by V gene sequence analysis (e.g. replacement mutation that prevent proper folding of the V domains or heavy and light chain pairing, or reduce affinity to the antigen), we speculated that HRS cells may derive from pre-apoptotic GC B cells as a rule [23]. There are a few cases of HL with unmutated V region genes [24]. These cases may derive from pre-GC B cells. However, GC founder B cells become apoptosis-prone al-

Tab. 14.2: Genetic features of HRS cells arguing for their B-cell origin*.

Feature	Argument	References
Clonal Ig heavy and light chain gene rearrangements in HRS cells	V_H and V_L gene rearrangements take place only in B lineage cells.	[9–12, 14]
Somatic mutations in Ig V genes of HRS cells of nearly all HL	Process of somatic hypermutation is restricted to antigen-activated B cells proliferating in GC.	[9–12, 14]
Aberrant somatic hypermutation of several proto-oncogenes in HRS cells of about half of HL	Aberrant somatic hypermutation is specific for several types of GC B-cell-derived lymphomas.	[17]
Class switch recombination in most HL cell lines and presumably also primary HRS cells	Class switch recombination takes place specifically in antigen-activated mature B cells.	[18, 19]
Clonally related composite lymphomas of a classical HL and a B-cell NHL with shared and distinct Ig V gene mutations	Clonal relationship of composite lymphomas with V gene mutation pattern suggesting their derivation from distinct members of a GC B-cell clone proves the B-cell origin of the HRS cells in these cases	[20]
Ig loci-associated chromosomal translocations in HRS cells of about 20 % of classical HL	Ig loci-associated translocations take place only in B cells as side-products of Ig gene-remodeling processes	[19]

*Adapted from ref. [21]

ready before the acquisition of the first mutations [25], so that potentially also these cases are derived from GC B cells. There is also indication that HRS cells have often undergone class switch recombination, a further genetic process that is typical for GC B cells (Tab. 14.2) [18, 19].

The expression of T-cell markers by HRS cells in a fraction of cases of classical HL also caused speculations about a potential T-cell origin of HRS cells. Consequently, several such cases were studied for rearranged T-cell receptor genes, and a few of these cases indeed carried clonal T-cell receptor rearrangements [26, 27]. However, most of the cases with T-cell marker expression lacked T-cell receptor rearrangements and were of B-cell origin [26, 27]. It is debated whether cases with a T-cell origin should be grouped together with the B-cell-derived cases as one type of lymphoma. However, with the standard diagnostic tools, the T-cell derived cases appear like typical classical HL. As there are very few cases of classical HL with a T-cell derivation, there is no information whether such lymphomas have a particular clinical behavior, or distinct gene expression pattern.

In NLPHL the LP cells show a B-cell phenotype (Tab. 14.1), so it was assumed that the LP cells are of B-cell origin. This was confirmed when isolated LP cells were studied for rearranged IgV region genes. All NLPHL that were analyzed carried IgV gene rearrangements [28–30]. The IgV genes were somatically mutated, functional, and a low level of ongoing hypermutation was detected [28, 29, 31]. This strongly suggests that LP cells represent transformed GC B cells that appear to be selected for BCR expression and still show low level hypermutation activity. This conclusion is supported by phenotypic features of LP cells, as they express, for example, the transcription factor BCL6, the master regulator of the GC B-cell program, and other typical GC B-cell factors (AICDA, SERPINA9 (GCET1), HGAL) (Tab. 14.1) [32–35].

The detection of clonal IgV gene rearrangements in HRS and LP cells also clarified the clonal nature of these cells, because these rearrangements represent ideal clonal markers due to their very high diversity. This was later further confirmed by the detection of clonal cytogenetic aberrations and other genetic lesions (see below). Thus, HRS and LP cells indeed represent the clonal tumor cells in HL.

14.3 Relationship of Hodgkin cells to Reed–Sternberg cells

As HRS cells express markers of various hematopoietic cell types, and as numerical chromosomal aberrations are frequently seen in these cells, it has been speculated that the HRS cells, or at least the multinucleated Reed–Sternberg cell variants, derive from fusions of different cells. However, a detailed analysis of antigen receptor loci in several cases provided strong evidence that HRS cells are not derived from such cell fusions [36]. The fact that the mononuclear Hodgkin and the multinuclear Reed–Sternberg cells show an identical immunophenotype already indicated that these two types of cells are closely linked and presumably together represent the tumor clone in this malignancy. In line with this idea, several studies of HL cell lines showed that the mononuclear Hodgkin cells can give rise to the RS cells, which themselves have little further proliferative activity [37–39]. These observations led to the idea that RS cells are generated by endomitosis or a process of acytokinetic mitosis. A recent timelapse analysis of HL cell lines provided further insight into this issue. It was shown that mononuclear Hodgkin cells undergo an incomplete cytokinesis followed by a refusion of the daughter cells [40]. Thus, the final separation step of the daughter cells after an attempted cell division seems to be frequently compromised in dividing Hodgkin cells.

Studies with HL cell lines indicated that there might be a small subpopulation of Hodgkin cells with particular proliferative and chemoresistant features that resemble in some aspects cancer stem cells [41, 42]. This Hodgkin cell subpopulation extrudes the Hoechst dye due to expression of multidrug receptors, including MDR1 and ABCG2, and hence are called side population cells in flow cytometric studies. The $CD30^+ CD20^-$ side population cells showed increased chemoresistance and were particularly efficient in reestablishing the mixed HRS cells population in subcloning experiments [41, 42]. However, not all HL cell lines investigated showed such cells [41, 42], and it is unclear whether also *in vivo*, the HRS cell clone is sustained by a small subpopulation of Hodgkin cells with characteristics of side population cells. One study also reported the existence of $CD20^+CD30^-$ and surface Ig-positive cells belonging to the HRS cell clone and potentially representing a reservoir or precursor population for the HRS cell clone [43]. However, that study was heavily criticized, so that the existence of such cells is still uncertain [44].

14.4 The lost B-cell identity of HRS cells

The studies of the rearranged IgV genes unequivocally identified mature B cells as the cellular origin of the HRS cells. However, HRS cells phenotypically no longer resemble B cells. Many typical B-cell transcription factors are downregulated in HRS cells, including PU.1, OCT2, BOB1, EBF1, and others [45–47]. HRS cells still express the B-cell master transcription factor PAX5 at low level, and also E2A is detectable [48, 49]. In addition, B-cell surface markers, such as CD19, CD20, and CD79a, and GC B-cell markers, including BCL6, are only rarely expressed or not at all (Tab. 14.1) [8]. Gene expression profiling studies showed a global downregulation of the B-cell phenotype in HRS cells [50]. Notably, a number of markers that are essential for an interaction of B cells with $CD4^+$ T cells are still expressed by HRS cells. These markers include CD40, CD80, CD86, CD54 and CD58 [51]. This indicates that HRS cells still need to interact with T helper cells. MHC class II expression, however, is often lacking, partly due to genetic lesions [52]. Hence, at least in a fraction of cases, a cognate interaction between HRS cells and the surrounding $CD4^+$ T cells through MHC class II-presented peptides is not occurring.

Multiple mechanisms contribute to the loss of the B-cell phenotype of HRS cells. First, the lack of expression of many B-cell transcription factors explains why many of the B-cell genes regulated by these factors are not expressed. Second, the activity of the B-cell identity master regulator E2A is impaired by aberrant expression of its inhibitors ABF1 and ID2 [49, 53]. ID2 is physiologically expressed by NK cells and promotes NK cell development at the expense of a B-cell fate in lymphoid precursors [54]. Third, HRS cells show constitutive activity of the T-cell transcription factor NOTCH1 [55, 56], which promotes T-cell development by suppressing B-cell genes in common lymphoid progenitors. Activation of NOTCH1 is presumably partly caused by interaction with its ligand Jagged-1, that is expressed by cells in the HL microenvironment [56], and by high level expression of the NOTCH1 coactivator MAML2 [57]. Additionally, the NOTCH inhibitor DELTEX1 is downregulated in HRS cells [55]. Fourth, HRS cells ectopically express two further transcription factors that suppress B-cell genes, namely STAT5 and IRF5 [58, 59]. Fifth, there is indication that in EBV-infected HRS cells, the EBV latent membrane proteins LMP1 and LMP2a moderately downmodulate B-cell genes [60, 61]. Sixth, HRS cells express several members of the polycomb repressive complexes 1 and 2, which dampen B-cell gene expression in hematopoietic and lymphoid precursors [62–64]. Seventh, many B-cell genes are epigenetically silenced in HRS cells by DNA methylation [65–67]. Taken together, a large number of factors contribute to the downmodulation of the B-cell program in HRS cells. However, so far no genetic lesion has been identified that could play a master role in this process, and it is also still unclear how the “dedifferentiation” of the HRS precursor cells is initiated in GC B cells. Since “dedifferentation” and the upregulation of ID2 and NOTCH1, as seen in HRS cells, are also hallmarks of hypoxic cells in various cancers, we wondered whether hypoxia might induce an initial “dedifferentiation” of B cells. Indeed, culturing of human B cells under hypoxic conditions induced a partial HRS cell phenotype, so that transient hypoxic conditions in GC reactions may represent an initial trigger for the cells to develop into the direction of HRS cells [68].

Considering the consistent downmodulation of the B-cell program in HRS cells and the many factors involved in orchestrating this process, including the aberrant expression of markers not typically expressed by B cells, it seems likely that this “reprogramming” is not a chance occurrence during classical HL pathogenesis, but a key factor for the generation of HRS cells. Perhaps this feature is intimately linked to the proposed origin of HRS cells from “crippled” GC B cells. As mentioned above, GC B cells are very stringently selected for expression of a high-affinity B-cell receptor, and an apoptosis program is pre-activated in these cells [22]. By losing their B-cell gene expression program, the HRS precursor cells may also get rid of the selectional pressure to either express a fitting B-cell receptor or undergo apoptosis. Hence, the “reprogramming” of the HRS cells may represent an escape mechanism from the tight selectional forces which regulate the survival of GC B cells.

14.5 Genetic lesions

Early cytogenetic studies revealed that HRS cells often have an aberrant karyotype with frequent aneuploidy [69–71]. However, whether these events recurrently affect specific genes or mostly reflect a genomic instability is unclear. There is evidence for genomic instability in the HRS cell clones, as HRS cells often show subclonal imbalances and other abnormal chromosomal features [69, 72]. Similar to many B-NHL, HRS cells carry translocations involving the Ig loci (in about 20 % of cases), but for most of these instances the partner genes were not identified [19]. Some cases involve *BCL2*, *BCL3*, *REL*, *BCL6* or *MYC* [19, 70, 73]. The relevance of these translocations in the established HRS cell clone remains enigmatic, because the silencing of B-cell receptor expression in HRS cells implies that the Ig enhancer, which typically drives the constitutive activation of proto-oncogenes translocated into the Ig loci [74], is not or only weakly active in HRS cells [47, 75]. Thus, it might be that many of these translocations were essential in early stages of HL pathogenesis, but became less important during later stages, when the HRS cells largely lost their B-cell gene expression program.

The search for specific mutations in proto-oncogenes or tumor suppressor genes was and still is very much hampered by the rareness of the HRS cells in the lymphoma tissue. In several instances, genetic lesions were initially detected in HL cell lines and their occurrence was then tested also in primary microdissected HRS cells. Such studies revealed frequent genetic lesions in numerous members of the NF-κB pathway. Deleterious mutations in the genes of the NF-κB inhibitors IκBα and IκBε are present in about 20 % and 10 % of cHL, respectively [76, 78–80]. Inactivating point mutations and deletions also affect the NF-κB signaling inhibitor A20, encoded by the *TNFAIP3* gene, in about 40 % of classical HL [80, 81]. Mutations in the negative NF-κB regulators CYLD and TRAF3 also occur, but they are rare [82, 83]. The NF-κB factor REL is affected by gains and amplifications in about 40 % of classical HL, and gains of *MAP3K4* (*NIK*) are present in the HRS cells in ca. 25 % of cases [84–86]. Gains or translocations affecting the *BCL3* gene, which encodes a coactivating factor of the noncanonical NF-κB pathway, have also been observed at low frequency [87, 88]. As is evident from this listing, members of both the canonical and noncanonical NF-κB pathway are affected by mutations, and studies of HL cell lines showed that often more than one of the genes is mutated in an HRS cell clone (Tab. 14.3). This indicates that a strong and multifactorial deregulation of NF-κB activity is essential for HRS cells.

A second signaling pathway for which several members show genetic lesions in HRS cells is the JAK/STAT pathway, the main mediator for cytokine signaling in immune cells. Gains of the *JAK2* kinase gene are present in about 30 % of classical HL, and SOCS1, the main negative regulator of JAK/STAT signaling, is affected by inactivating mutations in HRS cells of 40 % of classical HL [84, 89, 90]. Notably, gains of *JAK2* typically not only affect this gene but also encompass the neighboring *PD-L1*, *PD-L2* and *JMJD2C* genes [91, 92]. JMJD2C is a histone demethylase with a presumed role in the epigenetic regulation of HRS cells [92]. PD-L1 and PD-L2 play a role in

Tab. 14.3: Features of HL cell lines.

Cell line	HL subtype	Cell Type	EBV	Selected genetic lesions	Remarks
L428	Classical	B cell	–	*B2M*, *NFKBIA*, *NFKBIE*, *SOCS1*, *PTPN1*, *TP53*	
L540	Classical	T cell	–	*PTPN1*	
L591	Classical	B cell	+		EBV gene expression pattern distinct from the HL-typical latency II
L1236	Classical	B cell	–	*CD95*, *PTPN1*, *SOCS1*, *TNFAIP3*, *TP53*	Only line with proven origin from HRS cells
KMH2	Classical	B cell	–	*CTIIA* (translocation), *CYLD*, *NFKBIA*, *TNFAIP3*	
HDLM2	Classical	T cell	–	*SOCS1*,*TNFAIP3*, *TNFSF7*, *TNFSF9*, *TP53*, *UTX*	
SUP-HD1	Classical	B cell	–	*PTPN1*, *PTPN2*, *TNFAIP3*	
UHO-1	Classical	B cell	–	*PTPN1*, *TNFAIP3*, *TRAF3*	
DEV	NLPHL	B cell	–	*B2M*, *BCL6* (translocation), *SOCS1*	

inhibiting cytotoxic T cells, which express the PD1 receptor. Recently, also destructive mutations in the *PTPN1* gene were detected in HRS cells in 20 % of cases [93]. PTPN1 is a phosphatase that, among other functions, negatively regulates STAT activity.

HRS cells show a heterogenous pattern of major histocompatibility complex (MHC) class I and II expression [52, 94, 95]. The lack of expression of these molecules in a fraction of cases is partly due to genetic lesions. The MHC class I component β-2-microglobulin (B2M) is mutated in the HL cell line L428 [96] and in 7/10 cases of classical HL studied by exome sequencing [97]. Downmodulation of MHC class I is often a strategy of cancer cells to evade from attack by cytotoxic T cells. The gene of the MHC class II transactivator *CIITA*, which is an essential factor for MHC class II expression, shows chromosomal translocations with various partners in 15 % of classical HL [52]. These translocations inactivate the gene and lead to downmodulation of MHC class II expression. This might play a role in immune evasion of HRS cells. In this regard, it is also noteworthy that several HL cell lines carry inactivating mutations in the CD58 gene, which may aid in the escape of HRS cells from an attack by cytotoxic cells, as CD58 is important for the binding of cytotoxic cells to their target cells [98]. However, mutations in *CD58* are rare in primary HRS cells [98].

A number of additional candidate genes were studied for mutations in isolated HRS cells and/or HL cell lines, but these genes were either only rarely mutated (*TP53, CD95, ATM*), or no mutations were found (*BAD, FADD, CASP8, CASP10, NRAS*) [8].

Very little is known about genetic lesions in LP cells of NLPHL. LP cells harbor chromosomal translocations involving the *BCL6* gene in 40 % of cases [99–101]. In addition, *SOCS1* is mutated in about 40 % of cases of NLPHL [102]. Although LP cells show strong NF-κB activity [103], mutations in TNFAIP3 or NFKBIA, as they are frequent in HRS cells, were not found [104]. As there is also no indication for REL gains, it is still unclear whether genetic lesions cause the strong NF-κB activation in LP cells.

Genetic lesions (in particular genomic imbalances) as well as the clonal IgV gene rearrangements of HRS cells have recently also been used to search for HRS cell-specific DNA in the peripheral blood of HL patients by massively parallel sequencing [105, 106]. Such sequences were indeed found among cell-free DNA in the serum [105, 106], and it will be interesting to explore whether the detection of HRS cell-specific DNA in the serum can be used to monitor therapy and to serve as biomarker.

HL is one of the lymphomas with the strongest familial association [107]. This prompted studies to clarify whether germline mutations or polymorphisms play a role in HL pathogenesis. In a family with several members affected by HL, a constitutional translocation of the *KLHGC8B* gene was detected [108]. The function of this gene is largely unclear, but its downmodulation in a cell line caused increased generation of binucleated cells, indicating a potential involvement of the gene in the multinuclearity of Reed–Sternberg cells. In a family with several members affected by NLPHL, a germline mutation in the *NPAT* gene was found [109]. Additionally, a rare germline replacement mutation/polymorphism in *NPAT* was found at higher frequency in HL patients than in healthy individuals. The function of NPAT in LP or HRS cells is not yet clear.

Several genome-wide association studies have been performed for HL [110–113]. These studies revealed risk loci for HL at 2p16.1, 3p24.1, 6p21.32, 6q23.3, 8q24.21, 10p14, and 19p13.3. The polymorphism at 2p16.1 may relate to the *REL* gene, which is often altered by somatic lesions in HRS cells (see above); the risk allele at 6q23.3 localizes 5' of *EOMES*, which has a role in embyrogenesis and pluripotency. The risk locus at 10p14 is close to the *GATA3* gene, which is aberrantly expressed by HRS cells [114], and the susceptibility locus at 19p13.3 is within an intron of the *TCF3* gene, encoding the E2A transcription factor. The risk locus at 6p21 involves the MHC loci and points to an important role of HLA alleles in the risk for HL development. This was already indicated by earlier studies of HLA alleles and their link to HL [115]. There is also evidence for specific risk alleles for EBV-positive HL within the HLA class I locus [113, 116].

14.6 Role of EBV in HL pathogenesis

In about 30 % of cases of classical HL in the Western world, the HRS cells are infected by EBV [117]. In childhood cases of classical HL in Latin America, even up to 90 % of cases are EBV-positive [117]. Cases of mixed cellularity HL are more frequently EBV$^+$ than nodular sclerosis cases. LP cells in NLPHL are not EBV-infected. In the EBV-positive cases of classical HL, the virus is seen in all HRS cells (with very few exceptions [118]), and the EBV genome structure indicates a single infection event early in HL development [119]. EBV-infected HRS cells show a latency type II expression pattern with expression of the three EBV-encoded proteins EBV nuclear antigen 1 (EBNA1), latent membrane protein 1 (LMP1) and latent membrane protein 2a (LMP2a), as well as the viral nonprotein coding EBER RNAs [117]. EBNA1 is essential for the replication of the episomal viral genome and hence essential for all proliferating latently infected cells. LMP1 is a classical oncogene. It mimics an active CD40 receptor and hence causes activation of NF-κB and other signaling pathways [120]. LMP2a has a cytoplasmic domain that mimics the signaling component of the B-cell receptor [121]. Thus, LMP1 and LMP2a mimic two main survival signals in GC B cells – a productive interaction with T helper cells via CD40 signaling, and presence of a productive and active B-cell receptor [122]. We and others therefore speculated that EBV infection of GC B cells may be a way how crippled GC B cells can be rescued from apoptosis. It was indeed shown that EBV infection of primary human tonsillar GC B cells is able to rescue crippled GC B cells from apoptosis, supporting this scenario [123–125]. A restriction of the rescue studies is, however, that the EBV immortalized GC B cells with crippled B-cell receptor showed the latency III gene expression profile that is typical for *in vitro* EBV-infected B cells, and which includes further viral proteins in addition to those expressed in the HRS cell-typical latency II. A further strong argument for a decisive role of EBV in the development of classical HL with HRS cells lacking the ability to express a B-cell receptor at all is the observation that all cases of classical HL with destructive IgV gene mutations that impaired B-cell receptor expression were found to be EBV-positive [122]. It is likely that LMP2a expression in GC B cells that lose the capacity to express a B-cell receptor through somatic mutations is needed to prevent the immediate death of the cells by apoptosis. It is also notable that most cases of classical HL with *TNFAIP3* mutations are EBV-negative [81]. This indicates complementing roles of EBV (in particular LMP1) and *TNFAIP3* inactivation in contributing to strong NF-κB activity in HRS cells. In other words, in EBV$^+$ HRS cell clones, there is no need for inactivating *TNFAIP3* mutations, because LMP1 already provides strong NF-κB activation. Although EBV is likely a major pathogenetic factor in the EBV-positive cases, global gene expression profiling of primary HRS cells from EBV$^+$ and EBV$^-$ cases of classical HL revealed hardly any consistent EBV-associated changes in gene expression [126].

14.7 Deregulated signaling pathways and transcription factors

HRS cells show a deregulated constitutive activation of numerous signaling pathways and transcription factors. As is already evident from the discussion of the genetic lesions in HRS cells, constitutive NF-κB activation is a hallmark of HRS cells. Besides the genetic lesions and the role of EBV in EBV$^+$ cases, it is likely that also signaling events through surface receptors on HRS cells contribute to this activation. HRS cells express several members of the TNF receptor superfamily, including CD30, CD40, TNFRSF13B (TACI), TNFRSF17 (BCMA), and TNFRSF11A (RANK) [5, 127–129]. Signaling through these receptors typically activates the NF-κB pathway, and partly also other pathways. Cells expressing the ligands for these receptors are present in the microenvironment of HL and thus may activate these receptors. CD30 ligand is expressed by eosinophils and mast cells [130, 131], CD40 ligand by activated T cells [132], APRIL, which can bind to TACI and BCMA, is expressed by neutrophils [133]. BAFF, a further receptor for TACI and BCMA, is expressed by HRS cells themselves and by other microenvironmental cells [128]. RANK may be activated in an autocrine fashion, as HRS cells express its ligand [129].

Also the JAK/STAT pathway is not only activated by genetic lesions, but additionally by paracrine and autocrine mechanisms. HRS cells express the active forms of four STAT factors, STAT3, STAT5a, STAT5b, and STAT6 [59, 134–136]. The coexpression of IL13 and the IL13 receptor by HRS cells contributes to STAT6 activation in an autocrine manner [137]. Similarly, HRS cells coexpress IL21 and the IL21 receptor, so that also autocrine IL21 signaling contributes to a chronic STAT activity [59, 138]. Expression and activity of the STAT5 factors is additionally augmented by NF-κB [139]. Furthermore, STAT factors can be stimulated by receptor tyrosine kinases, several of which are active in HRS cells.

Receptor tyrosine kinases are involved in the regulation of multiple cellular processes. HRS cells aberrantly express numerous receptor tyrosine kinases that are not normally expressed by B cells, including platelet derived growth factor receptor alpha (PDGFRA), discoidin domain receptor tyrosine kinase 2 (DDR2), macrophage stimulating 1 receptor (MST1R, RON), ephrinB1 (EPHRB), tyrosine receptor kinase A and B (TRKA and TRKB), and colony-stimulating factor 1 receptor (CSF1R) [140, 141]. For some of these receptors, there is an indication that they are activated in an autocrine or paracrine manner in HRS cells (Fig. 14.1) [140, 142]. HRS cells also express the receptor tyrosine kinase MET, and its ligand hepatocyte growth factor (HGF) is produced by dendritic cells in the microenvironment [143]. Mutations in receptor tyrosine kinase genes, as is often seen for single receptor tyrosine kinases in other malignancies, seem not to play a major or any role in classical HL [140]. For *CSF1R* it has been shown that this gene is aberrantly expressed in HRS cells through an unusual mechanism involving derepression of an upstream endogenous long terminal repeat that functionally replaces the normal *CSF1R* promoter, which is not active in B cells [141].

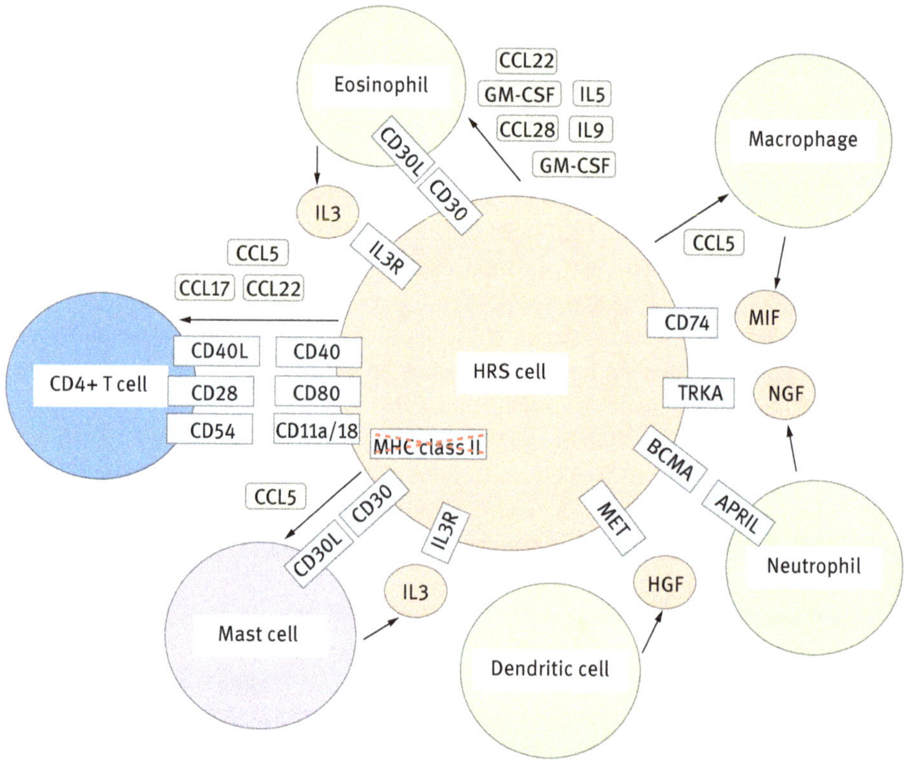

Fig. 14.1: **Supportive factors for HRS cells in their microenvironment.** Shown are main cellular interactions that presumably support the growth and/or survival of HRS cells. In addition, chemokines and cytokines secreted by HRS cells and attracting other cells or promoting their activation are indicated. MHC class II is downregulated in HRS cells in a large fraction of classical HL cases.

The PI3K/AKT pathway is chronically active in HRS cells [144, 145]. Signaling through CD30, CD40, RANK and receptor tyrosine kinases likely contributes to its activation. Also the ERK pathway is active in HRS cells. Again, CD30, CD40, and RANK may contribute to induce ERK activity [146, 147]. The ERK family members ERK1, ERK2 and ERK5 have been detected in HRS cells [146, 147].

The AP1 transcription factors function as dimers composed of JUN and FOS family members. In HRS cells, JUN and JUNB are highly expressed and they induce among many other genes CD30, which is a diagnostic marker for HRS cells [148, 149].

For all of the pathways mentioned above, there is evidence from functional studies with HL cell lines that their constitutive activity is essential for HRS cell survival and/or proliferation. Deregulated activity of signaling pathways is a hallmark of all cancers, but the multitude of signaling pathways and transcription factors that show a deregulation in HRS cells appears to be unique for classical HL among lymphoid malignancies.

14.8 Microenvironmental interactions

The microenvironment in classical HL is characterized by the infiltration of many different types of inflammatory cells, including T cells, B cells, plasma cells, neutrophils, eosinophils, mast cells, macrophages, and others, as well as stromal cells and fibroblasts. There is considerable variation in the composition of the microenvironment between the various HL subtypes [5]. It is likely that the rare HRS cells, which typically represent less than one or only a few percent of cells in the HL tissue, actively orchestrate the infiltration of other cells into the HL-involved lymph nodes. HRS cells produce numerous cytokines and chemokines with which they can attract and influence the other inflammatory cells [7, 150]. For example, they produce CCL5, CCL17 and CCL22, which attract T_H2 cells and regulatory T (Treg) cells (Fig. 14.1) [151–153]. Eosinophils are attracted through secretion of IL5, CCL5, CCL28 and granulocyte-macrophage colony-stimulating factor [152]. Activation and proliferation of fibroblasts, as seen particularly in nodular sclerosis HL, can be mediated by the HRS cells through secretion of IL13, TNFα, and fibroblast growth factors [152]. The activated fibroblasts can then contribute to eosinophil and T_H2 cell infiltration by secretion of CCL11 [154].

It is likely that HRS cells actively attract these other cells to obtain survival and proliferation signals from them. This includes activation of the receptor tyrosine kinase TKRA through NGF produced by neutrophils [142], and activation of MET through hepatocyte growth factor secreted by dendritic cells [143]. Several types of microenvironmental cells produce IL3, which stimulates the IL3R on HRS cells [155, 156]. Besides HRS cells themselves, also infiltrating macrophages produce macrophage migration inhibitory factor (MIF), which supposedly has stimulatory effects on HRS cells by binding to CD74, which is consistently expressed by HRS cells [157]. Further stimulatory signaling is mediated through members of the TNFR superfamily (e.g. CD30, BCMA) as mentioned above.

The largest population of cells in the HL microenvironment are T cells, including $CD8^+$ T cells, T_H cells and Treg cells [158, 159]. $CD4^+$ T cells typically are in direct contact with HRS cells and can interact with them for example through the receptor-ligand pairs CD40-CD40L, CD80-CD28 and CD54-CD11a/CD18 (Fig. 14.1). There are controversial data about whether the T_H cells are more of the T_H1 or T_H2 subtype [160, 161]. It seems that the T_H cells provide survival signals for the HRS cells (Fig. 14.1). The dependency of HRS cells on survival signals from other cells in the tissue is implicated also through several well-known features of HRS cells: (i) it is very difficult to grow HRS cells in culture without other cells and establish HL cell lines (see also below); (ii) primary HRS cells do not survive in immunodeficient mouse strains upon transplantation [162]; (iii) HRS cells are very rarely detectable in the peripheral blood [15], and even when HL metastasizes to other organs, such as the bone marrow, they are typically seen in their typical microenvironment. Remarkably, in patients with AIDS, the incidence of HL increased with improved therapy causing higher numbers of $CD4^+$

T cells, whereas the incidence of other B-cell lymphomas dropped, again pointing to a dependency of HRS cells on CD4$^+$ T cells [163].

The microenvironment in HL is, however, not the only factor supportive for HRS cells, and there are also multiple mechanisms by which these tumor cells try to evade an attack by cytotoxic T and NK cells (Fig. 14.2). As mentioned above, HRS cells attract Treg cells, which suppress cytotoxic cells by production of IL10 and presumably additional mechanisms [158, 159]. HRS cells also secrete immunosuppressive factors, including IL10, TGFβ, MIF and galectin-1 [152, 164–168]. There is also an indication that prostaglandin E2 is secreted in the HL microenvironment and contributes to the suppression of T-cell activity, but the cellular source of this factor has not yet been identified [169]. Furthermore, HRS cells show ectopic expression of CD137 and transfer this factor by trogocytosis to other HRS and antigen presenting cells, causing downregulation of CD137L in the target cells and, as a consequence, reduced T-cell stimulation

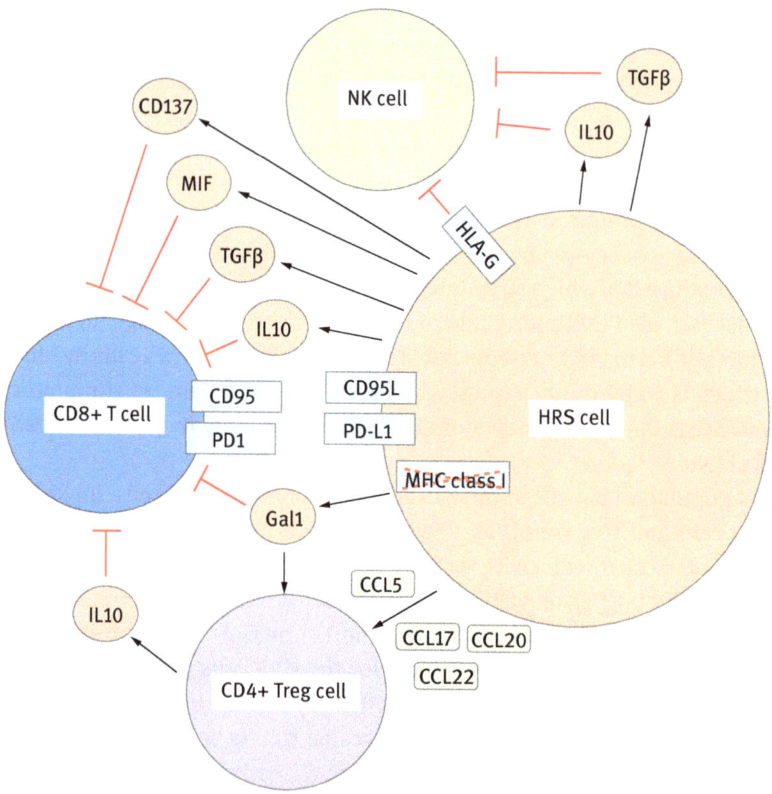

Fig. 14.2: **Immune escape mechanisms of HRS cells.** Shown are main mechanisms by which HRS cells escape from an attack by cytotoxic T and NK cells. The release of CD137 from HRS cells by trogocytosis functions indirectly on the inhibition of T-cell activity: CD137 is taken up by HRS and antigen presenting cells, causing downregulation of CD137L in these cells, and as a consequence impaired T-cell activation.

[170]. The frequent downmodulation of MHC class I and II molecules mentioned above may also contribute to an evasion from cytotoxic T cells. HRS cells express the non-classical MHC class I molecule HLA-G, which is a ligand for inhibitory receptors on NK and other cytotoxic cells, and thereby may escape from a control by such cells [171]. Furthermore, HRS cells express on their surface CD95L, PD-L1 and PD-L2, which can induce apoptosis of or functionally inhibit $CD8^+$ T cells [164, 172, 173]. An important role of PD-L in shielding HRS cells from killing by PD1-positive T cells is indicated by the recent observation that targeted therapy of HL patients with an antibody inhibiting the PD1-PD-L interaction has beneficial therapeutic effects [174].

Much less is known about microenvironmental interactions in NLPHL. However, LP cells are typically rosetted by $CD40L^+$ GC-type T_H cells and are found in networks of follicular dendritic cells that are an important component of primary B-cell follicles and GC [5, 175, 176]. This indicates that the LP cells as transformed GC B cells still depend on a GC-like microenvironment for their survival and proliferation.

14.9 Lessons learnt from composite lymphomas

In rare instances, two distinct types of lymphomas can occur simultaneously in the same patient. Such cases are called composite lymphomas. Often, they represent a combination of a classical HL and a B-NHL [20]. It is an intriguing question whether these cases represent the chance occurrence of two independent tumors in the same patient, or whether they share a common origin. To clarify this, a number of composite lymphomas were studied for a clonal relationship of both lymphoma components by molecular analysis of isolated lymphoma cells for rearranged IgV genes or other clonal markers [20, 177, 178]. In the majority of cases, the lymphomas were derived from a common precursor cell [20, 177, 178]. Typically, the rearranged IgV genes from the two related lymphomas carried shared as well as distinct somatic mutations [20, 177, 178]. This provides strong evidence that the common precursor was a mutating GC B cell. The B-NHL clonally related to the classical HL included follicular lymphomas, diffuse large B-cell lymphomas, chronic lymphocytic leukemia, mantle cell lymphomas, and others [20, 177–179].

In some patients, an HL and a B-NHL occur sequentially. Several such cases were studied for a clonal relationship of the lymphomas, and, strikingly, also in those sequential lymphomas, there was a common origin in more than half of the cases [20]. Often, the second clonally related lymphoma was diagnosed several years after the first lymphoma (up to 15 years), indicating that the second lymphoma or its precursor cells resided in the patients for extended time before it developed into a recognizable malignant tumor. Also in these clonally related sequential lymphomas, many cases showed shared and distinct IgV gene mutations, so that the same scenario applies as for the concurrent composite lymphomas [20].

Considering the derivation of two distinct types of lymphomas from a common GC B-cell precursor, it is likely that the common precursor already carried some transforming events, and that later further separate genetic lesions happened (mostly in the GC), so that two different lymphomas developed. In a candidate gene approach, examples for both shared as well as distinct transforming events were indeed identified for several composite and sequential lymphomas [73, 179]. For example, in combinations of classical HL and follicular lymphomas, the t(14;18) *BCL2*-IgH translocation was present not only in the follicular lymphomas, but also in the HRS cells [73].

In a number of clonally related composite lymphomas of a classical HL and a B-NHL with EBV$^+$ HRS cells of the classical HL cases, EBV was only detected in the HRS cells, but not in the B-NHL tumor cells [20]. This indicates that EBV infection was a rather late event, happening in a daughter cell of the (pre-malignant) common precursor, presumably often a GC B cell.

Overall, the molecular analysis of composite lymphomas provides further evidence for the GC B-cell origin of HRS cells in classical HL and a close relationship between B-NHL and classical HL (Tab. 14.2). Moreover, the genetic analysis of these cases supports the concept of a multistep transformation process in lymphomagenesis.

14.10 HL cell lines

From a very large number of attempts, only a few HL cell lines could be established. Presumably, HRS cells are strictly dependent on microenvironmental interactions, so that they normally cannot survive without other cells in culture. Indeed, all HL cell lines that are currently existing and available have been established from sources other than lymph nodes (bone marrow, peripheral blood, pleural effusions) [180–182], i.e. in situations where HRS cells apparently lost their dependency on the lymph node microenvironment already in the patient. Moreover, all lines were established from advanced or end-stage patients who received multiple cycles of therapy, so that one may speculate that the HRS cells had acquired additional genetic lesions already in the patient, rendering them less dependent on signals from surrounding cells. Overall, eight classical HL and one NLPHL cell line are currently used in the field (Tab. 14.3). Two of the lines (HDLM2 and L540) are of T-cell origin, so they may stem from the rare instances where a classical HL has a T-cell origin. Cell line L591 is the only EBV-infected one. Cell line CO was originally also reported as an HL cell line, but later turned out to be a cell culture contamination [183]. Moreover, although line HD-MYZ is sometimes still used, it should be pointed out that this line is of myeloid and not B- or T-cell origin [184]. As all informative primary classical HL cases are of B-cell or in rare instances a T-cell origin, the nonlymphoid origin of HD-MYZ is a strong argument that this line is not derived from HRS cells. Moreover, the phenotype of HD-MYZ cells does not fit to the phenotype of the HRS cells from the patient it was established from

[185]. Hence, this line should not be used as a model for HRS cells. Although a direct proof for a derivation from the HRS cells was provided only for line L1236 [14], the fact that all eight classical HL cell lines mentioned in Tab. 14.3 have a similar phenotype and (as far as it has been analyzed) global gene expression pattern (including the T-cell-derived HDLM2 cells) supports the idea that they originate from HRS cells [186, 187].

Due to the rarity of HRS cells and the technical challenge to obtain enough pure HRS cells for functional studies, and as there is still no mouse model for HL, the few classical HL cell lines represent essential tools for functional studies in HL. Moreover, many aspects of the pathophysiology and pathogenesis of classical HL were initially revealed with HL cell lines and later confirmed with primary HRS cells. It should, however, be kept in mind that the lines are adapted to the growth in culture and lost their interaction with other cells, as it occurs in the lymph node microenvironment. This is also reflected in many differences in gene expression between HL cell lines and primary HRS cells [126].

References

[1] Hodgkin T. On some morbid appearances of the absorbent glands and spleen. Med Chir Trans 1832;17:68–114.

[2] Reed D. On the pathological changes in Hodgkin's disease with special reference to its relation to tuberculosis. John Hopkins Hosp Rep 1902;10:133–193.

[3] Sternberg C. Über eine eigenartige unter dem Bilde der Pseudoleukämie verlaufende Tuberkolose des lymphatischen Apparates. Z Heilkunde 1898;19:21–90.

[4] Swerdlow SH, Campo E, Harris NL, et al. Classification of Tumours of Haematopoietic and Lymphoid Tissues 4th ed. Lyon: IARC Press; 2008.

[5] WHO Classification of Tumours of Haematopoietic and Lymphoid Tissues. Jaffe ES, Harris NL, Stein H, Vardiman JWe. Lyon, France: IARC Press; 2001.

[6] Engert A, Horning SJ. Hodgkin Lymphoma. Engert A, Horning SJ. Heidelberg: Springer; 2011.

[7] Steidl C, Connors JM, Gascoyne RD. Molecular pathogenesis of Hodgkin's lymphoma: increasing evidence of the importance of the microenvironment. J Clin Oncol 2011;29:1812–1826.

[8] Schmitz R, Stanelle J, Hansmann M-L, Küppers R. Pathogenesis of classical and lymphocyte-predominant Hodgkin lymphoma. Annu Rev Pathol 2009;4:151–174.

[9] Bräuninger A, Wacker HH, Rajewsky K, Küppers R, Hansmann ML. Typing the histogenetic origin of the tumor cells of lymphocyte-rich classical Hodgkin's lymphoma in relation to tumor cells of classical and lymphocyte-predominance Hodgkin's lymphoma. Cancer Res 2003;63:1644–1651.

[10] Kanzler H, Küppers R, Hansmann ML, Rajewsky K. Hodgkin and Reed-Sternberg cells in Hodgkin's disease represent the outgrowth of a dominant tumor clone derived from (crippled) germinal center B cells. J Exp Med 1996;184:1495–1505.

[11] Küppers R, Rajewsky K, Zhao M, et al. Hodgkin disease: Hodgkin and Reed-Sternberg cells picked from histological sections show clonal immunoglobulin gene rearrangements and appear to be derived from B cells at various stages of development. Proc Natl Acad Sci USA 1994;91:10962–10966.

[12] Marafioti T, Hummel M, Foss H-D, et al. Hodgkin and Reed-Sternberg cells represent an expansion of a single clone originating from a germinal center B-cell with functional immunoglobulin gene rearrangements but defective immunoglobulin transcription. Blood 2000;95:1443–1450.

[13] Irsch J, Nitsch S, Hansmann ML, et al. Isolation of viable Hodgkin and Reed-Sternberg cells from Hodgkin disease tissues. Proc Natl Acad Sci USA 1998;95:10117–10122.

[14] Kanzler H, Hansmann ML, Kapp U, et al. Molecular single cell analysis demonstrates the derivation of a peripheral blood-derived cell line (L1236) from the Hodgkin/Reed-Sternberg cells of a Hodgkin's lymphoma patient. Blood 1996;87:3429–3436.

[15] Vockerodt M, Soares M, Kanzler H, et al. Detection of clonal Hodgkin and Reed-Sternberg cells with identical somatically mutated and rearranged VH genes in different biopsies in relapsed Hodgkin's disease. Blood 1998;92:2899–2907.

[16] Küppers R, Zhao M, Hansmann ML, Rajewsky K. Tracing B cell development in human germinal centres by molecular analysis of single cells picked from histological sections. EMBO J 1993;12:4955–4967.

[17] Liso A, Capello D, Marafioti T, et al. Aberrant somatic hypermutation in tumor cells of nodular-lymphocyte-predominant and classic Hodgkin lymphoma. Blood 2006;108:1013–1020.

[18] Irsch J, Wolf J, Tesch H, Diehl V, Radbruch A, Staratschek-Jox A. Class switch recombination was specifically targeted to immunoglobulin (Ig)G4 or IgA in Hodgkin's disease-derived cell lines. Br J Haematol 2001;113:785–793.

[19] Martin-Subero JI, Klapper W, Sotnikova A, et al. Chromosomal breakpoints affecting immunoglobulin loci are recurrent in Hodgkin and Reed-Sternberg cells of classical Hodgkin lymphoma. Cancer Res 2006;66:10332–10338.

[20] Küppers R, Dührsen U, Hansmann ML. Pathogenesis, diagnosis, and treatment of composite lymphomas. Lancet Oncol 2014;15:e435–46.

[21] Küppers R, Hartmann S, Hansmann M-L. Hodgkin's lymphoma: Cell of origin, immunobiology, and pathogenesis. In: Orazi A, Weiss LM, Foucar K, Knowles DM, eds. Knowles' Neoplastic Hematopathology. 3rd ed. Philadelphia, USA: Lippincott Williams & Wilkins; 2014.

[22] Liu YJ, Joshua DE, Williams GT, Smith CA, Gordon J, MacLennan IC. Mechanism of antigen-driven selection in germinal centres. Nature 1989;342:929–931.

[23] Küppers R, Rajewsky K. The origin of Hodgkin and Reed/Sternberg cells in Hodgkin's disease. Annu Rev Immunol 1998;16:471–493.

[24] Müschen M, Küppers R, Spieker T, Bräuninger A, Rajewsky K, Hansmann ML. Molecular single-cell analysis of Hodgkin- and Reed-Sternberg cells harboring unmutated immunoglobulin variable region genes. Lab Invest 2001;81:289–295.

[25] Lebecque S, de Bouteiller O, Arpin C, Banchereau J, Liu YJ. Germinal center founder cells display propensity for apoptosis before onset of somatic mutation. J Exp Med 1997;185:563–571.

[26] Müschen M, Rajewsky K, Bräuninger A, et al. Rare occurrence of classical Hodgkin's disease as a T cell lymphoma. J Exp Med 2000;191:387–394.

[27] Seitz V, Hummel M, Marafioti T, Anagnostopoulos I, Assaf C, Stein H. Detection of clonal T-cell receptor gamma-chain gene rearrangements in Reed-Sternberg cells of classic Hodgkin disease. Blood 2000;95:3020–3024.

[28] Braeuninger A, Küppers R, Strickler JG, Wacker HH, Rajewsky K, Hansmann ML. Hodgkin and Reed-Sternberg cells in lymphocyte predominant Hodgkin disease represent clonal populations of germinal center-derived tumor B cells. Proc Natl Acad Sci USA 1997;94:9337–9342.

[29] Marafioti T, Hummel M, Anagnostopoulos I, et al. Origin of nodular lymphocyte-predominant Hodgkin's disease from a clonal expansion of highly mutated germinal-center B cells. N Engl J Med 1997;337:453–458.

[30] Ohno T, Stribley JA, Wu G, Hinrichs SH, Weisenburger DD, Chan WC. Clonality in nodular lymphocyte-predominant Hodgkin's disease. N Engl J Med 1997;337:459–465.

[31] Klein U, Goossens T, Fischer M, et al. Somatic hypermutation in normal and transformed human B cells. Immunol Rev 1998;162:261–280.

[32] Carbone A, Gloghini A, Gaidano G, et al. Expression status of BCL-6 and syndecan-1 identifies distinct histogenetic subtypes of Hodgkin's disease. Blood 1998;92:2220–2228.

[33] Greiner A, Tobollik S, Buettner M, et al. Differential expression of activation-induced cytidine deaminase (AID) in nodular lymphocyte-predominant and classical Hodgkin lymphoma. J Pathol 2005;205:541–547.

[34] Montes-Moreno S, Roncador G, Maestre L, et al. Gcet1 (centerin), a highly restricted marker for a subset of germinal center-derived lymphomas. Blood 2008;111:351–358.

[35] Natkunam Y, Lossos IS, Taidi B, et al. Expression of the human germinal center-associated lymphoma (HGAL) protein, a new marker of germinal center B-cell derivation. Blood 2005;105:3979–3986.

[36] Küppers R, Bräuninger A, Müschen M, Distler V, Hansmann ML, Rajewsky K. Evidence that Hodgkin and Reed-Sternberg cells in Hodgkin disease do not represent cell fusions. Blood 2001;97:818–821.

[37] Drexler HG, Gignac SM, Hoffbrand AV, Minowada J. Formation of multinucleated cells in a Hodgkin's-disease-derived cell line. Int J Cancer 1989;43:1083–1090.

[38] Ikeda J, Mamat S, Tian T, et al. Tumorigenic potential of mononucleated small cells of Hodgkin lymphoma cell lines. Am J Pathol 2010;177:3081–3088.

[39] Newcom SR, Kadin ME, Phillips C. L-428 Reed-Sternberg cells and mononuclear Hodgkin's cells arise from a single cloned mononuclear cell. Int J Cell Cloning 1988;6:417–431.

[40] Rengstl B, Newrzela S, Heinrich T, et al. Incomplete cytokinesis and re-fusion of small mononucleated Hodgkin cells lead to giant multinucleated Reed-Sternberg cells. Proc Natl Acad Sci USA 2013;110:20729–20734.

[41] Nakashima M, Ishii Y, Watanabe M, et al. The side population, as a precursor of Hodgkin and Reed-Sternberg cells and a target for nuclear factor-kappaB inhibitors in Hodgkin's lymphoma. Cancer Sci 2010;101:2490–2496.

[42] Shafer JA, Cruz CR, Leen AM, et al. Antigen-specific cytotoxic T lymphocytes can target chemoresistant side-population tumor cells in Hodgkin lymphoma. Leuk Lymphoma 2010;51:870–880.

[43] Jones RJ, Gocke CD, Kasamon YL, et al. Circulating clonotypic B cells in classic Hodgkin lymphoma. Blood 2009;113:5920–5926.

[44] Küppers R. Clonogenic B cells in classic Hodgkin lymphoma. Blood 2009;114:3970–3971.

[45] Kuzu I, Delsol G, Jones M, Gatter KC, Mason DY. Expression of the Ig-associated heterodimer (mb-1 and B29) in Hodgkin's disease. Histopathol 1993;22:141–144.

[46] Re D, Müschen M, Ahmadi T, et al. Oct-2 and Bob-1 deficiency in Hodgkin and Reed Sternberg cells. Cancer Res 2001;61:2080–2084.

[47] Stein H, Marafioti T, Foss HD, et al. Down-regulation of BOB.1/OBF.1 and Oct2 in classical Hodgkin disease but not in lymphocyte predominant Hodgkin disease correlates with immunoglobulin transcription. Blood 2001;97:496–501.

[48] Foss HD, Reusch R, Demel G, et al. Frequent expression of the B-cell-specific activator protein in Reed-Sternberg cells of classical Hodgkin's disease provides further evidence for its B-cell origin. Blood 1999;94:3108–3113.

[49] Mathas S, Janz M, Hummel F, et al. Intrinsic inhibition of transcription factor E2A by HLH proteins ABF-1 and Id2 mediates reprogramming of neoplastic B cells in Hodgkin lymphoma. Nature Immunol 2006;7:207–215.

[50] Schwering I, Bräuninger A, Klein U, et al. Loss of the B-lineage-specific gene expression program in Hodgkin and Reed-Sternberg cells of Hodgkin lymphoma. Blood 2003;101:1505–1512.

[51] Küppers R. New insights in the biology of Hodgkin lymphoma. Hematology Am Soc Hematol Educ Program 2012;2012:328–334.

[52] Steidl C, Shah SP, Woolcock BW, et al. MHC class II transactivator CIITA is a recurrent gene fusion partner in lymphoid cancers. Nature 2011;471:377–381.

[53] Renné C, Martin-Subero JI, Eickernjager M, et al. Aberrant expression of ID2, a suppressor of B-cell-specific gene expression, in Hodgkin's lymphoma. Am J Pathol 2006;169:655–664.

[54] Yokota Y, Mansouri A, Mori S, et al. Development of peripheral lymphoid organs and natural killer cells depends on the helix-loop-helix inhibitor Id2. Nature 1999;397:702–76.

[55] Jundt F, Acikgoz O, Kwon SH, et al. Aberrant expression of Notch1 interferes with the B-lymphoid phenotype of neoplastic B cells in classical Hodgkin lymphoma. Leukemia 2008;22:1587–1594.

[56] Jundt F, Anagnostopoulos I, Förster R, Mathas S, Stein H, Dörken B. Activated Notch 1 signaling promotes tumor cell proliferation and survival in Hodgkin and anaplastic large cell lymphoma. Blood 2002;99:3398–3403.

[57] Köchert K, Ullrich K, Kreher S, et al. High-level expression of Mastermind-like 2 contributes to aberrant activation of the NOTCH signaling pathway in human lymphomas. Oncogene 2011.

[58] Kreher S, Bouhlel MA, Cauchy P, et al. Mapping of transcription factor motifs in active chromatin identifies IRF5 as key regulator in classical Hodgkin lymphoma. Proc Natl Acad Sci USA 2014;111:E4513–4522.

[59] Scheeren FA, Diehl SA, Smit LA, et al. IL-21 is expressed in Hodgkin lymphoma and activates STAT5; evidence that activated STAT5 is required for Hodgkin lymphomagenesis. Blood 2008;111:4706–4715.

[60] Vockerodt M, Morgan SL, Kuo M, et al. The Epstein–Barr virus oncoprotein, latent membrane protein-1, reprograms germinal centre B cells towards a Hodgkin's Reed-Sternberg-like phenotype. J Pathol 2008;216:83–92.

[61] Portis T, Dyck P, Longnecker R. Epstein–Barr Virus (EBV) LMP2A induces alterations in gene transcription similar to those observed in Reed-Sternberg cells of Hodgkin lymphoma. Blood 2003;102:4166–4178.

[62] Dukers DF, van Galen JC, Giroth C, et al. Unique polycomb gene expression pattern in Hodgkin's lymphoma and Hodgkin's lymphoma-derived cell lines. Am J Pathol 2004;164:873–881.

[63] Raaphorst FM, van Kemenade FJ, Blokzijl T, et al. Coexpression of BMI-1 and EZH2 polycomb group genes in Reed-Sternberg cells of Hodgkin's disease. Am J Pathol 2000;157:709–715.

[64] Sanchez-Beato M, Sanchez E, Garcia JF, et al. Abnormal PcG protein expression in Hodgkin's lymphoma. Relation with E2F6 and NFkappaB transcription factors. J Pathol 2004;204:528–537.

[65] Ammerpohl O, Haake A, Pellissery S, et al. Array-based DNA methylation analysis in classical Hodgkin lymphoma reveals new insights into the mechanisms underlying silencing of B cell-specific genes. Leukemia 2012;26:185–188.

[66] Doerr JR, Malone CS, Fike FM, et al. Patterned CpG methylation of silenced B cell gene promoters in classical Hodgkin lymphoma-derived and primary effusion lymphoma cell lines. J Mol Biol 2005;350:631–640.

[67] Ushmorov A, Leithäuser F, Sakk O, et al. Epigenetic processes play a major role in B-cell-specific gene silencing in classical Hodgkin lymphoma. Blood 2006;107:2493–2500.

[68] Wein F, Otto T, Lambertz P, Fandrey J, Hansmann M-L, Küppers R. Potential role of hypoxia in early stages of Hodgkin lymphoma pathogenesis. Haematologica 2015; in press.

[69] Weber-Matthiesen K, Deerberg J, Poetsch M, Grote W, Schlegelberger B. Numerical chromosome aberrations are present within the CD30+ Hodgkin and Reed-Sternberg cells in 100 % of analyzed cases of Hodgkin's disease. Blood 1995;86:1464–1468.
[70] Poppema S, Kaleta J, Hepperle B. Chromosomal abnormalities in patients with Hodgkin's disease: evidence for frequent involvement of the 14q chromosomal region but infrequent bcl-2 gene rearrangement in Reed-Sternberg cells. J Natl Cancer Inst 1992;84:1789–1793.
[71] Falzetti D, Crescenzi B, Matteuci C, et al. Genomic instability and recurrent breakpoints are main cytogenetic findings in Hodgkin's disease. Haematologica 1999;84:298–305.
[72] Martin-Subero JI, Knippschild U, Harder L, et al. Segmental chromosomal aberrations and centrosome amplifications: pathogenetic mechanisms in Hodgkin and Reed-Sternberg cells of classical Hodgkin's lymphoma? Leukemia 2003;17:2214–2219.
[73] Schmitz R, Renné C, Rosenquist R, et al. Insight into the multistep transformation process of lymphomas: IgH-associated translocations and tumor suppressor gene mutations in clonally related composite Hodgkin's and non-Hodgkin's lymphomas. Leukemia 2005;19:1452–1458.
[74] Küppers R, Dalla-Favera R. Mechanisms of chromosomal translocations in B cell lymphomas. Oncogene 2001;20:5580–5594.
[75] Ushmorov A, Ritz O, Hummel M, et al. Epigenetic silencing of the immunoglobulin heavy-chain gene in classical Hodgkin lymphoma-derived cell lines contributes to the loss of immunoglobulin expression. Blood 2004;104:3326–3334.
[76] Cabannes E, Khan G, Aillet F, Jarrett RF, Hay RT. Mutations in the IκBα gene in Hodgkin's disease suggest a tumour suppressor role for IκBα. Oncogene 1999;18:3063–3070.
[77] Emmerich F, Theurich S, Hummel M, et al. Inactivating I kappa B epsilon mutations in Hodgkin/Reed-Sternberg cells. J Pathol 2003;201:413–420.
[78] Jungnickel B, Staratschek-Jox A, Bräuninger A, et al. Clonal deleterious mutations in the IkBa gene in the malignant cells in Hodgkin's disease. J Exp Med 2000;191:395–401.
[79] Lake A, Shield LA, Cordano P, et al. Mutations of NFKBIA, encoding IkappaBalpha, are a recurrent finding in classical Hodgkin lymphoma but are not a unifying feature of non-EBV-associated cases. Int J Cancer 2009;125:1334–1342.
[80] Kato M, Sanada M, Kato I, et al. Frequent inactivation of A20 in B-cell lymphomas. Nature 2009;459:712–716.
[81] Schmitz R, Hansmann ML, Bohle V, et al. TNFAIP3 (A20) is a tumor suppressor gene in Hodgkin lymphoma and primary mediastinal B cell lymphoma. J Exp Med 2009;206:981–989.
[82] Otto C, Giefing M, Massow A, et al. Genetic lesions of the TRAF3 and MAP3K14 genes in classical Hodgkin lymphoma. Br J Haematol 2012;157:702–708.
[83] Schmidt A, Schmitz R, Giefing M, et al. Rare occurrence of biallelic CYLD gene mutations in classical Hodgkin lymphoma. Genes Chromosomes Cancer 2010;49:803–809.
[84] Joos S, Menz CK, Wrobel G, et al. Classical Hodgkin lymphoma is characterized by recurrent copy number gains of the short arm of chromosome 2. Blood 2002;99:1381–1387.
[85] Martin-Subero JI, Gesk S, Harder L, et al. Recurrent involvement of the REL and BCL11A loci in classical Hodgkin lymphoma. Blood 2002;99:1474–1477.
[86] Steidl C, Telenius A, Shah SP, et al. Genome-wide copy number analysis of Hodgkin Reed-Sternberg cells identifies recurrent imbalances with correlations to treatment outcome. Blood 2010;116:418–427.
[87] Martin-Subero JI, Wlodarska I, Bastard C, et al. Chromosomal rearrangements involving the BCL3 locus are recurrent in classical Hodgkin and peripheral T-cell lymphoma. Blood 2006;108:401–402.
[88] Mathas S, Jöhrens K, Joos S, et al. Elevated NF-kappaB p50 complex formation and Bcl-3 expression in classical Hodgkin, anaplastic large-cell, and other peripheral T-cell lymphomas. Blood 2005;106:4287–4293.

[89] Joos S, Küpper M, Ohl S, et al. Genomic imbalances including amplification of the tyrosine kinase gene JAK2 in CD30+ Hodgkin cells. Cancer Res 2000;60:549–552.

[90] Weniger MA, Melzner I, Menz CK, et al. Mutations of the tumor suppressor gene SOCS-1 in classical Hodgkin lymphoma are frequent and associated with nuclear phospho-STAT5 accumulation. Oncogene 2006;25:2679–2684.

[91] Green MR, Monti S, Rodig SJ, et al. Integrative analysis reveals selective 9p24.1 amplification, increased PD-1 ligand expression, and further induction via JAK2 in nodular sclerosing Hodgkin lymphoma and primary mediastinal large B-cell lymphoma. Blood 2010;116:3268–3277.

[92] Rui L, Emre NC, Kruhlak MJ, et al. Cooperative epigenetic modulation by cancer amplicon genes. Cancer Cell 2010;18:590–605.

[93] Gunawardana J, Chan FC, Telenius A, et al. Recurrent somatic mutations of PTPN1 in primary mediastinal B cell lymphoma and Hodgkin lymphoma. Nat Genet 2014;46:329–335.

[94] van den Berg A, Visser L, Eberwine J, Dadvand L, Poppema S. Frequent lack of translation of antigen presentation-associated molecules MHC class I, CD1a and Beta(2)-microglobulin in Reed-Sternberg cells. Int J Cancer 2000;86:548–552.

[95] Oudejans JJ, Jiwa NM, Kummer JA, et al. Analysis of major histocompatibility complex class I expression on Reed-Sternberg cells in relation to the cytotoxic T-cell response in Epstein–Barr virus-positive and -negative Hodgkin's disease. Blood 1996;87:3844–3851.

[96] Liu Y, Abdul Razak FR, Terpstra M, et al. The mutational landscape of Hodgkin lymphoma cell lines determined by whole-exome sequencing. Leukemia 2014;28:2248–2251.

[97] Reichel J, Chadburn A, Rubinstein PG, et al. Flow-sorting and exome sequencing reveals the oncogenome of primary Hodgkin and Reed-Sternberg cells. Blood 2015;125:1061–1072.

[98] Schneider M, Schneider S, Zühlke-Jenisch R, et al. Alterations of the CD58 gene in classical Hodgkin lymphoma. Genes Chromosomes Cancer 2015; in press.

[99] Renné C, Martin-Subero JI, Hansmann ML, Siebert R. Molecular cytogenetic analyses of immunoglobulin loci in nodular lymphocyte predominant Hodgkin's lymphoma reveal a recurrent IGH-BCL6 juxtaposition. J Mol Diagn 2005;7:352–356.

[100] Wlodarska I, Nooyen P, Maes B, et al. Frequent occurrence of BCL6 rearrangements in nodular lymphocyte predominance Hodgkin lymphoma but not in classical Hodgkin lymphoma. Blood 2003;101:706–710.

[101] Wlodarska I, Stul M, De Wolf-Peeters C, Hagemeijer A. Heterogeneity of BCL6 rearrangements in nodular lymphocyte predominant Hodgkin's lymphoma. Haematologica 2004;89:965–972.

[102] Mottok A, Renné C, Willenbrock K, Hansmann ML, Bräuninger A. Somatic hypermutation of SOCS1 in lymphocyte-predominant Hodgkin lymphoma is accompanied by high JAK2 expression and activation of STAT6. Blood 2007;110:3387–3390.

[103] Brune V, Tiacci E, Pfeil I, et al. Origin and pathogenesis of nodular lymphocyte-predominant Hodgkin lymphoma as revealed by global gene expression analysis. J Exp Med 2008;205:2251–2268.

[104] Schumacher MA, Schmitz R, Brune V, et al. Mutations in the genes coding for the NF-kappaB regulating factors IkappaBalpha and A20 are uncommon in nodular lymphocyte-predominant Hodgkin's lymphoma. Haematologica 2010;95:153–157.

[105] Oki Y, Neelapu SS, Fanale M, et al. Detection of classical Hodgkin lymphoma specific sequence in peripheral blood using a next-generation sequencing approach. Br J Haematol 2015;169:689–693.

[106] Vandenberghe P, Wlodarska I, Tousseyn T, et al. Non-invasive detection of genomic imbalances in Hodgkin/Reed-Sternberg cells in early and advanced stage Hodgkin's lymphoma by sequencing of circulating cell-free DNA: a technical proof-of-principle study. Lancet Haematology 2015;2:e55–e65.

[107] Goldin LR, Bjorkholm M, Kristinsson SY, Turesson I, Landgren O. Highly increased familial risks for specific lymphoma subtypes. Br J Haematol 2009;146:91–94.
[108] Salipante SJ, Mealiffe ME, Wechsler J, et al. Mutations in a gene encoding a midbody kelch protein in familial and sporadic classical Hodgkin lymphoma lead to binucleated cells. Proc Natl Acad Sci USA 2009;106:14920–14925.
[109] Saarinen S, Aavikko M, Aittomaki K, et al. Exome sequencing reveals germline NPAT mutation as a candidate risk factor for Hodgkin lymphoma. Blood 2011;118:493–498.
[110] Cozen W, Timofeeva MN, Li D, et al. A meta-analysis of Hodgkin lymphoma reveals 19p13.3 TCF3 as a novel susceptibility locus. Nat Commun 2014;5:3856.
[111] Enciso-Mora V, Broderick P, Ma Y, et al. A genome-wide association study of Hodgkin's lymphoma identifies new susceptibility loci at 2p16.1 (REL), 8q24.21 and 10p14 (GATA3). Nat Genet 2010;42:1126–1130.
[112] Frampton M, da Silva Filho MI, Broderick P, et al. Variation at 3p24.1 and 6q23.3 influences the risk of Hodgkin's lymphoma. Nat Commun 2013;4:2549.
[113] Urayama KY, Jarrett RF, Hjalgrim H, et al. Genome-wide association study of classical Hodgkin lymphoma and Epstein–Barr virus status-defined subgroups. J Natl Cancer Inst 2012;104:240–253.
[114] Stanelle J, Döring C, Hansmann ML, Küppers R. Mechanisms of aberrant GATA3 expression in classical Hodgkin lymphoma and its consequences for the cytokine profile of Hodgkin and Reed/Sternberg cells. Blood 2010;116:4202–4211.
[115] Staratschek-Jox A, Shugart YY, Strom SS, Nagler A, Taylor GM. Genetic susceptibility to Hodgkin's lymphoma and to secondary cancer: workshop report. Ann Oncol 2002;13 Suppl 1:30–33.
[116] Diepstra A, Niens M, Vellenga E, et al. Association with HLA class I in Epstein–Barr-virus-positive and with HLA class III in Epstein–Barr-virus-negative Hodgkin's lymphoma. Lancet 2005;365:2216–2224.
[117] Kapatai G, Murray P. Contribution of the Epstein–Barr virus to the molecular pathogenesis of Hodgkin lymphoma. J Clin Pathol 2007;60:1342–1349.
[118] Tinguely M, Rosenquist R, Sundstrom C, et al. Analysis of a clonally related mantle cell and Hodgkin lymphoma indicates Epstein–Barr virus infection of a Hodgkin/Reed-Sternberg cell precursor in a germinal center. Am J Surg Pathol 2003;27:1483–1488.
[119] Anagnostopoulos I, Herbst H, Niedobitek G, Stein H. Demonstration of monoclonal EBV genomes in Hodgkin's disease and Ki-1- positive anaplastic large cell lymphoma by combined Southern blot and in situ hybridization. Blood 1989;74:810–816.
[120] Kilger E, Kieser A, Baumann M, Hammerschmidt W. Epstein–Barr virus-mediated B-cell proliferation is dependent upon latent membrane protein 1, which simulates an activated CD40 receptor. EMBO J 1998;17:1700–1709.
[121] Alber G, Kim KM, Weiser P, Riesterer C, Carsetti R, Reth M. Molecular mimicry of the antigen receptor signalling motif by transmembrane proteins of the Epstein–Barr virus and the bovine leukemia virus. Curr Biol 1993;3:333–339.
[122] Bräuninger A, Schmitz R, Bechtel D, Renné C, Hansmann M-L, Küppers R. Molecular biology of Hodgkin and Reed/Sternberg cells in Hodgkin's lymphoma. Int J Cancer 2006;118:1853–1861.
[123] Bechtel D, Kurth J, Unkel C, Küppers R. Transformation of BCR-deficient germinal-center B cells by EBV supports a major role of the virus in the pathogenesis of Hodgkin and post-transplantation lymphomas. Blood 2005; 106:4345–4350.
[124] Chaganti S, Bell AI, Begue-Pastor N, et al. Epstein–Barr virus infection in vitro can resue germinal centre B cells with inactivated immunoglobulin genes. Blood 2005; 106:4249–4252.
[125] Mancao C, Altmann M, Jungnickel B, Hammerschmidt W. Rescue of "crippled" germinal center B cells from apoptosis by Epstein–Barr virus. Blood 2005; 106:4339–4344.

[126] Tiacci E, Döring C, Brune V, et al. Analyzing primary Hodgkin and Reed-Sternberg cells to capture the molecular and cellular pathogenesis of classical Hodgkin lymphoma. Blood 2012;120:4609–4620.
[127] Carbone A, Gloghini A, Gattei V, et al. Expression of functional CD40 antigen on Reed-Sternberg cells and Hodgkin's disease cell lines. Blood 1995;85:780–789.
[128] Chiu A, Xu W, He B, et al. Hodgkin lymphoma cells express TACI and BCMA receptors and generate survival and proliferation signals in response to BAFF and APRIL. Blood 2007;109:729–739.
[129] Fiumara P, Snell V, Li Y, et al. Functional expression of receptor activator of nuclear factor kappaB in Hodgkin disease cell lines. Blood 2001;98:2784–2790.
[130] Molin D, Fischer M, Xiang Z, et al. Mast cells express functional CD30 ligand and are the predominant CD30L-positive cells in Hodgkin's disease. Br J Haematol 2001;114:616–623.
[131] Pinto A, Aldinucci D, Gloghini A, et al. Human eosinophils express functional CD30 ligand and stimulate proliferation of a Hodgkin's disease cell line. Blood 1996;88:3299–3305.
[132] Carbone A, Gloghini A, Gruss HJ, Pinto A. CD40 ligand is constitutively expressed in a subset of T cell lymphomas and on the microenvironmental reactive T cells of follicular lymphomas and Hodgkin's disease. Am J Pathol 1995;147:912–922.
[133] Schwaller J, Went P, Matthes T, et al. Paracrine promotion of tumor development by the TNF ligand APRIL in Hodgkin's disease. Leukemia 2007;21:1324–1327.
[134] Baus D, Pfitzner E. Specific function of STAT3, SOCS1, and SOCS3 in the regulation of proliferation and survival of classical Hodgkin lymphoma cells. Int J Cancer 2006;118:1404–1413.
[135] Kube D, Holtick U, Vockerodt M, et al. STAT3 is constitutively activated in Hodgkin cell lines. Blood 2001;98:762–770.
[136] Skinnider BF, Elia AJ, Gascoyne RD, et al. Signal transducer and activator of transcription 6 is frequently activated in Hodgkin and Reed-Sternberg cells of Hodgkin lymphoma. Blood 2002;99:618–626.
[137] Kapp U, Yeh WC, Patterson B, et al. Interleukin 13 is secreted by and stimulates the growth of Hodgkin and Reed-Sternberg cells. J Exp Med 1999;189:1939–1946.
[138] Lamprecht B, Kreher S, Anagnostopoulos I, et al. Aberrant expression of the Th2 cytokine IL-21 in Hodgkin lymphoma cells regulates STAT3 signaling and attracts Treg cells via regulation of MIP-3{alpha}. Blood 2008;112:3339–3347.
[139] Hinz M, Lemke P, Anagnostopoulos I, et al. Nuclear factor kappaB-dependent gene expression profiling of Hodgkin's disease tumor cells, pathogenetic significance, and link to constitutive signal transducer and activator of transcription 5a activity. J Exp Med 2002;196:605–617.
[140] Renné C, Willenbrock K, Küppers R, Hansmann M-L, Bräuninger A. Autocrine and paracrine activated receptor tyrosine kinases in classical Hodgkin lymphoma. Blood 2005;105:4051–4059.
[141] Lamprecht B, Walter K, Kreher S, et al. Derepression of an endogenous long terminal repeat activates the CSF1R proto-oncogene in human lymphoma. Nat Med 2010;16:571–579.
[142] Renne C, Minner S, Küppers R, Hansmann ML, Bräuninger A. Autocrine NGFbeta/TRKA signalling is an important survival factor for Hodgkin lymphoma derived cell lines. Leuk Res 2008;32:163–167.
[143] Teofili L, Di Febo AL, Pierconti F, et al. Expression of the c-met proto-oncogene and its ligand, hepatocyte growth factor, in Hodgkin disease. Blood 2001;97:1063–1069.
[144] Dutton A, Reynolds GM, Dawson CW, Young LS, Murray PG. Constitutive activation of phosphatidyl-inositide 3 kinase contributes to the survival of Hodgkin's lymphoma cells through a mechanism involving Akt kinase and mTOR. J Pathol 2005;205:498–506.

[145] Georgakis GV, Li Y, Rassidakis GZ, Medeiros LJ, Mills GB, Younes A. Inhibition of the phosphatidylinositol-3 kinase/Akt promotes G1 cell cycle arrest and apoptosis in Hodgkin lymphoma. Br J Haematol 2006;132:503–511.
[146] Nagel S, Scherr M, Quentmeier H, et al. HLXB9 activates IL6 in Hodgkin lymphoma cell lines and is regulated by PI3K signalling involving E2F3. Leukemia 2005;19:841–846.
[147] Zheng B, Fiumara P, Li YV, et al. MEK/ERK pathway is aberrantly active in Hodgkin disease: a signaling pathway shared by CD30, CD40, and RANK that regulates cell proliferation and survival. Blood 2003;102:1019–1027.
[148] Mathas S, Hinz M, Anagnostopoulos I, et al. Aberrantly expressed c-Jun and JunB are a hallmark of Hodgkin lymphoma cells, stimulate proliferation and synergize with NF-kappa B. EMBO J 2002;21:4104–4113.
[149] Watanabe M, Ogawa Y, Ito K, et al. AP-1 mediated relief of repressive activity of the CD30 promoter microsatellite in Hodgkin and Reed-Sternberg cells. Am J Pathol 2003;163:633–641.
[150] Liu Y, Sattarzadeh A, Diepstra A, Visser L, van den Berg A. The microenvironment in classical Hodgkin lymphoma: an actively shaped and essential tumor component. Semin Cancer Biol 2014;24:15–22.
[151] Aldinucci D, Lorenzon D, Cattaruzza L, et al. Expression of CCR5 receptors on Reed-Sternberg cells and Hodgkin lymphoma cell lines: involvement of CCL5/Rantes in tumor cell growth and microenvironmental interactions. Int J Cancer 2008;122:769–776.
[152] Skinnider BF, Mak TW. The role of cytokines in classical Hodgkin lymphoma. Blood 2002;99:4283–4297.
[153] van den Berg A, Visser L, Poppema S. High expression of the CC chemokine TARC in Reed-Sternberg cells. A possible explanation for the characteristic T-cell infiltration Hodgkin's lymphoma. Am J Pathol 1999;154:1685–1691.
[154] Jundt F, Anagnostopoulos I, Bommert K, et al. Hodgkin/Reed-Sternberg cells induce fibroblasts to secrete eotaxin, a potent chemoattractant for T cells and eosinophils. Blood 1999;94:2065–2071.
[155] Aldinucci D, Gattei V. Interleukin-3 receptors in Hodgkin's disease. Am J Pathol 2003;162:356–357.
[156] Aldinucci D, Gattei V. The role of interleukin-3 and stem cell factor in classical Hodgkin disease. Blood 2003;101:376–377.
[157] Sarker AB, Akagi T, Jeon HJ, et al. Bauhinia purpurea–a new paraffin section marker for Reed-Sternberg cells of Hodgkin's disease. A comparison with Leu-M1 (CD15), LN2 (CD74), peanut agglutinin, and Ber-H2 (CD30). Am J Pathol 1992;141:19–23.
[158] Gandhi MK, Lambley E, Duraiswamy J, et al. Expression of LAG-3 by tumor-infiltrating lymphocytes is coincident with the suppression of latent membrane antigen-specific CD8+ T-cell function in Hodgkin lymphoma patients. Blood 2006;108:2280–2289.
[159] Marshall NA, Christie LE, Munro LR, et al. Immunosuppressive regulatory T cells are abundant in the reactive lymphocytes of Hodgkin lymphoma. Blood 2004;103:1755–1762.
[160] Greaves P, Clear A, Owen A, et al. Defining characteristics of classical Hodgkin lymphoma microenvironment T-helper cells. Blood 2013;122:2856–2863.
[161] Poppema S. Immunobiology and pathophysiology of Hodgkin lymphomas. Hematology Am Soc Hematol Educ Program 2005:231–238.
[162] Kapp U, Wolf J, Hummel M, et al. Hodgkin's lymphoma-derived tissue serially transplanted into severe combined immunodeficient mice. Blood 1993;82:1247–1256.
[163] Biggar RJ, Jaffe ES, Goedert JJ, Chaturvedi A, Pfeiffer R, Engels EA. Hodgkin lymphoma and immunodeficiency in persons with HIV/AIDS. Blood 2006;108:3786–3791.

[164] Chemnitz JM, Eggle D, Driesen J, et al. RNA fingerprints provide direct evidence for the inhibitory role of TGFbeta and PD-1 on CD4+ T cells in Hodgkin lymphoma. Blood 2007;110:3226–3233.

[165] Gandhi MK, Moll G, Smith C, et al. Galectin-1 mediated suppression of Epstein–Barr virus specific T-cell immunity in classic Hodgkin lymphoma. Blood 2007;110:1326–1329.

[166] Juszczynski P, Ouyang J, Monti S, et al. The AP1-dependent secretion of galectin-1 by Reed Sternberg cells fosters immune privilege in classical Hodgkin lymphoma. Proc Natl Acad Sci USA 2007;104:13134–13139.

[167] Newcom SR, Gu L. Transforming growth factor beta 1 messenger RNA in Reed-Sternberg cells in nodular sclerosing Hodgkin's disease. J Clin Pathol 1995;48:160–163.

[168] Ma Y, Visser L, Roelofsen H, et al. Proteomics analysis of Hodgkin lymphoma: identification of new players involved in the cross-talk between HRS cells and infiltrating lymphocytes. Blood 2008;111:2339–2346.

[169] Chemnitz JM, Driesen J, Classen S, et al. Prostaglandin E2 impairs CD4+ T cell activation by inhibition of lck: implications in Hodgkin's lymphoma. Cancer Res 2006;66:1114–1122.

[170] Ho WT, Pang WL, Chong SM, et al. Expression of CD137 on Hodgkin and Reed-Sternberg cells inhibits T-cell activation by eliminating CD137 ligand expression. Cancer Res 2013;73:652–661.

[171] Diepstra A, Poppema S, Boot M, et al. HLA-G protein expression as a potential immune escape mechanism in classical Hodgkin's lymphoma. Tissue Antigens 2008;71:219–226.

[172] Yamamoto R, Nishikori M, Kitawaki T, et al. PD-1-PD-1 ligand interaction contributes to immunosuppressive microenvironment of Hodgkin lymphoma. Blood 2008;111:3220–3224.

[173] Metkar SS, Naresh KN, Redkar AA, Soman CS, Advani SH, Nadkarni JJ. Expression of Fas and Fas ligand in Hodgkin's disease. Leuk Lymphoma 1999;33:521–530.

[174] Ansell SM, Lesokhin AM, Borrello I, et al. PD-1 blockade with Nivolumab in relapsed or refractory Hodgkin's lymphoma. N Engl J Med 2015;372:311–319.

[175] Hansmann ML, Fellbaum C, Hui PK, Zwingers T. Correlation of content of B cells and Leu7-positive cells with subtype and stage in lymphocyte predominance type Hodgkin's disease. J Cancer Res Clin Oncol 1988;114:405–410.

[176] Kamel OW, Gelb AB, Shibuya RB, Warnke RA. Leu 7 (CD57) reactivity distinguishes nodular lymphocyte predominance Hodgkin's disease from nodular sclerosing Hodgkin's disease, T-cell-rich B-cell lymphoma and follicular lymphoma. Am J Pathol 1993;142:541–546.

[177] Bräuninger A, Hansmann ML, Strickler JG, et al. Identification of common germinal-center B-cell precursors in two patients with both Hodgkin's disease and Non-Hodgkin's lymphoma. N Engl J Med 1999;340:1239–1247.

[178] Marafioti T, Hummel M, Anagnostopoulos I, Foss HD, Huhn D, Stein H. Classical Hodgkin's disease and follicular lymphoma originating from the same germinal center B cell. J Clin Oncol 1999;17:3804–3809.

[179] Schneider S, Crescenzi B, Schneider M, et al. Subclonal evolution of a classical Hodgkin lymphoma from a germinal center B-cell-derived mantle cell lymphoma. Int J Cancer 2014;134:832–8343.

[180] Drexler HG. Recent results on the biology of Hodgkin and Reed-Sternberg cells. II. Continuous cell lines. Leuk Lymphoma 1993;9:1–25.

[181] Wolf J, Kapp U, Bohlen H, et al. Peripheral blood mononuclear cells of a patient with advanced Hodgkin's lymphoma give rise to permanently growing Hodgkin-Reed Sternberg cells. Blood 1996;87:3418–3428.

[182] Mader A, Bruderlein S, Wegener S, et al. U-HO1, a new cell line derived from a primary refractory classical Hodgkin lymphoma. Cytogenet Genome Res 2007;119:204–210.

[183] Drexler HG, Dirks WG, MacLeod RA. False human hematopoietic cell lines: cross-contaminations and misinterpretations. Leukemia 1999;13:1601–1607.
[184] Bargou RC, Mapara MY, Zugck C, et al. Characterization of a novel Hodgkin cell line, HD-MyZ, with myelomonocytic features mimicking Hodgkin's disease in severe combined immunodeficient mice. J Exp Med 1993;177:1257–1268.
[185] Stein H, Diehl V, Marafioti T, Jox A, Wolf J, Hummel M. The nature of Reed-Sternberg cells, lymphocytic and histiocytic cells and their molecular biology in Hodgkin's disease. In: Armitage JO, Diehl V, Hoppe RT, Weiss l, eds. Hodgkin's disease. Philadelphia: Lippincott Williams & Wilkins; 1999:121–137.
[186] Küppers R, Klein U, Schwering I, et al. Identification of Hodgkin and Reed-Sternberg cell-specific genes by gene expression profiling. J Clin Invest 2003;111:529–537.
[187] Willenbrock K, Küppers R, Renne C, et al. Common features and differences in the transcriptome of large cell anaplastic lymphoma and classical Hodgkin's lymphoma. Haematologica 2006;91:596–604.

Brunangelo Falini and Stefano Pileri

15 Anaplastic large cell lymphoma

15.1 Introduction

The term anaplastic large cell lymphoma (ALCL) was first coined by Stein et al. [1] in 1985 to describe a particular type of lymphoma characterized by a proliferation of neoplastic lymphoid cells showing anaplastic appearance, propensity to grow cohesively and invade lymph node sinuses, and strong expression of the Ki-1 molecule [2] (later renamed CD30), a 120 kD transmembrane cytokine receptor of the tumor necrosis factor (TNF) receptor family [3, 4].

Further progresses in the characterization of this pathological condition came from the observation that a proportion of ALCL cases, often referred to as "malignant histiocytosis" [5, 6], carried a nonrandom t(2;5) (p23;q35) translocation [7–9] that was later discovered to cause the fusion of the *NPM* (nucleophosmin) gene with a previously unrecognized gene named *ALK* (anaplastic lymphoma kinase) [10].

In the subsequent years, the development of monoclonal antibodies directed against fixative-resistant epitopes of the ALK protein [11, 12] had a major impact in separating ALCL into two major subgroups (ALK$^+$ and ALK$^-$) and also in helping to better define the borders between ALCL and other lymphoma entities, particularly Hodgkin's lymphoma and peripheral T-cell lymphoma (PTCL). Moreover, a large bulk of information has now accumulated on the key role of the ALK kinase in the molecular pathogenesis of ALK$^+$ ALCL [13] and on the genomics of ALK$^-$ ALCL.

15.2 Anaplastic large cell lymphoma, ALK-positive (ALK$^+$ ALCL)

In the fourth (2008) World Health Organization (WHO) classification of lymphohemopoietic neoplasms [14], ALK$^+$ ALCL is listed as a distinct disease entity within the spectrum of mature T-cell neoplasms, clearly separated from ALK$^-$ ALCL that is regarded as a provisional entity (see below).

ALK$^+$ ALCL is thought to be derived from the transformation of activated mature cytotoxic T cells as a consequence of the constitutive expression and activation of oncogenic ALK fusion proteins that are generated by rearrangements of the *ALK* gene with the *NPM1* gene or with other gene partners. Notably, ALCL carrying NPM1-ALK or other ALK variant fusion proteins exhibit close similarities in their clinicopathological findings [15], gene expression signature [16] and molecular pathogenesis. This justifies their inclusion in the WHO classification under the single term of ALK$^+$ ALCL.

15.2.1 Clinical features

ALK$^+$ ALCL is a rare pathological condition (about 3 % of adult and 10–20 % of childhood lymphomas) that shows male predominance (M:F ratio = 1.5:1) [14]. It mainly occurs in the first three decades of life and involves both lymph nodes (peripheral and/or abdominal) and extranodal sites [17]. Most patients present at diagnosis with advanced stage III-IV disease and systemic symptoms, especially high fever [17]. The most commonly affected extranodal sites are skin and bone followed by soft tissues, lung and liver [17] (Fig. 15.1 A, B). Involvement of the central nervous system and the gastrointestinal tract is very rare. Immunohistochemistry detects bone marrow infiltration by tumor cells in about 30 % of cases. Leukemic presentation with peripheral blood involvement has been mainly observed in patients with the "small cell variant" of ALK$^+$ ALCL and carries a poor prognosis [14].

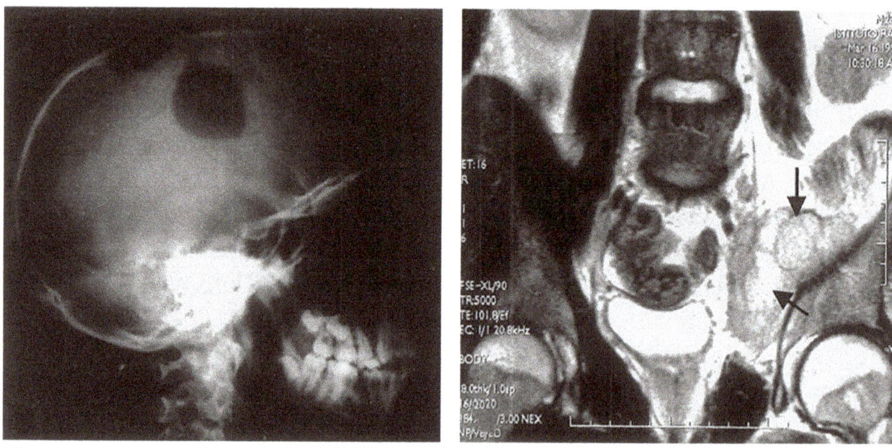

(a) (b)

Fig. 15.1: **Extranodal involvement in ALK$^+$ ALCL.** (a) Large osteolytic lesions of the skull in a 14-year-old boy. (b) Involvement of the right obturator muscle (arrows) in a 25-year-old man. This research was originally published in [20].

15.2.2 Morphology

ALK$^+$ ALCL is characterized by a broad morphological spectrum that is determined by a number of factors, including the pattern of lymphomatous infiltration (e.g. intrasinusoidal involvement), the cytological features of the dominant neoplastic population (ranging in size from small to large anaplastic), and the presence of accompanying reactive elements, especially histiocytes [12, 18–21]. Nevertheless, "hallmark cells" with eccentric nuclei and nuclear lobes surrounding the Golgi area – which appears as a clear, or more eosinophilic region – are detectable in all morphological variants of ALK$^+$ ALCL (see below) [19]. ALK activity may also contribute to anaplastic morphology since ALK expression controls the cell shape and F-actin filament assembly of ALCL cells through Cdc42 activation [22].

The most frequent morphological variant of ALK$^+$ ALCL (about 60 % of cases) is called “common type” and is mostly composed by large anaplastic tumor cells with abundant, often basophilic cytoplasm [1, 20, 23–25], and horseshoe or kidney-shaped nuclei containing multiple small basophilic nucleoli (Fig. 15.2(a)–(c)). These cells strongly express the CD30 and ALK proteins (Fig. 15.2(d) and Fig. 15.3).

The “small cell variant”, accounting for about 5–10 % of ALK$^+$ ALCL, is composed by a predominant population of small- to medium-sized tumor cells with irregular, generally not cerebriform, nuclei [24, 26]. These cells are ALK$^+$ but weakly positive or negative for the CD30 molecule and are often admixed with a minor population of CD30+/ALK+ large anaplastic tumor cells [12, 17, 19] (sometimes showing the fea-

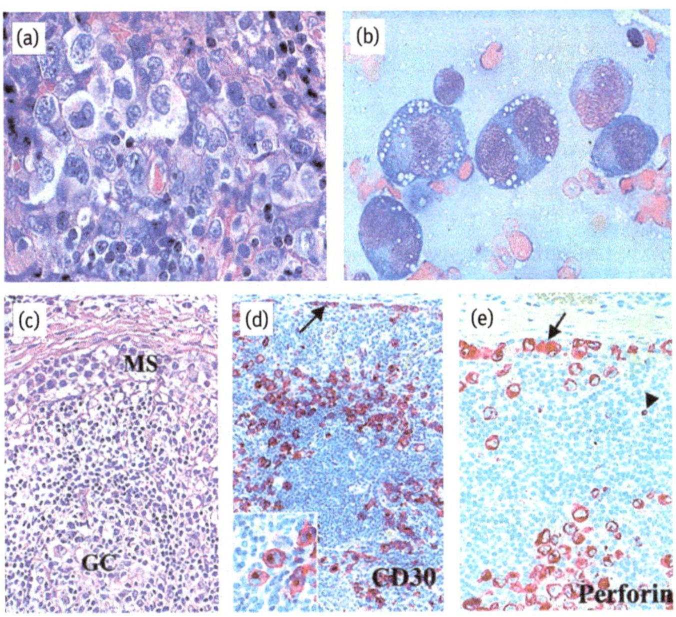

Fig. 15.2: **ALCL common type.** (a) Cytologic features of ALCL, common type, in a paraffin section (hematoxylin and eosin). (b) Cytologic features of ALCL, common type, in a touch imprint (May–Grünwald–Giemsa). (c) Dissemination of ALCL cells in a marginal sinus. MS indicates marginal sinus; GC, reactive germinal center (hematoxylin and eosin). (d) CD30 immunoreactivity of ALCL, common type. All tumor cells are CD30$^+$. Note the sinusoidal dissemination (arrow) and the perifollicular homing of the CD30$^+$ tumor cells (antibody clone Ber-H2; alkaline phosphatase-antialkaline phosphatase [APAAP]). (e) Immunoreactivity of an ALCL of common type for the cytotoxic molecule perforin. The sinusoidal dissemination of the tumor cells (arrow) and a small reactive cytotoxic cell (arrowhead) are indicated (antibody clone P1-8; APAAP). This research was originally published in [20].

tures of "hallmark cells") that are often located around blood vessels. Leukemic involvement is more frequent than in other morphological variants of ALK$^+$ ALCL and is characterized by the presence of atypical cells with flower-like appearance in the peripheral blood and/or bone marrow smears [14]. Bone marrow involvement in trephine biopsies is better detected by immunostaining for ALK protein [25].

The "lymphohistiocytic variant" [20, 23, 27] (about 10 % of ALK$^+$ ALCL) is composed of a population of small neoplastic cells (cytologically and immunophenotypically resembling those seen in the "small cell variant") admixed with a large number of pale reactive histiocytes (CD68+) which sometimes show signs of erythrophagocytosis. A minor population of CD30$^+$/ALK$^+$ large tumor cells, usually clustering around blood vessels, is also observed.

More rare morphological variants of ALK$^+$ ALCL include cases with giant cell rich [20, 23], Hodgkin-like [28], sarcomatoid [29, 30] or hypocellular [31] appearance. Immunohistochemistry is crucial for the correct diagnosis of these cases that otherwise can be misdiagnosed as atypical inflammatory lesions or malignant tumors other than ALCL.

15.2.3 Immunohistochemical markers

In addition to ALK (see below), other markers, such as CD30 and epithelial membrane antigen (EMA), can help in the immunohistochemical diagnosis of ALK$^+$ ALCL. Noteworthy, CD30 expression correlates with the size of tumor cells being a reliable assay for recognizing the "common type" but not the "small cell variant" of ALK+ ALCL that is usually negative. Characteristically, large anaplastic cells strongly express CD30 on the cell membrane and in the Golgi area [1, 4, 20] (Fig. 15.2(d) and Fig. 15.3(b)). ALK$^+$ ALCL cells frequently express EMA [32, 33], although in a variable percentage of tumor cells.

Most ALK$^+$ ALCL show a T-cell phenotype as defined by the expression of one or more T-cell antigens (usually CD2, CD5 and CD4). ALCL cases with a so-called "null phenotype" (due to negativity for pan-T antigens) [1, 34] are also thought to be T-cell derived since tumor cells carry clonally rearranged TCR genes [35] and often express the cytotoxic molecules perforin (Fig. 15.2(e)), granzyme B and TIA-1 [34, 35]. ALCL, ALK$^+$ is consistently negative for EBV markers [14, 36].

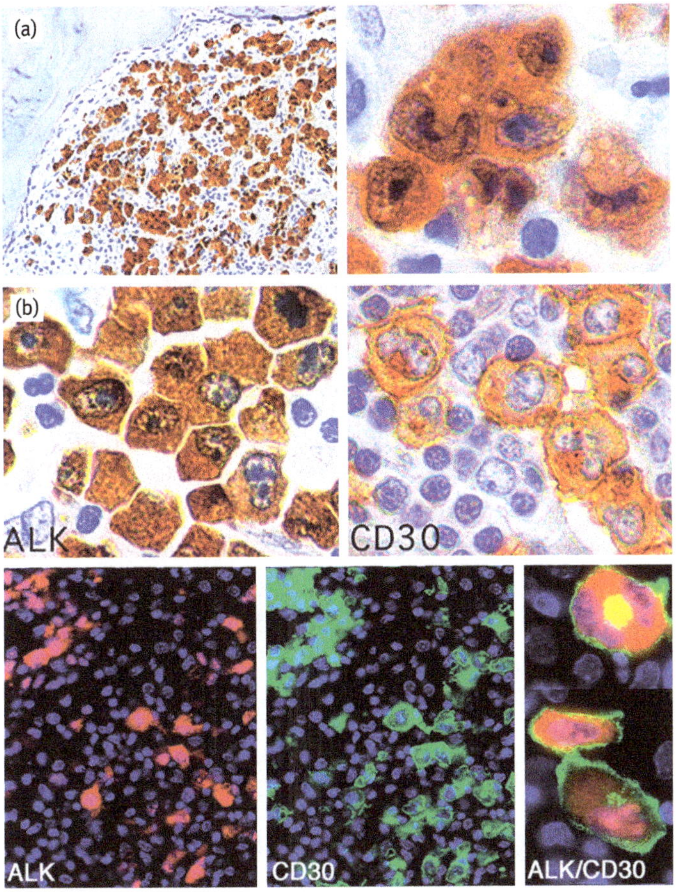

Fig. 15.3: **Immunostaining for ALK and CD30 in NPM-ALK positive ALCL.** (a) Labeling for NPM-ALK in the nucleus and cytoplasm in a lymphoma with anaplastic large cell morphologic characteristics. (b) A t(2;5)-positive large cell lymphoma showing the contrast between NPM-ALK expression (cytoplasmic with some nuclear staining) and CD30 staining (surface membrane and Golgi-associated labeling). The double fluorescent images at bottom show that the same cells express CD30 and ALK but in different patterns. Original magnification (a) left, ×250; (b) right, ×1000; (b) top, ×1000; (b) bottom (ALK and CD30, ×400; ALK/CD30, ×1000). This research was originally published in [21].

15.2.4 Patterns of ALK protein expression in ALK⁺ ALCL

The native ALK molecule (also known as CD246) belongs to the insulin receptor superfamily and plays a key role in neural development. This protein is not expressed in normal human tissues with the exception of rare cells in the nervous system [11]. Therefore, detection of ALK positivity in lymphoma cells [12, 20, 21] implies the presence of an ALK fusion product that has been generated by some kind of rearrangement of the *ALK* gene. Immunohistochemical staining for ALK has now supplanted FISH and molecular assays for the diagnosis of ALK$^+$ ALCL since it is highly specific, sensitive, is applicable to archival paraffin-embedded material, and provides important information on the identity of the labeled cells and their topographical distribution [37–39]. Subcellular expression of ALK protein also correlates with particular molecular variants of ALK$^+$ ALCL (see below).

15.2.4.1 ALCL expressing the NPM1-ALK fusion protein

ALK expression is detected in approximately 60 % of ALCLs. Most cases (about 85 %) harbor the NPM1-ALK fusion protein, i.e. the product of the t(2;5) translocation. The subcellular distribution of ALK protein mainly depends on the size of tumor cells. In particular, large tumor cells express ALK both in the cytoplasm (the expected location site of wild-type ALK) and the nucleus [15, 20, 21] (Fig. 15.3(a), Figs. 15.4 and 15.5). The "ectopic" nuclear expression of ALK is the result of the nucleo-cytoplasmic shuttling properties of nucleophosmin (NPM1). In fact, the ALK fusion protein is transported within the nucleus, through formation of NPM-ALK/wild-type NPM heterodimers [40].

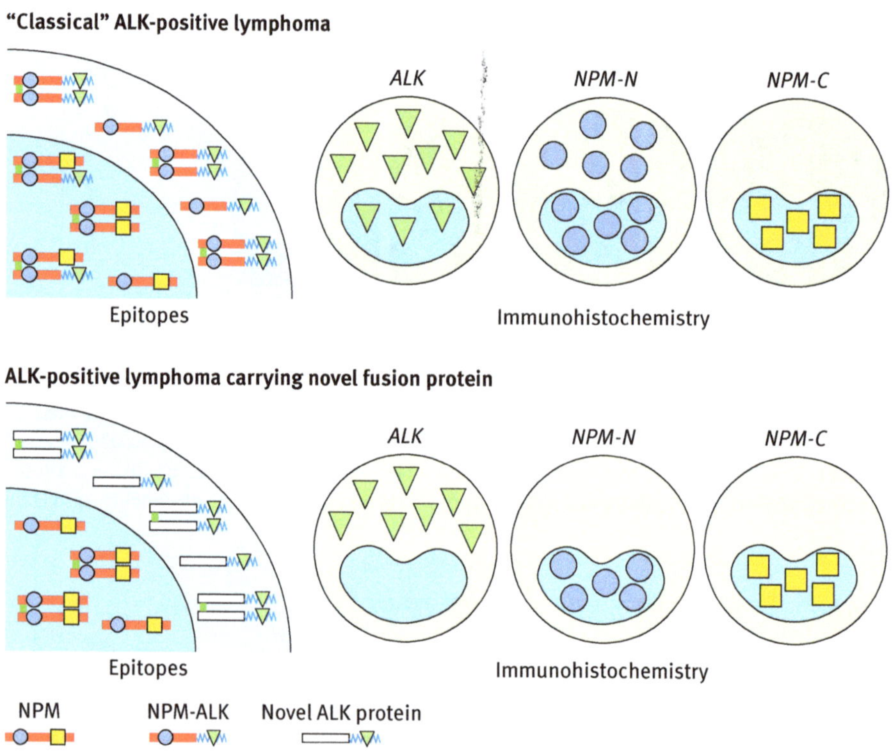

Fig. 15.4: Schematic illustration of the staining patterns for ALK and the N- and C-terminal portions of NPM in lymphomas carrying NPM-ALK or variant ALK fusion protein(s). This research was originally published in [15].

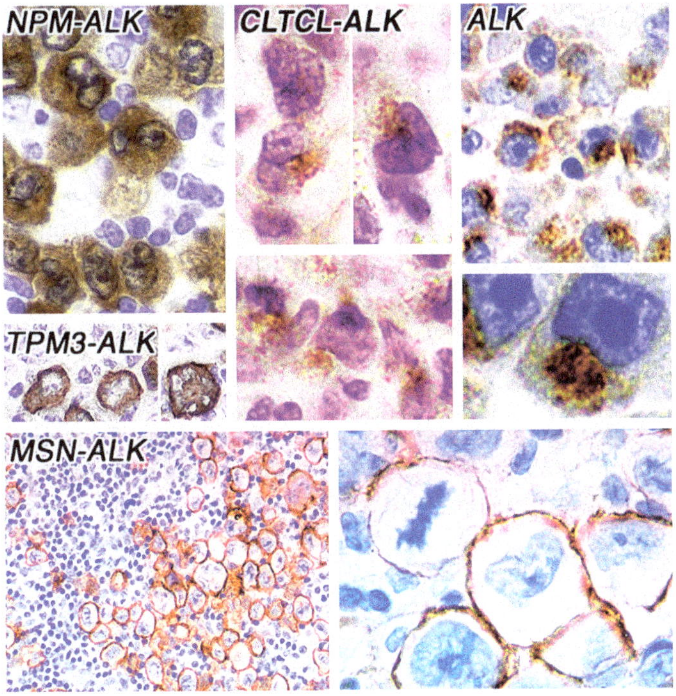

Fig. 15.5: Different patterns of ALK immunostaining in ALCL with t(2;5) and variant translocation. All tumors contain a hybrid ALK protein, except for that marked ALK, an example of the rare subtype of large B-cell lymphoma (CD30-negative and positive for epithelial membrane antigen) in which full-length wild-type ALK protein is expressed. Note the typical granular cytoplasmic positivity for ALK in tumor cells bearing the clathrin (CLTCL)-ALK (top panel) and TPM3-ALK (middle panel) translocation, and the surface expression for ALK in ALCL cells bearing moesin(MSN)-ALK translocation (bottom panels) (immunoperoxidase technique in paraffin sections; ×800 to 1,000). Original magnification: ×400 to ×1000. This research was originally published in [21].

In contrast, the small tumor cells characteristically observed in the "small cell" and "lympho-histiocytic" variants of ALK^+ ALCL, exhibit a nuclear-restricted ALK expression [12].

Immunostaining of ALCL harboring t(2;5) with antibodies specifically directed against the amino-terminal portion of NPM1 [41] (which is retained in the NPM1-ALK fusion protein) results into a unique pattern characterized by the expected expression of NPM1 in the nucleus and its aberrant accumulation in the cytoplasm, which is related to the presence in this cellular compartment of the NPM-ALK fusion protein [15, 21, 25, 41]. Extension of this immunohistochemical observation from ALCL to other human tumors then led to the discovery of aberrant cytoplasmic expression of nucleophosmin and of *NPM1* gene mutations in about one-third of acute myeloid leukemia [42].

15.2.4.2 ALCL expressing ALK fusion proteins other than NPM-ALK

About 15 % of ALK$^+$ ALCLs harbor cytogenetic abnormalities that lead to the formation of ALK fusion proteins other than NPM1-ALK (Tab. 15.1). These cases can be easily recognized by immunohistochemistry because the ALK variant fusion proteins are characterized by cytoplasmic-restricted ALK positivity [11, 12, 15, 19–21]. This staining pattern is clearly different from the cytoplasmic plus nuclear positivity typical of NPM1-ALK (Figs. 15.4 and 15.5). Western blotting of these cases reveals ALK fusion proteins with molecular weights different from that expected for NPM1-ALK [43] (i.e. 80 kDa).

In some cases, immunohistochemistry can even predict the type of the underlying cytogenetic abnormality. In fact, bioptic samples carrying the chimeric *CTCL-ALK* fusion protein generated by the t(2;17) (p23;q23) translocation [14] (Tab. 15.1) show a unique cytoplasmic positivity of granular type [44] (Fig. 15.5). This reflects the capability of the CTCL-ALK fusion protein to form clathrin coat on the surface of cytoplasmic vesicles. Moreover, immunohistochemistry can identify the MSN-ALK fusion protein generated by the very rare t(2;X) (p23;q11-12) translocation [45] that is characterized by surface membrane-restricted ALK positivity (since moesin is physiologically located on the cell surface) (Fig. 15.5).

15.2.5 Differential diagnosis

Immunohistochemistry plays a critical role in the distinguishing ALK$^+$ ALCL from other reactive or malignant conditions.

15.2.5.1 ALK$^+$ ALCL vs. reactive conditions

Because of the exuberant histiocytic proliferation that frequently masks the tumor cell component, the "lymphohistiocytic variant" of ALK$^+$ ALCL can be confused with atypical inflammatory lesions or with hemophagocytic syndrome [24, 27]. Moreover, ALK$^+$ ALCL with hypocellular pattern can be misdiagnosed with an inflammatory process in the lymph node [31]. Immunostaining for CD30 and ALK usually provides a correct diagnosis.

15.2.5.2 ALK$^+$ ALCL vs other lymphoma types

The "small cell variant" of ALK$^+$ ALCL can be misdiagnosed as peripheral T-cell lymphoma not otherwise specified (PTCL-NOS) [14, 24]. The diagnosis of ALK$^+$ ALCL is mainly based on demonstration of ALK nuclear positivity on the small atypical cells, which are usually CD30-negative [12, 20]. The distinction between these pathological conditions is clinically relevant, since ALK$^+$ ALCL has a better prognosis than PTCL-NOS. Expression of ALK protein also assists in the differential diagnosis between ALK$^+$

Tab. 15.1: Chromosomal translocations and ALK fusion proteins in ALK⁺ ALCL.

Chromosomal translocation	ALK partner*	ALK fusion protein (mWt)	Staining for ALK	Staining for NPM1**	Frequency
t(2;5)(p23;q35)	*NPM1*	80 kDa	Nuclear plus diffuse cytoplasmic	Nuclear plus diffuse cytoplasmic	~ 85%
t(1;2)(q25;p23)	*TPM3*	104 kDa	Diffuse cytoplasmic	Predicted nuclear	~ 13%
Inv(2)(p23q35)	*ATIC*	96 kDa	Diffuse cytoplasmic	Predicted nuclear	Rare
t(2;3)(p23;q21)	*TFG Xlong*	113 kDa	Diffuse cytoplasmic	Predicted nuclear	Rare
	TFG long	97 kDa	Diffuse cytoplasmic	Predicted nuclear	Rare
	TFG short	85 kDa	Diffuse cytoplasmic	Predicted nuclear	Rare
t(2;17)(p23;q23)	*CLTC*	250 kDa	Granular cytoplasmic	Predicted nuclear	Rare
t(X;2)(q11-12 ;p23)	*MSN*	125 kDa	Surface cell membrane	Predicted nuclear	Rare
t(2;19)(p23;p13.1)	*TPM4*	95 kDa	Diffuse cytoplasmic	Predicted nuclear	Rare
t(2;22)(p23;q11.2)	*MYH9*	220 kDa	Diffuse cytoplasmic	Predicted nuclear	Rare
t(2;17)(p23;q25)	*ALO17*	?	Diffuse cytoplasmic	Predicted nuclear	Rare

* More explanation about the characteristics of *ALK* fusion partner genes are given in the text.
** Using monoclonal antibodies against the N-terminus portion of nucleophosmin (NPM).

ALCL and Hodgkin's lymphoma (especially cases rich in neoplastic cells and with nodular sclerosis pattern) [20, 24, 46]. In fact, ALK positivity is only detected in ALCL but not in classic Hodgkin's lymphoma [12, 20].

ALK⁺ ALCL must be distinguished also from rare types of B-cell lymphomas and histiocytic proliferations expressing ALK fusion proteins (Fig. 15.6). Unlike ALK⁺ ALCL, ALK⁺ large B-cell lymphoma [47, 48] is characterized by immunoblastic/plasmablastic (rather than anaplastic) morphology, cytoplasmic expression of IgA and negativity or weak positivity for CD30 [48–50]. Chan et al. [51] recently reported 3 cases of ALK⁺ infant histiocytosis. These tumors differ from ALK⁺ ALCL because of the histiocytic appearance of tumor cells, their positivity for the macrophage-restricted form of CD68 [52], and the lack of CD30 expression.

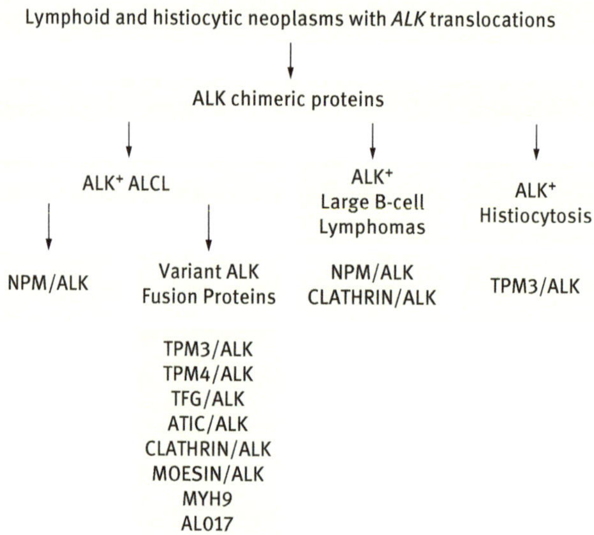

Fig. 15.6: Hemopoietic neoplasms carrying ALK gene alterations.

15.2.5.3 ALK$^+$ ALCL vs nonhematological malignancies

ALK$^+$ ALCL with lymphonodal intrasinusoidal or "cohesive" growth pattern may mimic a metastatic carcinoma. In these instances, the differential diagnosis is mainly based on the detection of ALK protein (only expressed in ALCL) and cytokeratins (only expressed in carcinomas). CD30 is a less reliable marker since it can be expressed both in ALCL and in some carcinomas [4].

Myofibroblastic tumors have been reported to express the ALK protein due to a variety of *ALK* rearrangements [53]. Diagnosis is mainly based on morphological criteria and staining for CD30 that is positive in ALCL but negative in ALK$^+$ inflammatory myofibroblastic tumors.

15.2.6 Genetic abnormalities and molecular pathogenesis of ALK$^+$ ALCL

Chromosomal translocations leading to the generation of ALK fusion proteins in ALCL are listed in Table 15.1. The t (2;5) is the most common event, being detectable in roughly 85 % of ALK$^+$ ALCLs. The fusion partner of ALK in the t(2;5) is the *NPM1* gene that encodes for a ubiquitously expressed nucleolar [41] phosphoprotein shuttling continuously between the nucleus and the cytoplasm [54]. NPM1 is involved in multiple functions [55], including the regulation of ribosome biogenesis, the maintenance of nucleolar integrity through its interaction with the oncosuppressor p19Arf and the control of centrosome duplication. In the NPM1-ALK fusion protein (80 kDa), the entire cytoplasmic region of ALK (corresponding to the last C-terminus 563 amino acids that contain the tyrosine kinase domain [10]) is fused with the amino-terminal portion of NPM1 (corresponding to the amino acids 1–116), that containing the oligomerization domain.

The second most frequent translocation is the t(1;2)(q25;p23) that causes the fusion of the *ALK* and *TPM3* (nonmuscle tropomyosin) genes [56, 57]. This translocation encodes for a 104 kDa *TPM3-ALK* fusion protein.

Other translocations occur at lower frequency. The *TFG* (tropomyosin receptor kinase (*TRK*)-fused gene)-*ALK* fusion gene [58] can generate three chimeric proteins with different molecular weights, i.e. 85 kDa (short TFG-ALK), 97 kDa (long TFG-ALK), and 113 kDa (extra-long TFG-ALK), corresponding to different breakpoints in the *TFG* gene [58, 59]. The TFG-ALK 97 kDa fusion protein is generated by the t(2;3)(p23;q21) translocation [60]. The inv(2)(p23;q35) fuses the *ALK* gene with the *ATIC* (5 aminoimidazole-4-carboxamide-1-beta-D-ribonucleotide transfomylase/inosine mono-phosphate (IMP) cyclohydrolase) gene that encodes for a bifunctional homodimeric enzyme catalyzing the penultimate and ultimate steps of the *de novo* purine nucleotide biosynthesis [61]. The encoded ATIC-ALK fusion protein is composed by the entire cytoplasmic portion of ALK and the amino-terminus of ATIC [62–65]. The t(2;17) (p23;q23) translocation results into the fusion of the *ALK* gene with the *CLTCL* (*clathrin heavy polypeptide-like*) gene, which encodes for the main structural protein of coated vesicles [14] (Tab. 15.1). This translocation can be predicted by immunohistochemistry [44] (see above). Other very rare ALK rearrangements in ALCL include: (i) the fusion of *ALK* with the *TPM4* gene [66], as a result of t(2;19) (p23;p13.1); (ii) the fusion of *ALK* with the *MYH9* (myosin heavy chain gene 9) gene, as a consequence of t(2;22) (p23;q11.2)(14); (iii) the fusion of *ALK* with the *ALO17* (*ALK lymphoma oligomerization partner on chromosome 17*) gene [14]; and (iv) the fusion of *ALK* with the *MSN* (moesin) gene, as a result of t(2;X) (p23;q11-12) [45]. This translocation can be also predicted by immunohistochemistry (see above).

NPM1-ALK and the less common "variant ALK fusion proteins" share two features that suggest they may contribute to lymphomagenesis through a similar mechanism. First, in all ALK rearrangements listed in Tab. 15.1 ALK is placed under the control of the strong promoter of the ALK fusion partner gene. This leads to the constitutive expression of the ALK chimeric RNA and the corresponding fusion protein. Second, all ALK fusion proteins contain at their N-terminus (corresponding to the ALK fusion partner protein moiety) an oligomerization domain capable to mediate homodimer formation [20, 25] through cross-linking with other ALK fusion proteins. Thus, the main role of *NPM1* and other fusion partner genes (*TPM3*, *ATIC*, etc.) in lymphomagenesis is to serve as an oligomerization domain for the chimera [67]. Self-association of the fusion proteins then results into the constitutive activation of the ALK tyrosine kinase catalytic domain. This event induces phosphorylation and aberrant activation of multiple, potentially tumorigenic, downstream substrates and signaling cascades. Activation of the ras-ERK pathway is mainly responsible for the uncontrolled cell-cycle progression and increased growth of ALCL cells that occurs through downregulation of CDKN1A (p21) and concomitant upregulation of cyclins D3 and A [68]. The ERK signaling pathway, in turn, results into the activation of the mechanistic target of rapamycin (mTOR), with consequent phosphorylation of the mTOR targets that promote cell-cycle progression.

The cell survival and phenotypic changes of ALCL cells are mainly attributed to the activation of the JAK3-STAT3 and PI3K-AKT pathways [68]. Activation of the signal transducer and activator of transcription 3 (STAT3) may occur through direct phosphorylation by the ALK fusion protein. Alternatively, the ALK chimera activates JAK3, which in turn can activate STAT3. Activation of STAT3 results into enhanced transcription of several anti-apoptotic molecules and cell cycle regulators such as BCL2, BCL2L1, CEBPB, MCL1 and various cyclins.

NPM1-ALK and other ALK variant fusion proteins can also bind to the regulatory p85 subunit of PI3K [68]. Activated PI3K results into the phosphorylation of its downstream effectors AKT1 and AKT2 that, through their inhibitory activity towards the function of pro-apoptotic proteins such as BAD, lead to enhanced survival of ALCL cells. Moreover, AKT1 and AKT2 can phosphorylate FOXO3, thus blocking FOXO3-mediated transcription of target genes that promote apoptosis, cell-cycle arrest and metabolic processes, forcing G1 phase cell-cycle block.

15.2.7 Prognostic factors and therapy

Several studies have reported a better overall survival of ALK$^+$ as compared to ALK$^-$ALCL [17, 69]. However, the younger age of ALK$^+$ALCL patients may account for this difference in outcome since no difference in terms of failure-free survival (FFS) and overall survival was observed between ALK$^+$ and ALK$^-$ ALCL if patients were stratified according to their median age at diagnosis [70]. The International Prognostic Index (IPI) continues to maintain its prognostic value even in the good prognostic category of ALK$^+$ ALCL [17].

About 60 % of adults and 80 % of children affected by ALK$^+$ ALCL are cured by chemotherapy using CHOP-based regimens [17]. Autologous hemopoietic stem cell transplantation (ASCT) is recommended for patients who relapse after CHOP-based regimens and remain chemosensitive to salvage chemotherapy. Allogeneic hemopoietic stem cell transplantation was reported to be an effective procedure for refractory ALK$^+$ ALCL [71] or patients relapsing after ASCT. New biological therapies based on the use of anti-CD30 monoclonal antibody Brentuximab [72–74] and drugs inhibiting NPM-ALK kinase activity [75–78] or interfering with the downstream signaling pathways [79–81] are expected to have an increasing role in the therapy of ALK$^+$ ALCL resistant to standard chemotherapy regimens.

15.3 Anaplastic large cell lymphoma, ALK-negative (ALK⁻ ALCL)

In the fourth edition (2008) of the WHO Classification [82], ALK⁻ ALCL was listed among peripheral T-cell lymphomas (PTCLs) as a provisional entity, since at that time the pathobiology of the tumor was still obscure and the borders with the not otherwise specified form (PTCL-NOS) not sharply defined. Nowadays, there is consensus to regard it as an accepted entity thanks to the many new pieces of information that in the meanwhile have become available. As such, it will be quoted in the update of the classification that is expected by mid-2016 (personal communication).

15.3.1 Epidemiology

ALK⁻ ALCL represents more than 5 % of all mature T-cell neoplasms worldwide [70]. However, its prevalence varies based on the geographic area. For instance, it is much commoner in South America than in Europe and USA [83].

15.3.2 Clinical features

ALK⁻ ALCL is a systemic disease that occurs in adults (from 40 to 65 years), conversely to ALCL ALK⁺ that has its highest prevalence in the first three decades of life [17]. There is a modest male preponderance (1.5:1) [70]. The tumor extensively involves the lymph nodes and less frequently the bone, soft tissue, and skin [70]. In case of disease presentation at the latter site, accurate staging procedures are mandatory to distinguish between a systemic tumor and a primary cutaneous ALCL, which has a completely different clinical course (see below). ALK⁻ ALCL should also be differentiated from the anaplastic variant of enteropathy-associated T-cell lymphoma [84]. Most patients present with widespread disease (stage III–IV) and B symptoms [70]. Immunohistochemistry with anti-CD30 antibodies should always be performed on staging bone marrow biopsies in order to detect possible occult involvement [85].

15.3.3 Morphology and phenotype

ALK⁻ ALCL is almost undistinguishable from the ALK⁺ form [82]. In particular, it consists of extremely large cells (measuring about 60 µm in diameter), with horseshoe or kidney-shaped nuclei, prominent inclusion-like nucleoli, and a large rim of cytoplasm, grayish-violet at Giemsa and provided with a paranuclear clear halo, corresponding to a well-developed Golgi apparatus (Fig. 15.7). They show a cohesive growth pattern and tend to diffuse through the sinuses. Because of these features, ALK⁻ ALCL was in the past misdiagnosed as undifferentiated carcinoma [86]. At times, the tumor mimics classical Hodgkin lymphoma of the nodular sclerosing type (Hodgkin-like variant) or contains a huge amount of histiocytes that can obscure the neoplastic population (lympho-histiocytic variant) [86]. In contrast to ALK$^+$ ALCL, no small variant is recognized, since the distinction from PTCL-NOS expressing CD30 would be actually impossible [86].

Similarly to ALK$^+$ ALCL, the neoplastic cells strongly express CD30 at the cytoplasmic membrane level as well as in the Golgi area [20] (Fig. 15.7). This staining pattern is the result of the synthesis of the molecule that is assembled in a proteic form of 90 kDa molecular weight in the Golgi apparatus and, following glycosylation – enhancing the weight to 120 kDa – moves to the cytoplasmic membrane, where it acquires a transmembranic location [20]. Part of the external domain of CD30 is shed into the peripheral blood where it can be quantitated [87]. Notably, the epitopes detected by the Ber-H2 and SGN30 monoclonal antibodies, although located on the external domain of CD30 are not released in the serum [86]. This is quite important since the two reagents represent tools employed for diagnostic and therapeutic purposes, respectively. In particular, SGN30 conjugated with mono-methyl auristatin E enters in the composition of Brentuximab-Vedotin, an antibody-drug conjugate approved for the treatment of refractory/relapsed classical Hodgkin lymphoma (cHL) and ALCL [88].

The tumor are negative for CD45 in about 30 % of cases [89] and more often expresses the epithelial membrane antigen (EMA) [86]. The latter can be more easily detected on Bouìn-fixed samples or, alternatively, by heat-based antigen retrieval techniques in high pH buffers, when formalin represents the fixative of choice [90]. Besides morphology, such findings contributed in the past to the misinterpretation of ALK$^-$ ALCL as undifferentiated carcinoma. Nevertheless, the usage of anti-cytokeratins antibodies (which produce no staining in ALK$^-$ ALCL) largely contributes to the correct diagnosis [86]. Among T-cell associated antigens, CD4 and CD43 are generally expressed, the staining for the remaining ones being erratic. The latter finding corresponds to a global downregulation of the T-cell program [86]. In most cases, an activated cytotoxic profile is recorded, with positivities for TIA-1, granzyme B and/or perforin [86]. Search for the ALK protein turns negative, as does the one for BSAP/PAX5 [86]. The latter is of paramount importance for the differential diagnosis with cases of cHL rich in neoplastic cells (as observed in lymphocyte depleted or nodular sclerosing type II), which consistently express the PAX5 protein, although at lower intensity in comparison to normal B lymphocytes [86]. A significant proportion of ALCLs Hodgkin-like of the REAL classification are nowadays diagnosed as cHL based on the results of the anti-PAX5 antibody and of gene rearrangement studies (see below).

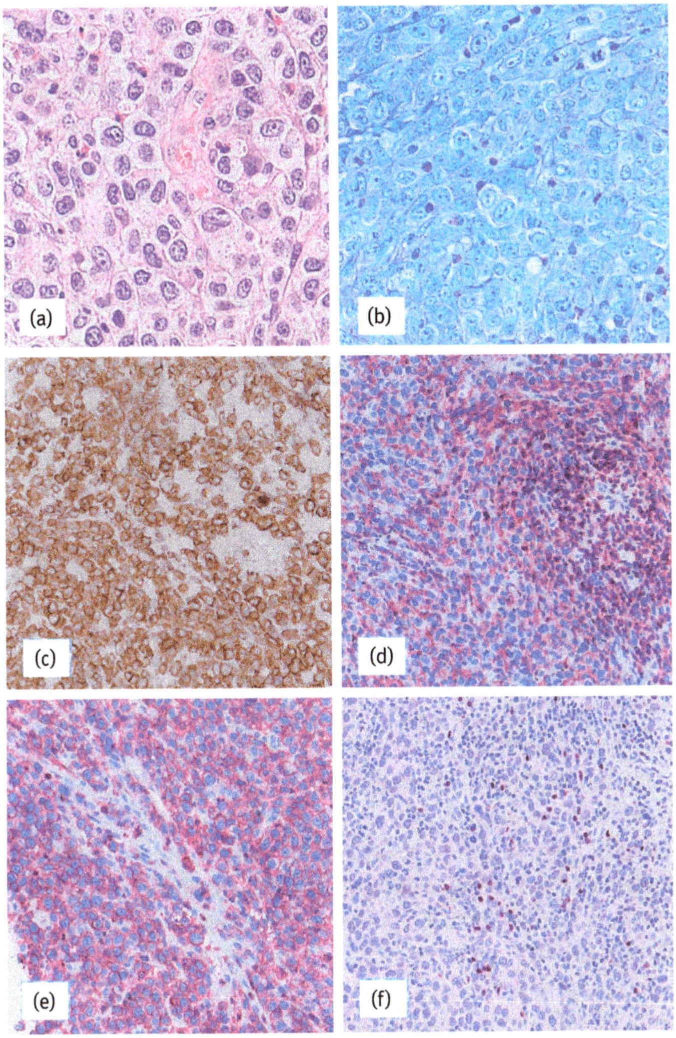

Fig. 15.7: **ALK-negative ALCL.** (a, b) Cytological details at hematoxylin and eosin, and Giemsa, respectively (×600). Please, note the size of the cells that are provided with kidney-shaped nuclei, prominent nucleoli and a large rim of cytoplasm, grayish-violet at Giemsa. (c) Neoplastic cells strongly express CD30 at the cytoplasmic membrane level as well as in a paranuclear dot-like fashion (Ber-H2 monoclonal antibody; immunoperoxidase Envision^{+} technique; hematoxylin nuclear counterstaining; ×300). (d) CD45 is expressed by reactive small lymphocytes but turns negative on lymphomatous elements (antibody clone 2B11 + PD7/26; immunoalkaline phosphatase technique; Gill's hematoxylin counterstaining; ×300). (e) Strong positivity at the determination of the epithelial membrane antigen (EMA) (antibody clone E29; immunoalkaline phosphatase technique; Gill's hematoxylin counterstaining; ×300). (f) Negativity for BSAP/PAX5, which is expressed by a few reactive small B-lymphocytes (antibody DAK-PAX5; immunoalkaline phosphatase technique; Gill's hematoxylin counterstaining; ×300).

15.3.4 Molecular studies

Molecular analysis of ALK$^-$ ALCL reveal clonal rearrangement of the genes encoding for the T-cell receptor (*TCR@*), different from cHL that stems from late or post-germinal center B lymphocytes [82]. A further aid to the distinction between the two diseases may be provided by EBER *in situ* hybridization, which shows EBV infection of the neoplastic cells in the setting of cHL only [82].

Important contributions to a better understand of ALK$^-$ ALCL have been provided by gene expression profiling (GEP) and next generation sequencing (NGS) studies. In particular, GEP carried out on mRNA extracted from both frozen and formalin-fixed, paraffin-embedded (FFPE) tissue samples revealed that ALK$^-$ ALCL has a signature close to the one of the ALK$^+$ form [91–95]. The main differences between the two entities are due to the overexpression of the transcription domain of the *ALK* gene, due to the occurrence of the t(2;5) translocation or its variants, causing the formation of a hybrid gene, encoding for a chimeric protein. Such similarity can explain the common morphology and phenotypic profile of ALK$^+$ and ALK$^-$ ALCLs. In particular, the JAK/STAT and AKT/mTOR pathways as well as the high expression of IRF4 causing deregulation of MYC target genes might represent the common functional link between the two diseases [96–101].

Importantly, ALK$^-$ ALCL has a gene signature distinct from PTCL/NOS, massively expressing CD30 [91–93, 95, 102]. The two conditions, in fact, have only one trait in common: the downregulation of the T-cell program [103]. Such distinction is not only relevant on histogenetic grounds, but has an important prognostic impact, as CD30+ PTCLs-NOS behave significantly worse than ALK$^-$ ALCL [70, 93, 95]. Interestingly, the results obtained by GEP studies have been reproduced by miRNA profiling [95, 98]. Once again, miRNA signatures have been identified, which contribute to clarifying the differences among PTCLs and underline the practical usefulness of the distinction between ALK$^-$ ALCL and CD30+ PTCL-NOS [95, 98]. Of note, *ad hoc* compounds can target some of these metabolic alterations [101]. Irrespective of the ALK status, ALCLs overexpress PDGFRα and PDGFRβ at both the gene and protein level [104, 105]. Importantly, both receptors are in their active phosphorylated form. The mechanism behind this phenomenon is largely unknown, although an autocrine loop has been suggested [105]. This causes sensitivity to TKI, as shown experimentally in an *ad hoc* developed mouse model [104].

NGS studies have significantly contributed to the better categorization of ALK$^-$ ALCL. In particular, they showed that a variable proportion of cases carry *DUSP22* rearrangement, while others are characterized by *TP53* or *TP63* mutations [100, 106, 107].

15.3.5 Prognosis

When treated according to conventional therapies, the prognosis of ALK⁻ ALCL (5-year-OS: 30–49 %) is worse than ALK⁺ ALCL (70–86 %) but better than PTCL-NOS (19–25 %), [93, 95, 108]. Cases carrying *DUSP22* rearrangement differ in their prognosis from patients with *TP53* or *TP63* mutations [100, 106, 107]. In fact, the former show a very good response to therapy, whereas the latter behave very aggressively (5-year-OS 90 % vs. 17 %) [100, 107].

Up-front autologous stem cell transplantation improves the outcome [109], as does the usage of Brentuximab-Vedotin in relapsed or refractory patients [110]. The usage of targeted therapies (e.g. JAK, STAT and TKI inhibitors) might further improve the present results [101, 104, 111].

15.4 Primary cutaneous ALCL

These tumors belong to a group of conditions collectively termed primary cutaneous CD30-positive T-cell lymphoproliferative disorders, which also include lymphomatoid papulosis (LyP) and some borderline lesions [112, 113]. In particular, LyP is a self-healing condition characterized by rhythmic paradoxical papular eruptions, morphologically malignant but clinically benign. In fact, the lesions consist of large atypical cells expressing CD30 and carrying both T-cell-associated antigens and clonal *TCR@* rearrangement, but undergo spontaneous regression by leaving a scar [113]. Based on the amount of admixed reactive elements and infiltration modalities, six subtypes of LyP have been reported in the literature [112–116]. Two of these are of special interest, termed C and F. The former consists of blasts only and is morphologically indistinguishable from primary cutaneous ALCL, the differential diagnosis being based on the different clinical course. The latter is characterized by *DASP22* rearrangement, like the more favorable form of ALCL/ALK⁻ [116].

Conversely, primary cutaneous ALCL presents with solitary or localized nodules, tumors or less frequently papules on the trunk, face, extremities, and buttocks that usually do not undergo spontaneous regression but often tend to ulceration [112, 113] (Fig. 15.8(a)). Multifocal lesions are encountered in about 20 % of patients and regional lymph node involvement in 10 %. Patients are more often males (2–3:1) in the seventh decade of life. No previous history of mycosis fungoides (MF) should be recorded, since the tumoral phase pf MF can be indistinguishable from primary cutaneous ALCL [112, 113]. Morphologically, the latter consists of large cells more often provided with anaplastic profile, although cases with immunoblastic or pleomorphic features can also be seen (Fig. 15.8(b),(c)). Immunohistochemistry reveals CD30 positivity in more than 75 % of neoplastic elements, defective T-cell phenotype, and expression of CD45 and cytotoxic markers [112, 113] (Fig. 15.8(d)). The search for EMA and ALK turns negative. *TCR@* genes are clonally rearranged [112, 113]. *DUSP22* abnormalities have been

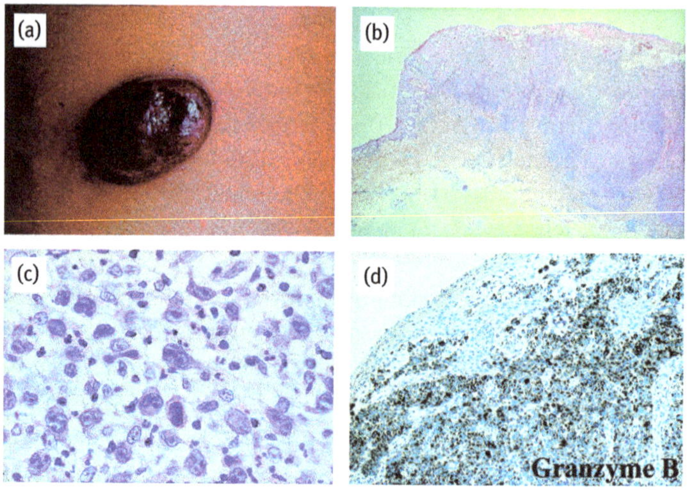

Fig. 15.8: **Primary cutaneous ALCL.** (a) Macroscopic appearance: ulcerated reddish-violet tumor of the skin. (b) Histologic appearance of the same lesion as in (a) at low magnification (hematoxylin and eosin). (c) Cytologic composition of the same lesion as in (a) and (b). Note the presence of many large anaplastic cells; they are strongly CD30$^+$ (not shown) (hematoxylin and eosin). (d) Immunoreactivity of the same lesion as in (a), (b), and (c) for the cytotoxic molecule granzyme B. All of the tumor cells are strongly positive (antibody clone GrB-7, avidin-biotin-peroxidase). This research was originally published in [20].

detected in 20–57 % of cases [117, 118]. The prognosis is usually favorable also in case of multifocal skin lesions or regional lymph node involvement, with a 10-year disease-related survival of approximately 90 % [112, 113]. No difference in clinical presentation, behavior or prognosis is found between cases with anaplastic and pleomorphic or immunoblastic morphology.

15.5 Breast-implant ALCL

This is a recently identified entity, which is not comprised in the WHO Classification and occurs in women who underwent a breast implant because of reconstructive or esthetic surgery [119–122]. So far, more than 170 cases have been reported in the literature [123] (Fig. 15.9).

The clinical presentation varies widely, from peri-implant seromas to a mass attached to the capsule or – more rarely – a fulminant, treatment-resistant widespread disease [123, 124]. There is no preference for saline or silicone fill or for cosmetic or reconstructive indications [123, 124]. Risk estimates range from one in 500,000 to one in 3 million women with implants [123, 124]. The tumor is not time dependent, since it can arise within months or up to 25 years after implantation, and delay from first symptoms to treatment in most patients does not seem to alter the course of the disease

[123, 124]. The natural history of the neoplasm remains obscure. The texturing of the silicone breast-implant surface suggests a causative role exerted by site- and material-specific chronic inflammation. A genetic predisposition and the role of biofilm organisms have also been questioned [123, 124].

Morphologically and phenotypically (Fig. 15.9), the process fulfills the criteria for the diagnosis of ALK$^-$ ALCL [123, 124]. In the few cases studied by molecular techniques, clonal *TCR@* rearrangement was detected [123, 124]. No studies based on high-throughput technologies have been so far carried out.

A careful staging evaluation should direct the appropriate treatment plan [123–125]. Although the fluid and scar capsule usually appear abnormal, they can seem grossly normal, and therefore it is recommended that all fluid and capsule tissues from patients with seromas are submitted for analysis [123–125]. Recently, it has been suggested that, if the tumor is restricted to the capsule, only bilateral implant removal with total capsulectomy should be considered with close follow-up, and without oncologic treatment [123–125]. Multiagent chemotherapy (CHOP-like regimens) with or without radiation has been successful for local extra-capsular involvement [126], but even ASCT was unsuccessful in most patients with aggressive disseminated disease [123, 124].

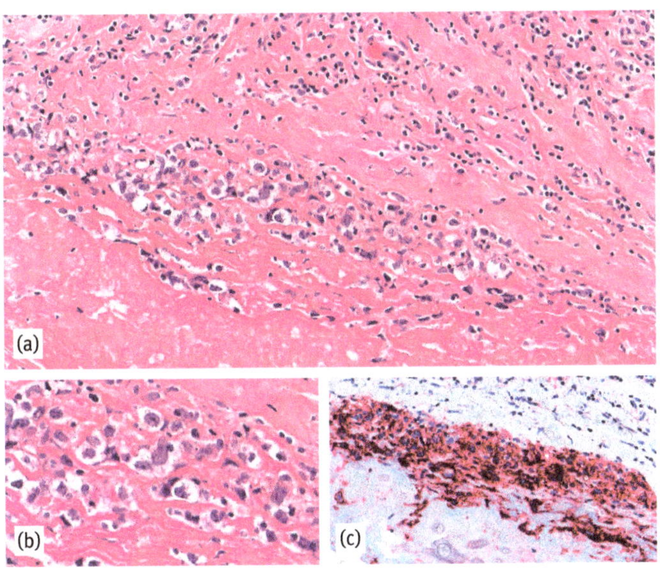

Fig. 15.9: **Breast-implant ALCL.** (a) An example of breast-implant ALCL; the tumor is limited to the inner portion of the capsule; please note the difference between the large anaplastic cells and the reactive lymphoid infiltrate, which is observed in the outer part of the capsule (hematoxylin and eosin, ×200). (b) Cytological details at higher magnification, underlying the bizarre shape of neoplastic cells, which are provided with prominent nucleoli (hematoxylin and eosin, x400). (c) Lymphomatous elements strongly express CD30 (Ber-H2 monoclonal antibody; immunoalkaline phosphatase technique; Gill's hematoxylin nuclear counterstaining; ×200).

References

[1] Stein H, Mason DY, Gerdes J, et al. The expression of the Hodgkin's disease associated antigen Ki-1 in reactive and neoplastic lymphoid tissue: evidence that Reed-Sternberg cells and histiocytic malignancies are derived from activated lymphoid cells. Blood 1985;66:848–858.

[2] Schwab U, Stein H, Gerdes J, et al. Production of a monoclonal antibody specific for Hodgkin and Sternberg- Reed cells of Hodgkin's disease and a subset of normal lymphoid cells. Nature 1982;299:65–67.

[3] Durkop H, Latza U, Hummel M, Eitelbach F, Seed B, Stein H. Molecular cloning and expression of a new member of the nerve growth factor receptor family that is characteristic for Hodgkin's disease. Cell 1992;68:421–427.

[4] Falini B, Pileri S, Pizzolo G, et al. CD30 (Ki-1) molecule: a new cytokine receptor of the tumor necrosis factor receptor superfamily as a tool for diagnosis and immunotherapy. Blood 1995;85:1–14.

[5] Benz-Lemoine E, Brizard A, Huret JL, et al. Malignant histiocytosis: a specific t(2;5)(p23;q35) translocation? Review of the literature. Blood 1988;72:1045–1047.

[6] Kaneko Y, Frizzera G, Edamura S, et al. A novel translocation, t(2;5)(p23;q35), in childhood phagocytic large T- cell lymphoma mimicking malignant histiocytosis. Blood 1989;73:806–813.

[7] Rimokh R, Magaud JP, Berger F, et al. A translocation involving a specific breakpoint (q35) on chromosome 5 is characteristic of anaplastic large cell lymphoma ('Ki-1 lymphoma'). Br J Haematol 1989;71:31–36.

[8] Bitter MA, Franklin WA, Larson RA, et al. Morphology in Ki-1(CD30)-positive non-Hodgkin's lymphoma is correlated with clinical features and the presence of a unique chromosomal abnormality, t(2;5)(p23;q35). Am J Surg Pathol 1990;14:305–316.

[9] Mason DY, Bastard C, Rimokh R, et al. CD30-positive large cell lymphomas ('Ki-1 lymphoma') are associated with a chromosomal translocation involving 5q35. Br J Haematol 1990;74:161–168.

[10] Morris SW, Kirstein MN, Valentine MB, et al. Fusion of a kinase gene, ALK, to a nucleolar protein gene, NPM, in non- Hodgkin's lymphoma. Science 1994;263:1281–1284.

[11] Pulford K, Lamant L, Morris SW, et al. Detection of anaplastic lymphoma kinase (ALK) and nucleolar protein nucleophosmin (NPM)-ALK proteins in normal and neoplastic cells with the monoclonal antibody ALK1. Blood 1997;89:1394–1404.

[12] Falini B, Bigerna B, Fizzotti M, et al. ALK expression defines a distinct group of T/null lymphomas ("ALK lymphomas") with a wide morphological spectrum. Am J Pathol 1998;153:875–886.

[13] Chiarle R, Voena C, Ambrogio C, Piva R, Inghirami G. The anaplastic lymphoma kinase in the pathogenesis of cancer. Nat Rev Cancer 2008;8:11–23.

[14] Delsol G, Falini B, Muller-Hermelink HK, et al. Anaplastic large cell lymphoma (ALCL), ALK-positive. In: Swerdlow SHea, ed. WHO Classification of Tumours of Haematopoietic and Lymphoid Tissues. Lyon: IARC; 2008:312-316.

[15] Falini B, Pulford K, Pucciarini A, et al. Lymphomas expressing ALK fusion protein(s) other than NPM-ALK. Blood 1999;94:3509–3515.

[16] Bohling SD, Jenson SD, Crockett DK, Schumacher JA, Elenitoba-Johnson KS, Lim MS. Analysis of gene expression profile of TPM3-ALK positive anaplastic large cell lymphoma reveals overlapping and unique patterns with that of NPM-ALK positive anaplastic large cell lymphoma. Leuk Res 2008;32:383–393.

[17] Falini B, Pileri S, Zinzani PL, et al. ALK^+ lymphoma: clinico-pathological findings and outcome. Blood 1999;93:2697–2706.

[18] Pileri SA, Pulford K, Mori S, et al. Frequent expression of the NPM-ALK chimeric fusion protein in anaplastic large-cell lymphoma, lympho-histiocytic type. Am J Pathol 1997;150:1207–1211.
[19] Benharroch D, Meguerian-Bedoyan Z, Lamant L, et al. ALK-positive lymphoma: a single disease with a broad spectrum of morphology. Blood 1998;91:2076–2084.
[20] Stein H, Foss HD, Durkop H, et al. CD30(+) anaplastic large cell lymphoma: a review of its histopathologic, genetic, and clinical features. Blood 2000;96:3681–3695.
[21] Falini B, Mason DY. Proteins encoded by genes involved in chromosomal alterations in lymphoma and leukemia: clinical value of their detection by immunocytochemistry. Blood 2002;99:409–426.
[22] Ambrogio C, Voena C, Manazza AD, et al. The anaplastic lymphoma kinase controls cell shape and growth of anaplastic large cell lymphoma through Cdc42 activation. Cancer Res 2008;68:8899–8907.
[23] Harris NL, Jaffe ES, Stein H, et al. A revised European-American classification of lymphoid neoplasms: a proposal from the International Lymphoma Study Group. Blood 1994;84:1361–1392.
[24] Kadin ME. Anaplastic large cell lymphoma and its morphological variants. Cancer Surv 1997;30:77–86.
[25] Falini B. Anaplastic large cell lymphoma: pathological, molecular and clinical features. Br J Haematol 2001;114:741–760.
[26] Kinney MC, Collins RD, Greer JP, Whitlock JA, Sioutos N, Kadin ME. A small-cell-predominant variant of primary Ki-1 (CD30)+ T-cell lymphoma. Am J Surg Pathol 1993;17:859–868.
[27] Pileri S, Falini B, Delsol G, et al. Lymphohistiocytic T-cell lymphoma (anaplastic large cell lymphoma CD30+/Ki-1 + with a high content of reactive histiocytes). Histopathology 1990;16:383–391.
[28] Vassallo J, Lamant L, Brugieres L, et al. ALK-positive anaplastic large cell lymphoma mimicking nodular sclerosis Hodgkin's lymphoma: report of 10 cases. Am J Surg Pathol 2006;30:223–229.
[29] Chan JK, Buchanan R, Fletcher CD. Sarcomatoid variant of anaplastic large-cell Ki-1 lymphoma. Am J Surg Pathol 1990;14:983–988.
[30] Pereira EM, Maeda SA, Reis-Filho JS. Sarcomatoid variant of anaplastic large cell lymphoma mimicking a primary breast cancer: a challenging diagnosis. Arch Pathol Lab Med 2002;126:723–726.
[31] Cheuk W, Hill RW, Bacchi C, Dias MA, Chan JK. Hypocellular anaplastic large cell lymphoma mimicking inflammatory lesions of lymph nodes. Am J Surg Pathol 2000;24:1537–1543.
[32] Delsol G, Al Saati T, Gatter KC, et al. Coexpression of epithelial membrane antigen (EMA), Ki-1, and interleukin-2 receptor by anaplastic large cell lymphomas. Diagnostic value in so-called malignant histiocytosis. Am J Pathol 1988;130:59–70.
[33] ten Berge RL, Snijdewint FG, von Mensdorff-Pouilly S, et al. MUC1 (EMA) is preferentially expressed by ALK positive anaplastic large cell lymphoma, in the normally glycosylated or only partly hypoglycosylated form. J Clin Pathol 2001;54:933–939.
[34] Krenacs L, Wellmann A, Sorbara L, et al. Cytotoxic cell antigen expression in anaplastic large cell lymphomas of T- and null-cell type and Hodgkin's disease: evidence for distinct cellular origin. Blood 1997;89:980–989.
[35] Foss HD, Anagnostopoulos I, Araujo I, et al. Anaplastic large-cell lymphomas of T-cell and null-cell phenotype express cytotoxic molecules. Blood 1996;88:4005–4011.
[36] Tan BT, Seo K, Warnke RA, Arber DA. The frequency of immunoglobulin heavy chain gene and T-cell receptor gamma-chain gene rearrangements and Epstein–Barr virus in ALK^+ and ALK^- anaplastic large cell lymphoma and other peripheral T-cell lymphomas. J Mol Diagn 2008;10:502–512.

[37] Shiota M, Fujimoto J, Takenaga M, et al. Diagnosis of t(2;5)(p23;q35)-associated Ki-1 lymphoma with immunohistochemistry. Blood 1994;84:3648–3652.

[38] Herbst H, Anagnostopoulos J, Heinze B, Durkop H, Hummel M, Stein H. ALK gene products in anaplastic large cell lymphomas and Hodgkin's disease. Blood 1995;86:1694–1700.

[39] Lamant L, Meggetto F, al Saati T, et al. High incidence of the t(2;5)(p23;q35) translocation in anaplastic large cell lymphoma and its lack of detection in Hodgkin's disease. Comparison of cytogenetic analysis, reverse transcriptase-polymerase chain reaction, and P-80 immunostaining. Blood 1996;87:284–2891.

[40] Mason DY, Pulford KA, Bischof D, et al. Nucleolar localization of the nucleophosmin-anaplastic lymphoma kinase is not required for malignant transformation. Cancer Res 1998;58:1057–1062.

[41] Cordell JL, Pulford KA, Bigerna B, et al. Detection of normal and chimeric nucleophosmin in human cells. Blood 1999;93:632–642.

[42] Falini B, Mecucci C, Tiacci E, et al. Cytoplasmic nucleophosmin in acute myelogenous leukemia with a normal karyotype. N Engl J Med 2005;352:254–266.

[43] Pulford K, Falini B, Cordell J, et al. Biochemical detection of novel anaplastic lymphoma kinase proteins in tissue sections of anaplastic large cell lymphoma. Am J Pathol 1999;154:1657–1663.

[44] Touriol C, Greenland C, Lamant L, et al. Further demonstration of the diversity of chromosomal changes involving 2p23 in ALK-positive lymphoma: 2 cases expressing ALK kinase fused to CLTCL (clathrin chain polypeptide-like). Blood 2000;95:3204–3207.

[45] Tort F, Pinyol M, Pulford K, et al. Molecular characterization of a new ALK translocation involving moesin (MSN-ALK) in anaplastic large cell lymphoma. Lab Invest 2001;81:419–426.

[46] Vassallo J, Lamant L, Brugieres L, et al. ALK-positive anaplastic large cell lymphoma mimicking nodular sclerosis Hodgkin's lymphoma: report of 10 cases. Am J Surg Pathol. 2006;30:223–229.

[47] Delsol G, Campo E, Gascoyne R. ALK-positive large B-cell lymphoma. In: Swerdlow SH ed al. (Eds.). WHO Classificationof Tumours of Haematopoietic and Lymphoid Tissues. Fourth ed. Lyon: IARC; 2008:254–255.

[48] Momose S, Tamaru J, Kishi H, et al. Hyperactivated STAT3 in ALK-positive diffuse large B-cell lymphoma with clathrin-ALK fusion. Hum Pathol 2009;40:75–82.

[49] Delsol G, Lamant L, Mariame B, et al. A new subtype of large B-cell lymphoma expressing the ALK kinase and lacking the 2; 5 translocation. Blood 1997;89:1483–1490.

[50] Gascoyne RD, Lamant L, Martin-Subero JI, et al. ALK-positive diffuse large B-cell lymphoma is associated with Clathrin-ALK rearrangements: report of 6 cases. Blood 2003;102:2568–2573.

[51] Chan JK, Lamant L, Algar E, et al. ALK^+ histiocytosis: a novel type of systemic histiocytic proliferative disorder of early infancy. Blood 2008;112:2965–2968.

[52] Falini B, Flenghi L, Pileri S, et al. PG-M1: a new monoclonal antibody directed against a fixative-resistant epitope on the macrophage-restricted form of the CD68 molecule. Am J Pathol 1993;142:1359–1372.

[53] Griffin CA, Hawkins AL, Dvorak C, Henkle C, Ellingham T, Perlman EJ. Recurrent involvement of 2p23 in inflammatory myofibroblastic tumors. Cancer Res 1999;59:2776–2780.

[54] Borer RA, Lehner CF, Eppenberger HM, Nigg EA. Major nucleolar proteins shuttle between nucleus and cytoplasm. Cell 1989;56:379–390.

[55] Grisendi S, Mecucci C, Falini B, Pandolfi PP. Nucleophosmin and cancer. Nat Rev Cancer 2006;6:493–505.

[56] Lamant L, Dastugue N, Pulford K, Delsol G, Mariame B. A new fusion gene TPM3-ALK in anaplastic large cell lymphoma created by a (1;2)(q25;p23) translocation. Blood 1999;93:3088–3095.

[57] Siebert R, Gesk S, Harder L, et al. Complex variant translocation t(1;2) with TPM3-ALK fusion due to cryptic ALK gene rearrangement in anaplastic large-cell lymphoma. Blood 1999;94:3614–3617.

[58] Hernandez L, Pinyol M, Hernandez S, et al. TRK-fused gene (TFG) is a new partner of ALK in anaplastic large cell lymphoma producing two structurally different TFG-ALK translocations. Blood 1999;94:3265–3268.

[59] Hernandez L, Bea S, Bellosillo B, et al. Diversity of genomic breakpoints in TFG-ALK translocations in anaplastic large cell lymphomas: identification of a new TFG-ALK(XL) chimeric gene with transforming activity. Am J Pathol 2002;160:1487–1494.

[60] Rosenwald A, Ott G, Pulford K, et al. t(1;2)(q21;p23) and t(2;3)(p23;q21): two novel variant translocations of the t(2;5)(p23;q35) in anaplastic large cell lymphoma. Blood 1999;94:362–364.

[61] Beardsley GP, Rayl EA, Gunn K, et al. Structure and functional relationships in human pur H. Adv Exp Med Biol 1998;431:221–226.

[62] Wlodarska I, De Wolf-Peeters C, Falini B, et al. The cryptic inv(2)(p23q35) defines a new molecular genetic subtype of ALK-positive anaplastic large-cell lymphoma. Blood 1998;92:2688–2695.

[63] Colleoni GW, Bridge JA, Garicochea B, Liu J, Filippa DA, Ladanyi M. ATIC-ALK: A novel variant ALK gene fusion in anaplastic large cell lymphoma resulting from the recurrent cryptic chromosomal inversion, inv(2)(p23q35). Am J Pathol 2000;156:781–789.

[64] Ma Z, Cools J, Marynen P, et al. Inv(2)(p23q35) in anaplastic large-cell lymphoma induces constitutive anaplastic lymphoma kinase (ALK) tyrosine kinase activation by fusion to ATIC, an enzyme involved in purine nucleotide biosynthesis. Blood 2000;95:2144–2149.

[65] Trinei M, Lanfrancone L, Campo E, et al. A new variant anaplastic lymphoma kinase (ALK)-fusion protein (ATIC-ALK) in a case of ALK-positive anaplastic large cell lymphoma. Cancer Res 2000;60:793–798.

[66] Meech S, Grump J, McGavran L, Meltesen L, Odom LF, Wei Q, Hunger SP. Tropomyosin 4 (TPM4) is fused to ALK by a t(2:19)(p23;p13) in an unusual childhood ALCL with the immunophenotype and functional properties of a natural killer cell malignancy. Blood 2000;92.

[67] Bischof D, Pulford K, Mason DY, Morris SW. Role of the nucleophosmin (NPM) portion of the non-Hodgkin's lymphoma- associated NPM-anaplastic lymphoma kinase fusion protein in oncogenesis. Mol Cell Biol 1997;17:2312–2325.

[68] Boi M, Zucca E, Inghirami G, Bertoni F. Advances in understanding the pathogenesis of systemic anaplastic large cell lymphomas. Br J Haematol 2015;168:771–783.

[69] Gascoyne RD, Aoun P, Wu D, et al. Prognostic significance of anaplastic lymphoma kinase (ALK) protein expression in adults with anaplastic large cell lymphoma. Blood 1999;93:3913–3921.

[70] Savage KJ, Harris NL, Vose JM, et al. ALK – anaplastic large-cell lymphoma is clinically and immunophenotypically different from both ALK^{+} ALCL and peripheral T-cell lymphoma, not otherwise specified: report from the International Peripheral T-Cell Lymphoma Project. Blood 2008;111:5496–5504.

[71] Liso A, Tiacci E, Binazzi R, et al. Haploidentical peripheral-blood stem-cell transplantation for ALK-positive anaplastic large-cell lymphoma. Lancet Oncol 2004;5:127–128.

[72] Younes A, Bartlett NL, Leonard JP, et al. Brentuximab vedotin (SGN-35) for relapsed CD30-positive lymphomas. N Engl J Med 2010;363:1812–1821.

[73] Eyre TA, Khan D, Hall GW, Collins GP. Anaplastic lymphoma kinase-positive anaplastic large cell lymphoma: current and future perspectives in adult and paediatric disease. European Journal of Haematology 2014;93:455–468.

[74] Heidegger S, Beer AJ, Geissinger E, et al. Combination therapy with brentuximab vedotin and cisplatin/cytarabine in a patient with primarily refractory anaplastic lymphoma kinase positive anaplastic large cell lymphoma. OncoTargets and Therapy 2014;7:1123–1127.
[75] Galkin AV, Melnick JS, Kim S, et al. Identification of NVP-TAE684, a potent, selective, and efficacious inhibitor of NPM-ALK. Proc Natl Acad Sci USA 2007;104:270–275.
[76] Li R, Morris SW. Development of anaplastic lymphoma kinase (ALK) small-molecule inhibitors for cancer therapy. Med Res Rev 2008;28:372–412.
[77] McDermott U, Iafrate AJ, Gray NS, et al. Genomic alterations of anaplastic lymphoma kinase may sensitize tumors to anaplastic lymphoma kinase inhibitors. Cancer Res 2008;68:3389–3395.
[78] Gambacorti-Passerini C, Messa C, Pogliani EM. Crizotinib in anaplastic large-cell lymphoma. N Engl J Med. 2011; 364:775–776.
[79] Bonvini P, Gastaldi T, Falini B, Rosolen A. Nucleophosmin-anaplastic lymphoma kinase (NPM-ALK), a novel Hsp90- client tyrosine kinase: down-regulation of NPM-ALK expression and tyrosine phosphorylation in ALK(+) CD30(+) lymphoma cells by the Hsp90 antagonist 17-allylamino,17-demethoxygeldanamycin. Cancer Res 2002;62:1559–1566.
[80] Gu L, Gao J, Li Q, et al. Rapamycin reverses NPM-ALK-induced glucocorticoid resistance in lymphoid tumor cells by inhibiting mTOR signaling pathway, enhancing G1 cell cycle arrest and apoptosis. Leukemia 2008;22:2091–2096.
[81] Cho-Vega JH, Vega F, Medeiros LJ. An attractive therapeutic target, mTOR pathway, in ALK$^+$ anaplastic large cell lymphoma. Adv Anat Pathol 2008;15:105–112.
[82] Mason DY, Harris NL, Delsol G, et al. Anaplastic large cell lymphoma, ALK-negative. In: Swerdlow SH et al. (Eds.). WHO Classification of Tumours of Haematopoietic and Lymphoid Tissues. Lyon: IARC; 2008:317-319.
[83] Bellei M, Chiattone CS, Luminari S, et al. T-cell lymphomas in South America and Europe. Revista brasileira de hematologia e hemoterapia 2012;34:42–47.
[84] Wang L, Liu Y, Lin XY, et al. A case of enteropathy-associated T-cell lymphoma (type I) arising in stomach without refractory celiac disease. Diagnostic Pathology 2012;7:172.
[85] Park SH, Chi HS, Cho YU, Jang S, Park CJ. Immunohistopathological features of anaplastic large-cell lymphoma according to anaplastic lymphoma kinase expression and bone marrow involvement pattern. Histopathology 2013;63:13–18.
[86] Ferreri AJ, Govi S, Pileri SA, Savage KJ. Anaplastic large cell lymphoma, ALK-negative. Critical Reviews in Oncology/Hematology 2013;85:206–215.
[87] Zinzani PL, Pileri S, Bendandi M, et al. Clinical implications of serum levels of soluble CD30 in 70 adult anaplastic large-cell lymphoma patients. Journal of Clinical Oncology: Official journal of the American Society of Clinical Oncology 1998;16:1532–1537.
[88] Kumar A, Younes A. Role of CD30 targeting in malignant lymphoma. Current treatment options in oncology 2014;15:210–225.
[89] Falini B, Pileri S, Stein H, et al. Variable expression of leucocyte-common (CD45) antigen in CD30 (Ki1)-positive anaplastic large-cell lymphoma: implications for the differential diagnosis between lymphoid and nonlymphoid malignancies. Hum Pathol 1990;21:624–629.
[90] Pileri SA, Roncador G, Ceccarelli C, et al. Antigen retrieval techniques in immunohistochemistry: comparison of different methods. J Pathol 1997;183:116–123.
[91] Piva R, Agnelli L, Pellegrino E, et al. Gene expression profiling uncovers molecular classifiers for the recognition of anaplastic large-cell lymphoma within peripheral T-cell neoplasms. J Clin Oncol 2010;28:1583–1590.
[92] Agnelli L, Mereu E, Pellegrino E, et al. Identification of a 3-gene model as a powerful diagnostic tool for the recognition of ALK-negative anaplastic large-cell lymphoma. Blood 2012;120:1274–1281.

[93] Piccaluga PP, Fuligni F, De Leo A, et al. Molecular profiling improves classification and prognostication of nodal peripheral T-cell lymphomas: results of a phase III diagnostic accuracy study. Journal of Clinical Oncology: Official Journal of the American Society of Clinical Oncology 2013;31:3019–3025.
[94] Kitahara S, Nakamura T. [Therapy and nursing of asthmatic patients]. Kango tenbo The Japanese Journal of Nursing Science 1977;2:55–60.
[95] Laginestra MA, Piccaluga PP, Fuligni F, et al. Pathogenetic and diagnostic significance of microRNA deregulation in peripheral T-cell lymphoma not otherwise specified. Blood Cancer Journal 2014;4:259.
[96] Feldman AL, Law M, Remstein ED, et al. Recurrent translocations involving the IRF4 oncogene locus in peripheral T-cell lymphomas. Leukemia 2009;23:574–580.
[97] Meier C, Hoeller S, Bourgau C, et al. Recurrent numerical aberrations of JAK2 and deregulation of the JAK2-STAT cascade in lymphomas. Mod Pathol 2009;22:476–487.
[98] Liu C, Iqbal J, Teruya-Feldstein J, et al. MicroRNA expression profiling identifies molecular signatures associated with anaplastic large cell lymphoma. Blood 2013;122:2083–2092.
[99] Spaccarotella E, Pellegrino E, Ferracin M, et al. STAT3-mediated activation of microRNA cluster 17~ 92 promotes proliferation and survival of ALK-positive anaplastic large cell lymphoma. Haematologica 2014;99:116–124.
[100] Boi M, Rinaldi A, Kwee I, et al. PRDM1/BLIMP1 is commonly inactivated in anaplastic large T-cell lymphoma. Blood 2013;122:2683–2693.
[101] Crescenzo R, Abate F, Lasorsa E, et al. Convergent mutations and kinase fusions lead to oncogenic STAT3 activation in anaplastic large cell lymphoma. Cancer Cell 2015;27:516–532.
[102] Iqbal J, Wright G, Wang C, et al. Gene expression signatures delineate biological and prognostic subgroups in peripheral T-cell lymphoma. Blood 2014;123:2915–2923.
[103] Agostinelli C, Rizvi H, Paterson J, et al. Intracellular TCR-signaling pathway: novel markers for lymphoma diagnosis and potential therapeutic targets. Am J Surg Pathol 2014;38:1349–1359.
[104] Laimer D, Dolznig H, Kollmann K, et al. PDGFR blockade is a rational and effective therapy for NPM-ALK-driven lymphomas. Nat Med 2012;18:1699–1704.
[105] Piccaluga PP, Rossi M, Agostinelli C, et al. Platelet-derived growth factor alpha mediates the proliferation of peripheral T-cell lymphoma cells via an autocrine regulatory pathway. Leukemia 2014;28:1687–1697.
[106] Feldman AL, Dogan A, Smith DI, et al. Discovery of recurrent t(6;7)(p25.3;q32.3) translocations in ALK-negative anaplastic large cell lymphomas by massively parallel genomic sequencing. Blood 2011;117:915–919.
[107] Parrilla Castellar ER, Jaffe ES, Said JW, et al. ALK-negative anaplastic large cell lymphoma is a genetically heterogeneous disease with widely disparate clinical outcomes. Blood 2014;124:1473–1480.
[108] Vose J, Armitage J, Weisenburger D. International peripheral T-cell and natural killer/T-cell lymphoma study: pathology findings and clinical outcomes. Journal of clinical oncology : official journal of the American Society of Clinical Oncology 2008;26:4124–4130.
[109] d'Amore F, Relander T, Lauritzsen GF, et al. Up-front autologous stem-cell transplantation in peripheral T-cell lymphoma: NLG-T-01. Journal of clinical oncology : official journal of the American Society of Clinical Oncology 2012;30:3093–3099.
[110] Gisselbrecht C, Borchmann P, D'Amore F, Illidge TM, Zinzani PL. Challenging CD30-positive lymphomas–current challenges, new insights and future directions: joining a conversation on CD30+ lymphomas. Leuk Res 2014;38:1003.
[111] Coiffier B, Federico M, Caballero D, et al. Therapeutic options in relapsed or refractory peripheral T-cell lymphoma. Cancer Treatment Reviews 2014;40:1080–1088.

[112] Ralfkier E, Willemze R, Paulli M, Kadin ME. Primary cutaneous CD30-positive T-cell lymphoproliferative disorders. In: Swerdlow SHea, ed. WHO Classification of Tumours of Haematopoietic and Lymphoid Tissues. Lyon: IARC; 2008:300–301.
[113] Kempf W, Kazakov DV, Kerl K. Cutaneous lymphomas: an update. Part 1: T-cell and natural killer/t-cell lymphomas and related conditions. The American Journal of Dermatopathology 2014;36:105–123.
[114] Saggini A, Gulia A, Argenyi Z, et al. A variant of lymphomatoid papulosis simulating primary cutaneous aggressive epidermotropic CD8+ cytotoxic T-cell lymphoma. Description of 9 cases. Am J Surg Pathol 2010;34:1168–1175.
[115] Kempf W, Kazakov DV, Scharer L, et al. Angioinvasive lymphomatoid papulosis: a new variant simulating aggressive lymphomas. Am J Surg Pathol 2013;37:1–13.
[116] Karai LJ, Kadin ME, Hsi ED, et al. Chromosomal rearrangements of 6p25.3 define a new subtype of lymphomatoid papulosis. Am J Surg Pathol 2013;37:1173–1181.
[117] Samols MA, Su A, Ra S, et al. Intralymphatic cutaneous anaplastic large cell lymphoma/lymphomatoid papulosis: expanding the spectrum of CD30-positive lymphoproliferative disorders. Am J Surg Pathol 2014;38:1203–1211.
[118] Xing X, Feldman AL. Anaplastic large cell lymphomas: ALK positive, ALK negative, and primary cutaneous. Adv Anat Pathol 2015;22:29–49.
[119] Roden AC, Macon WR, Keeney GL, Myers JL, Feldman AL, Dogan A. Seroma-associated primary anaplastic large-cell lymphoma adjacent to breast implants: an indolent T-cell lymphoproliferative disorder. Modern pathology : an official journal of the United States and Canadian Academy of Pathology, Inc 2008;21:455–463.
[120] de Jong D, Vasmel WL, de Boer JP, et al. Anaplastic large-cell lymphoma in women with breast implants. JAMA 2008;300:2030–2035.
[121] Bishara MR, Ross C, Sur M. Primary anaplastic large cell lymphoma of the breast arising in reconstruction mammoplasty capsule of saline filled breast implant after radical mastectomy for breast cancer: an unusual case presentation. Diagnostic Pathology 2009;4:11.
[122] Kim B, Roth C, Chung KC, et al. Anaplastic large cell lymphoma and breast implants: a systematic review. Plastic and Reconstructive Surgery 2011;127:2141–2150.
[123] Brody GS, Deapen D, Taylor CR, et al. Anaplastic large cell lymphoma occurring in women with breast implants: analysis of 173 cases. Plastic and Reconstructive Surgery 2015;135:695–705.
[124] Peters W. Update on anaplastic large cell lymphoma in women with breast implants. Plast Surg (Oakv) 2014;22:267–269.
[125] Kim B, Predmore ZS, Mattke S, van Busum K, Gidengil CA. Breast implant-associated anaplastic large cell lymphoma: updated results from a structured expert consultation process. Plastic and Reconstructive Surgery Global Open 2015;3:e296.
[126] Estes CF, Zhang D, Reyes R, Korentager R, McGinness M, Lominska C. Locally advanced breast implant-associated anaplastic large-cell lymphoma: a case report of successful treatment with radiation and chemotherapy. Frontiers in Oncology 2015;5:26.

José Rodríguez Cortés, Adolfo Ferrando, and Teresa Palomero

16 Molecular pathogenesis of malignant lymphomas: PTCL, NOS and AITL

16.1 Introduction

Peripheral T-cell lymphomas (PTCLs) are a heterogeneous group of non-Hodgkin lymphomas frequently associated with poor prognosis. Overall, PTCLs account for 10–15 % of non-Hodgkin lymphomas (NHL) in North America. This diagnosis covers a heterogeneous group of clinical and pathologic entities characterized by the presence of malignant post-thymic T and NK cells. Compared with B-cell lymphomas, PTCLs are less common, more difficult to diagnose and classify, more aggressive, and have inferior outcomes with current treatments. They are also less understood in terms of their normal T-cell counterparts, and lack specific cell signaling pathways for the development of targeted therapies.

There are currently more than 20 subtypes of PTCL and NK T-cell lymphomas in the WHO classification, which can be grouped according to their clinical, histological and immunophenotypic features at presentation [1, 2]. However, the varied morphology of these lesions and the lack of definitive diagnostic markers for most of the subtypes make the diagnosis and classification of PTCLs challenging. Indeed, about 30 % to 50 % of PTCL cases lack clear morphological features and are not classifiable with current immunophenotypic and molecular markers. These cases are categorized as PTCL not otherwise specified (PTCL, NOS) making it the most common group of PTCLs.

Among specific diagnostic subtypes, angioimmunoblastic T-cell lymphomas (AITL) are the most frequent diagnosis, accounting for 18.5 % of PTCLs. The diagnosis of AITL is based on characteristic histopathological features, immunophenotype and T-cell receptor (*TCR*) clonality. Morphologically, AITLs show partial effacement of lymph node architecture, marked expansion of the paracortex with proliferation of arborizing high endothelial venules, small clusters of neoplastic cells around follicles, high endothelial venules and a prominent paracortical infiltrate of small to intermediate sized lymphocytes, plasma cells, eosinophils, histiocytes and proliferating follicular dendritic cells. Despite this cellular heterogeneity, tumor cells are characteristically CD4 positive (CD4+) and show monoclonal *TCR* rearrangements [3, 4]. Immunophenotypically, neoplastic T cells express most pan-T-cell antigens (CD2, CD3 and CD5), CD4 and T follicular helper cell (Tfh) markers including CD10, CXCR5, CXCL13 and programmed death-1 (PD1) [5–7]. In addition, there is a characteristic expansion of follicular dendritic cells with expression of CD21, CD23 and CD35 around the high endothelial venules. Gene profiling studies as well as immunohistochemistry approaches have shown marked expression of Tfh genes in AITL, including the chemokine *CXCL13*, *ICOS* (a *CD28* homologue with costimulatory function in T-cell activation and expansion), *CD154*, *CD40L* and *NFATC1*, which supports the concept that germinal center T helper cells may represent the cell of origin for this disease [8, 9]. Another cytoplasmic protein, SAP (SLAM-associated protein), has been recently validated as a novel marker of both normal germinal center T cells and AITL [10].

Clinically, PTCL, NOS and AITL are usually diagnosed at advanced stage (stages II-IV) and typically affect elderly adults, with a median age of 55–66 years and a slightly higher prevalence observed in males [11, 12]. Both diseases are clinically heterogeneous, the most prominent manifestation being a generalized lymphadenopathy associated with B symptoms [13–15]. In addition, most common symptoms include hepatomegaly, splenomegaly, pruritus, skin rash, pleural effusion, edema and ascites [14–16] and bone marrow involvement [17]. A high percentage of patients present autoimmune manifestations, including lupus, arthritis or autoimmune thyroiditis [13, 18], suggesting abnormalities in the regulation of the immune system. Both PTCL, NOS and AITL are aggressive tumors associated with a similar dismal prognosis of 30 % survival at 5 years [19].

16.2 Follicular helper T cells in the pathogenesis of AITL

A major advance in the understanding of the pathophysiology of AITL and several subtypes of PTCL, NOS was the discovery of the cellular derivation of these lymphomas from T follicular helper (Tfh) cells, a subset of T helper cells residing in the follicular centers involved in promoting B-cell differentiation and survival and in the development of high-affinity long-lived plasma cells and memory B cells.

Tfh cells were initially described as a separate CD4+ T-cell population characterized by high levels of expression of the chemokine receptor CXCR5 [20–22]. Since then, Tfh have emerged as a distinct subset of effector T helper cells with a characteristic gene expression signature and functionally separated from other known CD4 T-cell subsets (T_H1, T_H2, and T_H17) [23–25]. Besides CXCR5, Tfh cells are defined by high levels of expression of inducible T-cell co-stimulator (ICOS), programmed cell death 1 (PD1), the transcriptional repressor B-cell lymphoma 6 (BCL6) and the CXCL13 chemokine. In addition, Tfh cells express high levels of cytokines that influence B-cell differentiation and antibody production such as interleukin 21 (IL21) and interleukin 4 (IL4) [26–28]. The relationship between Tfh cells and AITL tumor cells was initially proposed based on the expression of CXCL13 in AITL samples [6, 9], and was further substantiated by gene expression studies which established a close similarity between the expression signatures of Tfh cells and AITL tumor biopsies [8, 29]. Based on these findings, current models propose Tfh cells as the normal counterpart and the cell of origin of AITL. This hypothesis is consistent with additional features of AITL lesions such as their association with germinal centers in early stages of the disease and the accompanying expansion of B cells and proliferating follicular dendritic cells [16, 30].

In this context, Tfh development represents the conceptual framework to understand the origins and mechanisms of AITL pathogenesis.

Tfh cells differentiation is initiated by the activation of a CD4 naïve T lymphocyte by dendritic cells [31]. This interaction involves the activation of the ICOS in the T cell [32, 33], which activates phosphatidylinositol-3-OH kinase (PI(3)K) signaling inducing the expression of BCL6, a critical transcription factor required for Tfh development [34–36]. Upon activation, the CD4 T cells upregulate the expression of PD1 and CXCR5 in addition to BCL6, becoming Tfh precursors, which will subsequently migrate to the border of the B-cell follicle to engage in a second cell-cell interaction involving Tfh precursors and antigen-specific B cells that is critical for the generation of germinal center Tfh cells [26, 37]. As antigen stimulation builds up a germinal center reaction, Tfh precursor cells complete their maturation and acquire their definitive phenotype characterized by expression of high levels of CXCR5, PD1, BCL6, MAF and SAP [26].

The critical role of BCL6 in Tfh development is substantiated by the abrogation of Tfh cell differentiation in absence of BCL6 [37, 38], and the activation of a Tfh program in T cells upon BCL6 overexpression [37, 39]. Although the precise mechanisms operating downstream of BCL6 are not fully clarified yet, this transcriptional repressor seems to participate in the restriction of alternative cell fates during Tfh cell development via transcriptional downregulation of critical factors implicated in T_H1 (T-bet), T_H2 (GATA3) and T_H17 (RORγt) cell development [36, 37, 40, 41]. Other transcription factors also required for Tfh cell differentiation in addition to BCL6 include ASCL2, c-MAF, IRF4, and AP-1 [41–44].

In addition to ICOS engagement, Tfh cell differentiation depends on multiple cytokine signals. In mice, activation of CD4+ T cells in presence of Il6 can induce some Tfh characteristics *in vitro* [45], probably as the result of Stat3-mediated activation of the *Bcl6* gene [46, 47]. In addition, *Il21* KO mice have impaired Tfh differentiation [45] and Il21-induced Stat3 activation has been reported to participate in Tfh differentiation *in vivo* [48]. Notably, Il21 is induced by Il6/Stat3, suggesting a role of this cytokine axis in the reinforcement in Tfh differentiation [49]. In humans, in addition to IL6 and IL21, the induction of STAT4 activation by IL12 seems to play a prominent role in Tfh cell development. Thus, IL12 induces high levels of expression of *IL21*, *ICOS*, *CXCR5* and *BCL6* [50–52]. Moreover, the loss of expression of a functional IL12 receptor β1 chain (IL12Rβ1) results in altered germinal center formation in lymph nodes and reduced levels of memory Tfh cells and memory B cells in blood [51]. Notably, and in contrast to the positive role of IL6/IL21-STAT3 and IL12-STAT4 signaling in Tfh development, the IL2-STAT5 signaling axis plays a negative role in Tfh cell development. Thus, STAT5 blockade increases Tfh cell differentiation, whereas overexpression or constitutive active STAT5 reduces Tfh *in vivo* [53–55] (Fig. 16.1).

The prominent role of cytokine signaling in Tfh development could play an active role in the pathogenesis and clinical manifestations of AITL. Thus, the cytokine microenvironment is thought to be important in the early stages of lymphoma progression and in establishing the abundant inflammatory component of AITL tumor

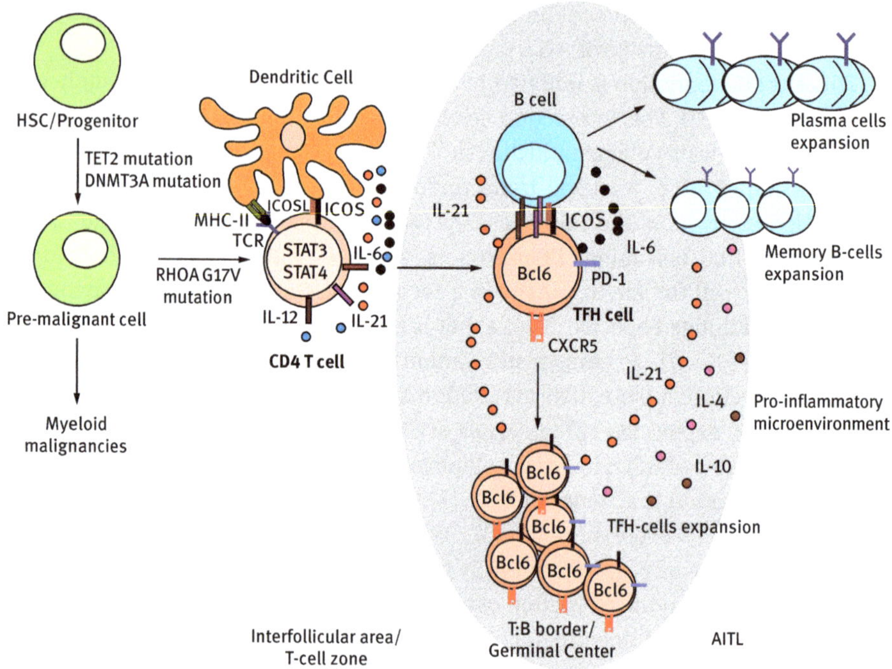

Fig. 16.1: **Current model for AITL development.** Hematopoietic stem cells (HSC) /progenitor cells acquire TET2 and/or DNMT3A premalignant mutations followed by second-hit RHOA G17V mutation to achieve malignant transformation into AITL. Tfh (T follicular helper) cell differentiation is initiated by activation of CD4 naïve T cells by dendritic cells in presence of IL6, IL21 and IL12 to induce STAT3/STAT4 activation. Dendritic cells provide signals (ICOS-ICOSL) that upregulate BCL6 and CXCR5 in CD4$^+$ T cells allowing them to migrate to B-cell follicles to induce germinal centers formation. Ongoing stimulation of Tfh cells and antigen presentation by B cells drive full development of Tfh cells, supporting the germinal center reaction and providing help to B cells to facilitate the generation of long-lived plasma cells and memory B cells. Deregulated expansion and/or function of Tfh could induce the generation of cytokines (IL4, IL6, IL21 and IL10), which play a prominent role in the early stages of lymphoma progression and in setting the abundant inflammatory component of AITL tumor lesions.

lesions, and could contribute to the frequent co-occurrence of autoimmune alterations in this disease (Fig. 16.1).

16.3 Gene expression profiling insights in the classification and biology of PTCL, NOS and AITL

Peripheral T-cell lymphomas are a very heterogeneous group of mature T-cell malignancies. Different morphologies and lack of diagnostic markers make classification of the different subtypes a challenge that affects clinical management and precludes a better knowledge of this disease. In this context, microarray gene expression profiling studies have been instrumental in the characterization of PTCLs, allowing improved diagnosis and providing critical novel information on the pathobiology of these tumors [8, 29, 56–62]. Transcriptional characterization of PTCL tumor samples and normal lymphocyte populations established the unique molecular profile of AITLs and demonstrated a close relationship between these lymphomas and follicular T helper cells [8, 29]. Moreover, expression analysis in PTCL, NOS samples can identify an AITL-like signature in about 20 % of cases, supporting that morphology-based classification frequently fails to adequately assign an AITL diagnosis. Further support to the intrinsic value of genomic profiling was provided in a recent study, where 14 % of morphologically diagnosed PTCL, NOS cases were reclassified as AITLs by molecular signatures in a large cohort of PTCL cases [62]. In all, 37 % of PTCL, NOS in this study were assigned to a different diagnostic group based on their pattern of gene expression. PTCL, NOS cases not related to other diagnostic categories showed two major groups characterized by the expression of either GATA3 (33 %) or TBX21 (49 %). Notably the PTCL, NOS GATA3 positive cases were significantly associated with poor overall survival [62].

Analysis of the expression signatures of AITL cases reflects the important contribution of the microenvironment in these tumors. Thus, in addition to showing enrichment for Tfh genes, AITL gene expression programs are also enriched in B-cell, follicular dendritic cell, and angiogenesis-related genes from the microenvironment [63]. The B-cell signature seems to be associated with a better outcome [62]. Moreover, analysis of cytokine gene expression patterns shows a rich cytokine milieu including angiogenic (*IL8*), proinflammatory (*IL6, IL18*), immunosuppressive (*IL10*), and proliferation-inducing (*IL21*) interleukins, as well as chemokines and their receptors. Notably, the high expression of IL4, IL6, and IL21 could be linked with the characteristic expansion of B cells and plasma cells present in these tumors and with the hypergammaglobulinemia and co-occurrence of autoimmune disorders frequently associated with the diagnosis of AITL. Finally, gene expression profiling of AITL samples shows high levels of expression of genes promoting angiogenesis which may reflect a pathogenic role of VEGF not only promoting angiogenesis but also by stimulating lymphoma cell growth via an autocrine/paracrine loop [64].

16.4 The genetic landscape of AITL and PTCL, NOS

Molecular cytogenetic profiling of AITL and PTCL, NOS has implicated numerous chromosomal regions as targets for recurrent deletions and amplifications in this disease, with gains exceeding the losses [65–67]. There is little concordance in the different profiling studies that have been published, which might be due to the heterogeneity of the disease itself and the different genomic platforms used for the analysis [2]. Altogether, the genetic aberrations in PTCL are complex and no association between specific chromosomal abnormalities and overall survival has been found yet.

A recurrent chromosomal translocation t(5;9)(q33;q22), resulting in an interleukin-2–inducible T-cell kinase (ITK)–spleen tyrosine kinase (SYK) fusion tyrosine kinase has been described as a recurrent pathogenic event in PTCL [68, 69]. This rearrangement fuses the PH-TH domain of ITK with most of the interdomain-B and the kinase domain of SYK, to generate a chimeric kinase protein that associates constitutively with lipid rafts in T cells and triggers antigen-independent phosphorylation of T cell receptor–proximal proteins [70]. The pathogenic role of the ITK-SYK fusion is evidenced by the development of aggressive PTCLs in mice upon expression of this oncogene [70, 71].

Genomic profiling studies have started to dissect the repertoire of genetic alterations driving the pathogenesis of AITL and PTCL, NOS tumors. These studies have already uncovered a major role for mutations in *RHOA* and epigenetic factor genes such as *TET2*, *DNMT3A* and *IDH2* in the pathogenesis of these tumors, and pointed to additional relevant pathways in these diseases (Fig. 16.2).

RHOA is a small GTPase that regulates multiple biological processes, including cytoskeleton remodeling, cell adhesion, migration, proliferation and survival [72, 73]. RHOA alternates between an active GTP-bound configuration and an inactive GDP-bound state. The transition between both states is catalyzed by RHO GEFs (guanine nucleotide exchange factors), which facilitate the incorporation of GTP and transition to the active configuration, and RHOA GAPs (GTPase activating proteins), which stimulate the hydrolysis of GTP leading the protein to the inactive state [72, 73]. RHOA has been shown to play important roles during thymocyte development, as well as in mature T cells. Initial studies using RHOA dominant negatives or inactivation of RhoA by using C3 transferase in transgenic mice suggested a role of RHOA in T-cell adhesion, polarization, migration and TCR signaling [74–78]. Even more, RHOA inactivation has been shown to cause developmental block during thymocyte progression [79–81], and inactivation of RhoA in the thymus has been linked to development of T-cell lymphomas [82]. Recently, the development of a conditional knockout model where *RhoA* is specifically deleted in T cells has demonstrated a role for RhoA in multiple steps during thymocyte development, including TCRB-selection, positive selection, early single positive lineage commitment, thymocyte proliferation and survival, and notably, mitochondrial function [83]. Blocking RhoA activation in mature T cells leads to diminished ERK1/2 activation, calcium responses, and IL2 production [84] and interestingly, Rho GTPase-mediated pathways have been shown to play a role in the development of systemic autoimmunity [85].

In recent years, whole exome sequencing approaches have identified *RHOA* heterozygous missense mutations as one of the most frequent events in PTCL [86–88]. The mutation is a recurrent substitution of valine for glycine at residue 17 of the protein (RHOA Gly17Val or G17V). The *RHOA G17V* mutation is present in close to 70 % of AITL patients and 20 % of PTCL, NOS [86–88]. Biochemical analysis of the mutant protein demonstrated that the RHOA G17V mutant does not bind GTP and acts as a dominant negative protein [86–88], most probably by sequestering the RHOA GEFs that activate

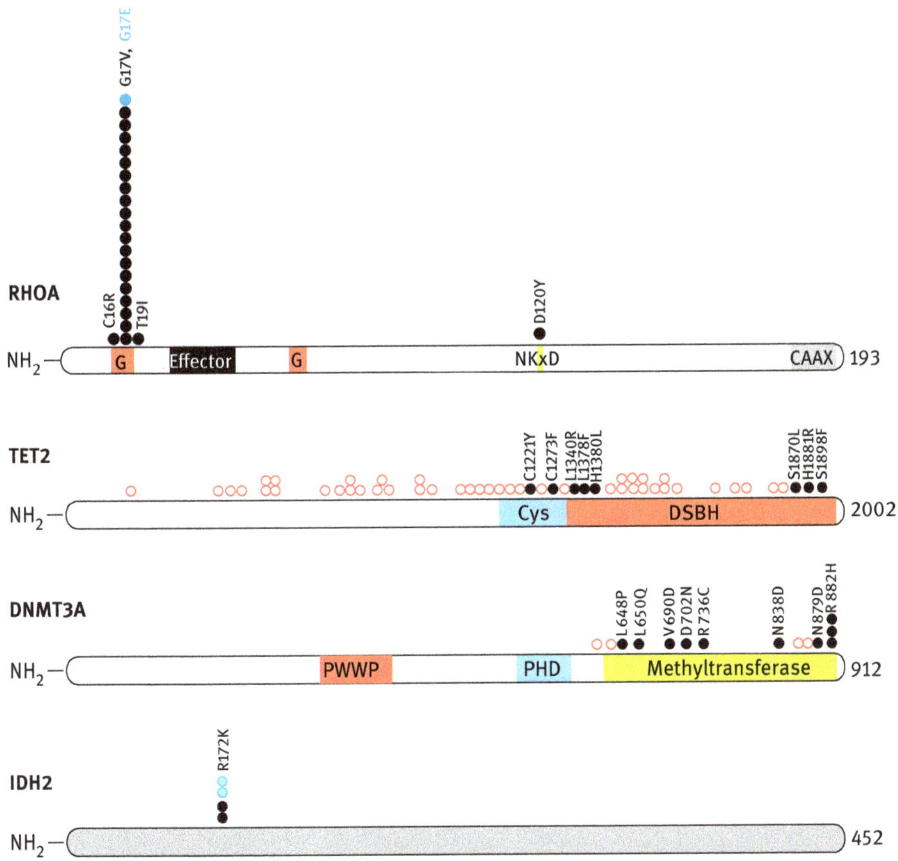

Fig. 16.2: **Most frequent mutations identified in AITL and PTCL, NOS.** Schematic representation of the structure of the most frequently mutated proteins in AITL and PTCL, NOS: RHOA, TET2, DNMT3A and IDH2, detected via exome sequencing (N = 12) and targeted resequencing (N = 125). Filled circles represent predicted amino acid substitutions while open red circles indicate truncating mutations. G:GTP/GDP binding domain; Effector: effector interaction domain; NKXD: NKXD GTP-binding domain; CAAX: CAAX box prenylation domain; Cys: cysteine-rich domain; DSBH: double-stranded beta helix fold domain; PWWP: PWWP domain; PHD: plant homeodomain.

the wild-type protein [86]. After the identification of the RHOA G17V allele, recurrent mutations in different RHOA residues have been identified in Burkitt lymphoma and gastric carcinoma [89–91]

The ten-eleven translocation 2 (*TET2*) encodes a 2 oxoglutarate/Fe^{2+}–dependent oxygenase that participates in the epigenetic control of gene expression through the oxidation of methylated cytosines and DNA demethylation by catalyzing the oxidation of DNA 5-methylcytosine to 5-hydroxymethylcytosine [92–94]. First identified as a tumor suppressor in myeloid malignancies [95, 96], recent studies have shown a high prevalence of loss-of-function mutations in *TET2* in PTCLs [97–99]. *TET2* mutations in

PTCL are particularly prevalent in AITL (~ 50 %) and PTCL, NOS (~ 40 %), and were mostly absent in all other PTCL entities except for a couple of mutations described in enteropathy-associated T-cell lymphoma [99]. In the AITL and PTCL, NOS groups, *TET2* mutations were associated with advanced stage disease, adverse clinical parameters at presentation and shorter progression free survival [99].

TET2 mutations found in PTCL are primarily frameshift and nonsense loss-of-function alleles and seem to be a founder lesion, as they can be identified not only in the malignant CD4+ T-cell population, but also in accompanying normal reactive cells [100]. The role of *Tet2* has been studied in multiple loss-of-function mouse models, showing that the loss of *Tet2* caused increased frequency of hematopoietic stem cells, enhanced competitive repopulation capacity and biased differentiation towards the myeloid lineage [100–107]. Loss of *Tet2* in mice has also been associated with the development of both myeloid malignancies [100–102] and AITL like T-cell lymphoma [108].

The DNA (cytosine-5) methyltransferase 3A (*DNMT3A*) gene encodes a methyltransferase involved in the epigenetic regulation of gene expression via methylation of cytosines in the DNA. Mutations in *DNMT3A* gene were originally described in acute myeloid leukemia [109, 110], where they represent one of the most frequently mutated genes and are associated with adverse overall survival [111]. Analysis of *DNMT3A* mutations in PTCL identified the presence of recurrent genetic lesions in this gene in 10 % of AITL samples [98]. In this case, mutations in *DNMT3A* are frequently associated with the presence of mutations in *TET2*, and it seems that the *DNMT3A* changes may even predate the occurrence of a *TET2* mutation in the natural history of the disease [98].

The isocitrate dehydrogenase 2 (*IDH2*) gene, mutated in about 30–40 % of AITL cases [112], encodes a metabolic mitochondrial enzyme physiologically involved in the oxidative decarboxylation of isocitrate to 2-oxoglutarate. Recurrent mutations in this gene can typically be found in acute myeloid leukemia, glioma, chondrosarcoma and intrahepatic cholangiocarcinoma in addition to AITL, and typically involve residue Arg172 as single amino acid substitutions (R172K, R172S) [113]. Notably, the resulting mutant enzymes have a neomorphic enzymatic activity catalyzing the conversion of alpha ketoglutarate to 2 hydroxyglutarate (2-HG), an oncometabolite that inhibits the activity of alpha ketoglutarate dependent enzymes including the TET family enzymes [114].

Notably, and despite the proposed converging mechanisms of action of *TET2*, *DNMT3A* and *IDH2* mutations in AITL, these genetic alterations can be frequently found co-occurring in the same tumor in this disease, highlighting the central role of epigenetic deregulation in the pathogenesis of these lymphomas and suggesting a multistep model for the development of AITL, where the immature blood cells would first acquire premalignant mutations, mostly *TET2* and/or *DNMT3A* mutations, followed by *RHOA G17V* as a second-hit that would lead to the specific development of AITL [115]. It has been postulated that defects in RHOA may alter Tfh cell motility and drive increased cell numbers into the follicular environment, where they are retained and receive signals promoting lymphomagenesis [116]. More studies are required to validate the multistep model and to understand the role of RHOA G17V in the development of AITL.

Finally, recurrent – albeit less frequent – genetic defects have been described in *FYN* in AITL and PTCL, NOS cases [86]. The FYN tyrosine kinase is, together with LCK, the predominant SRC family kinase found in T lymphocytes, and plays an important role in T-cell activation upon T-cell receptor (TCR) stimulation [117]. Strikingly, FYN mutations found in PTCL, NOS and AITL cases specifically disrupt the intramolecular inhibitory interaction of the FYN SH2 domain with C-terminal SRC kinase (CSK)-phosphorylated FYN Tyr531, resulting in increased tyrosine kinase signaling. Mutant active FYN proteins encoded by these mutations can be effectively inhibited with dasatinib, a multikinase inhibitor with activity against ABL1 and SRK kinases [86].

16.5 Lessons from animal models

The development of accurate animal models that recapitulate the natural history of AITL and PTCL, NOS is of fundamental importance to dissect the specific mechanisms driving these diseases and to accelerate the development and testing of novel therapies.

A specific mouse model of PTCL has been derived via expression of the ITK-SYK fusion protein [70]. This fusion tyrosine kinase oncogene, resulting from the t(5;9)(q33;q22) translocation [68], mimics constitutive TCR signaling and drives lymphoma when expressed in T cells. In this model, the human ITK-SYK fusion cDNA was inserted into the Rosa26 locus preceded by a loxP-flanked stop cassette. Then, the Rosa26$^{ITK\text{-}SYK}$ mouse was crossed with a CD4-Cre transgenic mouse to activate ITK-SYK expression specifically in T cells. The expression of ITK-SYK in T cells induces a malignancy that resembles PTCL with a latency of 20 weeks. ITK-SYK–positive T cells are mostly CD4+CD8-, with a high proliferative condition due to spontaneous and chronic TCR activation. Malignant cells are found infiltrated in the skin, spleen, lymph nodes, and other organs, in accordance with the phenotype seen in some human samples [118]. The lymphoma cells are highly transplantable, inducing the same type of disease with clonal origin. Interestingly, the authors reported that the ITK-SYK–expressing T cells are responsive to SYK inhibitors, paving the way to target PTCLs with SYK inhibitors.

Similar results have been found using transduction of bone marrow progenitors with retroviruses carrying the ITK-SYK fusion gene [71]. Forced expression of ITK-SYK in the bone marrow caused a T-cell lymphoproliferative disease with a short latency of 8 weeks after transplantation. In concordance with the Rosa26 ITK-SYK knockin model [70], retroviral expression of ITK-SYK induces an accumulation of CD4+ and CD4–CD8– T cells, and complete loss of CD8+ T cells. This malignancy is serially transplantable, inducing a clonal/biclonal disease. In addition, the development of PTCL in this model comes with an inflammatory condition that could be related to autoimmune diseases found in some PTCL, NOS and AITL patients [63, 119].

The role of inflammation in PTCL has been addressed in a different model by enforcing expression of Lin28b, which in turn induces a PTCL that develops into an inflammatory tumor microenvironment characterized by elevated levels of Il6 and NFκB [120]. Lin28a and Lin28b are RNA-binding proteins expressed in undifferentiated cells whose expression has been linked to malignant transformation *in vitro* [121]. In this model, murine *Lin28b* cDNA was inserted into a vector containing *Vav*-regulatory elements, inducing the expression of Lin28b in hematopoietic tissues. First, mice display aberrant T-cell development with lymphopenia and increased number of mature T cells in the thymus. Later on, they develop an aggressive fatal T-cell malignancy characterized by lymphadenopathy, splenomegaly, and increase of neutrophils and eosinophils in peripheral blood. The overexpression of Lin28b induces activation of NF-κB and overexpression of pro-inflammatory cytokines such as Il6, leading to development of systemic inflammation that favors the onset of PTCL. It is worth noting that, although most mice present lymphopenia, not all of them develop PTCL, indicating that additional unidentified mutations are needed for malignant transformation. Phenotypically, the malignant cells are closely related to Tfh cells, as they are CD4+ PD1+ and express some Tfh cell markers including *Bcl6*, *Cxcl13*, *Icos* and *Il21*. However, tumor cells are negative for Cxcr5, a key marker for Tfh. Thus, although the lymphomas that developed in the Lin28b-transgenic mice have some features in common with AITL, they are better classified within the PTCL, NOS group.

Another PTCL mouse model has been established by specific deletion of the tumor suppressor *Snf5* during T-cell development [122]. Snf5 (also known as Smarcb1) is a subunit of the SWI/SNF complex whose function is to regulate chromatin structure, and plays an important role in epigenetic regulation of gene expression [123]. Inactivating mutations in the SWI/SNF complex have been found in a number of human cancer types, including malignant rhabdoid tumors and meningiomas [124–126]. In this model, deletion of *Snf5* during T-cell development using the Lck-Cre or CD4-Cre recombinase induces an aggressive monoclonal/oligoclonal CD8+ mature form of PTCL, whereas earlier loss of *Snf5* in early stem/progenitor cells, due to expression of the Cre recombinase under the CD2 promoter, induces abnormal cell development, but not T-cell tumors [122]. The mature malignant cells have a memory phenotype (CD8+ $CD44^{high}$ $CD122^{low}$) and are found exclusively in peripheral organs, mostly in the spleen. Interestingly, the expansion of the lymphoma is dependent on TCR-MHC interactions, suggesting that, at least for some types of PTCLs, TCR blockade might constitute a therapeutic opportunity.

Currently, the mouse model that most closely recapitulates the features of AITL is the *sanroque* (*san*) model, generated by ENU mutagenesis and identified in a screening for anti-nuclear antibodies [127]. *san* mice have a homozygous point mutation in the *Rc3h1* gene that substitutes a methionine residue at position 199 in the Rc3h1 protein by arginine (Met199Arg), resulting in an aberrant expression of *Icos*. Rc3h1 is a RING-type E3 ubiquitin ligase that binds *Icos* mRNA through its ROQ domain and regulates its stability. In *sanroque* mice, mutated Rc3h1 is unable to promote *Icos* mRNA degradation, resulting in the overexpression of Icos on CD4+ T cells, which in turn leads to excessive accumulation of Tfh cells and development of a lupus erythematosus-like autoimmune disease [28]. Mice heterozygous for the *Rc3h1* allele (*Rc3h1*$^{+/san}$) also have deregulated Tfh cells, but do not develop the full-blown lupus-like autoimmunity [127]. Instead, heterozygosity for *Rc3h1*$^{+/san}$ induces a nonlethal lymphadenopathy with histopathological features reminiscent of AITL, including accumulation of clonal Tfh cells, large B cells and a significant increase in the frequency of germinal center B cells. The dependence on Tfh cells was demonstrated by genetic blockade of Tfh development through deletion of *Cd28*, *Icos* or *Sap*, resulting in prevention of AITL-like tumor development. However, alterations in the *Rc3h1* gene have not been found in humans and, despite the generalized presence of deregulated Tfh cells, only 53 % of *Rc3h1* heterozygous mice develop AITL-like disease, suggesting that additional mutagenic events are required for neoplastic transformation.

Recently, *Tet2* loss-of-function models have been also under study as a model for PTCL, due to the frequent *TET2* loss-of-function mutations found in AITL and PTCL, NOS [99, 108, 128]. It has been shown that reduced Tet2 function in mice induces a late latency T-cell lymphoma with Tfh-like features [108]. In this model, *Tet2* knockdown results in 80 % reduction of *Tet2* mRNA levels, accumulation of Tfh-like cells in the spleen, and the development of T-cell lymphoma with a latency of 67 weeks. However, in spite of their Tfh origin, Tet2 knockdown tumors lack most of the features of AITL, suggesting that this model recapitulates PTCL, NOS rather than AITL.

Finally, a study proposed that intraperitoneal injection of primary human AITL cells into immunodeficient NOG mice can replicate human AITL in mice [129]. Following transplant of human AITL cells, mice develop marked splenomegaly and mild hepatomegaly with splenic infiltration by T cells and B cells of human origin. The infiltrate contained a small fraction of PD1+ T cells, together with a large number of B- and T-reactive lymphocytes and polyclonal plasma cells, and immunohistological analysis demonstrated it closely resembled the original tumor. The AITL transplanted NOG mice produce polyclonal human immunoglobulins, including IgG, indicating that AITL cells function as Tfh cells, regulating B-cell mediated humoral immunity [129]. Serial transplantations resulted in the reduction of the reactive components, such as B cells or CD8+ T cells with concomitant expansion of neoplastic cells. Thus, after several rounds of transplants, most of cells were CD4+ only, and human IgG production was absent [129].

16.6 Concluding remarks

PTCL is a very heterogeneous disease, with the most frequent subgroups being PTCL, NOS and AITL, both associated with poor prognosis. Neoplastic cells in AITL and a subset of PTCL, NOS have been shown to derive from Tfh cells, a subset of T cells normally present in germinal centers with a helper function to germinal center B cells. The heterogeneous presentation of PTCLs makes them often challenging to diagnose and classify. Gene expression profiling has been used to improve the classification of PTCLs, demonstrating the association of AITL with a Tfh-like gene expression signature, and the clustering of PTCL, NOS into two different groups associated with GATA3 or TBX21 gene signatures and with different survival outcomes. Genetic alterations in PTCL, NOS and AITL are frequent and characterized by complex karyotypes with recurrent chromosomal losses and gains, including a recurrent chromosomal translocation ITK-SYK. High throughput sequencing analyses have identified several genes frequently mutated in AITL and PTCL, NOS, including epigenetic regulators such as *TET2*, *DNMT3A* and *IDH2* and the small GTPase *RHOA*, which is recurrently mutated in 70 % of AITL cases and 20 % of PTCL, NOS. The specific mechanisms that elicit T-cell transformation downstream of these genes are still unknown; however, transformation seems to follow a multistep model with *TET2* and/or *DNMT3A* mutations occurring first at the hematopoietic progenitor stage, followed by specification into the Tfh lineage guided by RHOA G17V mutation. Significant efforts are underway to develop and characterize mouse models for AITL and PTCL, NOS. These models will provide powerful tools for the study of the pathogenesis of PTCLs and for the identification and development of molecularly targeted therapies to treat this disease.

References

[1] Swerdlow SH, Campo E, Harris NL, Jaffe ES, Pileri SA, Stein H, Thiele J, Vardiman JW Pathology and genetics of tumours of hematopoietic and lymphoid tissues. World Health Organization classification of tumours. International Agency for Research on Cancer Press, Lyon, 2008;(11):270–299.

[2] de Leval L, et al. Molecular classification of T-cell lymphomas. Crit Rev Oncol Hematol, 2009;72(2):125–143.

[3] Willenbrock K, et al. Analysis of T-cell subpopulations in T-cell non-Hodgkin's lymphoma of angioimmunoblastic lymphadenopathy with dysproteinemia type by single target gene amplification of T cell receptor- beta gene rearrangements. Am J Pathol 2001;158(5):1851–1857.

[4] Willenbrock K, et al. In angioimmunoblastic T-cell lymphoma, neoplastic T cells may be a minor cell population. A molecular single-cell and immunohistochemical study. Virchows Arch 2005;446(1):15–20.

[5] Attygalle AD, et al. CD10 expression in extranodal dissemination of angioimmunoblastic T-cell lymphoma. Am J Surg Pathol 2004;28(1):54–61.

[6] Dupuis J, et al. Expression of CXCL13 by neoplastic cells in angioimmunoblastic T-cell lymphoma (AITL): a new diagnostic marker providing evidence that AITL derives from follicular helper T cells. Am J Surg Pathol 2006;30(4):490–494.
[7] Dorfman DM, et al. Programmed death-1 (PD-1) is a marker of germinal center-associated T cells and angioimmunoblastic T-cell lymphoma. Am J Surg Pathol 2006;30(7):802–810.
[8] de Leval L, et al. The gene expression profile of nodal peripheral T-cell lymphoma demonstrates a molecular link between angioimmunoblastic T-cell lymphoma (AITL) and follicular helper T (Tfh) cells. Blood 2007;109(11):4952–4963.
[9] Grogg KL, et al. Expression of CXCL13, a chemokine highly upregulated in germinal center T-helper cells, distinguishes angioimmunoblastic T-cell lymphoma from peripheral T-cell lymphoma, unspecified. Mod Pathol 2006;19(8):1101–1107.
[10] Roncador G, et al. Expression of two markers of germinal center T cells (SAP and PD-1) in angioimmunoblastic T-cell lymphoma. Haematologica 2007;92(8):1059–1066.
[11] Campo E, et al. Report of the European Task Force on Lymphomas: workshop on peripheral T-cell lymphomas. Ann Oncol 1998;9(8):835–843.
[12] Vose J, et al. International peripheral T-cell and natural killer/T-cell lymphoma study: pathology findings and clinical outcomes. J Clin Oncol 2008;26(25):4124–4130.
[13] Iannitto E, et al. Angioimmunoblastic T-cell lymphoma. Crit Rev Oncol Hematol 2008;68(3):264–271.
[14] Mosalpuria K, Bociek RG, Vose JM. Angioimmunoblastic T-cell lymphoma management. Semin Hematol 2014;51(1):52–58.
[15] Siegert W, et al. Angioimmunoblastic lymphadenopathy (AILD)-type T-cell lymphoma: prognostic impact of clinical observations and laboratory findings at presentation. The Kiel Lymphoma Study Group. Ann Oncol 1995;6(7):659–664.
[16] de Leval L, Gisselbrecht C, Gaulard P. Advances in the understanding and management of angioimmunoblastic T-cell lymphoma. Br J Haematol 2010;148(5):673–689.
[17] Cho YU, et al. Distinct features of angioimmunoblastic T-cell lymphoma with bone marrow involvement. Am J Clin Pathol 2009;131(5):640–66.
[18] Dogan A, Attygalle AD, Kyriakou C. Angioimmunoblastic T-cell lymphoma. Br J Haematol 2003;121(5):681–691.
[19] Mourad N, et al. Clinical, biologic, and pathologic features in 157 patients with angioimmunoblastic T-cell lymphoma treated within the Groupe d'Etude des Lymphomes de l'Adulte (GELA) trials. Blood 2008;111(9):4463–4470.
[20] Schaerli P, et al. CXC chemokine receptor 5 expression defines follicular homing T cells with B cell helper function. J Exp Med 2000;192(11):1553–1562.
[21] Kim CH, et al. Subspecialization of CXCR5+ T cells: B helper activity is focused in a germinal center-localized subset of CXCR5+ T cells. J Exp Med 2001;193(12):1373–1381.
[22] Breitfeld D, et al. Follicular B helper T cells express CXC chemokine receptor 5, localize to B cell follicles, and support immunoglobulin production. J Exp Med 2000;192(11):1545–1552.
[23] Chtanova T, et al. T follicular helper cells express a distinctive transcriptional profile, reflecting their role as non-T_H1/T_H2 effector cells that provide help for B cells. J Immunol 2004;173(1):68–78.
[24] Ueno H, Banchereau J, Vinuesa CG. Pathophysiology of T follicular helper cells in humans and mice. Nat Immunol 2015;16(2):142–152.
[25] Vinuesa CG, et al. A RING-type ubiquitin ligase family member required to repress follicular helper T cells and autoimmunity. Nature 2005;435(7041):452–458.
[26] Choi YS, et al. ICOS receptor instructs T follicular helper cell versus effector cell differentiation via induction of the transcriptional repressor Bcl6. Immunity 2011;34(6):932–946.

[27] King C. New insights into the differentiation and function of T follicular helper cells. Nat Rev Immunol 2009;9(11):757–766.

[28] Vinuesa CG, et al. Follicular B helper T cells in antibody responses and autoimmunity. Nat Rev Immunol 2005;5(11):853–865.

[29] Piccaluga PP, et al. Gene expression analysis of angioimmunoblastic lymphoma indicates derivation from T follicular helper cells and vascular endothelial growth factor deregulation. Cancer Res 2007;67(22):10703–10710.

[30] Gaulard P, de Leval L. Pathology of peripheral T-cell lymphomas: where do we stand? Semin Hematol 2014;51(1):5–16.

[31] Goenka R, et al. Cutting edge: dendritic cell-restricted antigen presentation initiates the follicular helper T cell program but cannot complete ultimate effector differentiation. J Immunol 2011;187(3):1091–1095.

[32] Stone EL, et al. ICOS coreceptor signaling inactivates the transcription factor FOXO1 to promote Tfh cell differentiation. Immunity 2015;42(2):239–251.

[33] Weber JP, et al. ICOS maintains the T follicular helper cell phenotype by down-regulating Kruppel-like factor 2. J Exp Med 2015;212(2):217–233.

[34] Leavenworth JW, et al. A p85alpha-osteopontin axis couples the receptor ICOS to sustained Bcl-6 expression by follicular helper and regulatory T cells. Nat Immunol 2015;16(1):96–106.

[35] Crotty S. T follicular helper cell differentiation function, and roles in disease. Immunity 2014;41(4):529–542.

[36] Liu X, Nurieva RI, Dong C. Transcriptional regulation of follicular T-helper (Tfh) cells. Immunol Rev 2013;252(1):139–145.

[37] Nurieva RI, et al. Bcl6 mediates the development of T follicular helper cells. Science 2009;325(5943):1001–1005.

[38] Yu D, et al. Lineage specification and heterogeneity of T follicular helper cells. Curr Opin Immunol 2009;21(6):619–625.

[39] Johnston RJ, et al. Bcl6 and Blimp-1 are reciprocal and antagonistic regulators of T follicular helper cell differentiation. Science 2009;325(5943):1006–1010.

[40] Qi H, et al. Tfh cell differentiation and their function in promoting B-cell responses. Adv Exp Med Biol 2014;841:153–180.

[41] Hatzi K, et al. BCL6 orchestrates Tfh cell differentiation via multiple distinct mechanisms. J Exp Med 2015;212(4):539–553.

[42] Liu X, et al. Transcription factor achaete-scute homologue 2 initiates follicular T-helper-cell development. Nature 2014;507(7493):513–518.

[43] Kroenke MA, et al. Bcl6 and Maf cooperate to instruct human follicular helper CD4 T cell differentiation. J Immunol 2012;188(8):3734–3744.

[44] Kwon H, et al. Analysis of interleukin-21-induced Prdm1 gene regulation reveals functional cooperation of STAT3 and IRF4 transcription factors. Immunity 2009;31(6):941–952.

[45] Nurieva RI, et al. Generation of T follicular helper cells is mediated by interleukin-21 but independent of T helper 1, 2, or 17 cell lineages. Immunity 2008;29(1):138–149.

[46] Liao W, et al. Modulation of cytokine receptors by IL-2 broadly regulates differentiation into helper T cell lineages. Nat Immunol 2011;12(6):551–559.

[47] Choi YS, et al. Bcl6 expressing follicular helper CD4 T cells are fate committed early and have the capacity to form memory. J Immunol 2013;190(8):4014–4026.

[48] Nurieva RI, Dong C. (IL-)12 and 21: a new kind of help in the follicles. Immunol Cell Biol 2009;87(8):577–578.

[49] Suto A, et al. Development and characterization of IL-21-producing CD4+ T cells. J Exp Med 2008;205(6):1369–1379.

[50] Ma CS, et al. Early commitment of naive human CD4(+) T cells to the T follicular helper (T(FH)) cell lineage is induced by IL-12. Immunol Cell Biol 2009;87(8):590–600.
[51] Schmitt N, et al. IL-12 receptor beta1 deficiency alters in vivo T follicular helper cell response in humans. Blood 2013;121(17):3375–3385.
[52] Schmitt N, et al. Human dendritic cells induce the differentiation of interleukin-21-producing T follicular helper-like cells through interleukin-12. Immunity 2009;31(1):158–169.
[53] Walker SR, et al. STAT5 outcompetes STAT3 to regulate the expression of the oncogenic transcriptional modulator BCL6. Mol Cell Biol 2013;33(15):2879–2890.
[54] Johnston RJ, et al. STAT5 is a potent negative regulator of Tfh cell differentiation. J Exp Med 2012;209(2):243–250.
[55] Nurieva RI, et al. STAT5 protein negatively regulates T follicular helper (Tfh) cell generation and function. J Biol Chem 2012;287(14):11234–11239.
[56] Martinez-Delgado B, et al. Differential expression of NF-kappaB pathway genes among peripheral T-cell lymphomas. Leukemia 2005;19(12):2254–22563.
[57] Ballester B, et al. Gene expression profiling identifies molecular subgroups among nodal peripheral T-cell lymphomas. Oncogene 2006;25(10):1560–1570.
[58] Martinez-Delgado B. Peripheral T-cell lymphoma gene expression profiles. Hematol Oncol 2006;24(3):113–119.
[59] Piccaluga PP, et al. Gene expression analysis of peripheral T cell lymphoma, unspecified, reveals distinct profiles and new potential therapeutic targets. J Clin Invest 2007;117(3):823–834.
[60] Cuadros M, et al. Identification of a proliferation signature related to survival in nodal peripheral T-cell lymphomas. J Clin Oncol 2007;25(22):3321–3329.
[61] Iqbal J, et al. Molecular signatures to improve diagnosis in peripheral T-cell lymphoma and prognostication in angioimmunoblastic T-cell lymphoma. Blood 2010;115(5):1026–1036.
[62] Iqbal J, et al. Gene expression signatures delineate biological and prognostic subgroups in peripheral T-cell lymphoma. Blood 2014;123(19):2915–2923.
[63] Gaulard P, de Leval L. The microenvironment in T-cell lymphomas: emerging themes. Semin Cancer Biol 2014;24:49–60.
[64] Piccaluga PP, Tabanelli V, Pileri SA. Molecular genetics of peripheral T-cell lymphomas. Int J Hematol 2014;99(3):219–226.
[65] Zettl A, et al. Genomic profiling of peripheral T-cell lymphoma, unspecified, and anaplastic large T-cell lymphoma delineates novel recurrent chromosomal alterations. Am J Pathol 2004;164(5):1837–1848.
[66] Thorns C, et al. Chromosomal aberrations in angioimmunoblastic T-cell lymphoma and peripheral T-cell lymphoma unspecified: A matrix-based CGH approach. Genes Chromosomes Cancer 2007;46(1):37–44.
[67] Nelson M, et al. Cytogenetic abnormalities and clinical correlations in peripheral T-cell lymphoma. Br J Haematol 2008;141(4):461–469.
[68] Streubel B, et al. Novel t(5;9)(q33;q22) fuses ITK to SYK in unspecified peripheral T-cell lymphoma. Leukemia 2006;20(2):313–318.
[69] Rigby S, et al. The lymphoma-associated fusion tyrosine kinase ITK-SYK requires pleckstrin homology domain-mediated membrane localization for activation and cellular transformation. J Biol Chem 2009;284(39):26871–26881.
[70] Pechloff K, et al. The fusion kinase ITK-SYK mimics a T cell receptor signal and drives oncogenesis in conditional mouse models of peripheral T cell lymphoma. J Exp Med 2010;207(5):1031–1044.
[71] Dierks C, et al. The ITK-SYK fusion oncogene induces a T-cell lymphoproliferative disease in mice mimicking human disease. Cancer Res 2010;70(15):6193–6204.

[72] Jaffe AB, Hall A. Rho GTPases: biochemistry and biology. Annu Rev Cell Dev Biol 2005;21:247–269.
[73] Boulter E, et al. Off the beaten paths: alternative and crosstalk regulation of Rho GTPases. FASEB J 2012;26(2):469–479.
[74] del Pozo MA, et al. Rho GTPases control migration and polarization of adhesion molecules and cytoskeletal ERM components in T lymphocytes. Eur J Immunol 1999;29(11):3609–3620.
[75] Heasman SJ, et al. Coordinated RhoA signaling at the leading edge and uropod is required for T cell transendothelial migration. J Cell Biol 2010;190(4):553–563.
[76] Heasman SJ, Ridley AJ. Multiple roles for RhoA during T cell transendothelial migration. Small GTPases 2010;1(3):174–179.
[77] Mou F, et al. The Mst1 and Mst2 kinases control activation of rho family GTPases and thymic egress of mature thymocytes. J Exp Med 2012;209(4):741–759.
[78] Corre I, et al. Analysis of thymocyte development reveals that the GTPase RhoA is a positive regulator of T cell receptor responses in vivo. J Exp Med 2001;194(7):903–914.
[79] Henning SW, et al. The GTPase Rho has a critical regulatory role in thymus development. EMBO J 1997;16(9):2397–2407.
[80] Galandrini R, Henning SW, Cantrell DA. Different functions of the GTPase Rho in prothymocytes and late pre-T cells. Immunity 1997;7(1):163–174.
[81] Cleverley S, Henning S, Cantrell D. Inhibition of Rho at different stages of thymocyte development gives different perspectives on Rho function. Curr Biol 1999;9(12):657–660.
[82] Cleverley SC, et al. Loss of Rho function in the thymus is accompanied by the development of thymic lymphoma. Oncogene 2000;19(1):13–20.
[83] Zhang S, et al. Gene targeting RhoA reveals its essential role in coordinating mitochondrial function and thymocyte development. J Immunol 2014;193(12):5973–5982.
[84] Angkachatchai V, Finkel TH. ADP-ribosylation of rho by C3 ribosyltransferase inhibits IL-2 production and sustained calcium influx in activated T cells. J Immunol 1999;163(7):3819–3825.
[85] Pernis AB. Rho GTPase-mediated pathways in mature CD4+ T cells. Autoimmun Rev 2009;8(3):199–203.
[86] Palomero T, et al. Recurrent mutations in epigenetic regulators, RHOA and FYN kinase in peripheral T cell lymphomas. Nat Genet 2014;46(2):166–170.
[87] Sakata-Yanagimoto M, et al. Somatic RHOA mutation in angioimmunoblastic T cell lymphoma. Nat Genet 2014;46(2):171–175.
[88] Yoo HY, et al. A recurrent inactivating mutation in RHOA GTPase in angioimmunoblastic T cell lymphoma. Nat Genet 2014;46(4):371–375.
[89] Rohde M, et al. Recurrent RHOA mutations in pediatric Burkitt lymphoma treated according to the NHL-BFM protocols. Genes Chromosomes Cancer 2014;53(11):911–916.
[90] Kakiuchi M, et al. Recurrent gain-of-function mutations of RHOA in diffuse-type gastric carcinoma. Nat Genet 2014;46(6):583–587.
[91] Wang K, et al. Whole-genome sequencing and comprehensive molecular profiling identify new driver mutations in gastric cancer. Nat Genet 2014;46(6):573–582.
[92] Tahiliani M, et al. Conversion of 5-methylcytosine to 5-hydroxymethylcytosine in mammalian DNA by MLL partner TET1. Science 2009;324(5929):930–935.
[93] Ito S, et al. Role of Tet proteins in 5mC to 5hmC conversion ES-cell self-renewal and inner cell mass specification. Nature 2010;466(7310):1129–1133.
[94] Cimmino L, et al. TET family proteins and their role in stem cell differentiation and transformation. Cell Stem Cell 2011;9(3):193–204.
[95] Pronier E, Delhommeau F. Role of TET2 mutations in myeloproliferative neoplasms. Curr Hematol Malig Rep 2012;7(1):57–64.

[96] Abdel-Wahab O, et al. Genetic characterization of TET1, TET2, and TET3 alterations in myeloid malignancies. Blood 2009;114(1):144–147.
[97] Vose JM. Update on T-cell lymphoma. Ann Oncol 2008;19 Suppl 4:iv74-6.
[98] Couronne L, C Bastard, and OA Bernard, TET2 and DNMT3A mutations in human T-cell lymphoma. N Engl J Med 2012;366(1):95–96.
[99] Lemonnier F, et al. Recurrent TET2 mutations in peripheral T-cell lymphomas correlate with Tfh-like features and adverse clinical parameters. Blood 2012;120(7):1466–1469.
[100] Quivoron C, et al. TET2 inactivation results in pleiotropic hematopoietic abnormalities in mouse and is a recurrent event during human lymphomagenesis. Cancer Cell 2011;20(1):25–38.
[101] Moran-Crusio K, et al. Tet2 loss leads to increased hematopoietic stem cell self-renewal and myeloid transformation. Cancer Cell 2011;20(1):11–24.
[102] Li Z, et al. Deletion of Tet2 in mice leads to dysregulated hematopoietic stem cells and subsequent development of myeloid malignancies. Blood 2011;118(17):4509–4518.
[103] Ko M, et al. Ten-Eleven-Translocation 2 (TET2) negatively regulates homeostasis and differentiation of hematopoietic stem cells in mice. Proc Natl Acad Sci USA 2011;108(35):14566–14571.
[104] Ko M, Rao A. TET2: epigenetic safeguard for HSC. Blood 2011;118(17):4501–4503.
[105] Koh KP, et al. Tet1 and Tet2 regulate 5-hydroxymethylcytosine production and cell lineage specification in mouse embryonic stem cells. Cell Stem Cell 2011;8(2):200–213.
[106] Kunimoto H, et al. Tet2 disruption leads to enhanced self-renewal and altered differentiation of fetal liver hematopoietic stem cells. Sci Rep 2012;2:273.
[107] Shide K, et al. TET2 is essential for survival and hematopoietic stem cell homeostasis. Leukemia 2012;26(10):2216–2223.
[108] Muto H, et al. Reduced TET2 function leads to T-cell lymphoma with follicular helper T-cell-like features in mice. Blood Cancer J 2014;4:e264.
[109] Ley TJ, et al. DNMT3A mutations in acute myeloid leukemia. N Engl J Med 2010;363(25):2424–2433.
[110] Shah MY, Licht JD. DNMT3A mutations in acute myeloid leukemia. Nat Genet 2011;43(4):289–290.
[111] Thiede C. Mutant DNMT3A: teaming up to transform. Blood 2012;119(24):5615-7.
[112] Cairns RA, et al. IDH2 mutations are frequent in angioimmunoblastic T-cell lymphoma. Blood 2012;119(8):1901–1903.
[113] Dang L, Jin S, Su SM. IDH mutations in glioma and acute myeloid leukemia. Trends Mol Med 2010;16(9):387–397.
[114] Parker SJ, Metallo CM. Metabolic consequences of oncogenic IDH mutations. Pharmacol Ther 2015;152:54–62.
[115] Sakata-Yanagimoto M. Multistep tumorigenesis in peripheral T cell lymphoma. Int J Hematol 2015;102(5):523–527.
[116] Ahearne MJ, et al. Follicular helper T-cells: expanding roles in T-cell lymphoma and targets for treatment. Br J Haematol 2014;166(3):326–335.
[117] Palacios EH, Weiss A. Function of the Src-family kinases, Lck and Fyn, in T-cell development and activation. Oncogene 2004;23(48):7990–8000.
[118] Cotta CV, Hsi ED. Pathobiology of mature T-cell lymphomas. Clin Lymphoma Myeloma 2008;8 Suppl 5:S168–179.
[119] Carbone A, et al. The role of inflammation in lymphoma. Adv Exp Med Biol 2014;816:315–333.
[120] Beachy SH, et al. Enforced expression of Lin28b leads to impaired T-cell development, release of inflammatory cytokines, and peripheral T-cell lymphoma. Blood 2012;120(5):1048–1059.

[121] Viswanathan SR, et al. Lin28 promotes transformation and is associated with advanced human malignancies. Nat Genet 2009;41(7):843–848.
[122] Wang X, et al. TCR-dependent transformation of mature memory phenotype T cells in mice. J Clin Invest 2011;121(10):3834–3845.
[123] Roberts CW, Orkin SH. The SWI/SNF complex–chromatin and cancer. Nat Rev Cancer 2004;4(2):133–142.
[124] Margol AS, Judkins AR. Pathology and diagnosis of SMARCB1-deficient tumors. Cancer Genet 2014;207(9):358–364.
[125] Schmitz U, et al. INI1 mutations in meningiomas at a potential hotspot in exon 9. Br J Cancer 2001;84(2):199–201.
[126] Sevenet N, et al. Spectrum of hSNF5/INI1 somatic mutations in human cancer and genotype-phenotype correlations. Hum Mol Genet 1999;8(13):2359–2368.
[127] Ellyard JI, et al. Heterozygosity for Roquinsan leads to angioimmunoblastic T-cell lymphoma-like tumors in mice. Blood 2012;120(4):812–821.
[128] Nguyen TB, et al. Double somatic mosaic mutations in TET2 and DNMT3A–origin of peripheral T cell lymphoma in a case. Ann Hematol 2015;94(7):1221–1223.
[129] Sato F, et al. Angioimmunoblastic T-cell lymphoma mice model. Leuk Res 2013;37(1):21–27.

Wing C. Chan

17 Extranodal NK/T-cell lymphoma

17.1 Introduction

Extranodal NK/T-cell lymphoma (ENKTCL) is a group of rare lymphomas derived mostly from NK cells but also including cases of T-cell origin. These lymphomas share several immunophenotypic markers and clinical features and are often managed similarly. While modern technologies allow differentiation of the cell of origin of these tumors, these tools have not been employed uniformly in clinical practice, and the term ENKTCL is still applied to these cases in general.

17.2 General clinicopathologic features

The prototypical ENKTCL is an aggressive lymphoma occurring in the nasal and paranasal region including the hard palate [1]. The tumor is locally invasive and can cause extensive destruction of the midline facial structures. Morphologically, the tumor infiltrate is typically angiocentric and angiodestructive, with extensive tissue necrosis. The cytology of the cells varies among cases and can range from mainly small, mildly irregular cells to predominantly large pleomorphic cells. Most of the cases are derived from NK cells, although a minority of cases (approximately 20 %) are of T-cell origin, the majority of which are γδ T cells [2, 3]. The tumor cells are almost invariably Epstein–Barr virus (EBV) positive, regardless of the cell of origin. The clinical behavior also appears to be similar and cases are managed similarly.

The tumor cells are cytotoxic lymphocytes and show cytoplasmic azurophilic granules on Giemsa staining. Cytotoxic molecules such as TIA1, granzyme B, granzyme M and perforin [4, 5] can be demonstrated by immunohistochemistry. The tumor cells express surface CD2, CD7, sometimes CD8 and frequently CD56. Both NK- and T-cell tumors express cytoplasmic CD3ε, but surface CD3 is expressed only by tumors of the T-cell lineage. The expression of γδ or αβ T-cell receptor can be immunohistochemically demonstrated by recently available monoclonal antibodies but T-cell cases demonstrating dual reactivity or dual negativity have been observed. A recent study suggests that the expression of CD16, CXCL13 and OCT2 is much more commonly detected in tumors of NK cell origin, while PD-1 marks T-lineage cases. CD56 and IRF4 are also more frequently expressed by the former [3]. The presence of EBV is best demonstrated by *in situ* hybridization for EBV-encoded small RNAs (EBERs), which are present in a large quantity in the nuclei of infected cells. The EBV infection in ENKTCL shows a latency type II pattern with expression of EBNA1, LMP1, and LMP2 [6]. However, immunostaining for LMP1 shows variable results and is much less sensitive than EBER *in situ* hybridization for diagnostic purposes.

While the most frequent sites of involvement are the nasal, paranasal and upper aerodigestive tract, a very similar tumor can present in extranasal sites including the skin, gastrointestinal tract and the testis. Such cases are designated as ENKTCL-nasal type to indicate their similarity to the prototypical nasal tumor. It is possible that some of these tumors actually represent dissemination from a nasal tumor that may not be evident initially. Primary lymph node involvement is very rare, but regional lymph nodes may be involved secondarily while bone marrow involvement is uncommon. Even without direct bone marrow involvement, cases may be complicated by a hemophagocytic syndrome that could be rapidly fatal [7]. For patients presenting with extensive multiorgan, bone marrow and peripheral blood involvement, the disease overlaps with aggressive NK cell leukemia and differential diagnosis may be difficult [8].

Early stage disease is responsive to and occasionally cured by radiation therapy. However, relapse is common and combination of radiation therapy and chemotherapy is more appropriate [9]. The prognosis is dismal with disseminated disease. A more recently designed regimen consisting of dexamethasone, methotrexate, ifosfamide, L-Asparaginase and etoposide (SMILE) appears to be much more effective for disseminated or relapsed disease [10]. For patients presenting with more advanced disease, hematopoietic stem cell transplant may be indicated if a remission is achieved [9].

17.3 Receptor expression

NK cells do not express a single dominant receptor like T cells and B cells, but express multiple members of inhibitory and stimulatory receptors belonging to the killer immunoglobin-like receptor (KIR) or killer cell lectin-like receptors (KLR) family (Fig. 17.1). Since the profile of expression of these molecules, particularly KIR, is different among NK cells, a clonal proliferation is expected to show a restricted pattern of KIR expression reflecting the profile of KIR expression on the founder cells, and this profile may be used as the clonal marker for NK cell proliferative disorders [11, 12]. In ENKTCL, we have found that KIR transcripts are markedly downregulated in almost all cases with the exception of KIR2DL4. For T-cell derived tumors, the T-cell receptor is expected to be expressed, but occasionally it is not detectable, and sometimes both γδ and αβ receptors are expressed. Since every mature T cell has undergone rearrangement of the TCRγ locus, specific rearrangement of the TCRγ locus can be used as an indication of T-cell lineage differentiation as well as a clonal marker.

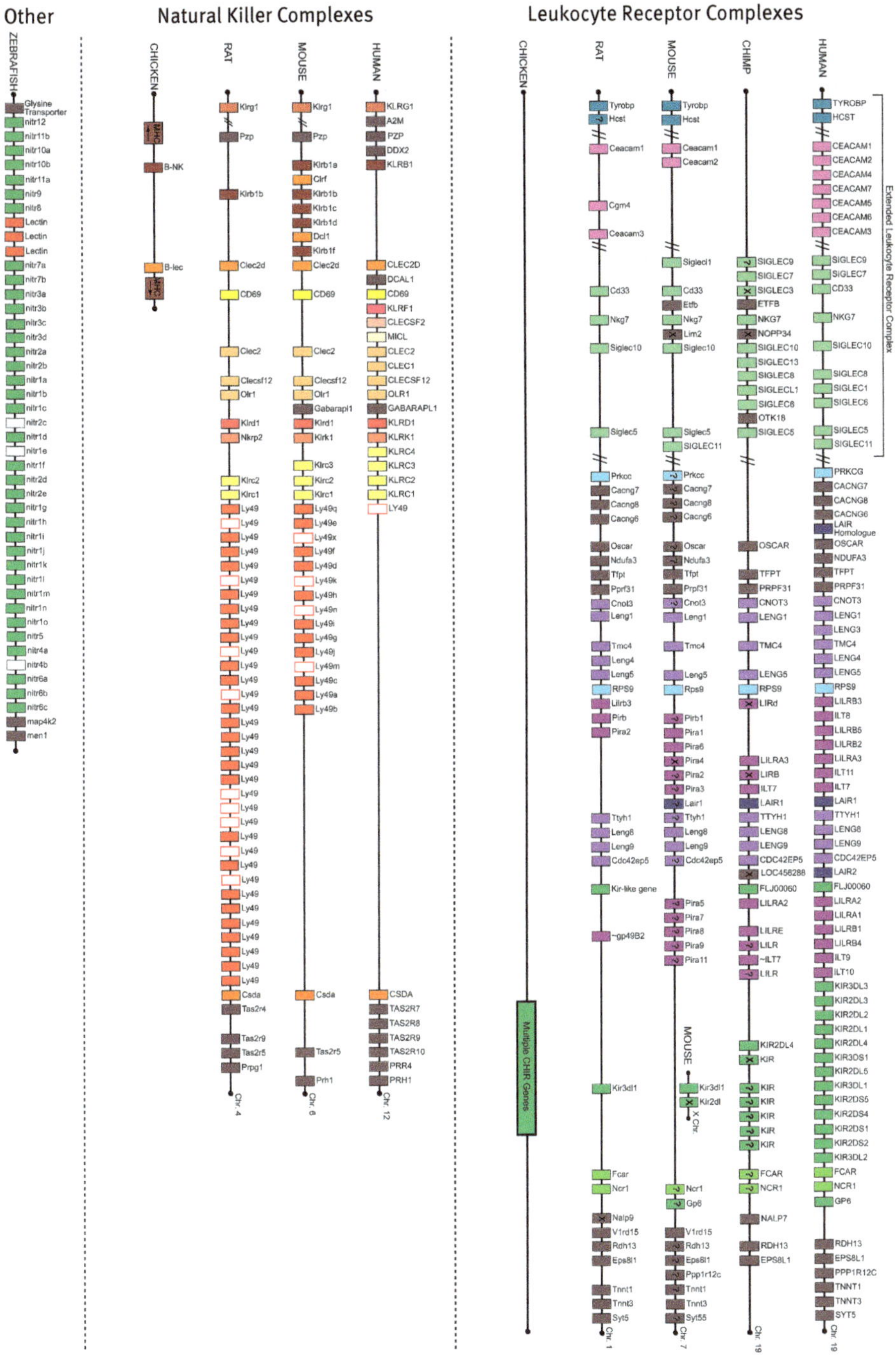

Fig. 17.1: Comparative genomic maps of NK cell receptor gene clusters. There are individual differences in the complement of receptor genes present in the genome. Reprinted with permission: [88, Figure 2].

17.4 The role of EBV in pathogenesis

The role of EBV in the pathogenesis of ENKTCL is not entirely clear. There is evidence indicating that the EBV genome is clonal, consistent with infection prior to the clonal expansion of the tumor cells [13]. LMP1 has been demonstrated to be an important transforming protein in B cells and is able to activate both the canonical and non-canonical NF-κB pathways (Fig. 17.2) [14, 15]. EBNA1 is essential for the maintenance of the viral episomes. However, there is evidence that EBNA1 may also have a role in altering the host cells [16], e.g. by reducing TP53 levels, as elaborated in the next paragraph, and thus may also have oncogenic properties. EBNA2 is also implicated in B-cell transformation in altering cellular mRNA & miRNA profiles to promote cellular proliferation and survival [17–19]. EBER1 and 2 are abundantly expressed in type 2 latency and they may also play a role in tumorigenesis by inhibiting apoptosis, enhancing cell cycle progression and inducing the production of cytokines such as interleukin 9, 6 and 10, which may have autocrine and/or paracrine functions [20, 21]. While these functions have not been directly demonstrated in ENKTCL, their roles in

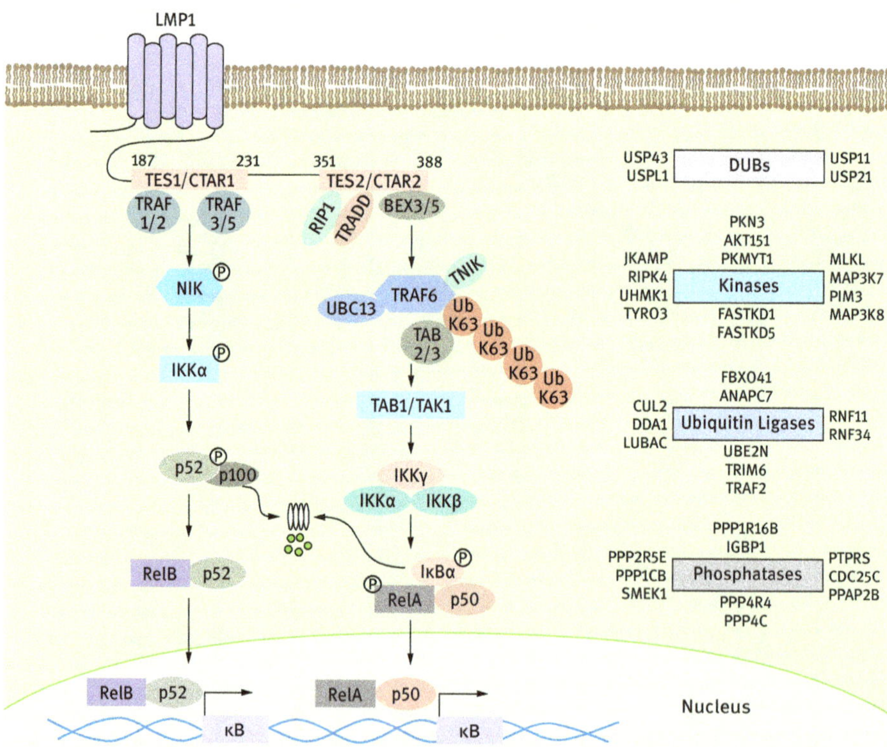

Fig. 17.2: Schematics of the activation of the canonical and alternative NF-κB pathway by LMP1. Reprinted with permission from [15] (Figure 2).

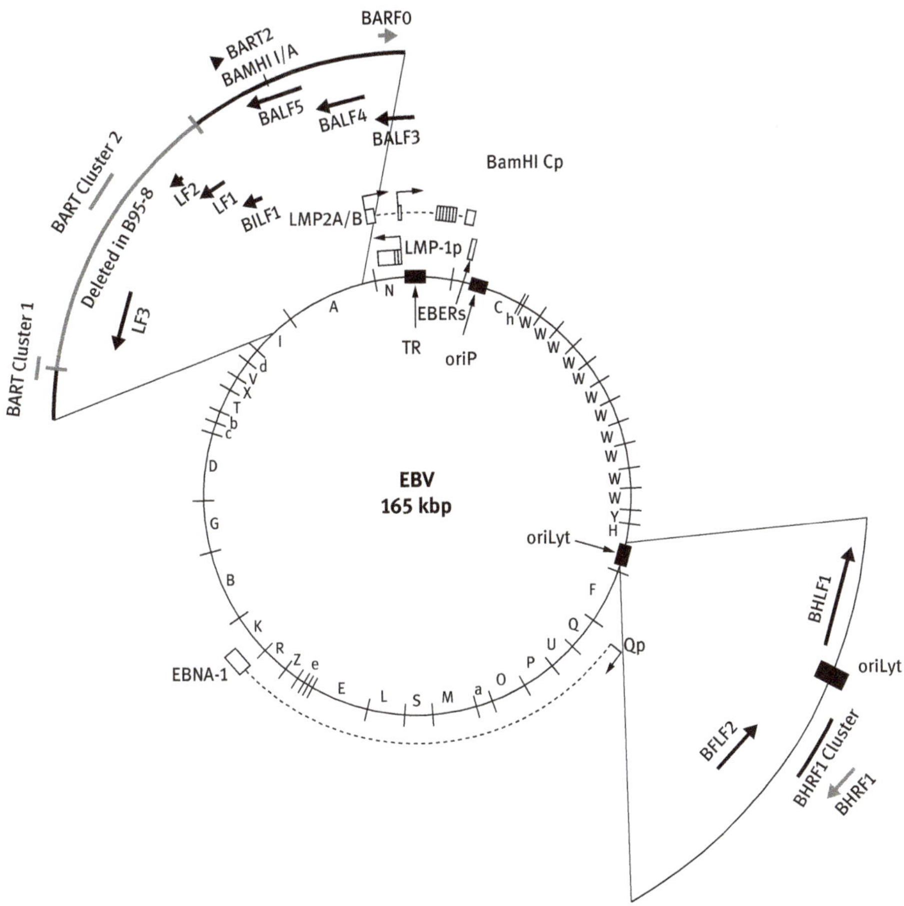

Fig. 17.3: Diagram of the EBV genome, showing two regions that are associated with microRNA transcripts.

this lymphoma should be investigated. In addition, EBV has been reported to produce a number of microRNAs from 2 primary transcripts [22, 23]. Only the BART locus associated miRNAs but not the BHRF1 cluster are expressed with type 1 or 2 latency (Fig. 17.3). There is evidence that at least some of these miRNAs promote transformation in B lymphocytes [24]. A recent report [25] suggests that miR-bart 20-5p & miR-BART8 may repress the IFN-STAT1 pathway with decreased expression of TP53 in ENKTCL. The expression profile and implications of these microRNAs in ENKTCL pathogenesis warrant further investigation [26].

17.5 Genetic and epigenetic alterations

Cytogenetic information on ENKTCL is limited but a number of studies on chromosomal copy number changes and loss of heterozygosity (LOH) have been reported [27–32]. There are a number of frequent gains and losses including gains of 1q21-q44 and frequent losses of 6q21 and 17p11-p13 (summarized in Tab. 17.1). The minimal common region of del 6q21 includes *PRDM1*, *ATG5*, *AIM1* and *HACE1*. Additional studies have found inactivating mutations affecting PRDM1 in cell lines. However, such mutations are infrequent in patient samples. Instead, frequent DNA methylation that reduces the transcription of this gene is detected (Fig. 17.4). *In vitro* functional studies also support the possibility that *PRDM1* is a tumor suppressor gene in ENKTCL [33, 34]. There is also evidence implicating *HACE1* [35], an E3 ubiquitin ligase reported to be a tumor suppressor gene in other systems [36]. These studies suggest that 6q21 may be a highly important region harboring one or more tumor suppressor genes. The 17p deletion includes the *TP53* gene which has also been found to be mutated in many ENKTCL [37]. It is interesting that EBNA1 has been found to decrease TP53 expression, probably through the interruption of the interaction of USP7 (HAUSP) with TP53. These observations, together with the possibility of decreased TP53 expression due to miR-BART20-5p and 8, indicate that the TP53 pathway is likely to be impaired in many ENKTCL lymphomas through multiple mechanisms [38–40].

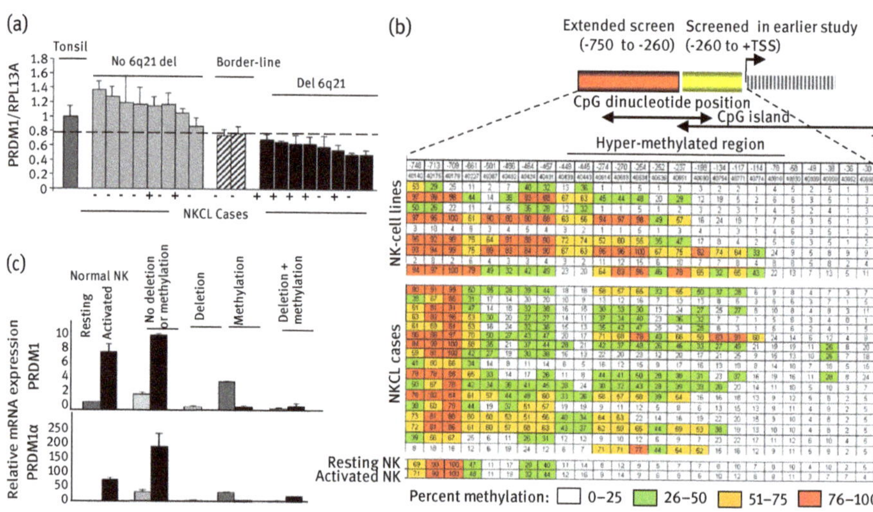

Fig. 17.4: **PRDM1 is frequently deleted as part of the del6q21 abnormality.** The PRDM1 promoter region is also frequently methylated, resulting in lower expression of the gene. Reprinted with permission from [34] (figure 5).

As *PRDM1* appears to be inactivated mainly by DNA methylation, there are a few important genes that have also been reported to be frequently methylated such as *TP73*

Tab. 17.1: Recurrent chromosomal gains and losses in ENKTCL. Modified from [31] (Table 1b).

Chromosomal location	Frequency (%)							
	Cell line	Samples	Ref. [90]	Ref. [92]	Ref. [89]	Ref. [91]	Ref. [95]	Ref. [94]
Recurring gains								
1q21.1	62.5	14.3	30	10				
1q23.1–q23.3	62.5	14.3	30	20				
1q31.3–q32.1	75	28.5	50	10				
1q32.2–q41	75	28.5	50					
1q41–q42.1	75	28.5	50					
1qter–q41	74	28.5	50					
2p16.3–p21	37.5	14.3	17					
2q13–q14	25	28.5	24		28.6			
2q31.1–q32.2	25	14.3	24		42.8			
6p25.2–p25.3	37.5	14.3	40					
7q11.2	37.5	14.3	24	20				
7q21.13–q21.3	50	14.3		20				
7q31.1–q31.2	50	14.3	24	20				
7q35	50	28.5		40				
16q22.2	25	14.3		20				
17q21.1	62.5	43		40	28.6			
20q11.1	62.5	28.5		50				
21q11.2–q22.11	25	14.3			28.6			
Recurring losses[a]								
1p36.32	37.5	14.3	58		57			
1p36.23–p36.31	37.5	14.3	30					
1p32.3–34.1	37.5	14.3		28	28.6			
1p21.3–p22.3	37.5	14.3		10				
3q21.3	–	14.3	30					
6q21	62.5	43	35	20		31	100	80
6q23.2	62.5	28.5	47	20				
6q25.3	62.5	14.3	29	30		67	58	
6q26	62.5	14.3	35	10		46		
7p22.1	25	43.5	40	10				
9p22.1–p22.3	37.5	43.5		10				
9cent–p23	37.5	14.3		10				
10p14–p15	37.5	14.3		10				
11q23.1	12.5	14.3	29	20		31		
11q24.3–q25	37.5	14.3		20				
13q14.11	25	28.5		25	14.3	60		
13q14.11–q14.2	12.5	14.3		30				
15q13.1	37.5	14.3	29					
17p13–p13.1	62.5	28.5	40	20	43	40		

Abbreviation: CGH, comparative genomic hybridization.
The "chromosomal location" shown reflects the minimal common region (MCR) compiled from all the studies and may not be identical to the MCR derived from our study.
[a] References [89, 90, 92] include CGH data and other references include LOH data on selected chromosomes e.g. [91] on chromosome 6, 11, 13 and 17 and [92, 93] on chromosome 6 only. Only references with at least five cases are included in the table.
Bold: recurrent abnormalities observed in at least two (gains) or three (losses) previous studies.

and *SHP1* [41, 42]. We have performed a genome-wide methylation analysis and found frequent global hypermethylation in ENKTCL. This study confirmed the frequent DNA methylation of several previously reported genes, and revealed additional potentially important methylated loci, including some that have been reported as methylated or inactivated by mutation in other tumors, or are known tumor suppressor genes. These include BIM, a crucial pro-apoptotic BH3 only protein; *SOCS6*, an inhibitor of the JAK/STAT pathway; *DDX3X*, and *DAPK1* [43]. One interesting finding is the frequent methylation of *Asparagine Synthetase (ASNS)*. Its inactivation has implications on the sensitivity of the tumor cells to Asparaginase treatment [44].

TP73 is a member of the TP53 family and may regulate a similar set of target genes as TP53, although it may also have functions other than that of a transactivator. Inactivating *TP73* by DNA methylation may cause similar loss of tumor suppressor function as loss of *TP53* particularly when TP53 function is compromised [45, 46]. *SHP1* is frequently inactivated in a variety of cancers. It is one of the major negative regulators of NK cell activation [47] through dephosphorylation of receptor/adaptor phosphotyrosine. In addition, SHP1 also regulates STAT3 activation in several systems although the exact mechanism has not been elucidated [48–52]; as such, SHP1 could play a similar role in ENKTCL where its loss may contribute with loss of SOSC6 and other genetic abnormalities to activate the JAK/Stat3 pathway (see below).

There have been reports of frequent *FAS* mutations in ENKTCL [53, 54]. In addition, survivin (BIRC5), an inhibitor of apoptosis, has been reported to be overexpressed in ENKTCL, probably through TP53 inactivation. These changes, together with *BIM* methylation and upregulation of BCLXL and BCL2 through STAT3 or STAT5 activation, represent some of the mechanisms for ENKTCL to resist apoptosis.

The STAT3 pathway is frequently activated in ENKTCL. This may be achieved largely through autocrine and paracrine activation of STAT3 via the cytokine receptor/JAK/STAT pathway [55]. Inactivation of negative regulators may play a role as mentioned above; however, STAT3 mutations affecting the SH2 domain have also been detected and rarely STAT5B SH2 mutation is observed (Fig. 17.5) [56]. These mutations may prolong and enhance the activation of STAT3, and NK cell lines with STAT3 mutants demonstrate higher levels of tyrosine phosphorylated protein; moreover, cell lines with transduced STAT3 mutants displayed growth advantage [44]. These STAT3 mutants may also promote tumor development through the production of tumor-promoting inflammation and an immunosuppressive tumor microenvironment as a result of the production of cytokines such as VEGF and IL10 [57–59]. As STAT3 activation is dependent on the cytokines/JAK/Stat3 axis, interruption of this pathway through a JAK-inhibitor is expected to impair tumor proliferation/survival, and this is indeed observed both for WT and mutant cell lines in vitro. There were two reports of frequent JAK3 mutations in ENKTCL [60, 61]. However, other studies failed to confirm these findings [44, 62]. A recent mutation analysis using whole exome sequencing and subsequently more focused sequencing have demonstrated the mutational profile of ENKTCL [63]. Interestingly, the most frequent mutations involved *DDX3X*, an RNA he-

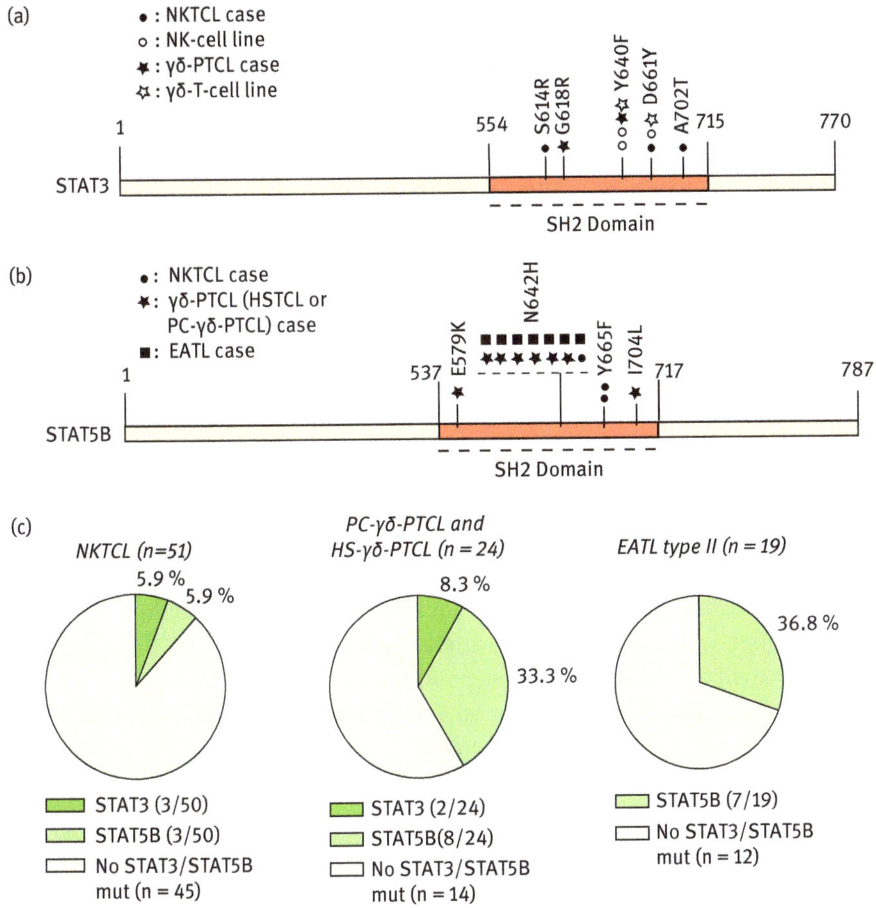

Fig. 17.5: STAT3 and STAT5B SH2 domain mutations. STAT3 mutations are more common in NK cell lymphoma, whereas STAT5B mutations are far more common in γδ T-cell lymphoma, and the most common mutation is the N642H mutation. Reprinted with permission [53] (figure 1).

licase that has also been reported to be mutated in other tumors. The mutants have reduced RNA helicase activity but the precise roles of DDX3X in ENKTCL need further exploration. This study also confirms the frequent *TP53* and *STAT3* mutations and mutations affecting epigenetic modifiers.

The frequent angiocentric and angiodestructive infiltrates of the tumor are expected to produce hypoxia and thereby activation of hypoxia induced factors (HIFs). Indeed, activation of the HIF1α pathway is observed in gene expression profiling (GEP) studies, and this pathway has been shown to also promote tumorigenesis and angiogenesis. The platelet derived growth factor receptor (PDGFR) signaling pathway has also been reported to be activated in ENKTCL. Immunohistochemistry confirmed the expression of phosphorylated PDGFRα. If this is a clinically important pathway, it is possible that imatinib could be an effective treatment. GEP studies also revealed activation of the NOTCH pathway, and tumor cell lines with NOTCH pathway activation were highly sensitive to γ-secretase inhibitors [32, 64].

It is interesting that aurora kinase A (AURKA) is frequently upregulated in ENKTCL, and NK cell lines are highly sensitive to AURKA inhibitors. Aside from being an important regulator of mitosis, AURKA also downregulates the TP53 pathway, with an increase in survivin expression [65] that is reversed by the inhibitor.

17.6 γδ T-cell lymphoma

17.6.1 Cell of origin

The T-cell antigen receptor is a heterodimer consisting of either an α and β chain or a γ and δ chain transcribed from specifically rearranged genes in the α, β, γ or δ loci [66]. The γδ T cells belong to a more primitive branch of T cells, closer in development and function to the innate immune system, and the γδ receptor has less diversity compared with the αβ receptor. These cells do not show major histocompatibility complex (MHC) restriction and can respond to a variety of antigens such as the CD1d-lipid antigen [67] and small phosphorylated antigens without MHC presentation [68, 69]. These cells can also use the NK cell receptors for recognition, and can be activated to produce cytokines such as IL17 without TCR ligation [68–70]. They can also mediate MHC unrestricted cytotoxic activities.

γδ T cells are a minor subset of T cells, present at low percentages in the lymph node, spleen and peripheral blood, but more frequently found in mucosal surfaces and the skin. Together with NK cells, γδ T cells are part of the innate immune system that is frequently the first line of defense against infectious agents [71]. γδ T cells are typically CD4 and CD8 negative and express many NK cell associated molecules, including the cytotoxic molecules TIA1, perforin, granzyme B and granzyme M. Depending on the δ gene used, these cells can be divided into two major subpopulations: cells expressing the δ2 chain include most circulating γδ T cells and are also the prevalent population in tonsils, lymph nodes and skin, while δ1 expressing cells predominate in intestinal mucosa as well as in the spleen. Further heterogeneity involves surface antigen expression and cytokine production, such that subsets of these cells may show certain functional characteristics resembling T-cell subsets in the adaptive immune system.

17.6.2 Subtypes of lymphomas derived from γδ T cells

These are uncommon lymphomas that mostly develop in extranodal sites [72–74] and can be broadly divided into two categories: hepatosplenic γδ T-cell lymphoma and mucocutaneous γδ T-cell lymphoma. Tumors arising in certain sites tend to demonstrate unique clinicopathologic features as well as biology, and the tumor is named after their predilected site of involvement (e.g., hepatosplenic γδ T-cell lymphoma). On the other hand, different types of lymphoma can overlap in the sites of presentation; for example, NK cell lymphoma can present in extranasal sites, overlapping with γδ T-cell lymphoma, while γδ T-cell lymphoma can present in the nasal region and form part of the spectrum of ENKTCL [75, 76]. Therefore, while presentation site alone can guide diagnosis, other features have to be taken into consideration.

17.6.3 Hepatosplenic γδ T-cell lymphoma

This is an aggressive lymphoma with male predominance and tends to affect younger adults with a median age of about thirty-five years. The predominant site of involvement is the liver and the spleen, as reflected in its name; however, the bone marrow (BM) is also very frequently involved, and occasional abnormal cells may be seen in the peripheral blood smear [72]. The tumor cells are predominantly in the sinuses of the liver and BM but, in the spleen, the cords are prominently involved as well. Lymphadenopathy is not obvious; however, lymph nodes, such as splenic hilar nodes, may show involvement in the sinuses. The tumor cells are small to medium sized cells with some degree of nuclear irregularity but no prominent nucleolus. They are typically CD3 positive, CD4 and CD8 negative, CD5 negative, CD57 negative, and CD56 positive. CD16 may also be expressed. The cytotoxic marker TIA1 is almost invariably present but granzyme B and perforin are generally absent.

Isochromosome 7q is a frequent and characteristic abnormality. Loss of the Y chromosome and gain of chromosome 8 are common, and increase in 7q signals may be observed in progressive disease. Rare cases of hepatosplenic T-cell lymphoma express the αβ rather than γδ receptor, and these cases are similar to the γδ counterpart in clinicopathologic features except in female predominance and wider age distribution; the characteristic isochromosome 7q abnormality is also detected in the αβ hepatosplenic T-cell lymphomas. Association with EBV seems to be more frequent compared with the typical γδ type [77]. Interestingly, gene expression profiling studies indicated a similar gene expression profile for both αβ and γδ HSTCL [78]. HSTCL shares the expression of many NK cell associated transcripts and FOS, VAV3, S1PR5 and SYK tend to be upregulated. For tumors with αβ T cells, it is unclear whether the tumor is derived from a small subset of αβ T cells that have differentiated as a component of the innate immune system or from a cell originally belonging to the adaptive immune system, which developed features of innate T cells during transformation due to genetic and/or microenvironmental influences.

Hepatosplenic γδ T-cell lymphomas quite often arise in the setting of chronic antigenic stimulation or immune suppression: increase in γδ T cells has been reported in the peripheral blood of some transplant patients, and development of γδ T-cell lymphoma has been observed in solid organ transplant patients [79] as well as in patients with Crohn's disease treated with thiopurine or thiopurine and anti-TNF monoclonal antibodies [80, 81].

The outcome of patients treated with cyclophosphamide, doxorubicin, vincristine and prednisone (CHOP) or CHOP-like regimens has been poor. It seems that chemotherapy regimens that contain high dose cytarabine and platinum are more effective, suggesting that more intensive chemotherapy followed by stem cell transplantation may be more successful [73].

17.6.4 Primary cutaneous γδ T-cell lymphoma

Morphologically, the tumor can resemble mycosis fungoides in having a dermal infiltration with epidermotropism. The majority of cases, however, show histological features similar to subcutaneous panniculitis-like T-cell lymphoma with a subcutaneous infiltration among and surrounding adipocytes. Some lesions may show involvement of all three layers of the skin (epithelial, dermal and subcutaneous), while different patterns may be present in different lesions from the same patient. The tumor cells are γδ T cells and, as expected, are negative for CD4 and CD8, although rare CD8αα positive cases have been described. There is variable expression of CD7 and CD56, but strong expression of cytotoxic molecules including TIA1, granzyme B, and perforin. These tumors tend to express Vδ2 [72, 73]. This entity is typically EBV negative although rare EBV positive cases have been reported [82].

Clinically, patients most frequently present with involvement of the lower extremities followed by the torso and arms, with multiple deep plaques or nodules, some of which may show ulceration. Lymphadenopathy and bone marrow involvement are uncommon. The prognosis of this entity is worse than the TCR αβ positive panniculitis-like subcutaneous T-cell lymphoma, which is its mainly differential diagnosis and the major distinguishing features are listed in Tab. 17.2. Hemophagocytic syndrome is a rapidly fatal condition that occurs at a high frequency in the γδ lymphoma, although it has also been reported in more indolent αβ cases [83]. Treatment with CHOP or CHOP-like regimens has not been very successful. Intensive chemotherapies with stem cell transplantation may produce durable remissions and may be the preferred approach for patients with this lymphoma.

17.6.5 Mucosal γδ T-cell lymphoma

γδ T-cell lymphoma can affect any mucosal sites of the body. This may be related to the relatively high frequency of γδ T cells in the epithelium and mucosa. The most frequent sites of involvement are in the nasal, paranasal and nasopharyngeal region as well as the small intestine. The prototypical lesion of the ENKTCL, nasal type af-

Tab. 17.2: Features of subcutaneous panniculitis-like T-cell lymphoma (SPTCL) vs. primary cutaneous γδ T-cell lymphoma (PCGDTCL).

SPTCL	PCGDTCL
Mostly confined to the subcutaneous tissue Usually lower extremities	Typically also involving the dermis and epidermis Usually presenting with multiple lesions in the extremities and buttock
Expresses αβ TCR CD8+; CD56-	Expresses γδ TCR Double negative or CD8+, CD5-, CD56+
Clinically indolent (5 yr OS ~ 80 %). Recurrences tend to remain in the subcutaneous tissue Hemophagocytic syndrome uncommon	Clinically aggressive Frequent dissemination to other extranodal sites Hemophagocytic syndrome common

fecting the nasal and paranasal region has been described above. Most of these cases arise from NK cells, but a minority of cases are of γδ T-cell origin, and rarely the tumor may express αβ T-cell receptor. The NK cell lesion will not express TCR- γδ or αβ receptor and the TCRγ locus is in a germline configuration. It does not express a complete CD3 γ,δ and ε complex but expresses the epsilon chain of the CD3 and therefore show cytoplasmic staining with anti CD3-epsilon antibody by immunohistochemistry. The prognosis of this entity is relatively good if the disease is confined to the nasal region, as the tumor is highly responsive to radiation therapy with or without chemotherapy. For disseminated lesions or lesions arising in extranasal region or relapsed disease, prognosis is much poorer. In extranasal mucosal sites, γδ T-cell derived lymphomas are much more frequent than NK cell derived tumors. The lineage can generally be assigned with a combination of immunohistochemistry using antibodies against TCR-γδ and αβ, as well as molecular analysis of TCRγ-locus rearrangement.

17.6.6 γδ T-cell lymphoma in the intestine

A variety of PTCL may involve the gastrointestinal tract, but a primary intestinal lymphoma believed to be derived from intraepithelial T lymphocytes and often associated with some degree of enteropathic changes has been defined as enteropathy associated T-cell lymphoma (EATL). This entity was initially defined in European populations and the cases are associated with celiac disease or gluten sensitive enteropathy. Other primary intestinal lymphoma, seen mainly in Asia, while having overlapping clinical and pathologic features to the originally defined EATL, also show substantial differences leading to their segregation into a separate subtype EATL-Type II. Type II cases have a dense mucosal infiltrate that tends to be monomorphic and shows epitheliotropism. Geographically, type II EATL is much more common in Asia and, according to one report, cases are derived mostly from γδ T cells (78 %) although some cases may be both γδ and αβ TCR positive and a few cases show no demonstrable TCR expression [83]. However, another study reported mostly αβ T cells (ratio αβ: γδ = 1.6:1) for type II EATL [84]. The usual phenotype is CD3+, CD5-, CD4-, CD8+, CD56+, EBER-, TIA+ and most cases are also granzyme B+. CD20 is quite often expressed aberrantly. Most cases express the CD8 homodimer and have been reported to express nuclear megakaryocytic associated tyrosine kinase (MATK) [84]. It has been proposed that type II EATL should be separate from EATL based on the different geographic distribution, absence of evidence for gluten-sensitive enteropathy, frequent differences in morphology and immunophenotype and high γδ vs. αβ usage. However, there is some overlap in the genetic abnormalities with frequent gains of 9q31-qTER and loss of 16q12. Genetic differences are also observed, such as the frequent gain of 8q24 involving the *MYC* oncogene in type II EATL. It is generally considered that EATL, primary cutaneous γδ T-cell lymphoma and hepatosplenic T-cell lymphoma are EBV negative, and EBV status is useful for distinguishing these cases from ENKTCL, nasal type. However, rare EBV positive cases have been reported in each of these entities, and the γδ T-cell cases within ENKTCL, nasal type are frequently EBV positive. Thus, EBV positivity cannot be used as a defining criterion for the diagnosis of ENKTCL involving sites outside of the nasal region, although EBV positivity should raise concern in the diagnosis of hepatosplenic T-cell lymphoma, cutaneous γδ T-cell lymphoma and EATL [74].

Clinically, EATL type II cases show a prominent male predominance and frequently present with weight loss, abdominal pain and diarrhea, and bowel perforation is common. Overall survival is poor. Patients have not been uniformly treated, but long term survival has been observed in a few patients with high dose therapy and autologous stem cell transplant, suggesting that this approach may be effective in patients who are able to tolerate such aggressive treatment [84–86].

17.7 Gene expression profiling analysis

The strong similarity between the gene expression profiles of γδ T-cell lymphoma and NK cell lymphoma is clearly demonstrated in two recent gene expression profiling studies [2, 87]. Many of the genes expressed by resting and activated NK cells are also expressed in both types of lymphoma (Fig. 17.6). Using a group of genes that are associated with the γδ T-cell receptor, the NK and γδ cases can be separated from one another. As mentioned above for hepatosplenic T-cell lymphoma, it was found that the typical γδ T-cell cases and the rare αβ T-cell cases have very similar gene expression profiles [78]. This indicates that the type of T-cell receptor expression does not reliably predict the biology of the lymphoid cells, and some αβ expressing T-cell lymphomas have a fundamental similarity in biology to that of γδ T cells and NK cells. It is also clear that CD4 or CD8 expression may not predict helper vs cytotoxic T cell and therefore again, it is important to examine the biology by more comprehensive global

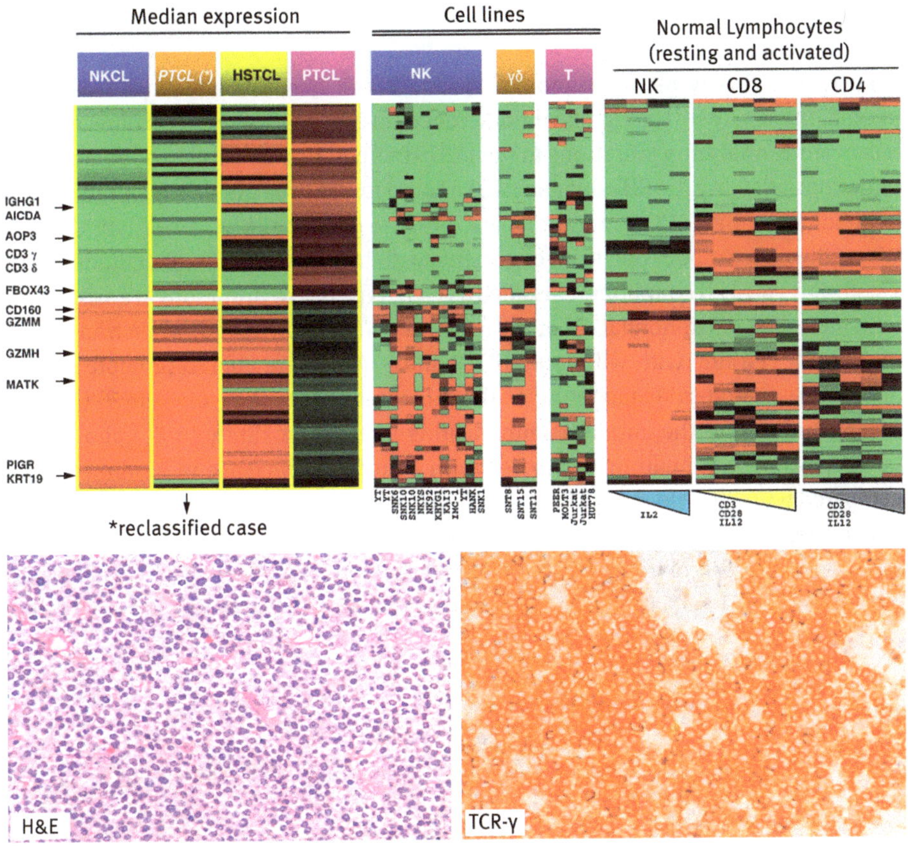

Fig. 17.6: The gene expression profiles of NK cell lymphoma and γδ T-cell lymphomas are very similar and also have substantial similarity to that of normal resting and activated NK cells. Inset: a case of γδ T-cell lymphoma stained with antibody against the T-cell receptor γ chain, showing strong positivity. Reprinted with permission from [62] (figure 2).

techniques. Moreover, there are cases that express both γδ and αβ receptors as well as cases where the T-cell antigen receptor expression cannot be determined at least by immunohistochemistry [74]. In these cases, perhaps the gene expression profile can indicate whether the cells resemble the T/NK cells of the innate immune system or are more similar to T cells of the adaptive immune system. The relevant oncogenic pathways that are activated can also be determined by GEP. This has implications not only in diagnosis but also in treatment approaches.

17.8 JAK/STAT pathway activation

Even within closely related tumors such as NK cell lymphoma and γδ T-cell lymphoma, there may still be significant differences in the pathogenetic mechanisms. In a recent study analyzing mutations in the JAK/STAT pathway in these lymphoma types, mutations affecting the SH2 domain of *STAT3* are found to be more frequent in NK cell lymphoma, while *STAT5B* mutations are far more common in γδ T-cell lymphomas, where they affect about one third of patients with hepatosplenic T-cell, cutaneous γδ T-cell, and γδ T-cell lymphoma classified as EATL type II. The mutations are associated with increased phosphorylation of the respective STAT protein and, when transduced into an NK cell line with wild-type STAT3 and STAT5, were able to promote growth and survival. Molecular modeling of the most frequent mutant of *STAT5B* (N642H) demonstrated that the mutated amino acid is in close proximity to the phosphorylated tyrosine of the pairing monomer, with a predicted increase in binding affinity compared with the wild-type. This prediction was confirmed by surface plasmon resonance spectroscopy [56]. Our current model (Fig. 17.7) suggests that the mutant confers marked increase in binding affinity between the phosphorylated dimer. This would lead to increased persistence of the dimer and prolonged/enhanced transactivation

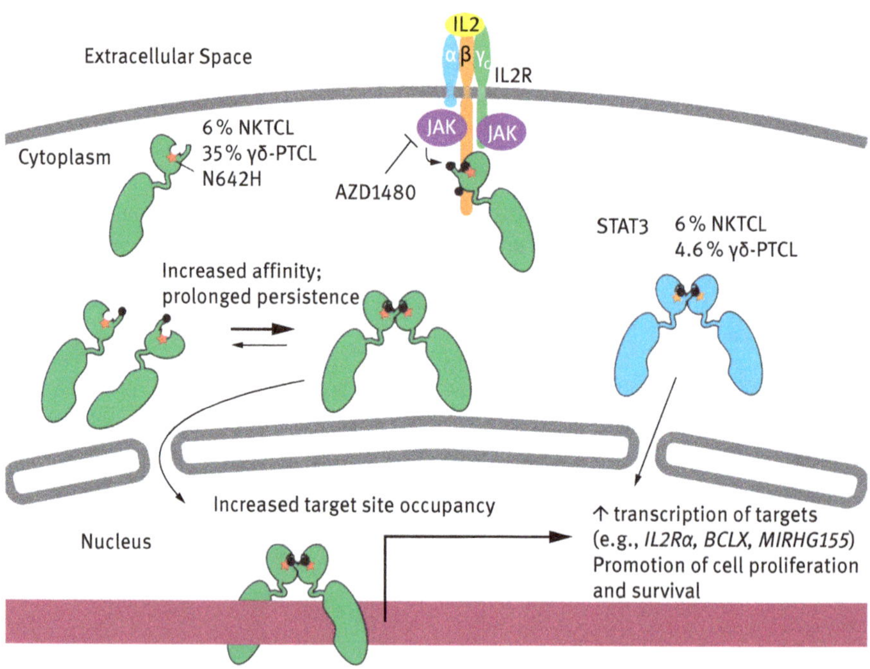

Fig. 17.7: Diagram illustrating the mechanism of increased STAT5B and STAT3 activation and enhanced transcription of target genes as a result of oncogenic mutations in the SH2 domain (Courtesy: Dr. Tim McKeithan).

of target genes that promote neoplastic transformation. Interruption of cytokine signaling using a JAK1/2 inhibitor was able to reduce the extent of STAT activation and its downstream consequences, even in the presence of the STAT mutants. This finding may have immediate therapeutic implications in the treatment of patients with γδ T-cell lymphoma, as JAK 1/2 inhibitors are already employed in the management of myeloproliferative neoplasms.

17.9 Concluding remarks

ENKTCL, nasal type, and the various defined subtypes of γδ T-cell lymphoma are rare tumors with very poor prognosis except for localized disease in the nasal region. Further understanding of the pathogenesis of these lymphomas and novel approaches for extranasal or disseminated disease is sorely needed. Recent studies have started to unravel its pathogenesis and identify possible pathways that may be amenable to targeted therapy. These tumors are rare and large multi-institutional, international collaborative studies will greatly facilitate the progress of research efforts and the management of this group of diseases. With such concerted efforts, it is anticipated that, in the future, additional important molecular lesions and pathways will be discovered, and effective targeted therapies will be found to improve the outcome of this group of patients.

References

[1] Chan JKC, Quintanilla-Martinez L, Ferry JA, Peh SC. Extranodal NK/T-cell lymphoma, nasal type p285-88 in WHO classification of tumours of haematopoietic and lymphoid tissues. SH Swerdlow, E Campo, NL Harris et al. Lyon, France, International Agency for Research on Cancer 2008:285–288.

[2] Iqbal J, Weisenburger DD, Chowdhury A, et al. Natural killer cell lymphoma shares strikingly similar molecular features with a group of non-hepatosplenic gammadelta T-cell lymphoma and is highly sensitive to a novel aurora kinase A inhibitor in vitro. Leukemia 2011;25(2):348–358.

[3] Pongpruttipan T, Sukpanichnant S, Assanasen T, et al. Extranodal NK/T-cell lymphoma, nasal type, includes cases of natural killer cell and alphabeta, gammadelta, and alphabeta/gammadelta T-cell origin:a comprehensive clinicopathologic and phenotypic study. Am J Surg Pathol 2012;36(4):481–499.

[4] Boulland ML, Kanavaros P, Wechsler J, et al. Cytotoxic protein expression in natural killer cell lymphomas and in alpha beta and gamma delta peripheral T-cell lymphomas. J Pathol 1997;183(4):432–439.

[5] Krenacs L, Smyth MJ, Bagdi E, et al. The serine protease granzyme M is preferentially expressed in NK-cell, gamma delta T-cell, and intestinal T-cell lymphomas:evidence of origin from lymphocytes involved in innate immunity. Blood 2003;101(9):3590–3593.

[6] Kis LL, Takahara M, Nagy N, et al. Cytokine mediated induction of the major Epstein–Barr virus (EBV)-encoded transforming protein, LMP-1. Immunol Lett 2006;104(1-2):83–88.

[7] Cheung MM, Chan JK, Lau WH, et al. Primary non-Hodgkin's lymphoma of the nose and nasopharynx:clinical features, tumor immunophenotype, and treatment outcome in 113 patients. J Clin Oncol 1998;16(1):70–77.
[8] Chan JKC, Jaffe ES, Ralfkiaer E, Ko KH. Aggressive NK-cell leukaemia. WHO classification of tumours of haematopoietic and lymphoid tissues. S. H. Swerdlow, E. Campo, N. L. Harris et al. Lyon, France, International Agency for Research on Cancer 2008:276–277.
[9] Tse E, Kwong YL. How I treat NK/T-cell lymphomas. Blood 2013;121(25):4997–5005.
[10] Yamaguchi M, Kwong YL, Kim WS, et al. Phase II study of SMILE chemotherapy for newly diagnosed stage IV, relapsed, or refractory extranodal natural killer (NK)/T-cell lymphoma, nasal type:the NK-Cell Tumor Study Group study. J Clin Oncol 2011;29(33):4410–4416.
[11] Morice WG, Kurtin PJ, Leibson PJ, et al. Demonstration of aberrant T-cell and natural killer-cell antigen expression in all cases of granular lymphocytic leukaemia. Br J Haematol 2003;120(6):1026–1036.
[12] Pascal V, Schleinitz N, Brunet C, et al. Comparative analysis of NK cell subset distribution in normal and lymphoproliferative disease of granular lymphocyte conditions. Eur J Immunol 2004;34(10):2930–2940.
[13] Minarovits J, Hu LF, Imai S, et al. Clonality, expression and methylation patterns of the Epstein–Barr virus genomes in lethal midline granulomas classified as peripheral angiocentric T cell lymphomas. J Gen Virol 1994;75 (Pt 1):77–84.
[14] Soni V, Cahir-McFarland E, Kieff E. LMP1 TRAFficking activates growth and survival pathways. Adv Exp Med Biol 2007;597:173–187.
[15] Ersing I, Bernhardt K, Gewurz BE. NF-kappaB and IRF7 pathway activation by Epstein–Barr virus Latent Membrane Protein 1. Viruses 2013;5(6):1587–1606.
[16] Frappier L. Contributions of Epstein–Barr nuclear antigen 1 (EBNA1) to cell immortalization and survival. Viruses 2012;4(9):1537–1547.
[17] Rosato P, Anastasiadou E, Garg N, Lenze D, Boccellato F, Vincenti S, Severa M, Coccia EM, Bigi R, Cirone M, Ferretti E, Campese AF, Hummel M, Frati L, Presutti C, Faggioni A, Trivedi P. *Differential regulation of miR-21 and miR-146a by Epstein-Barr virus-encoded EBNA2.* Leukemia. 2012;26(11):2343–52.
[18] Kohlhof H, Hampel F, Hoffmann R, Burtscher H, Weidle UH, Hölzel M, Eick D, Zimber-Strobl U, Strobl LJ. *Notch1, Notch2, and Epstein-Barr virus-encoded nuclear antigen 2 signaling differentially affects proliferation and survival of Epstein-Barr virus-infected B cells.* Blood. 2009;113(22):5506–15.
[19] Portal D, Zhao B, Calderwood MA, Sommermann T, Johannsen E, Kieff E. *EBV nuclear antigen EBNALP dismisses transcription repressors NCoR and RBPJ from enhancers and EBNA2 increases NCoR-deficient RBPJ DNA binding.* Proc Natl Acad Sci USA. 2011 ;108(19):7808–13.
[20] Samanta M, Takada K. Modulation of innate immunity system by Epstein–Barr virus-encoded non-coding RNA and oncogenesis. Cancer Sci 2010;101(1):29–35.
[21] Banerjee AS, Pal AD, Banerjee S. Epstein–Barr virus-encoded small non-coding RNAs induce cancer cell chemoresistance and migration. Virology 2013;443(2):294–305.
[22] Pratt ZL, Kuzembayeva M, Sengupta S, Sugden B. The microRNAs of Epstein–Barr Virus are expressed at dramatically differing levels among cell lines. Virology 2009;386(2):387–397.
[23] Klinke O, Feederle R, Delecluse HJ. Genetics of Epstein–Barr virus microRNAs. Semin Cancer Biol (2014;26:52–59.
[24] Vereide DT, Seto E, Chiu YF, et al. Epstein–Barr virus maintains lymphomas via its miRNAs. Oncogene 2014;33(10):1258–1264.
[25] Huang WT, Lin CW. EBV-encoded miR-BART20-5p and miR-BART8 inhibit the IFN-gamma-STAT1 pathway associated with disease progression in nasal NK-cell lymphoma. Am J Pathol 2014;184(4):1185–1197.

[26] Motsch N, Alles J, Imig J, et al. MicroRNA profiling of Epstein–Barr virus-associated NK/T-cell lymphomas by deep sequencing. PLoS One 2012;7(8):e42193.
[27] Siu LL, Wong KF, Chan JK, Kwong YL. Comparative genomic hybridization analysis of natural killer cell lymphoma/leukemia. Recognition of consistent patterns of genetic alterations. Am J Pathol 1999;155(5):1419–1425.
[28] Siu, LL, Chan V, Chan JK, et al. Consistent patterns of allelic loss in natural killer cell lymphoma. Am J Pathol 2000;157(6):1803–1809.
[29] Ko YH, Choi KE, Han JH, et al. Comparative genomic hybridization study of nasal-type NK/T-cell lymphoma. Cytometry 2001;46(2):85–91.
[30] Nakashima Y, Tagawa H, Suzuki R, et al. Genome-wide array-based comparative genomic hybridization of natural killer cell lymphoma/leukemia:different genomic alteration patterns of aggressive NK-cell leukemia and extranodal Nk/T-cell lymphoma, nasal type. Genes Chromosomes Cancer 2005;44(3):247–255.
[31] Iqbal J, Kucuk C, Deleeuw RJ, et al. Genomic analyses reveal global functional alterations that promote tumor growth and novel tumor suppressor genes in natural killer-cell malignancies. Leukemia 2009;23(6):1139–1151.
[32] Huang Y, de Reynies A, de Leval L, et al. Gene expression profiling identifies emerging oncogenic pathways operating in extranodal NK/T-cell lymphoma, nasal type. Blood 2010;115(6):1226–1237.
[33] Karube K, Nakagawa M, Tsuzuki S, et al. Identification of FOXO3 and PRDM1 as tumor-suppressor gene candidates in NK-cell neoplasms by genomic and functional analyses. Blood 2011;118(12):3195–3204.
[34] Kucuk C, Iqbal J, Hu X, et al. PRDM1 is a tumor suppressor gene in natural killer cell malignancies. Proc Natl Acad Sci USA 2011;108(50):20119–20124.
[35] Kucuk C, Hu X, Iqbal J, et al. HACE1 is a tumor suppressor gene candidate in natural killer cell neoplasms. Am J Pathol 2013;182(1):49–55.
[36] Zhang L, Anglesio MS, O'Sullivan M, et al. The E3 ligase HACE1 is a critical chromosome 6q21 tumor suppressor involved in multiple cancers. Nat Med 2007;13(9):1060–1069.
[37] Quintanilla-Martinez L, Kremer M, Keller G, et al. p53 Mutations in nasal natural killer/T-cell lymphoma from Mexico:association with large cell morphology and advanced disease. Am J Pathol 2001;159(6):2095–2105.
[38] Li M, Chen D, Shiloh A, et al. Deubiquitination of p53 by HAUSP is an important pathway for p53 stabilization. Nature (2002;416(6881):648–653.
[39] Saridakis V, Sheng Y, Sarkari F, et al. Structure of the p53 binding domain of HAUSP/USP7 bound to Epstein–Barr nuclear antigen 1 implications for EBV-mediated immortalization. Mol Cell 2005;18(1):25–36.
[40] Frappier L. Contributions of Epstein–Barr nuclear antigen 1 (EBNA1) to cell immortalization and survival. Viruses 2012;4(9):1537–1547.
[41] Oka T, Ouchida M, Koyama M, et al. Gene silencing of the tyrosine phosphatase SHP1 gene by aberrant methylation in leukemias/lymphomas. Cancer Res 2002;62(22):6390–6394.
[42] Siu LL, Chan JK, Wong KF, Kwong YL. Specific patterns of gene methylation in natural killer cell lymphomas: p73 is consistently involved. Am J Pathol 2002;160(1):59–66.
[43] Hu X, Kucuk C, Jang B, et al. Global methylation analysis reveals novel candidate tumor suppressor genes in natural killer cell lymphomas. 2015 (in press).
[44] Kucuk C, Hu X, Jiang B, et al. Global promoter methylation analysis reveals novel candidate tumor suppressor genes in natural killer cell lymphoma. Clin Cancer Res 2015;21(7):1699–711.
[45] Deyoung MP, Ellisen LW. p63 and p73 in human cancer:defining the network. Oncogene 2007;26(36):5169–5183.

[46] Candi E, Agostini M, Melino G, Bernassola F. How the TP53 family proteins TP63 and TP73 contribute to tumorigenesis:regulators and effectors. Hum Mutat 2014;35(6):702–714.
[47] Lanier LL. Up on the tightrope:natural killer cell activation and inhibition. Nat Immunol 2008;9(5):495–502.
[48] Chim CS, Fung TK, Cheung WC, et al. SOCS1 and SHP1 hypermethylation in multiple myeloma:implications for epigenetic activation of the Jak/STAT pathway. Blood 2004;103(12):4630–4635.
[49] Han Y, Amin HM, Frantz C, et al. Restoration of shp1 expression by 5-AZA-2'-deoxycytidine is associated with downregulation of JAK3/STAT3 signaling in ALK-positive anaplastic large cell lymphoma. Leukemia 2006;20(9):1602–1609.
[50] Kim DJ, Tremblay ML Digiovanni J. Protein tyrosine phosphatases, TC-PTP, SHP1, and SHP2, cooperate in rapid dephosphorylation of Stat3 in keratinocytes following UVB irradiation. PLoS One 2010;5(4):e10290.
[51] Chen KF, Su JC, Liu CY, et al. A novel obatoclax derivative, SC-2001, induces apoptosis in hepatocellular carcinoma cells through SHP-1-dependent STAT3 inactivation. Cancer Lett 2012;321(1):27–35.
[52] Gupta SC, Phromnoi K, Aggarwal BB. Morin inhibits STAT3 tyrosine 705 phosphorylation in tumor cells through activation of protein tyrosine phosphatase SHP1. Biochem Pharmacol 2013;85(7):898–912.
[53] Shen L, Liang AC, Lu L, et al. Frequent deletion of Fas gene sequences encoding death and transmembrane domains in nasal natural killer/T-cell lymphoma. Am J Pathol 2002;161(6):2123–2131.
[54] Takakuwa T, Dong Z, Nakatsuka S, et al. Frequent mutations of Fas gene in nasal NK/T cell lymphoma. Oncogene (2002;21(30):4702–4705.
[55] Coppo P, Gouilleux-Gruart V, Huang Y, et al. STAT3 transcription factor is constitutively activated and is oncogenic in nasal-type NK/T-cell lymphoma. Leukemia 2009;23(9):1667–1678.
[56] Kucuk C, Bei J, Xioazhou H, et al. Activating mutations of STAT5B and STAT3 in lymphomas derived from γδ-T or NK cells. nature communications 2015 (in press).
[57] Wang T, Niu G, Kortylewski M, et al. Regulation of the innate and adaptive immune responses by Stat-3 signaling in tumor cells. Nat Med 2004;10(1):48–54.
[58] Yu H, Kortylewski M,Pardoll D. Crosstalk between cancer and immune cells:role of STAT3 in the tumour microenvironment. Nat Rev Immunol 2007;7(1):41–51.
[59] Yu H, Pardoll D, Jove R. STATs in cancer inflammation and immunity:a leading role for STAT3. Nat Rev Cancer 2009;9(11):798–809.
[60] Koo GC, Tan SY, Tang T, et al. Janus kinase 3-activating mutations identified in natural killer/T-cell lymphoma. Cancer Discov 2012;2(7):591–597.
[61] Bouchekioua A, Scourzic L, de Wever O, et al. JAK3 deregulation by activating mutations confers invasive growth advantage in extranodal nasal-type natural killer cell lymphoma. Leukemia 2014;28(2):338–348.
[62] Kimura H, Karube K, Ito Y, et al. Rare occurrence of JAK3 mutations in natural killer cell neoplasms in Japan. Leuk Lymphoma 2014;55(4):962–963.
[63] Jiang L, Gu ZH, Yan ZX, et al. Exome sequencing identifies somatic mutations of DDX3X in natural killer/T-cell lymphoma. Nat Genet 2015;47(9):1061–1066.
[64] Iqbal J,Kucuk C, Deleeuw RJ, et al. Genomic analyses reveal global functional alterations that promote tumor growth and novel tumor suppressor genes in natural killer-cell malignancies. Leukemia 2009;23(6):1139–1151.
[65] Iqbal J, Weisenburger DD, Chowdhury A, et al. Natural killer cell lymphoma shares strikingly similar molecular features with a group of non-hepatosplenic gammadelta T-cell lymphoma

and is highly sensitive to a novel aurora kinase A inhibitor in vitro. Leukemia 2011;25(2):348–358.

[66] van Dongen JJ, Wolvers-Tettero IL. Analysis of immunoglobulin and T cell receptor genes. Part I:Basic and technical aspects. Clin Chim Acta 1991;198(1-2):1–91.

[67] Uldrich AP, Le Nours J, Pellicci DG, et al. CD1d-lipid antigen recognition by the gammadelta TCR. Nat Immunol 2013;14(11):1137–1145.

[68] Hayday AC. [gamma][delta] cells:a right time and a right place for a conserved third way of protection. Annu Rev Immunol 2000;18:975–1026.

[69] Carding SR, Egan PJ. Gammadelta T cells:functional plasticity and heterogeneity. Nat Rev Immunol 2002;2(5):336–345.

[70] Kapsenberg ML. Gammadelta T cell receptors without a job. Immunity 2014;31(2):181–183.

[71] Bonneville M, O'Brien RL, Born WK. Gammadelta T cell effector functions:a blend of innate programming and acquired plasticity. Nat Rev Immunol 2010;10(7):467–478.

[72] Gaulard P, Belhadj K, Reyes F. Gammadelta T-cell lymphomas. Semin Hematol 2003;40(3):233–243.

[73] Tripodo C, Iannitto E, Florena AM, et al. Gamma-delta T-cell lymphomas. Nat Rev Clin Oncol 2009;6(12):707–717.

[74] Swerdlow SH, Jaffe ES, Brousset P, et al. International Lymphoma Study. Cytotoxic T-cell and NK-cell Lymphomas:Current Questions and Controversies. Am J Surg Pathol 2014;38(10):e60–71.

[75] Pang DJ, Neves JF, Sumaria N, Pennington DJ. Understanding the complexity of gammadelta T-cell subsets in mouse and human. Immunology 2012;136(3):283–290.

[76] Caccamo N, Todaro M, Sireci G, et al. Mechanisms underlying lineage commitment and plasticity of human gammadelta T cells. Cell Mol Immunol 2013;10(1):30–34.

[77] Macon WR, Levy NB, Kurtin PJ et al. Hepatosplenic alphabeta T-cell lymphomas:a report of 14 cases and comparison with hepatosplenic gammadelta T-cell lymphomas. Am J Surg Pathol 2001;25(3):285–296.

[78] Travert M, Huang Y, de Leval L, et al. Molecular features of hepatosplenic T-cell lymphoma unravels potential novel therapeutic targets. Blood 2012;119(24):5795–5806.

[79] Khan WA, Yu L, Eisenbrey AB, et al. Hepatosplenic gamma/delta T-cell lymphoma in immunocompromised patients. Report of two cases and review of literature. Am J Clin Pathol 2001;116(1):41-50.

[80] Mackey AC, Green L, Liang LC, et al. Hepatosplenic T cell lymphoma associated with infliximab use in young patients treated for inflammatory bowel disease. J Pediatr Gastroenterol Nutr 2001;44(2):265–267.

[81] Kotlyar DS, Osterman MT, Diamond RH, et al. A systematic review of factors that contribute to hepatosplenic T-cell lymphoma in patients with inflammatory bowel disease. Clin Gastroenterol Hepatol 2011;9(1):36–41 e31.

[82] Guitart J, Weisenburger DD, Subtil A, et al. Cutaneous gammadelta T-cell lymphomas:a spectrum of presentations with overlap with other cytotoxic lymphomas. Am J Surg Pathol 2012;36(11):1656–1665.

[83] Magro CM, Wang X. Indolent primary cutaneous gamma/delta T-cell lymphoma localized to the subcutaneous panniculus and its association with atypical lymphocytic lobular panniculitis. Am J Clin Pathol 2012;138(1):50–56.

[84] Chan JK, Chan AC, Cheuk W, et al. Type II enteropathy-associated T-cell lymphoma:a distinct aggressive lymphoma with frequent gammadelta T-cell receptor expression. Am J Surg Pathol 2011;35(10):1557–1569.

[85] Tan SY, Chuang SS, Tang T, et al. Type II EATL (epitheliotropic intestinal T-cell lymphoma):a neoplasm of intra-epithelial T-cells with predominant CD8alphaalpha phenotype. Leukemia 2013;27(8):1688–1696.
[86] Iqbal J, Weisenburger DD, Greiner TC, et al. Molecular signatures to improve diagnosis in peripheral T-cell lymphoma and prognostication in angioimmunoblastic T-cell lymphoma. Blood 2010;115(5):1026–1036.
[87] Iqbal J, G. Wright, C. Wang, et al. Lymphoma Leukemia Molecular Profiling Gene expression signatures delineate biological and prognostic subgroups in peripheral T-cell lymphoma. Blood 2014;123(19):2915–2923.
[88] Kelley et al. Comparative Genomics of Natural Killer Cell Receptor Gene Clusters. PLoS Genetics. 2005;1(2):e27.
[89] Ko YH, Choi KE, Han JH, Kim JM, Ree HJ. Comparative genomic hybridization study of nasal-type NK/T-cell lymphoma. Cytometry 2001;46:85–91.
[90] Nakashima Y, Tagawa H, Suzuki R, Karnan S, Karube K, Ohshima K et al. Genome-wide array-based comparative genomic hybridization of natural killer cell lymphoma/leukemia: different genomic alteration patterns of aggressive NK-cell leukemia and extranodal Nk/T-cell lymphoma, nasal type. Genes Chromosomes Cancer 2005;44:247–255.
[91] Siu LL, Chan V, Chan JK, Wong KF, Liang R, Kwong YL. Consistent patterns of allelic loss in natural killer cell lymphoma. Am J Pathol 2000;157:1803–1809.
[92] Siu LL, Wong KF, Chan JK, Kwong YL. Comparative genomic hybridization analysis of natural killer cell lymphoma/leukemia. Recognition of consistent patterns of genetic alterations. Am J Pathol 1999;155:1419–1425.
[93] Wong KF. Genetic changes in natural killer cell neoplasms. Leuk Res 2002;26:977–978.
[94] Yoon J, Ko YH. Deletion mapping of the long arm of chromosome 6 in peripheral T and NK cell lymphomas. Leuk Lymphoma 2003;44:2077–2082.
[95] Simon R, Peng A. BRB-ArrayTools User Guide, version 3.6.0. Biometric Research Branch, National Cancer Institute.

Ethel Cesarman

18 Lymphomas associated with viral infection

18.1 Introduction

The ability of transmissible agents to cause cancer in a number of animal species has been recognized for more than 100 years, and even in humans Domenico Rigoni-Sterns observed in 1842 that nuns developed cervical cancer less frequently than married women, although the identification of the causal *Human Papilloma Virus* came many years later (reviewed in [1]). Regarding hematological malignancies, in 1907 the first observation that a viral infection could lead to the development of leukemia was made by Ellerman and Bang [2]. Since this time, many associations between viruses and leukemia/ lymphoma have been proposed, but only a few were firmly demonstrated to occur in humans. To conclude that a viral agent is causally associated with a human cancer, several Koch-like postulates must be fulfilled, which have been updated to account for a number of molecular techniques [3, 4]. These include: (i) epidemiologic evidence; (ii) consistent presence or persistence of the viral nucleic acids in the tumor cells; (iii) oncogenic effects of viral genome or genes *in vitro*; and (iv) demonstration that the malignant features of tumor cells depend on functions conferred by persistence of the viral genomes. Three human viruses meet these rigorous criteria: *Epstein–Barr virus* (EBV, also called human herpesvirus 4, HHV-4), *Kaposi's sarcoma herpesvirus* (KSHV/human herpesvirus 8, HHV-8), and *human T-cell lymphotropic virus 1* (HTLV-1).

While these viruses have documented oncogenicity, their prevalence in the normal population is quite high (for example > 90 % for EBV). A small proportion of infected individuals develop a virus-associated malignancy, and usually a long time after primary infection. However, the likelihood of this occurring is greatly increased if the individual has an underlying immunodeficiency, at least for EBV and KSHV. This indicates that (i) oncogenic viruses are necessary, but not sufficient for certain malignancies to occur, and represent only one component of a multistep process, as proposed for other cancers; and (ii) the immune system plays a critical role in keeping these infection agents from leading to malignant transformation.

In the following sections, the current understanding and recent developments concerning the pathobiology of EBV, KSHV, and HTLV-1 and their role in lymphomagenesis are reviewed. We also briefly summarize the evidence implicating *hepatitis C virus* (HCV) and human immunodeficiency virus (HIV) infections in the development of lymphomas.

18.2 Epstein–Barr virus (EBV)

The consistent association of EBV with specific lymphoid malignancies indicates a role in their etiopathogenesis. EBV infection occurs in > 90 % of healthy adults worldwide, so it is clearly not sufficient for the development of a lymphoproliferative disorder. However, there is extensive epidemiologic and experimental data indicating that EBV is an oncovirus. An early key piece of information was the finding that EBV can infect and transform normal human B cells *in vitro*, resulting in their "immortalization" and leading to continuously growing lymphoblastoid cell lines (LCLs) [5].

18.2.1 Patterns of EBV latency

Primary EBV infection is most frequently asymptomatic, but can also cause infectious mononucleosis, especially when it occurs in older children or young adults. After this primary infection, the virus establishes latency in resting memory B cells and remains throughout the entire life of the host, only very rarely causing disease in immunocompetent individuals. Latency is promoted in B cells by expression of PAX5, which blocks the function of the EBV protein ZEBRA (also called Z or ZTA), that is essential for expression of lytic viral genes [5]. These resting memory B cells can reactivate the virus (lytic cycle) when they become plasma cells as a consequence (at least in part) of PAX5 downregulation. This happens commonly in the oral cavity, leading to viral shedding in the saliva and transmission to other hosts [6]. Epithelial cells in the oral mucosa are thought to amplify viral replication and shedding [7]. A germinal center cell model has been proposed by Throley-Lawson and colleagues, based on careful analysis of expression patterns in different tissues from EBV-infected immunocompetent individuals. This model proposes five different transcription programs that explain the mechanisms used by the virus to establish an infection while avoiding recognition by the immune system (Fig. 18.1) (reviewed in [8–10]). When EBV first infects a naïve B cell in lymphoid tissue within the oral cavity, a transient "growth program" is established, where EBV expresses EBNA 1-6, as well as LMP1, LMP2A and LMP2B. These proteins force the infected cells to become proliferating B-cell blasts, probably allowing EBV infection to be propagated. *In vitro* generated LCLs have this growth program indefinitely. However, since many of these proteins are antigenic, this state is very transient in immunocompetent individuals. As soon as an immune response is established, most of the cells with this program are eliminated, or otherwise switch to a "default program" of EBV expression, where only EBNA1, LMP1 and LMP2A are expressed. This stage is also temporary because LMP1 and LMP2A mimic CD40 and antigen receptor signaling, thereby inducing the B cells in peripheral lymphoid organs to behave like germinal center B cells and differentiate into resting memory B cells [11, 12]. These infected cells in turn switch to a "latency program" where no viral genes are expressed, allowing lifetime persistence of EBV. Dividing peripheral blood infected

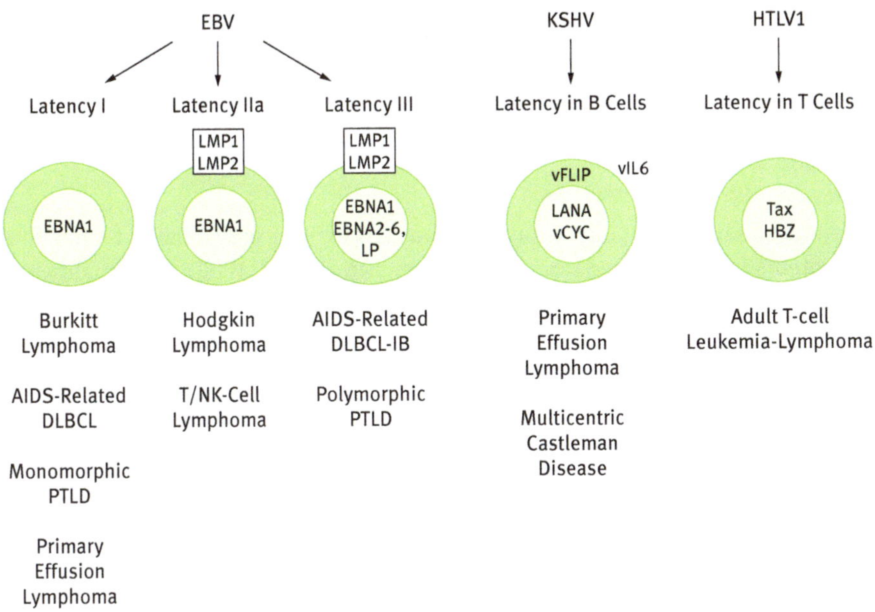

Fig. 18.1: **Major patterns of EBV and KSHV latent gene expression in lymphoproliferative disorders.** The main EBV latency patterns and the most common lymphoproliferative disorders in which these patterns are seen are illustrated. The KSHV proteins that are uniformly expressed in PEL and MCD are shown on the right. Modified from the previously published figure in [155].

memory B cells express only EBNA1, which is not immunogenic but allows the EBV episome to segregate and be propagated in dividing cells.

Some of the same transcriptional programs are broadly recapitulated in lymphomas and lymphoproliferative disease, although there is evidence that EBV latency in tumors is more heterogeneous than previously thought, both at the RNA and protein level [13, 14]. Three broad major patterns of EBV expression have been described in infected cells in lymphoproliferative disorders: Types I, II, and III, illustrated in Fig. 18.1 [5, 15]. Type III latency corresponds to the “growth program” and involves the unrestricted expression of all nine latent genes, including six EBV-encoded nuclear antigens (EBNA1-6) [15] and three latent membrane proteins (LMP1, LMP2A, and LMP2B). At the other end of the spectrum, in Type I latency, EBNA1 is the major viral protein produced. Type II latency corresponds to an intermediate form representing the “default program”, and consists of expression of EBNA1 and varying amounts LMP1 and LMP2 proteins.

In general, since EBNA proteins are immunogenic, with the exception of EBNA1, a feature of Type III latency is the recognition and elimination of the EBV-infected cells by the immune system. Thus, lymphomas with unrestricted EBV latency are generally encountered in people with immunodeficiency. In contrast, most lymphomas in immunocompetent, or at least largely immunocompetent hosts will have Type I or II latencies, because downregulation of the immunogenic EBNA proteins is an important mechanism of immune evasion by EBV [5].

18.2.2 EBV mediated lymphomagenesis

Most studies indicate that infected lymphoma cells express only a limited number of viral proteins in one of three canonical latency patterns [6]. While, in actuality, the pattern of EBV gene expression can be quite heterogenous at the tissue and single cell level [13, 14], this simplified classification can help understand the relationship between EBV infection and B-cell lymphoma development. The major proteins and RNAs expressed and their functions have been well described [16] (Tab. 18.1). As a basic concept, Type I latency tumors tend to only express EBNA-1 and EBERSs, which are not potent viral oncogenes; so, a number of additional genetic alternations are seen in tumors with this type of latency. The best example to illustrate this is Burkitt lymphoma, where a translocation of MYC is almost invariably found, in addition to activation of tonic B-cell receptor (BCR) signaling by TCF3 activation or inactivation of its negative regulator ID3 [17], plus a number of other cellular genetic alterations. On the other end of the spectrum, Latency III involves the unrestricted expression of all nine latent genes. A number of these proteins are highly transforming (Tab. 18.1), so in general, cellular genetic alterations are more variable.

EBV also expresses noncoding RNAs. Among these are the EBERs, which are expressed ubiquitously and abundantly in infected cells, making them a very useful diagnostic technique for detecting EBV infection in clinical specimens by *in situ* hybridization. Viral microRNAs (miRs) are also expressed in latently infected tumor cells, and at least 22 precursors and 44 mature EBV miRs have been described [18, 19]. Expression of the individual miRs varies depending on cell type and experimental context. The BHRF1 cluster of miRs is expressed in B cells with Type II latency and, at low levels, in cells with Type I latency [20, 21]. miRs in the *BHRF1* region inhibit apoptosis and enhance cell cycle progression and proliferation during early infection of human primary B cells [22].

EBV also has global effects on the epigenome. EBV infection of B cells leads to the appearance of large-scale hypomethylated blocks in about two thirds of the genome; these regions overlap with hypomethylated blocks seen in general in cancer and are associated with cellular gene expression hypervariability [23].

Tab. 18.1: Major EBV genes and their function expressed during latency.

EBV GENE	MAIN PROTEIN FUNCTIONS	Expression pattern Lat I	Lat II	Lat III
EBNA-1	Binds to the origin of replication of the EBV episome, which is required for replication of the virus DNA and its distribution to the daughter cells in proliferating cells. Not presented by MHC class I molecules, avoiding recognition by cytotoxic T cells. Transgenic mice have increased incidence of lymphoma alone, and in cooperation with MYC. May protect cells from P53 mediated apoptosis.			
EBER-1 and EBER-2	Short nonpolyadenylated RNAs. Can induce the secretion of interleukin-10 (IL-10), which might stimulate growth of infected B cells and suppress cytotoxic T cells. Might mediate resistance of EBV-positive Burkitt lymphoma cells to interferon-alpha. Because of their abundance, *in situ* hybridization for EBER is used for histological detection of EBV.			
LMP-1	Strongly activates the NF-kB pathway, important in lymphomagenesis. Functions as an oncogene; transgenic mice expressing the gene in B cells develop lymphomas. Inhibits apoptosis by upregulating expression of a number of cytokines and apoptosis inhibitory proteins. Induces AID upregulation and genomic instability.			
LMP-2A	Can mimic BCR signaling through its ITAM motifs. Can also inhibit BCR signaling by binding and sequestering tyrosine kinases from the BCR. This can prevent unwanted antigen-triggered activation of EBV-positive B cells that would cause entry into the lytic cycle.			
EBNA-2	First virus protein to be expressed after infection of B cells *in vitro*. Essential for cellular transformation. Interacts with the transcription factor RBP-Jk, mimicking the Notch signaling pathway. Regulates several virus genes, including LMP1 and LMP2A , and many cellular genes including MYC. Represses BIK, a pro-apoptotic molecule, thereby protecting infected cells from TGF-β1-induced apoptosis.			
EBNA-3A-C	EBNA-3A and 3C are essential for viral transformation of B cells. EBNA-3 A, B and C associate with RBP-Jk and alter transcription. EBNA-3C together with EBNA2 activates the LMP1 promoter. EBNA-3C can affect ubiquitination of RB1 and p27 affecting cell cycle control. EBNA3C associates with p73 and may inhibit its function.			
EBNA-LP	Cooperates with EBNA2 to activate gene expression.			

Table adapted from [155].

18.2.3 Lymphomas associated with EBV infection

A specific group of lymphomas are associated with EBV infection, and the majority are of B-cell origin and include:

18.2.3.1 Diffuse large B-cell lymphomas (DLBCL)

These lymphomas occur in both HIV-infected and uninfected individuals, and while they are considered the same disease, there are some important differences among these two groups. EBV infection is much more common in DLBCL occurring in HIV-infected individuals, and occurs in about 30 % of cases. In contrast, in HIV-negative patients, EBV is present in < 5 % of DLBCLs, and these are more common in older individuals. In addition, some of the EBV+ DLBCLs in HIV+ patients have an immunoblastic morphology, which is rare in HIV-negative patients. These include the majority of those presenting as primary in the central nervous system (CNS). In this immunoblastic type, EBV most commonly has a Type III latency pattern of viral gene expression, and occurs in people with more advanced immunodeficiency because the viral proteins expressed, including LMP1 and EBNA2, are oncogenic but also immunogenic (reviewed in [16]). The widespread implementation of combined antiretroviral therapy (CART) to control HIV has led to a decrease in frequency of immunoblastic DLBCL. In contrast, cases with centroblastic morphology occur in patients with or without AIDS, and most commonly have a EBV Type I latency [14, 24–26]. In HIV infected patients, the germinal center subtype is more common [24] and there is a higher proliferation rate, with approximately half of the cases expressing Ki67 in more than 80 % of the tumor cells, as compared to 13 % in HIV-negative patients, consistent with a more frequent high grade [27]. Another difference of AIDS-DLBCLs is a more common frequency of *MYC* rearrangements, lower numbers of CD4+ and FOXP3+ T cells, increased numbers of activated cytotoxic cells, and higher angiogenesis with increased vascular density. Within AIDS-related DLBCL, the presence of EBV has been reported to be associated with a worse clinical outcome [28, 29], but this is not observed in every cohort of patients [24].

18.2.3.2 Burkitt lymphomas (BL)

The vast majority of cases of endemic BL (occurring in Africa) are associated with EBV infection, while only approximately 30 % of sporadic or AIDS related cases are EBV-positive. One study on sporadic BL in the United States showed the presence of EBV in 29 % of the cases, but also reported that infection varies with age, being more common in young adults than in children or older individuals, and an association with race (more common in Blacks and Hispanics) and with HIV status (64 % in HIV+ vs. 22 % in HIV-) [30].

BL typically displays EBV Type I latency, but a different latency pattern, called Wp-restricted latency, was also reported to occur in approximately 15 % of BL, specifically in cases that have a mutant EBV genome with deletions that include EBNA2, leading to expression of EBNA 3A, 3B, 3C proteins, in addition to EBNA1, without expression of EBNA2 and the LMPs [31]. BL cells with this deleted version of EBV have a higher resistance to apoptosis due to expression of a viral BCL2 homologue, BHRF1, which is normally a lytic gene but is expressed in cells that are latently infected by this EBV variant [32]. In the context of deregulated MYC expression, a molecular hallmark of BL [33–36], MYC protein overexpression would lead to apoptosis without protective mechanisms such as deletion of TP53 (another common molecular alteration in BL) or BHRF1 (EBV BCL2) expression [37]. RNA sequencing of AIDS-related and sporadic BL has also shown a much more heterogeneous transcriptional pattern than previously thought, and, even at the protein level, proteins like LMP2, which should not be expressed in Type I latency, can be found by immunohistochemistry in 11 % of HIV-associated BL and 5 % of sporadic BL cases [14].

18.2.3.3 Classical Hodgkin's lymphoma (cHL)

This disease occurs in both HIV infected and uninfected individuals, but EBV is much more common in the former (approximately 90 % vs. 30 %). A related difference is the more common mixed cellularity and lymphocyte depleted histologies in HIV+ patients, as opposed to the more common nodular sclerosis in immunocompetent individuals [38, 39]. This is consistent with the generally more frequent presence of EBV infection in the mixed cellularity histological subtype. The number of tumor-associated macrophages has been found to be an adverse prognostic factor, and is higher in both the mixed cellularity subtype and in EBV+ cases of [40]. Specific HLA types have been reported in EBV+ vs. negative cHL, where HLA-DR2 and HLA-DR5 were found to be increased in the latter, while the allele frequencies of HLA-A1, HLA-B37 and HLA-DR10 were significantly increased in the EBV+ cHL population, suggesting specific genetic predispositions for different subtypes [41].

The tumor cells of cHL are the Hodgkin and Reed–Sternberg (HRS) cells, and these only comprise a small proportion of the overall tumor mass. EBV is present in HRS cells, and has a Type II latency, with expression of EBV LMP-1 and LMP-2A proteins. LMP-1 leads to constitutive NF-κB activation, a pathway shown to be activated in cHL regardless of EBV association. Loss of function mutations of TNFAIP3 (A20), an inhibitor of NF-κB, are frequently found in CHL but are less common in EBV+ lymphomas, where NF-κB activity is a consequence of LMP1 expression [42, 43], an observation also made in other EBV+ lymphomas types [44]. Another recently reported activity of LMP-1 is upregulation of discoidin domain receptor 1 (DDR1), a receptor tyrosine kinase that acts as a pro-survival factor and is activated by collagen [45]. This activity could be related to the presence of collagen in EBV+ cHL lymph nodes, which may promote the survival of HRS cells, and may explain the inability of these cells to survive in cell culture.

One of the interesting features of HRS cells is that in spite of a B-cell origin, BCRs are not expressed, including immunoglobulin heavy and light chains, and the alpha and beta chains (CD79A/B) involved in signaling. One model suggests that in the EBV-positive cases, LMP-2A replaces BCR by signaling though an ITAM motif, leading to activation of the same downstream pathways. A problem with this model is that BCR activation of EBV-infected cells leads to EBV lytic reactivation, raising the question of how LMP2A can signal while maintaining latency. An explanation may be provided by a study that showed that HRS cells do not express a set of genes that are induced by LMP-2A and BCR activation in other cell types, such as EGR1, a zinc-finger protein required for BCR-induced EBV replication [46]. Therefore, while the precise mechanism of this differential gene expression in the presence of the same signaling protein is not clear, this observation may explain why EBV can remain latent in HRS cells even in the context of LMP-2A expression.

18.2.3.4 Post-transplant lymphoproliferative disorders (PTLD)

This is the prototypical disease associated with EBV infection, as this virus is present in the majority of cases. Consistent with the immune-deficient state of the host, most lesions harbor a Type III EBV latency. One type of PTLD represents an exception; these are the cases with monomorphic histology, which may be EBV negative or have a latency pattern that corresponds to the lymphoma subtype they most resemble, such as Latency I in DLBCL. A gene expression profiling study of post-transplant DLBCL showed viral and inflammatory response signatures that segregated with the EBV status and latency subtypes, confirming that there are biological differences among these cases that may impact on clinical behavior and response to specific therapies [47].

18.2.3.5 Plasmablastic lymphoma (PBL)

This is an aggressive malignancy that was first reported in the oral cavity of HIV-infected individuals [48], although cases can also occur in other sites, and in immunocompetent individuals; these cases are recognized as PBL by the WHO classification, as long as they have morphological and immunophenotypic features consistent with PBL, which include plasma cell differentiation [49]. EBV is present in the vast majority of cases presenting in the oral cavity, while in other sites EBV can be found in approximately 75 % of cases. A Type I latency has been described in PBL. A report of five cases with a review of the literature described 248 PBL cases, out of which 157 were in HIV-positive patients and 43 % were outside the oral cavity, with EBV being identified in 86 % of patients [50].

18.2.3.6 EBV-associated T- and NK cell malignancies

While EBV is most commonly present in lymphomas of B-cell origin, there is a strong association between EBV infection and several non-B-cell malignancies (reviewed in [51]). Aggressive NK cell leukemia, a rare disease that is more common in Asians, shows EBV infection in over 90 % of cases. Extranodal NK-T-cell lymphoma, nasal type, are also associated with EBV infection. This malignancy has also been called angiocentric T-cell lymphoma and lethal midline granuloma. It is characterized by vascular damage, necrosis and a cytotoxic T-cell phenotype of the tumor cells. It most commonly occurs in the upper respiratory tract but can also present in other extranodal and nodal sites, with a higher incidence in Asians and Native Latin Americans. Two different pediatric T-cell lymphoproliferative disorders have been found to be associated with EBV infection: (i) systemic EBV-positive lymphoproliferative disease of childhood, which is characterized by proliferation of EBV-infected T cells with an activated cytotoxic phenotype; and (ii) hydroa vacciniforme-like lymphoma, an EBV-positive cutaneous T-cell lymphoma associated with sun sensitivity.

18.3 Kaposi's sarcoma-associated herpesvirus (KSHV)

Kaposi's sarcoma-associated herpesvirus is always found in Kaposi's sarcoma [52], and compelling evidence indicates that it is an etiologic agent for this disease [53]. In contrast to EBV, which is present in a number of lymphoma subtypes, KSHV is present in a more limited number of lymphoproliferative disorders, two of which have been best described and documented: primary effusion lymphoma (PEL) and multicentric Castleman's disease (MCD) [54, 55]. A third type of KSHV-associated lymphoproliferative disease is MCD-associated plasmablastic lymphoma [56], but this entity is more rare and controversial.

18.4 KSHV-mediated lymphomagenesis

KSHV encodes an astonishing number of proteins with cellular homologues and documented oncogenic properties. However, many of these are expressed during lytic replication, and are thus absent in the majority of lymphoma cells. An example of a potential KSHV-encoded oncogene only expressed during the lytic phase is the viral BCL2 homolog, which may be important to keep the infected cells alive long enough to complete the viral life cycle. In KSHV-associated lymphoma (i.e. PEL), cells have largely a single major latency pattern of virus infection, with some minor variations, while different latency patterns may be observed in KS, with possibly more lytic replication and heterogenous expression. In PEL, only a handful of viral proteins are made, which are important for viral episome maintenance, cell proliferation, cell survival and immune evasion. Three viral proteins are clearly abundantly expressed in all latently infected PEL cells: LANA, vCYC and vFLIP. The major function of LANA is tethering viral episomes during cell division, but it can also bind a number of tumor suppressor genes. vCYC is a cyclin D homolog, which can affect cell cycle regulation. vFLIP can induce NF-κB and suppress apoptosis as well as autophagy [57–62]. These three viral proteins are expressed under the control of the same regulatory elements in tri-cistronic and bi-cistronic transcripts, so they may act in concert. Other viral gene products are expressed in a variable proportion of cells in PEL or MCD, including Kaposin B, vIL-6, and vIRF3. The main feature of these proteins and their role in oncogenesis is summarized in Tab. 18.2 [63–71]. At least 12 pre-miRNAs are also encoded by KSHV and expressed during latency [72–76]. Deletion of a cluster of 10 precursor miRs from the KSHV genome rendered it unable to transform *in vitro* infected cells, and infection caused cell cycle arrest and apoptosis instead, revealing that these miRs play a protective role on the infected cell [77]. Specific miR functions include targeting of IκBα, thereby activating NF-κB by miR-K1, which strengthens NF-κB activity induced by vFLIP and promotes cellular survival [78]. miR-K1 also targets and depletes the cellular mRNAs encoding cyclin-dependent kinase inhibitor CDKN1a (p21), thereby blocking cell cycle arrest and promoting cellular proliferation [79]. Another interesting microRNA encoded by KSHV is miR-K12-11, a viral ortholog of cellular miR-155. Both the viral and cellular counterparts have been shown to enhance B-cell expansion [80–82]. Viral microRNAs have been found in virions, suggesting that they can be functional immediately after *de novo* infection [83]. Unbiased global analysis of microRNA function using PAR-CLIP and Ago HITS-CLIP has revealed that KSHV microRNAs target 1,000–2,000 cellular mRNAs in PEL cell lines. Targets are enriched for genes involved in a number of pathways relevant to KSHV lymphomagenesis, and include apoptosis, cell cycle regulation, lymphocyte proliferation, and immune evasion [84, 85].

18.4.1 Lymphomas associated with KSHV infection

Primary effusion lymphoma (PEL). This type of lymphoma has been defined by the presence of KSHV, but EBV is also an important feature of this lymphoma, as it is present in roughly 90 % of cases, which are dually infected by both human gamma-herpesviruses [39, 54, 86]. KSHV expresses a handful of proteins, which have clear oncogenic properties and drive the disease and tumor cell survival (Tab. 18.2). In contrast, EBV has a Type I latency in PEL, with little or no expression of the major EBV transforming proteins [87, 88], so it is thought that KSHV is the major driving virus, although both viruses appear to play a pathogenic role. Both KSHV and EBV microRNAs are also expressed in PEL [84, 89], providing evidence that EBV has contributory functions to PEL pathogenesis. In PEL, KSHV miRNAs directly target more than 2,000 cellular mRNAs, including those involved in transcriptional regulation and signaling, and 58 % of these mRNAs are also targeted by EBV miRNAs, via distinct binding sites [84].

Tab. 18.2: Major KSHV genes expressed during latency. Adapted from [155].

KSHV GENE	MAIN PROTEIN FUNCTIONS
LANA	Essential for episome maintenance. Interacts with may nucleosomal proteins including core nucleosomal histones as well as with histone H1 and HMGs. Binds and inactivates RB1,TP53 and GSK3B. Has transcriptional effects on viral and cellular genes, including IL-6, hTERT, Pim1, and the TGFbeta type II receptor. Interacts with full-length isoform TAp73a and its dominant negative regulator DNp73a, affecting TAp73a stability, subnuclear localization and TAp73a-mediated transcriptional activation of target genes. Upregulates expression and phosphorylation of survivin, upregulating the activities of histone acetyltransferases and deacetylases and leading to an increase in viral copy number in infected B cells. Antibodies useful for immunohistochemical diagnosis of KS, PEL and MCD.
vCYC	Functional cyclin that can associate with CDK6 and induce phosphorylation of RB1 overcoming cell-cycle arrest. Differs from the cellular cyclin D in its ability to promote degradation of the CDK inhibitor CDKN1B (p27Kip) when complexed with CDK6. vCYC transgenic mice develop lymphomas only when bred with *TP53* knock-out mice, and T-cell lymphomas on its own. vCYC–CDK6 complexes can induce phosphorylation of nucleophosmin (NPM1) leading to genomic instability.
vFLIP	Homologous to the cellular FLICE/caspase 8 inhibitory proteins (cFLIPs), which can inhibit death receptor mediated apoptosis. Activates NF-κB though both the classical and alternative pathways by binding to IKKγ/NEMO Induces expression of numerous genes, including anti-apoptotic genes and growth factors. Protects PEL cells from spontaneous apoptosis. Protects B cells from autophagy by binding to Atg3. Induces IRF4, which in turn supports latency by inhibiting KSHV lytic reactivation. vFLIP transgenic and conditional knock-in mice develop B cell malignancies when expressed in B cells, and accelerates the development of MYC induced lymphomas.
vIL6	Viral homolog of cellular IL6. Differs from IL6 in that it is selectively glycosylated and can bind IL6ST (previously called IL-6 receptor beta, or gp130) in the absence of IL6R (the high affinity IL6 receptor, also called alpha or gp80) to activate IL6-responsive genes and promote B-cell survival.

Tab. 18.3: (continued)

KSHV GENE	MAIN PROTEIN FUNCTIONS
vIL6	Can interact with IL6ST within the ER. Expressed by a variable but significant proportion of latently-infected PEL cells as well as in MCD, and expressed at low levels in during true latency. Secreted protein with paracrine effects on neoplastic and tumor-infiltrating cells. Present in sera of patients with KSHV-associated MCD and KSHV inflammatory cytokine syndrome (KICS).
vIRF-3	Does not bind DNA but can be recruited to the interferon promoters via its interaction with cellular IRF-3 and IRF-7, stimulating their transcriptional activity. Inhibits TP53 function by associating with the DNA-binding domain of p53, inhibiting p53 phosphorylation on serine residues S15 and S20, antagonizing TP53 oligomerization and reducing DNA binding affinity. Expressed in PEL but not KS. Inhibits transcription of CIITA and MHC Class II, leading to decreased recognition of infected cells by CD4+ T cells.
Kaposins	Complex region potentially encoding three proteins, called Kaposin A, B, and C. Kaposin A may be transforming. Kaposin A signals by recruiting cytohesin 1 to the plasma membrane. Expression of Kaposin A protein not known, but Kaposin B seems to be expressed in KS and some PEL cell lines, but not others. Kaposin B may have transforming functions, including activation of the p38 pathway. Kaposin B can stabilize cytokine mRNAs through activation of MK2 blocking the decay of mRNAs with AU-rich elements (AREs) in their 3′ untranslated region; AREs are present in cytokine mRNAs, so Kaposin B increases the production of pro-inflammatory cytokines. Kaposin B can activate STAT3.

KSHV can also be found in the tumor cells of AIDS-related NHL with no evidence of body cavity involvement. These have been called extracavitary or solid PEL and represent approximately 5 % of all AIDS-NHLs. They frequently resemble immunoblastic lymphomas histologically but they frequently lack of expression of B-cell antigens (a feature of PEL) and are also usually co-infected with EBV [90].

While KSHV is considered part of the definition of PEL, cases of primary lymphomatous effusions that lack KSHV have been documented. These have been called KSHV- or HHV-8-negative PEL. They generally occur in HIV-negative and older individuals more frequently than KSHV-associated PEL, so it is thought that they represent a somewhat different lymphoma type [91–96].

Multicentric Castleman's disease (MCD) and associated lymphomas. Approximately half of MCD cases from immunocompetent individuals and almost all individuals with HIV and MCD are associated with infection by KSHV [55]. This virus is present in B cells in the mantle zones, which can be sparse or numerous and coalesce, even forming microlymphomas or frank lymphomas, although the latter are extremely rare. Infected cells are monotypic but polyclonal, and express IgMλ [97]. Lymphomas that are found in association with MCD are different from PEL in that they are negative for EBV, express IgMλ and are thought to arise from naive B cells rather than from terminally differentiated B cells, as they do not have somatic hypermutation of the immunoglobulin genes [39, 56]. Another very rare B-cell lymphoma entity associated with KSHV is a germinotropic lymphoproliferative disorder, where germinal center B cells are co-infected with EBV and KSHV [98].

A feature of MCD is the presence of several constitutional symptoms that are thought to be due to production of excess cytokines, including interleukin-6 (IL-6) which is markedly elevated in this disease. Since KSHV encodes an IL-6 homolog, called vIL-6, which is also expressed in MCD, it may contribute to pathogenesis through this mechanism [99–101]. In addition, expression of vIL6 has been reported to confer a worse prognosis [102]. The expression of other KSHV lytic proteins in KSHV-associated MCD suggests that there is uncontrolled viral proliferation in this disease. An IL-6-related systemic inflammatory syndrome has also been described, called KSHV inflammatory cytokine syndrome (KICS). This syndrome is seen in patients with HIV and KS, but without a pathologic diagnosis of MCD [103, 104].

18.5 Human T-cell leukemia virus-1 (HTLV-1)

HTLV-1 is the first described human lymphotropic retrovirus [105], and is the etiologic agent for adult T-cell leukemia/lymphoma (ATL).

18.5.1 HTLV-1 mediated oncogenesis

In contrast to other animal retroviruses that transform cells by transduction of oncogenes or activation of oncogenes adjacent to insertion sites, HTLV-1 transforms T cells by a rather complex mechanism that is mediated by its transactivator protein Tax and the HTLV-1 bZIP factor (HBZ) [106]. Evidence supporting an oncogenic role for Tax stems from its ability to induce T-cell immortalization [107, 108] and mesenchymal tumors in transgenic mice [109]. Transgenic mice in which the expression of Tax was specifically targeted to T cells developed a T-cell lymphocytic leukemia [110]. An important function of HTLV-1 Tax is activation of classical and alternative NF-κB, but at least the classical NF-κB pathway is inhibited by another viral protein, called HBZ [111]. The major activities of these two proteins have been described and are summarized in Tab. 18.4 [105].

18.5.2 Adult T-cell leukemia-lymphoma (ATL)

ATL is an often aggressive malignancy of mature CD4+ lymphocytes. There are four different clinical subtypes: acute, chronic, lymphomatous, and smoldering. The typical acute form of ATL is characterized by generalized lymph node, peripheral blood, and skin involvement by pleomorphic tumor cells with hyperlobated nuclei, lytic bone lesions, hypercalcemia, rapidly progressive clinical course, and relatively short survival [112]. The virus integrates randomly into the cellular genome, so clonality can be monitored based on integration sites. Soon after infection of T cells by HTLV-1 clones arise, which are selected *in vivo* and can be seen in individuals with chronic infection. Expansion of specific clones is thought to set the stage for ATL development.

The molecular events induced by HTLV-1 that lead to ATL have been reviewed recently [113]. Tax is able to activate NF-κB and plays a crucial role in immortalization and persistent infection, so for a long time it has been considered the main viral oncogene in ATL. However, this protein is generally not expressed in ATL, and is actually frequently deleted from the integrated provirus [114–116]. A major breakthrough was made with the discovery of HBZ [117]. HBZ is ubiquitously expressed in ATL cells and HTLV-1 infected cells *in vivo*, and this protein can also promote cellular proliferation [118]. Interestingly, HBZ and Tax have opposite effects on a number of signaling pathways evaluated [113]. Among these pathways are NF-κB, AP1, NFAT, Wnt, and CREB, which are activated by Tax but inhibited by HBZ [113]. On the other hand, the TGF-

Tab. 18.4: HTLV-1 proteins involved in transformation and expressed during latency.

HTLV-1 GENE	MAIN PROTEIN FUNCTIONS
Tax	Indispensable for immortalization/transformation and for persistent HTLV-1 infection *in* vivo. Activates NF-kB by interacting with IKKγ/NEMO and activating IKKα and IKKβ. Activates PI3K/AKT pathway by directly binding to p85α, the inhibitory subunit of PI3K, and releasing the active catalytic p110α. Activates CDK2 and CDK4 promoting cell cycle progression. Induces genomic instability by interacting with the TP53 coactivator CBP/P300 reducing that transcriptional activating function of TP53. Activates telomerase by transcriptionally upregulating hTERT. Induces centrosome aberrations by inactivating TAX1BP1. Induces defects in DNA repair. Can induce cell cycle arrest at G1 or G2-M in some experimental conditions. Stimulates generation of reactive oxygen species (ROS). Increases production of cytokines and chemokines promoting inflammation, including IL2, and adhesion molecules. Induces constitutive AP-1 activity and related gene expression.
HBZ	Suppresses Tax-mediated viral gene transcription via interaction with CREB2 and JUN. Interacts with host factors with bZIP domains, such as CREB, JunB, ATF-1 and ATF-3, inhibiting their transcriptional activity. Interaction with JunD activates transcription of target genes. Enhances TGFβ pathway by interacting with SMAD2/3 and p300. Selectively inhibits the classical NF-kB pathway by preventing DNA binding of p65, and promoting its degradation. Induces FoxP3 expression in T cells, but interacts with FoxP3 impairing it functionally.

β/Smad pathway is inhibited by Tax but activated by HBZ. An explanation has been that, while Tax promotes cellular proliferation, it can also induce cellular senescence by induction of p21 and p27, which is prevented by HBZ [119]. Thus, it appears that Tax is important for the establishment of HTLV-1 infection of T cells, leading to proliferation (which also involves cytokines like IL2), while for ATL development HBZ appears to play a role in supporting ongoing T-cell leukemogenesis. In addition, ATLL tumor cells have oncogenic alterations in the TP53 [120, 121] and CDKN2/p16 tumor suppressor, genes as well as chromosomal aberrations [122, 123], the complexity of which correlates with tumor aggressiveness [124]; therefore, these tumors may have become at least partially independent of HTLV-1.

18.6 Hepatitis C virus

There is epidemiological and experimental evidence suggesting a role for *hepatitis C virus* (HCV) infection in the development of B-cell lymphomas as well as mixed cryoglobulinemia. A large number of epidemiologic studies have been published, evaluating the frequency of HCV infection in patients with lymphoma, and this evidence has been recently reviewed [125]. The relative risk of B-cell NHL among HCV-positive individuals as compared to HCV-negative individuals ranges from 2 to 5, although not all studies have reported a positive association between HCV and lymphoma, and geographical differences seem to be important (reviewed in [126]). One multicenter case-control study showed positive associations with specific lymphoma types, in particular follicular, marginal zone and mucosa-associated lymphoid tissue (MALT) non-Hodgkin's lymphomas [127]. One study found HCV-RNA in 37 % of the lymphoma tissue samples, but this high incidence has not been confirmed by others and remains controversial [128]. Nevertheless the possibility that on occasion lymphomas can contain the HCV genome is suggested by the report of a single B-cell lymphoma cell line containing the replicating viral genome [129]. In this cell line, genetic instability was observed and a hit-and-run mechanism has been proposed [130]. While HCV does not directly transform human B cells *in vitro*, the ability of HCV to induce B-cell lymphomas was shown in transgenic mice expressing the full HCV genome in B cells [131].

Most evidence suggests an indirect role resulting from chronic B-cell antigenic stimulation. This is supported by the finding that lymphomas in patients with HCV express a limited number of immunoglobulin sequences, consistent with an antigenic drive, and some lymphomas have been identified that actually express an antigen receptor that recognizes HCV [132]. Treatment of HCV infection with antiviral agents has resulted in regression of a significant proportion of low-grade lymphomas in HCV (+) cases but not in HCV (–) patients [133–136], strongly implicating antigen stimulation in the maintenance of the tumor.

18.7 HIV and lymphomagenesis

While EBV, KSHV and HTLV-1 are directly transforming viruses, HIV is also considered a carcinogen in humans [137]. The increased incidence of B-cell lymphomas in HIV infected patients is believed to be related to disrupted immune surveillance to tumor antigens, genetic alterations, chronic antigenic stimulation, and cytokine dysregulation [138–142]. There is evidence supporting a role for HIV beyond immunosuppression. A large study conducted by the Centers for AIDS Research (CFAR) Network of Integrated Clinical Systems found that the incidence of NHL is increased even in patients treated with combined antiretroviral therapy, and that HIV viremia is a continuous variable associated with NHL with a hazard ratio of 1.42 per $\log_{10}$ copies/mL [143]. This increased incidence of NHL was observed even in patients with CD4 cell counts of > 200 cells/μL. Another study also documented the association of detectable HIV viremia with the incidence of cHL [144]. Consistent with a role of HIV is a report documenting the development of B-cell tumors in transgenic mice with a 7.4-kb pNL4-3 HIV-1 provirus expressing p17, gp120 and nef, but lacking parts of the gag-pol region (rendering it replication incompetent) [145].

HIV itself does not infect B cells but there is also biological plausibility that this virus promotes lymphoma development through secreted viral proteins [137]. HIV-induces chronic B-cell stimulation, which precedes the development of lymphoma and may be critical [146–148]. Several reports have implicated specific HIV proteins in this B-cell hyperplasia, including Tat [149], Nef [150, 151], Env gp120 [152], Env gp130 and Gag p17 [153]. Tat has inhibitory effects on DNA repair mechanisms in B cells [154]. These effects may be compounded with the rapidly proliferating B cells in the germinal centers (where HIV is present) that also have ongoing rearrangements and somatic hypermutation, making these cells more likely to transform into a lymphoma.

Interactions between HIV viral proteins and other oncogenic herpesviral proteins have been reported, involving secretion or transmission of HIV proteins through structures like nanotubules or exosomes affecting properties of pre-lymphomatous or lymphoma B cells (reviewed in [137] and [155]). Two studies have documented interesting interactions between HIV Nef and KSHV: (i) Nef can synergize with KSHV vIL-6 and activate the AKT pathway leading to angiogenesis and tumor formation in a model of KS [156] and (ii) Nef can regulate KSHV latency by inducing cellular miRNA 1258 (hsa-miR-1258), which directly targets a seed sequence in the 3' untranslated region of the mRNA encoding the major lytic switch protein of KSHV (RTA) [157]. HIV Nef has been shown to be secreted in exosomes [158] and also penetrate bystander, uninfected B cells [151], so it may have an effect in maintaining latency of EBV or KSHV-infected B cells that eventually become transformed.

18.8 Concluding remarks

EBV, KSHV and HTLV-1 are associated with specific types of malignant lymphomas, and extensive experimental evidence indicates that they play a direct etiopathogenic role in their development. While these viruses are quite different, they all encode proteins that subvert the host molecular machinery to deregulate cell growth and immortalize infected lymphoid cells, and promoting the development of further cellular genetic alterations to take place so that the cells will progress to full malignancy. There is evidence that HIV and HCV are indirect carcinogens. The ability of all these viruses to transform cells and to lead to lymphomagenesis is highly dependent on host factors, since infection of none of these viruses is sufficient for cancer development; the fraction of infected individuals that develop lymphoma is fortunately relatively small. Our increasing understanding of the oncogenic and immunologic mechanisms involved in the pathobiology of virus-associated lymphomas is important for the eventual development of targeted therapeutic and preventive approaches.

References

[1] Weiss RA, Vogt PK. 100 years of Rous sarcoma virus. J Exp Med. 2011 Nov 21;208(12):2351-5.

[2] Ellermann V, Bang O. Experimentelle Leukämie bei Hühnern. Zentralbl Bakteriol Parasitenkd Infectionskr Hyg Abt Orig. 1908;46:595–609.

[3] Fredricks DN, Relman DA. Sequence-based identification of microbial pathogens: a reconsideration of Koch's postulates. Clin Microbiol Rev. 1996;9(1):18–33.

[4] zur Hausen H. Viruses in human cancers. Eur J Cancer. 1999 Dec;35(14):1878–1885.

[5] Rickinson AB, Kieff E. Epstein–Barr virus. In: Fields BN, Knipe DM, Howley PM, editors. Virology. Philadelphia: Lippincott-Raven Publishers; 1996. p. 2397–2446.

[6] Laichalk LL, Thorley-Lawson DA. Terminal differentiation into plasma cells initiates the replicative cycle of Epstein–Barr virus in vivo. J Virol. 2005 Jan;79(2):1296–1307.

[7] Hadinoto V, Shapiro M, Sun CC, Thorley-Lawson DA. The dynamics of EBV shedding implicate a central role for epithelial cells in amplifying viral output. PLoS Pathog. 2009 Jul;5(7):e1000496.

[8] Thorley-Lawson DA. Epstein–Barr virus: exploiting the immune system. Nat Rev Immunol. 2001 Oct;1(1):75–82.

[9] Thorley-Lawson DA, Gross A. Persistence of the Epstein–Barr virus and the origins of associated lymphomas. N Engl J Med. 2004 Mar 25;350(13):1328–1337.

[10] Thorley-Lawson DA, Hawkins JB, Tracy SI, Shapiro M. The pathogenesis of Epstein–Barr virus persistent infection. Curr Opin Virol. 2013 Jun;3(3):227–232.

[11] Babcock GJ, Hochberg D, Thorley-Lawson AD. The expression pattern of Epstein–Barr virus latent genes in vivo is dependent upon the differentiation stage of the infected B cell. Immunity. 2000 Oct;13(4):497–506.

[12] Laichalk LL, Hochberg D, Babcock GJ, Freeman RB, Thorley-Lawson DA. The dispersal of mucosal memory B cells: evidence from persistent EBV infection. Immunity. 2002 May;16(5):745–754.

[13] Niedobitek G, Agathanggelou A, Rowe M, Jones EL, Jones DB, Turyaguma P, et al. Heterogeneous expression of Epstein–Barr virus latent proteins in endemic Burkitt's lymphoma. Blood. 1995;86:659–665.

[14] Arvey A, Ojesina AI, Pedamallu CS, Ballon G, Jung J, Duke F, et al. The tumor virus landscape of AIDS-related lymphomas. Blood. 2015 May 14;125(20):e14–e22.

[15] Kieff E. Epstein–Barr virus and its replication. In: Fields BN, Knipe DM, Howley PM, editors. Virology. Philadelphia: Lippincott-Raven Publishers; 1996. p. 2343–2396.

[16] Longnecker RM, Kieff E, Cohen JI. Epstein–Barr virus. In: Knipe DM, Howley PM, editors. Fields Virology. 6th ed. Philadelphia: Lippincott, Williams, and Wilkins; 2013. p. 1898–1959.

[17] Schmitz R, Young RM, Ceribelli M, Jhavar S, Xiao W, Zhang M et al. Nature. 2012 Oct 4;490(7418):116–120. doi: 10.1038/nature11378.

[18] Swaminathan S. Noncoding RNAs produced by oncogenic human herpesviruses. J Cell Physiol. 2008 Aug;216(2):321–326.

[19] Lopes LF, Ruiz Miyazawa KW, de Almeida ER, Serafim KG, de Almeida Gualtieri K, Costa IC, et al. Epstein–Barr virus (EBV) microRNAs: involvement in cancer pathogenesis and immunopathology. Int Rev Immunol. 2013 Jun;32(3):271–281.

[20] Cai X, Schafer A, Lu S, Bilello JP, Desrosiers RC, Edwards R, et al. Epstein–Barr virus microRNAs are evolutionarily conserved and differentially expressed. PLoS Pathog. 2006 Mar;2(3):e23.

[21] Xing L, Kieff E. Epstein–Barr virus BHRF1 micro- and stable RNAs during latency III and after induction of replication. J Virol. 2007 Sep;81(18):9967–9975.

[22] Seto E, Moosmann A, Gromminger S, Walz N, Grundhoff A, Hammerschmidt W. Micro RNAs of Epstein–Barr virus promote cell cycle progression and prevent apoptosis of primary human B cells. PLoS Pathog. 2010;6(8).

[23] Hansen KD, Sabunciyan S, Langmead B, Nagy N, Curley R, Klein G, et al. Large-scale hypomethylated blocks associated with Epstein–Barr virus-induced B-cell immortalization. Genome Res. 2014 Feb;24(2):177–184.

[24] Chadburn A, Chiu A, Lee JY, Chen X, Hyjek E, Banham AH, et al. Immunophenotypic analysis of AIDS-related diffuse large B-cell lymphoma and clinical implications in patients from AIDS Malignancies Consortium clinical trials 010 and 034. J Clin Oncol. 2009 Oct 20;27(30):5039–5048.

[25] Dunleavy K, Wilson WH. Role of molecular subtype in predicting outcome of AIDS-related diffuse large B-cell lymphoma. J Clin Oncol. 2010;28(16):e260.

[26] Chadburn A, Noy A, Lee JY, Hyjek E, Banham AH, Sparano JA, et al. Reply to K. Dunleavy et al. J Clin Oncol. 2010;28(16):e261–e2.

[27] Liapis K, Clear A, Owen A, Coutinho R, Greaves P, Lee AM, et al. The microenvironment of AIDS-related diffuse large B-cell lymphoma provides insight into the pathophysiology and indicates possible therapeutic strategies. Blood. 2013 Jul 18;122(3):424–433.

[28] Kaplan LD, Shiramizu B, Herndier B, Hahn J, Meeker TC, Ng V, et al. Influence of molecular characteristics on clinical outcome in human immunodeficiency virus-associated non-Hodgkin's lymphoma: identification of a subgroup with favorable clinical outcome. Blood. [Research Support, Non-U.S. Gov't Research Support, U.S. Gov't, Non-P.H.S. Research Support, U.S. Gov't, P.H.S.]. 1995 Apr 1;85(7):1727–1735.

[29] Chao C, Silverberg MJ, Martinez-Maza O, Chi M, Abrams DI, Haque R, et al. Epstein–Barr virus infection and expression of B-cell oncogenic markers in HIV-related diffuse large B-cell Lymphoma. Clin Cancer Res. [Research Support, N.I.H., Extramural]. 2012 Sep 1;18(17):4702–4712.

[30] Mbulaiteye SM, Pullarkat ST, Nathwani BN, Weiss LM, Rao N, Emmanuel B, et al. Epstein–Barr virus patterns in US Burkitt lymphoma tumors from the SEER residual tissue repository during 1979-2009. Apmis. [Research Support, N.I.H., Extramural Research Support, N.I.H., Intramural Research Support, Non-U.S. Gov't Research Support, U.S. Gov't, P.H.S.]. 2014 Jan;122(1):5–15.

[31] Kelly G, Bell A, Rickinson A. Epstein–Barr virus-associated Burkitt lymphomagenesis selects for downregulation of the nuclear antigen EBNA2. Nat Med. 2002 Oct;8(10):1098–10104.

[32] Kelly GL, Long HM, Stylianou J, Thomas WA, Leese A, Bell AI, et al. An Epstein–Barr virus anti-apoptotic protein constitutively expressed in transformed cells and implicated in burkitt lymphomagenesis: the Wp/BHRF1 link. PLoS Pathog. 2009 Mar;5(3):e1000341.

[33] Cesarman E, Dalla-Favera R, Bentley D, Groudine M. Mutations in the first exon are associated with altered transcription of c-myc in Burkitt lymphoma. Science. 1987;238(4831):1272–1275.

[34] Rabbitts TH, Hamlyn PH, Baer R. Altered nucleotide sequences of a translocated c-myc gene in Burkitt lymphoma. Nature. 1983 Dec 22-1984 Jan 4;306(5945):760–765.

[35] Burns BF, Wood GS, Dorfman RF. The varied histopathology of lymphadenopathy in the homosexual male. Am J Surg Pathol. [Research Support, U.S. Gov't, P.H.S.]. 1985 Apr;9(4):287–297.

[36] Bhatia K, Huppi K, Spangler G, Siwarski D, Iyer R, Magrath I. Point mutations in the c-Myc transactivation domain are common in Burkitt's lymphoma and mouse plasmacytomas. Nature Genet. 1993;5:56–61.

[37] Allday MJ. How does Epstein–Barr virus (EBV) complement the activation of Myc in the pathogenesis of Burkitt's lymphoma? Semin Cancer Biol. 2009 Dec;19(6):366–376.

[38] Carbone A, Gloghini A, Serraino D, Spina M. HIV-associated Hodgkin lymphoma. Curr Opin HIV AIDS. 2009 Jan;4(1):3–10.

[39] Raphaël M, Said J, Borisch B, Cesarman E, Harris NL. Lymphomas associated with HIV infection. In: Swerdlow SH, Campo, E., Harris, N.L., Jaffe, E.S., Pileri, S.A., Stein, H., Thiele, J., editor. WHO Classification of Tumours of Haematopoietic and Lymphoid Tissues. Fourth ed. Lyon, France: IARC Press; 2008. p. 340–342.

[40] Barros MH, Hassan R, Niedobitek G. Tumor-associated macrophages in pediatric classical Hodgkin lymphoma: association with Epstein–Barr virus, lymphocyte subsets, and prognostic impact. Clin Cancer Res. [Research Support, Non-U.S. Gov't]. 2012 Jul 15;18(14):3762–3771.

[41] Huang X, Kushekhar K, Nolte I, Kooistra W, Visser L, Bouwman I, et al. HLA associations in classical Hodgkin lymphoma: EBV status matters. PLoS One. [Research Support, Non-U.S. Gov't]. 2012;7(7):e39986.

[42] Schmitz R, Hansmann ML, Bohle V, Martin-Subero JI, Hartmann S, Mechtersheimer G, et al. TNFAIP3 (A20) is a tumor suppressor gene in Hodgkin lymphoma and primary mediastinal B cell lymphoma. J Exp Med. [Research Support, Non-U.S. Gov't]. 2009 May 11;206(5):981–989.

[43] Reichel J, Chadburn A, Rubinstein PG, Giulino-Roth L, Tam W, Liu Y, et al. Flow-sorting and exome sequencing reveals the oncogenome of primary Hodgkin and Reed-Sternberg cells. Blood. 2014 Dec 8.

[44] Giulino L, Mathew S, Ballon G, Chadburn A, Barouk S, Antonicelli G, et al. A20 (TNFAIP3) genetic alterations in EBV-associated AIDS-related lymphoma. Blood. 2011 May 5;117(18):4852–4854.

[45] Cader FZ, Vockerodt M, Bose S, Nagy E, Brundler MA, Kearns P, et al. The EBV oncogene LMP1 protects lymphoma cells from cell death through the collagen-mediated activation of DDR1. Blood. [Research Support, Non-U.S. Gov't]. 2013 Dec 19;122(26):4237–4245.

[46] Vockerodt M, Wei W, Nagy E, Prouzova Z, Schrader A, Kube D, et al. Suppression of the LMP2A target gene, EGR-1, protects Hodgkin's lymphoma cells from entry to the EBV lytic cycle. J Pathol. [Research Support, Non-U.S. Gov't]. 2013 Aug;230(4):399–409.

[47] Morscio J, Dierickx D, Ferreiro JF, Herreman A, Van Loo P, Bittoun E, et al. Gene expression profiling reveals clear differences between EBV-positive and EBV-negative posttransplant lymphoproliferative disorders. Am J Transplant. [Comparative Study Research Support, Non-U.S. Gov't]. 2013 May;13(5):1305–1316.

[48] Delecluse HJ, Anagnostopoulos I, Dallenbach F, Hummel M, Marafioti T, Schneider U, et al. Plasmablastic lymphomas of the oral cavity: a new entity associated with the human immunodeficiency virus infection. Blood. 1997 Feb 15;89(4):1413–1420.

[49] Colomo L, Loong F, Rives S, Pittaluga S, Martinez A, Lopez-Guillermo A, et al. Diffuse large B-cell lymphomas with plasmablastic differentiation represent a heterogeneous group of disease entities. Am J Surg Pathol. [Research Support, Non-U.S. Gov't]. 2004 Jun;28(6):736–747.

[50] Corti M, Villafane MF, Bistmans A, Campitelli A, Narbaitz M, Bare P. Oral cavity and extra-oral plasmablastic lymphomas in AIDS patients: report of five cases and review of the literature. Int J STD AIDS. [Case Reports]. 2011 Dec;22(12):759–763.

[51] Suzuki R. Pathogenesis and treatment of extranodal natural killer/T-cell lymphoma. Semin Hematol. [Review]. 2014 Jan;51(1):42–51.
[52] Chang Y, Cesarman E, Pessin MS, Lee F, Culpepper J, Knowles DM, et al. Identification of herpesvirus-like DNA sequences in AIDS-associated Kaposi's sarcoma. Science. 1994;266:1865–1869.
[53] Mesri EA, Cesarman E, Boshoff C. Kaposi's sarcoma and its associated herpesvirus. Nature reviews Cancer. [Research Support, N.I.H., Extramural Research Support, Non-U.S. Gov't Review]. 2010 Oct;10(10):707–719.
[54] Cesarman E, Chang Y, Moore PS, Said JW, Knowles DM. Kaposi's Sarcoma-associated Herpesvirus-like DNA sequences in AIDS-related body cavity-based lymphomas. N Eng J Med. 1995;332:1186–1191.
[55] Soulier J, Grollet L, Oksenhendler E, Cacoub P, Cazals-Hatem D, Babinet P, et al. Kaposi's sarcoma-associated herpesvirus-like DNA sequences in multicentric Castleman's disease. Blood. 1995;86:1275–1280.
[56] Dupin N, Diss TL, Kellam P, Tulliez M, Du MQ, Sicard D, et al. HHV-8 is associated with a plasmablastic variant of Castleman disease that is linked to HHV-8-positive plasmablastic lymphoma. Blood. 2000;95(4):1406–1412.
[57] Ballestas ME, Chatis PA, Kaye KM. Efficient persistence of extrachromosomal KSHV DNA mediated by latency-associated nuclear antigen. Science. 1999;284(5414):641–644.
[58] Friborg J, Jr., Kong W, Hottiger MO, Nabel GJ. p53 inhibition by the LANA protein of KSHV protects against cell death. Nature. 1999;402(6764):889–894.
[59] Li M, Lee H, Yoon D-W, Albrecht J-C, Fleckenstein B, Jung J. Kaposi's sarcoma-associated herpesvirus encodes a functional cyclin. J Virol. 1997;71:1984–1991.
[60] Guasparri I, Keller SA, Cesarman E. KSHV vFLIP Is Essential for the Survival of Infected Lymphoma Cells. J Exp Med. 2004 Apr 5;199(7):993–1003.
[61] Sun Q, Matta H, Chaudhary PM. The human herpes virus 8 encoded viral FLICE inhibitory protein protects against growth factor withdrawal-induced apoptosis via NF- {kappa}B activation. Blood. 2002;24:24.
[62] Lee JS, Li Q, Lee JY, Lee SH, Jeong JH, Lee HR, et al. FLIP-mediated autophagy regulation in cell death control. Nat Cell Biol. 2009 Nov;11(11):1355–1362.
[63] Verma SC, Cai Q, Kreider E, Lu J, Robertson ES. Comprehensive analysis of LANA interacting proteins essential for viral genome tethering and persistence. PLoS One. [Research Support, N.I.H., Extramural]. 2013;8(9):e74662.
[64] Santag S, Jager W, Karsten CB, Kati S, Pietrek M, Steinemann D, et al. Recruitment of the tumour suppressor protein p73 by Kaposi's Sarcoma Herpesvirus latent nuclear antigen contributes to the survival of primary effusion lymphoma cells. Oncogene. [Research Support, Non-U.S. Gov't]. 2013 Aug 8;32(32):3676–3685.
[65] Bentz GL, Bheda-Malge A, Wang L, Shackelford J, Damania B, Pagano JS. KSHV LANA and EBV LMP1 induce the expression of UCH-L1 following viral transformation. Virology. [Research Support, N.I.H., Extramural]. 2014 Jan 5;448:293–302.
[66] Lu J, Jha HC, Verma SC, Sun Z, Banerjee S, Dzeng R, et al. Kaposi's sarcoma-associated herpesvirus-encoded LANA contributes to viral latent replication by activating phosphorylation of survivin. J Virol. [Research Support, N.I.H., Extramural]. 2014 Apr;88(8):4204–4217.
[67] Forero A, Moore PS, Sarkar SN. Role of IRF4 in IFN-stimulated gene induction and maintenance of Kaposi sarcoma-associated herpesvirus latency in primary effusion lymphoma cells. J Immunol. [Research Support, N.I.H., Extramural]. 2013 Aug 1;191(3):1476–1485.
[68] Cousins E, Nicholas J. Role of human herpesvirus 8 interleukin-6-activated gp130 signal transducer in primary effusion lymphoma cell growth and viability. J Virol. [Research Support, N.I.H., Extramural]. 2013 Oct;87(19):10816–10827.

[69] Chen D, Gao Y, Nicholas J. Human herpesvirus 8 interleukin-6 contributes to primary effusion lymphoma cell viability via suppression of proapoptotic cathepsin D, a cointeraction partner of vitamin K epoxide reductase complex subunit 1 variant 2. J Virol. [Research Support, N.I.H., Extramural]. 2014 Jan;88(2):1025–1038.

[70] Zuo J, Hislop AD, Leung CS, Sabbah S, Rowe M. Kaposi's sarcoma-associated herpesvirus-encoded viral IRF3 modulates major histocompatibility complex class II (MHC-II) antigen presentation through MHC-II transactivator-dependent and -independent mechanisms: implications for oncogenesis. J Virol. [Research Support, Non-U.S. Gov't]. 2013 May;87(10):5340–5350.

[71] King CA. Kaposi's sarcoma-associated herpesvirus kaposin B induces unique monophosphorylation of STAT3 at serine 727 and MK2-mediated inactivation of the STAT3 transcriptional repressor TRIM28. J Virol. 2013 Aug;87(15):8779–8791.

[72] Cai X, Lu S, Zhang Z, Gonzalez CM, Damania B, Cullen BR. Kaposi's sarcoma-associated herpesvirus expresses an array of viral microRNAs in latently infected cells. Proc Natl Acad Sci USA. [Research Support, N.I.H., Extramural Research Support, Non-U.S. Gov't Research Support, U.S. Gov't, P.H.S.]. 2005 Apr 12;102(15):5570–5575.

[73] Samols MA, Hu J, Skalsky RL, Renne R. Cloning and identification of a microRNA cluster within the latency-associated region of Kaposi's sarcoma-associated herpesvirus. J Virol. [Research Support, N.I.H., Extramural Research Support, Non-U.S. Gov't Research Support, U.S. Gov't, P.H.S.]. 2005 Jul;79(14):9301–9305.

[74] Pfeffer S, Sewer A, Lagos-Quintana M, Sheridan R, Sander C, Grasser FA, et al. Identification of microRNAs of the herpesvirus family. Nat Methods. [Comparative Study Evaluation Studies Research Support, Non-U.S. Gov't Research Support, U.S. Gov't, P.H.S.]. 2005 Apr;2(4):269–276.

[75] Grundhoff A, Sullivan CS, Ganem D. A combined computational and microarray-based approach identifies novel microRNAs encoded by human gamma-herpesviruses. Rna. [Research Support, N.I.H., Extramural]. 2006 May;12(5):733–750.

[76] Gottwein E. Kaposi's Sarcoma-Associated Herpesvirus microRNAs. Front Microbiol. 2012;3:165.

[77] Moody R, Zhu Y, Huang Y, Cui X, Jones T, Bedolla R, et al. KSHV microRNAs mediate cellular transformation and tumorigenesis by redundantly targeting cell growth and survival pathways. PLoS Pathog. [Research Support, N.I.H., Extramural]. 2013;9(12):e1003857.

[78] Lei X, Bai Z, Ye F, Xie J, Kim CG, Huang Y, et al. Regulation of NF-kappaB inhibitor IkappaBalpha and viral replication by a KSHV microRNA. Nat Cell Biol. 2010 Feb;12(2):193–199.

[79] Gottwein E, Cullen BR. A human herpesvirus microRNA inhibits p21 expression and attenuates p21-mediated cell cycle arrest. J Virol. 2010 May;84(10):5229–5237.

[80] Skalsky RL, Samols MA, Plaisance KB, Boss IW, Riva A, Lopez MC, et al. Kaposi's sarcoma-associated herpesvirus encodes an ortholog of miR-155. J Virol. [Research Support, N.I.H., Extramural Research Support, Non-U.S. Gov't]. 2007 Dec;81(23):12836–12845.

[81] Gottwein E, Mukherjee N, Sachse C, Frenzel C, Majoros WH, Chi JT, et al. A viral microRNA functions as an orthologue of cellular miR-155. Nature. [Research Support, N.I.H., Extramural Research Support, Non-U.S. Gov't]. 2007 Dec 13;450(7172):1096–1099.

[82] Dahlke C, Maul K, Christalla T, Walz N, Schult P, Stocking C, et al. A microRNA encoded by Kaposi sarcoma-associated herpesvirus promotes B-cell expansion in vivo. PLoS One. [Research Support, Non-U.S. Gov't]. 2012;7(11):e49435.

[83] Lin X, Li X, Liang D, Lan K. MicroRNAs and unusual small RNAs discovered in Kaposi's sarcoma-associated herpesvirus virions. J Virol. [Research Support, Non-U.S. Gov't]. 2012 Dec;86(23):12717–12730.

[84] Gottwein E, Corcoran DL, Mukherjee N, Skalsky RL, Hafner M, Nusbaum JD, et al. Viral microRNA targetome of KSHV-infected primary effusion lymphoma cell lines. Cell Host Microbe. [Research Support, N.I.H., Extramural Research Support, Non-U.S. Gov't]. 2011 Nov 17;10(5):515–526.
[85] Haecker I, Gay LA, Yang Y, Hu J, Morse AM, McIntyre LM, et al. Ago HITS-CLIP expands understanding of Kaposi's sarcoma-associated herpesvirus miRNA function in primary effusion lymphomas. PLoS Pathog. [Research Support, N.I.H., Extramural Research Support, Non-U.S. Gov't]. 2012;8(8):e1002884.
[86] Nador RG, Cesarman E, Chadburn A, Dawson DB, Ansari MQ, Said J, et al. Primary effusion lymphoma: a distinct clinicopathologic entity associated with the Kaposi's sarcoma-associated herpesvirus. Blood. 1996;88:645–656.
[87] Horenstein MG, Nador RG, Chadburn A, Hyjek EM, Inghirami G, Knowles DM, et al. Epstein–Barr virus latent gene expression in primary effusion lymphomas containing Kaposi's sarcoma-associated herpesvirus human herpesvirus-8. Blood. 1997;90(3):1186–1191.
[88] Szekely L, Chen F, Teramoto N, Ehlin-Henriksson B, Pokrovskaja K, Szeles A, et al. Restricted expression of Epstein–Barr virus (EBV)-encoded, growth transformation-associated antigens in an EBV- and human herpesvirus type 8-carrying body cavity lymphoma line. General Virology. 1998;79:1445–1452.
[89] Xia T, O'Hara A, Araujo I, Barreto J, Carvalho E, Sapucaia JB, et al. EBV microRNAs in primary lymphomas and targeting of CXCL-11 by ebv-mir-BHRF1-3. Cancer Res. [Research Support, N.I.H., Extramural Research Support, Non-U.S. Gov't]. 2008 Mar 1;68(5):1436–1442.
[90] Chadburn A, Hyjek E, Mathew S, Cesarman E, Said J, Knowles DM. KSHV-positive solid lymphomas represent an extra-cavitary variant of primary effusion lymphoma. Am J Surg Pathol. 2004 Nov;28(11):1401–1416.
[91] Ashihara E, Shimazaki C, Hirai H, Inaba T, Hasegawa G, Mori S, et al. Human herpes virus 8-negative primary effusion lymphoma in a patient with a ventriculoperitoneal shunt tube. Int J Hematol. [Case Reports Review]. 2001 Oct;74(3):327–332.
[92] Paner GP, Jensen J, Foreman KE, Reyes CV. HIV and HHV-8 negative primary effusion lymphoma in a patient with hepatitis C virus-related liver cirrhosis. Leuk Lymphoma. [Case Reports Review]. 2003 Oct;44(10):1811–1814.
[93] Matsumoto Y, Nomura K, Ueda K, Satoh K, Yasuda N, Taki T, et al. Human herpesvirus 8-negative malignant effusion lymphoma: a distinct clinical entity and successful treatment with rituximab. Leuk Lymphoma. [Case Reports Review]. 2005 Mar;46(3):415–419.
[94] Tsagarakis NJ, Argyrou A, Gortzolidis G, Kentrou N, Papadhimitriou SI, Tzanetou K, et al. Report of an HIV and HHV-8 negative case of primary effusion lymphoma with idiopathic T4 lymphocytopenia. Int J Hematol. [Case Reports]. 2009 Jul;90(1):94–98.
[95] Saini N, Hochberg EP, Linden EA, Jha S, Grohs HK, Sohani AR. HHV8-Negative Primary Effusion Lymphoma of B-Cell Lineage: Two Cases and a Comprehensive Review of the Literature. Case Rep Oncol Med. 2013;2013:292301.
[96] Xiao J, Selvaggi SM, Leith CP, Fitzgerald SA, Stewart J, 3rd. Kaposi sarcoma herpesvirus/human herpesvirus-8-negative effusion-based lymphoma: report of 3 cases and review of the literature. Cancer Cytopathol. [Case Reports Review]. 2013 Nov;121(11):661–669.
[97] Du MQ, Liu H, Diss TC, Ye H, Hamoudi RA, Dupin N, et al. Kaposi sarcoma-associated herpesvirus infects monotypic (IgM lambda) but polyclonal naive B cells in Castleman disease and associated lymphoproliferative disorders. Blood. 2001 Apr 1;97(7):2130–2136.
[98] Du MQ, Diss TC, Liu H, Ye H, Hamoudi RA, Cabecadas J, et al. KSHV- and EBV-associated germinotropic lymphoproliferative disorder. Blood. 2002 Nov 1;100(9):3415–3418.

[99] Parravicini C, Corbellino M, Paulli M, Magrini U, Lazzarino M, Moore PS, et al. Expression of a virus-derived cytokine, KSHV vIL-6, in HIV-seronegative Castleman's disease. Am J Pathol. 1997;151(6):1517–1522.

[100] Cannon JS, Nicholas J, Orenstein JM, Mann RB, Murray PG, Browning PJ, et al. Heterogeneity of viral IL-6 expression in HHV-8-associated diseases. Journal of Infectious Diseases. 1999;180:824–828.

[101] Staskus KA, Sun R, Miller G, Racz P, Jaslowski A, Metroka C, et al. Cellular tropism and viral interleukin-6 expression distinguish human herpesvirus 8 involvement in Kaposi's sarcoma, primary effusion lymphoma, and multicentric Castleman's disease. J Virol. 1999;73(5):4181–4187.

[102] Menke DM, Chadbum A, Cesarman E, Green E, Berenson J, Said J, et al. Analysis of the human herpesvirus 8 (HHV-8) genome and HHV-8 vIL-6 expression in archival cases of castleman disease at low risk for HIV infection. Am J Clin Pathol. 2002 Feb;117(2):268–275.

[103] Uldrick TS, Wang V, O'Mahony D, Aleman K, Wyvill KM, Marshall V, et al. An interleukin-6-related systemic inflammatory syndrome in patients co-infected with Kaposi sarcoma-associated herpesvirus and HIV but without Multicentric Castleman disease. Clin Infect Dis. 2010 Aug 1;51(3):350–358.

[104] Polizzotto MN, Uldrick TS, Hu D, Yarchoan R. Clinical Manifestations of Kaposi Sarcoma Herpesvirus Lytic Activation: Multicentric Castleman Disease (KSHV-MCD) and the KSHV Inflammatory Cytokine Syndrome. Front Microbiol. 2012;3:73.

[105] Fujii M, Matsuoka M. Human T-Cell Leukemia Virus Types 1 and 2. In: D. M. Knipe PMH, editor. Fields Virology. 6th ed. Philadelphia: Lippincott Williams & Wilkins; 2013, 1474–501.

[106] Matsuoka M, Jeang KT. Human T-cell leukemia virus type 1 (HTLV-1) and leukemic transformation: viral infectivity, Tax, HBZ and therapy. Oncogene. 2011 Mar 24;30(12):1379–1389.

[107] Grassmann R, Dengler C, Muller-Fleckenstein I, Fleckenstein B, McGuire K, Dokhelar MC, et al. Transformation to continuous growth of primary human T lymphocytes by human T-cell leukemia virus type I X-region genes transduced by a Herpesvirus saimiri vector. Proc Natl Acad Sci USA. [Research Support, Non-U.S. Gov't Research Support, U.S. Gov't, P.H.S.]. 1989 May;86(9):3351–3355.

[108] Grassmann R, Berchtold S, Radant I, Alt M, Fleckenstein B, Sodroski JG, et al. Role of human T-cell leukemia virus type 1 X region proteins in immortalization of primary human lymphocytes in culture. J Virol. [Research Support, Non-U.S. Gov't Research Support, U.S. Gov't, P.H.S.]. 1992 Jul;66(7):4570–4575.

[109] Nerenberg M, Hinrichs SH, Reynolds RK, Khoury G, Jay G. The tat gene of human T-lymphotropic virus type 1 induces mesenchymal tumors in transgenic mice. Science. 1987 Sep 11;237(4820):1324–1329.

[110] Grossman WJ, Kimata JT, Wong FH, Zutter M, Ley TJ, Ratner L. Development of leukemia in mice transgenic for the tax gene of human T-cell leukemia virus type I. Proc Natl Acad Sci USA. [Research Support, Non-U.S. Gov't Research Support, U.S. Gov't, P.H.S.]. 1995 Feb 14;92(4):1057–1061.

[111] Sun SC, Cesarman E. NF-kappaB as a target for oncogenic viruses. Curr Top Microbiol Immunol. 2011;349:197–244.

[112] Watanabe T. HTLV-1-associated diseases. Int J Hematol. [Research Support, Non-U.S. Gov't Review]. 1997 Oct;66(3):257–278.

[113] Matsuoka M, Yasunaga J. Human T-cell leukemia virus type 1: replication, proliferation and propagation by Tax and HTLV-1 bZIP factor. Curr Opin Virol. 2013 Dec;3(6):684–691.

[114] Reitz MS, Jr., Popovic M, Haynes BF, Clark SC, Gallo RC. Relatedness by nucleic acid hybridization of new isolates of human T-cell leukemia-lymphoma virus (HTLV) and demonstration of provirus in uncultured leukemic blood cells. Virology. 1983 Apr 30;126(2):688–692.

[115] Yoshida M, Seiki M, Yamaguchi K, Takatsuki K. Monoclonal integration of human T-cell leukemia provirus in all primary tumors of adult T-cell leukemia suggests causative role of human T-cell leukemia virus in the disease. Proc Natl Acad Sci USA. [Research Support, Non-U.S. Gov't]. 1984 Apr;81(8):2534–2537.

[116] Yoshida M, Hattori S, Seiki M. Molecular biology of human T-cell leukemia virus associated with adult T-cell leukemia. Curr Top Microbiol Immunol. [Research Support, Non-U.S. Gov't]. 1985;115:157–175.

[117] Gaudray G, Gachon F, Basbous J, Biard-Piechaczyk M, Devaux C, Mesnard JM. The complementary strand of the human T-cell leukemia virus type 1 RNA genome encodes a bZIP transcription factor that down-regulates viral transcription. J Virol. 2002 Dec;76(24):12813–12822.

[118] Satou Y, Yasunaga J, Yoshida M, Matsuoka M. HTLV-I basic leucine zipper factor gene mRNA supports proliferation of adult T cell leukemia cells. Proc Natl Acad Sci USA. 2006 Jan 17;103(3):720–725.

[119] Zhi H, Yang L, Kuo YL, Ho YK, Shih HM, Giam CZ. NF-kappaB hyper-activation by HTLV-1 tax induces cellular senescence, but can be alleviated by the viral anti-sense protein HBZ. PLoS Pathog. 2011 Apr;7(4):e1002025.

[120] Cesarman E, Chadburn A, Inghirami G, Gaidano G, Knowles DM. Structural and functional analysis of oncogenes and tumor suppressor genes in adult T-cell leukemia/lymphoma shows frequent p53 mutations. Blood. [Research Support, Non-U.S. Gov't Research Support, U.S. Gov't, P.H.S.]. 1992 Dec 15;80(12):3205–3216.

[121] Pise-Masison CA, Radonovich M, Sakaguchi K, Appella E, Brady JN. Phosphorylation of p53: a novel pathway for p53 inactivation in human T-cell lymphotropic virus type 1-transformed cells. J Virol. 1998 Aug;72(8):6348–6355.

[122] Maruyama K, Fukushima T, Kawamura K, Mochizuki S. Chromosome and gene rearrangements in immortalized human lymphocytes infected with human T-lymphotropic virus type I. Cancer Res. 1990 Sep 1;50(17 Suppl):5697S–5702S.

[123] Fukuhara S, Hinuma Y, Gotoh YI, Uchino H. Chromosome aberrations in T lymphocytes carrying adult T-cell leukemia-associated antigens (ATLA) from healthy adults. Blood. [Research Support, Non-U.S. Gov't]. 1983 Jan;61(1):205–207.

[124] Sanada I, Tanaka R, Kumagai E, Tsuda H, Nishimura H, Yamaguchi K, et al. Chromosomal aberrations in adult T cell leukemia: relationship to the clinical severity. Blood. [Research Support, Non-U.S. Gov't]. 1985 Mar;65(3):649–654.

[125] Paydas S. Hepatitis C virus and lymphoma. Crit Rev Oncol Hematol. 2015 Mar;93(3):246–256.

[126] Negri E, Little D, Boiocchi M, La Vecchia C, Franceschi S. B-cell non-Hodgkin's lymphoma and hepatitis C virus infection: a systematic review. Int J Cancer. 2004 Aug 10;111(1):1–8.

[127] Engels EA, Chatterjee N, Cerhan JR, Davis S, Cozen W, Severson RK, et al. Hepatitis C virus infection and non-Hodgkin lymphoma: results of the NCI-SEER multi-center case-control study. Int J Cancer. 2004 Aug 10;111(1):76–80.

[128] Paydas S, Ergin M, Tanriverdi K, Yavuz S, Disel U, Kilic NB, et al. Detection of hepatitis C virus RNA in paraffin-embedded tissues from patients with non-Hodgkin's lymphoma. Am J Hematol. 2004 Jul;76(3):252–257.

[129] Sung VM, Shimodaira S, Doughty AL, Picchio GR, Can H, Yen TS, et al. Establishment of B-cell lymphoma cell lines persistently infected with hepatitis C virus in vivo and in vitro: the apoptotic effects of virus infection. J Virol. 2003 Feb;77(3):2134–2146.

[130] Machida K, Cheng KT, Sung VM, Shimodaira S, Lindsay KL, Levine AM, et al. Hepatitis C virus induces a mutator phenotype: enhanced mutations of immunoglobulin and protooncogenes. Proc Natl Acad Sci USA. 2004 Mar 23;101(12):4262–4267.

[131] Kasama Y, Sekiguchi S, Saito M, Tanaka K, Satoh M, Kuwahara K, et al. Persistent expression of the full genome of hepatitis C virus in B cells induces spontaneous development of B-cell lymphomas in vivo. Blood. 2010 Dec 2;116(23):4926–4933.
[132] Weng WK, Levy S. Hepatitis C virus (HCV) and lymphomagenesis. Leuk Lymphoma. 2003 Jul;44(7):1113–1120.
[133] Hermine O, Lefrere F, Bronowicki JP, Mariette X, Jondeau K, Eclache-Saudreau V, et al. Regression of splenic lymphoma with villous lymphocytes after treatment of hepatitis C virus infection. N Engl J Med. 2002 Jul 11;347(2):89–94.
[134] Saadoun D, Suarez F, Lefrere F, Valensi F, Mariette X, Aouba A, et al. Splenic lymphoma with villous lymphocytes, associated with type II cryoglobulinemia and HCV infection: a new entity? Blood. 2005 Jan 1;105(1):74–76.
[135] Hartridge-Lambert SK, Stein EM, Markowitz AJ, Portlock CS. Hepatitis C and non-Hodgkin lymphoma: the clinical perspective. Hepatology. 2012 Feb;55(2):634–641.
[136] Vallisa D, Bernuzzi P, Arcaini L, Sacchi S, Callea V, Marasca R, et al. Role of anti-hepatitis C virus (HCV) treatment in HCV-related, low-grade, B-cell, non-Hodgkin's lymphoma: a multicenter Italian experience. J Clin Oncol. 2005 Jan 20;23(3):468–473.
[137] A review of human carcinogens, Part B: Biological agents. Lyon, France: World Health Organization International Agency for Research on Cancer; 2012.
[138] Carbone A. Emerging pathways in the development of AIDS-related lymphomas. Lancet Oncol. 2003 Jan;4(1):22–29.
[139] Gaidano G, Carbone A, Dalla-Favera R. Genetic basis of acquired immunodeficiency syndrome-related lymphomagenesis. J Natl Cancer Inst Monogr. 1998(23):95–100.
[140] Said J, Cesarman E, Knowles D. Lymphadenopathy and the lymphoid neoplasms associated with HIV infection and other causes of immunosuppression. In: Orazzi A WL, Foucar K, Knowles DM, editor. Knowles' Neoplastic Hematopathology. Third ed. Baltimore: Wolters Kuwer/Williams and Wilkins; 2014. p. 584–658.
[141] Hussain SK, Zhu W, Chang SC, Breen EC, Vendrame E, Magpantay L, et al. Serum levels of the chemokine CXCL13, genetic variation in CXCL13 and its receptor CXCR5, and HIV-associated non-hodgkin B-cell lymphoma risk. Cancer Epidemiol Biomarkers Prev. [Comparative Study Multicenter Study Research Support, N.I.H., Extramural Research Support, Non-U.S. Gov't]. 2013 Feb;22(2):295–307.
[142] Hussain SK, Hessol NA, Levine AM, Breen EC, Anastos K, Cohen M, et al. Serum biomarkers of immune activation and subsequent risk of non-hodgkin B-cell lymphoma among HIV-infected women. Cancer Epidemiol Biomarkers Prev. 2013 Nov;22(11):2084–2093.
[143] Achenbach CJ, Buchanan AL, Cole SR, Hou L, Mugavero MJ, Crane HM, et al. HIV viremia and incidence of non-Hodgkin lymphoma in patients successfully treated with antiretroviral therapy. Clin Infect Dis. [Research Support, N.I.H., Extramural]. 2014 Jun;58(11):1599–1606.
[144] Kowalkowski MA, Mims MP, Amiran ES, Lulla P, Chiao EY. Effect of immune reconstitution on the incidence of HIV-related Hodgkin lymphoma. PLoS One. [Research Support, N.I.H., Extramural]. 2013;8(10):e77409.
[145] Curreli S, Krishnan S, Reitz M, Lunardi-Iskandar Y, Lafferty MK, Garzino-Demo A, et al. B cell lymphoma in HIV transgenic mice. Retrovirology. [Research Support, N.I.H., Extramural]. 2013;10:92.
[146] Pelicci PG, Knowles DM, 2nd, Arlin ZA, Wieczorek R, Luciw P, Dina D, et al. Multiple monoclonal B cell expansions and c-myc oncogene rearrangements in acquired immune deficiency syndrome-related lymphoproliferative disorders. Implications for lymphomagenesis. J Exp Med. [Research Support, Non-U.S. Gov't Research Support, U.S. Gov't, P.H.S.]. 1986 Dec 1;164(6):2049–2060.

[147] Przybylski GK, Goldman J, Ng VL, McGrath MS, Herndier BG, Schenkein DP, et al. Evidence for early B-cell activation preceding the development of Epstein–Barr virus-negative acquired immunodeficiency syndrome-related lymphoma. Blood. [Research Support, U.S. Gov't, P.H.S.]. 1996 Dec 15;88(12):4620–4629.
[148] Martinez-Maza O, Breen EC. B-cell activation and lymphoma in patients with HIV. Curr Opin Oncol. [Research Support, Non-U.S. Gov't Research Support, U.S. Gov't, P.H.S. Review]. 2002 Sep;14(5):528–532.
[149] Lefevre EA, Krzysiek R, Loret EP, Galanaud P, Richard Y. Cutting edge: HIV-1 Tat protein differentially modulates the B cell response of naive, memory, and germinal center B cells. J Immunol. [Research Support, Non-U.S. Gov't]. 1999 Aug 1;163(3):1119–1122.
[150] Swingler S, Brichacek B, Jacque JM, Ulich C, Zhou J, Stevenson M. HIV-1 Nef intersects the macrophage CD40L signalling pathway to promote resting-cell infection. Nature. 2003 Jul 10;424(6945):213–219.
[151] Qiao X, He B, Chiu A, Knowles DM, Chadburn A, Cerutti A. Human immunodeficiency virus 1 Nef suppresses CD40-dependent immunoglobulin class switching in bystander B cells. Nat Immunol. 2006 Mar;7(3):302–310.
[152] He B, Qiao X, Klasse PJ, Chiu A, Chadburn A, Knowles DM, et al. HIV-1 envelope triggers polyclonal Ig class switch recombination through a CD40-independent mechanism involving BAFF and C-type lectin receptors. J Immunol. [Research Support, N.I.H., Extramural Research Support, Non-U.S. Gov't]. 2006 Apr 1;176(7):3931–3941.
[153] Giagulli C, Marsico S, Magiera AK, Bruno R, Caccuri F, Barone I, et al. Opposite effects of HIV-1 p17 variants on PTEN activation and cell growth in B cells. PLoS One. [Research Support, Non-U.S. Gov't]. 2011;6(3):e17831.
[154] Nunnari G, Smith JA, Daniel R. HIV-1 Tat and AIDS-associated cancer: targeting the cellular anti-cancer barrier? J Exp Clin Cancer Res. [Research Support, N.I.H., Extramural Research Support, Non-U.S. Gov't Review]. 2008;27:3.
[155] Cesarman E. How do viruses trick B cells into becoming lymphomas? Curr Opin Hematol. 2014 Jul;21(4):358–368.
[156] Zhu X, Guo Y, Yao S, Yan Q, Xue M, Hao T, et al. Synergy between Kaposi's sarcoma-associated herpesvirus (KSHV) vIL-6 and HIV-1 Nef protein in promotion of angiogenesis and oncogenesis: role of the AKT signaling pathway. Oncogene. [Research Support, Non-U.S. Gov't]. 2014 Apr 10;33(15):1986–1996.
[157] Yan Q, Ma X, Shen C, Cao X, Feng N, Qin D, et al. Inhibition of Kaposi's sarcoma-associated herpesvirus lytic replication by HIV-1 Nef and cellular microRNA hsa-miR-1258. J Virol. [Research Support, N.I.H., Extramural Research Support, Non-U.S. Gov't]. 2014 May;88(9):4987–5000.
[158] Lenassi M, Cagney G, Liao M, Vaupotic T, Bartholomeeusen K, Cheng Y, et al. HIV Nef is secreted in exosomes and triggers apoptosis in bystander CD4+ T cells. Traffic. [Research Support, N.I.H., Extramural]. 2010 Jan;11(1):110–122.

Loic Ysebaert, Roch Houot, and Franck Morschhauser

19 Translating science into therapy of lymphoma

19.1 Introduction

Until recently, therapeutic progresses in the field of lymphoma have largely been made by empirical observations or even by serendipity. The two major therapeutic advances were the introduction of anthracyclines in the seventies and the development of immunotherapy with anti-CD20 monoclonal antibodies (MoAb) between 1990 and 2010, as a consequence of better knowledge in cancer cell immunology and in the engineering of MoAb. Recently, better-engineered antibody drug conjugates have entered the clinical arena, as well as novel cancer immunotherapies with immune checkpoints inhibitors. However, we now have the opportunity to design "personalized therapeutic strategies" by a better understanding of the multistep process of lymphomagenesis [1]. Significant progress has been made since the description of the first chromosomal translocations involving the IGH genes and the oncogenes BCL2, BCL6 and MYC, recently through the input of chemists, systems biologists and image specialists who contributed to the development of novel tools allowing the measurement of interindividual disparities and intrapatient tumor heterogeneity. This is best illustrated by the advances in genomic analyses, which allowed (i) to recognize the existence of multiple phenotypic subtypes of diffuse large B-cell lymphoma (DLBCL) based on gene expression profiling [2, 3], (ii) to better understand the role of epigenetic changes [4] and (iii) to discover recurrent mutations affecting key signaling pathways or cellular functions [5]. The most comprehensive molecular assessment of lymphoma genetics came from whole genome and whole exome next generation sequencing (NGS) studies, which were performed on the major histological lymphoma subtypes: DLBCL [6–8], Burkitt [9, 10], follicular (FL) [11], mantle cell (MCL) [12], splenic marginal zone [13] and peripheral T-cell lymphoma [14, 15]. Only recently, an increasing knowledge of intracellular signaling cascades (Fig. 19.1) has paved the way for more rational drug designs and to envision synthetic lethality by targeting different pathways at stake in lymphoma survival.

In this chapter, we will focus on most recent "science-driven" therapeutic successes, namely targeting of the B-cell receptor (BCR) signaling pathway, molecules regulating apoptosis, epigenetic regulators, and innate/acquired immune checkpoints.

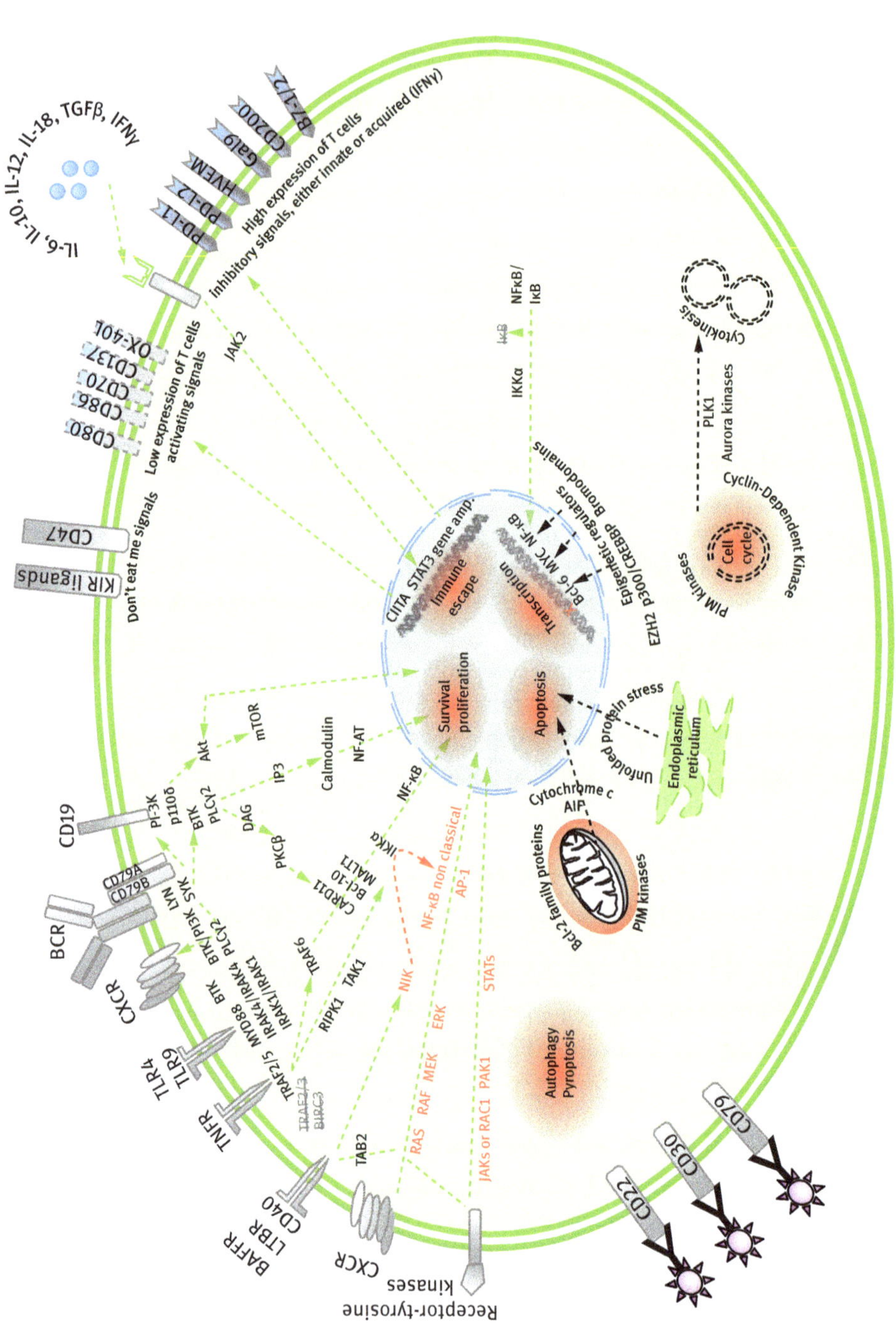

Fig. 19.1: Druggable pathways in Bcell malignancies. In red are highlighted suggested pathways of resistance to ibrutinib.

19.2 Inhibition of the B-cell receptor signaling pathway

Signal transduction through the BCR (reviewed in [16–18]) plays a pivotal role during physiologic B-cell development, and has been implicated in the pathogenesis of a variety of lymphoid malignancies, where it is often aberrantly activated. The BCR can be engaged through two main modes of activation, referred to as "tonic" (i.e. antigen (Ag)-independent, largely mediated by PI3K) or "active" (Ag-dependent, mediated by several tyrosine kinases and adaptor molecules, including Bruton's tyrosine kinase (BTK) [19]. Additionally, ABC DLBCLs are addicted to a "chronic active" form of BCR signaling that is in part explained by somatic mutations in the *CD79B* co-receptor molecules.

In recent years, drugs targeting the BCR signaling have become the subject of intense investigations. There are multiple aspects that should be considered when assessing the utility of BCR inhibitors for the treatment of a given lymphoma entity: (i) the maturation stage of the cell of origin, (ii) the predominant mode of BCR signaling ("tonic" vs "Ag-dependent" vs "chronic active"), and consequently the preferred downstream pathways perverted in the lymphoma cells; (iii) the transcription factor(s) that are engaged downstream of the BCR; (iv) the genetic status of the individual tumor (i.e. presence of mutations upstream or downstream the targeted kinase); and (v) the presence of other cell surface receptors that can hijack BCR-driven signaling hubs or drive pro-survival pathways not triggered by BCR activation, and would be left untouched by BCR-targeting agents.

19.2.1 Historical overview of BTK/PI3K inhibitors

Currently, two agents targeting BCR signaling have been approved: ibrutinib (-inib = inhibitor of a protein kinase) and idelalisib (-isib = inhibitor of lipid kinase).

BTK has been known since the 1960s but only recently has its role in BCR signaling been studied in inflammatory diseases, using RNA interference or pharmacological inhibition (1999–2001). Ibrutinib was synthesized to target the Cys481 residue of BTK (a neutrophilic site able to form a bond with the ATP binding site) and was shown to irreversibly block the catalytic site of BTK and 10 others tyrosine kinases, especially Tec family kinases [20]. The first demonstration of activity was made in a mouse clinical model of rheumatoid arthritis in 2007, but evidence was rapidly brought into a dog model of aggressive B-cell lymphoma (2010) [21]. BTK inhibition strategies in human DLBCL cell lines have been pioneered by the scientific group of Louis Staudt and have led to testing ibrutinib in B-cell malignancies. Only four years later, approval for CLL and MCL therapy was obtained.

PI3 kinases (PI3K) were discovered in the mid-1980s, and their activity as lipid kinases in mammalian cells has been known since 1990–1992 [22, 23]. LY294002, the first pan-PI3K inhibitor, was synthesized in 1994 and helped understand how PI3K/AKT was activated in many cancers, regulating cell survival, metabolism and proliferation.

Among PI3Ks, 3 classes are recognized, but only class I is able to transform PI-(4,5)P2 into PI-(3,4,5)P3. This family includes 4 catalytic isoforms, with class Ia (P110α,β,δ) being expressed ubiquitously in many tissues and organs, and class Ib (p110γ) being expressed primarily in leukocytes [24]. In 2002, knock-out mice models revealed p110δ and γ as the two major isotypes expressed in normal B cells, leading to the development of the first specific inhibitors in 2005 [25]. Targeting the δ isoform primarily expressed in leucocytes was the preferred option in B-cell malignancies to prevent effects in normal cells [26]. Idelalisib (inhibitor of delta lipid kinase) was synthesized based on the resolution of the crystal structure of the δ isoform in 2010 and approved for CLL and FL in 2014 based on outstanding clinical results.

19.2.2 Tumor hyperlymphocytosis

BCR stimulation is quickly followed by the regulation of adhesion receptors (e.g. integrin α4β1) and chemokine receptors (CXCR-4, CXCR-5, CCR-7), which are also crucial for adequate positioning of B cells within their microenvironment in lymph nodes and bone marrow [27, 28]. PI3Kδ and BTK inhibition lead to disruption of B α4β1 dependent adhesion, thereby inhibiting homing to lymphoid organs [29]. The ability of BTK/PI3K inhibitors to disrupt tumor-microenvironment interactions is evidenced in patients by rapid nodal shrinkage due to both *in situ* apoptosis and egress from niches leading to transient (median, 12–24 weeks) tumoral hyperlymphocytosis. This phenomenon is observed in CLL (> 80 % of cases), where cells are released from tissue niches and the bone marrow [30], and MCL (roughly 35 % of patients) where cells are released mostly from the bone marrow [31]. This phenomenon is observed to a much lesser extent in FL, MZL, WM or DLBCL, hinting that apoptosis is potentially the main mechanism of action in these diseases [32]. This finding led to the creation of a new response criterion: partial response (PR) with lymphocytosis > 5,000/μl [33]. Even in lymphoma with tumor circulating cells, the occurrence of tumor lysis syndrome is extremely rare with both compounds, reflecting their cytostatic rather than cytotoxic properties. Both ibrutinib and idelalisib share on-target effects against other immune effector cells (maturation, cytokine production, migration), platelets (collagen aggregation defects), vasculature and muscles (impaired calcium signaling) explaining some of the side effects most frequently observed.

19.2.3 Diffuse large B-cell lymphoma

DLBCL comprises two major molecular subtypes, called germinal center B-cell-like (GCB) and activated B-cell-like (ABC) (see also Chapter 8). Chronic active BCR signaling and constitutive activation of the NF-κB transcription complex are the pathogenetic hallmark of ABC DLBCL [34].

Ibrutinib demonstrated encouraging responses in patients with relapsed/refractory (R/R) ABC DLBCL, but almost no effect in GCB DLBCL, in line with preclinical data obtained in ABC DLBCL cell lines. Response rate varied in part according to the tumor mutational landscape, reaching 56 % (5/9) in patients with CD79A/B mutations (found in roughly 20 % of ABC DLBCL patient samples) and 31 % (9/29) in patients that lacked BCR mutations [35]. Also, 80 % (4/5) of patients with both mutant CD79B and MYD88 (myeloid differentiation factor 88) responded, whereas patients with wild-type CD79B and mutant MYD88 (detectable in 25–40 % of ABC DLBCLs) did not respond, suggesting a MYD88-independent pathway for NF-κB activation [36]. CARD11 activating mutations (detectable in approximately 10 % of ABC DLBCLs) and loss of A20 (24 % of ABC DLBCL) were associated with ibrutinib resistance, as expected based on their downstream activity [19].

Although activation of the PI3 kinase pathway is observed in various DLBCL cell lines [37], clinical results with idelalisib were disappointing, yielding no objective responses. Activation of PI3 kinase in DLBCL can be caused by loss of PTEN or by chronic B-cell receptor signaling [37, 38]. The fact that catalytic subunits can compensate for one another (especially p110α for δ) may in part explain why blocking a single isoform such as PI3Kδ failed in DLBCL [40–42]. Dual (or pan) targeting of PI3K isoforms in NHL may prevent the compensatory activity of various class I PI3K catalytic subunits in tumor cells and bystander stromal cells (α effect), but may also change the safety profile.

19.2.4 Follicular lymphoma

Genetic lesions leading to constitutional NF-κB or BCR signaling activation are observed at lower frequency in FL patients. However, a role for BCR signaling in the expansion and survival of FL cells has been evoked by several observations. Roughly 25 % of FL cases express mannosylated IgM BCRs that recognize mannose-binding lectins expressed by macrophages, thereby crosslinking the BCR and potentially initiating BCR signaling through chronic self-antigenic drive [44]. GEP studies in FL suggest that NF-κB addiction is BTK-independent, and results from a convergence of BAFF-R and CD40 signaling together with the loss of FoxP1-dependent NF-κB repression [45, 46]. This may explain the modest antitumor activity of ibrutinib in a phase 2 study in relapsed or refractory FL [43].

The importance of the PI3K pathway in FL is supported by the evidence of constitutive phosphorylation of AKT and PTEN loss (at least in 20 % of FL cases) across preclinical studies [47]. A model of genetically p110δ-mutant mice devoid of germinal center B cells due to a lack of T follicular helper cells (TFH) orchestrating GC reaction, suggests PI3Kδ inhibition may have pleiotropic consequences of potential therapeutic value in FL patients [48]. Idelalisib yielded a 56 % ORR (14 %CR) and a median PFS of 11 months in the major subgroup of 72 heavily pre-treated FL cases whose disease was refractory to rituximab plus alkylating agents enrolled in study 101-09 [49]. Preliminary results with duvelisib, which inhibits both γ and δ isoforms, showed similar encouraging results in FL, based on limited published phase I data [50].

19.2.5 Waldenström's macroglobulinemia (WM)

WM arises from IgM+ memory B cells and is characterized by the same activating MYD88 L265P mutation that was reported in roughly 30 % of ABC DLBCLs [51]. More recently, a S338X mutation in CXCR4, generating WHIM-like syndrome (warts, hypogammaglobulinemia, infections, myelokathexis), has been described in 28–36 % of WM cases associated with marrow infiltration of lymphoplasmacytic cells and cytopenia [52]. Interestingly, in WHIM disease, myelokathexis (retention of myeloid cells) accompanies MAPKinase activation. Similarly, cell lines expressing CXCR4 WHIM-like mutations showed impaired receptor internalization upon CXCL-12 exposure, prolonged AKT and ERK phosphorylation, and decreased response to ibrutinib [53]. This rescue effect was absent when direct MYD88 inhibitors were used, suggesting that IRAK1-4 inhibitors should overcome pro-survival benefits of CXCR4 mutations and synergize with ibrutinib.

A phase 2 trial in 69 R/R patients receiving ibrutinib reported a 90.5 % ORR with no CR, but major reduction (> 75 %) of IgM spike in 73.0 % of the patients [54]. In keeping with our recent understanding of the biology of the disease, the mutational status of MYD88 and CXCR4 had an impact on outcomes: ORR/Major PR were 100 %/91.2 % for patients with MYD88 L265P and CXCR4 wild-type, 85.7/61.9 % in cases of CXCR4 WHIM, and 71.4 %/28.6 % in cases with both MYD88 wild-type and CXCR4 wild-type. The estimated 2-year progression-free and overall survival (OS) rates among all patients were 69.1 % and 95.2 %, respectively. Among the 19 patients MW treated with idelalisib in study 101-02 [50] and 101-09 [49], 80 % achieving partial response and a median PFS of 33.3 mo (101-02) and 22.1 mo (101-09). Similarly to ibrutinib, patients with MYD88 L265P mutations respond to idelalisib while CXCR4 mutations confers resistance to this agent [53].

19.2.6 Marginal zone lymphomas

Marginal zone lymphoma (MZL) encompasses three different entities, namely nodal MZL, splenic MZL, and extranodal MZL, including mucosa-associated lymphoid tissue (MALT) lymphoma. BCR signaling through antigenic stimulation (e.g. hepatitis C virus infection) or inflammation receptors (e.g. Helicobacter *pylori*) is particularly relevant in this disease, as demonstrated by the observation that pathogen eradication using antibiotics at early stages of the disease (e.g. *Helicobacter pylori* in gastric MALT) leads to clinical responses. However, transition from Ag-driven dependence to a genetically acquired independence has been recently demonstrated, which may explain the failure of antimicrobial agents to efficiently eradicate the disease at later stages of lymphomagenesis. Indeed, NF-κB addiction of MZL cells may be promoted by balanced chromosomal translocations and recurrent somatic mutations in multiple NF-κB pathway components (see Chapter 6). Results of a larger phase 2 study (Clinical Trials.gov Identifier: NCT01980628) in all types of relapsed/refractory MZL with measurable disease patients are expected soon to get more insight into the efficiency of ibrutinib in MZL. Involvement of the PI3K pathway is suggested by spontaneous occurrence of MZL in a mouse model genetically modified to generate a loss of activity of both PTEN and SHIP1, the 3-PI and 5-PI phosphatases antagonizing PI3K activity [55]. Clinical data with idelalisib in MZL are limited: ORR was 47 % in 15 patients with double-refractory MZL enrolled in study 101-09 [49].

19.2.7 Chronic lymphocytic leukemia

The presence of multiple distinct sets of stereotyped BCRs with conserved CDR3 sequences has indicated a central role for the antigen in promoting CLL [56]. Consistently, ibrutinib yielded unprecedented response rates of 71 % in the first phase Ib/II trial of CLL patients, with durable remissions and manageable toxicity profile [57]. These excellent results were confirmed in separate trials in elderly patients [58] and in patients with TP53 disruption [59]. Of note, a landmark analysis performed later in patients with persistent hyperlymphocytosis (now classified as partial response with lymphocytosis, PRlym) showed similar outcomes compared to CR/PR patients i.e. 75 % progression-free survival (PFS) and OS at 26 months of follow-up while median PFS usually was less than 18 months with conventional immuno-chemotherapy (CIT) [30]. Similar response rates of 71 % and 96 % PFS and OS, respectively, after a median follow-up of 24 months were obtained with ibrutinib in the frontline setting in elderly patients urging therapy (median age 71 years) and unfit for CIT. These results, together with the superiority of ibrutinib over the anti-CD20 monoclonal antibody ofatumumab [60] in the RESONATE phase 3 randomized trial, led to FDA approval in 2014.

Phase I results with single-agent idelalisib showed significant responses in 54 R/R CLL [61]. The lower PFS rates compared to ibrutinib may be due to grade 3 adverse events occurring late (e.g. diarrhea), which urged halting of therapy and subsequent relapse. A significant improvement of ORR (81 % vs. 13 %) and median PFS (19.4 vs. 7.3 months) was shown in the randomized phase III trial 0116 [49], where rituximab plus idelalisib were compared to rituximab and placebo in 220 R/R CLL patients with comorbidities and enriched in IgHV unmutated genes (83 %) or del(17p) (42 %). Upon disease progression, cross-over was allowed to the idelalisib arm or to increased dose. Still, median OS was 20.8 months in the placebo arm, whereas median OS has not been reached yet in the idelalisib treated patients. The safety profile suggests that the frequency of grade 3/4 toxicity increases with time, especially diarrhea/colitis (16 % to

42 %), pneumonia (8 % to 13 %) and infections or neutropenia (42 %). CLL patients with del(17p), a critical unmet medical need population, treated with idelalisib as frontline therapy had 100 % ORR (CR 33 %) and 83 % PFS at 3 years. This was the first report ever of almost 100 % CLL patients responding to a chemotherapy-free approach [62]. Such excellent results, together with a manageable safety profile, led to market approval of idelalisib in combination with rituximab in R/R CLL and as first-line treatment for CLL patients with del(17p). Despite the encouraging *in vitro* activity of the p110α/δ inhibitor copanlisib against CLL cells [63], ORR in a phase I trial was 50 %, i.e. slightly lower than with idelalisib monotherapy. A similar 55 % ORR was reported with duvelisib in a phase I clinical trial [64], suggesting that selective p110δ inhibition is appropriate in CLL.

19.2.8 Mantle Cell Lymphoma

In addition to genetic lesions, antigenic drive has been proposed to play an important role in the pathogenesis of MCL, which is characterized by a remarkably restricted IgHV gene repertoire, including mostly four genes (IGHV3-21, IGHV4-34, IGHV1-8, and IGHV3-23) and being significantly different from that of normal B-cell subpopulations [65]. *In vitro*, ibrutinib was able to induce apoptosis in cell lines at low doses [66], and a phase II clinical trial evaluating 111 heavily pre-treated MCL patients showed response rates of 68 %, including 21 % CR and 47 % PR (but no PRlym by contrast to CLL patients) [67]. The higher CR rates compared to CLL may reflect better disease clearance in the bone marrow [31, 68]. With 58 % of patients surviving at 18 months and a median PFS of 13 months, ibrutinib is now licensed under this indication. The relatively short median PFS is explained by the rapid development of alternative pathways to overcome BTK dependence. Findings in primary tumor samples are consistent with those in MCL cell lines: activation of the alternative NF-κB pathway may be induced by NIK (NF-κB inducing kinase) upregulated in 15–20 % of ibrutinib-refractory cases through inactivating mutations of its negative regulators TRAF2/3 and BIRC3 [69].

Activation of the PI3K/AKT and mTOR axes due to inactivation of PTEN is a well-recognized finding in MCL cells [70, 71]. Everolimus, an mTOR inhibitor, has been approved in R/R MCL [71]. Idelalisib yielded 40 % response rate in 40 R/R patients with a very short median PFS of 3.7 months; this was much less than observed with ibrutinib, but 22 % of patients had ongoing responses at 12 months [72]. This heterogeneity in efficacy and response duration might be due to compensatory feedback loops (mTOR reactivating PDK1 thus keeping AKT phosphorylation intact), but also to the high expression of the p110α isoform in MCL cells still retaining a p110δ-dependence for adequate chronic BCR signaling. Activation of PIP3 despite PI3Kδ inhibition is expected to be the consequence of PI3Kα isoform upregulation in MCL, and is favored by concomitant PTEN loss occurring in 16 % of MCL cases with high α/δ ratio [73]. These studies suggest that dual α/δ targeting represents an option to improve response rates.

Indeed, the P110α/δ inhibitor copanlisib showed 71 % ORR in phase I [42]. Also, GDC-0941, a predominantly p110α/δ inhibitor, was significantly more active compared with idelalisib against MCL cell lines and primary samples [74].

19.3 Inducing the mitochondrial apoptosis pathway in lymphoma

Apoptosis is a well-organized process of preprogrammed cell death executed by cysteine proteases called caspases. Caspase activity can be initiated either via cell-surface death-receptors – the so-called death receptor or extrinsic pathway – or through the mitochondria, also called intrinsic pathway of apoptosis. Most stimuli induce apoptosis via the mitochondrial pathway, a process largely regulated by members of the BCL2 protein family [75]. This family can be subdivided into pro-apoptotic effector proteins (BAX and BAK), pro-apoptotic BH3-only proteins (BID, BIM, PUMA, Noxa, HRK, BIK, BMF, BAD) and anti-apoptotic BCL2 proteins (BCL2, BCL-xL, MCL1, A1, BCL-B, BCL-w).

Upregulation of the intrinsic apoptosis pathway participates to clonal survival in many mature B-cell malignancies (explaining the name BCL given to members of the family). High levels of BCL2 proteins may be the consequence of genetic events, such as balanced chromosomal translocations between IgH and the *BCL2* locus in FL, CLL, DLBCL; loss of miRNA15a-16 in CLL (deletion 13q being the most common genetic lesion in this disease and leading to the development of CLL-like lymphoma in a mouse model); gene amplification in one third of DLBCLs, or epigenetic modifications (hypomethylation of the BCL2 encoding gene) [76–78]. High levels of BCL2/BCL-xL/MCL1 may also be induced by microenvironmental cues delivered by stromal cells, tumor-associated macrophages, making drug targeting of this pathway particularly worthwhile [79].

The BH3-only protein family can be subdivided into activators – that are able to directly activate BAX/BAK – or sensitizers – that neutralize the apoptotic brake offered by anti-apoptotic BCL2 proteins. Competitive disruption of these interactions, leading to sensitization or induction of apoptosis, forms the basis of BH3 mimetic biological activity. Information obtained from NMR structural studies of BCL-xL complexed with the BH3 domain of BAK and BAD facilitated the design of BH3 domain peptidomimetics [80]. Most BH3 mimetics developed in the clinics are *sensitizers*, and induce apoptosis through blocking of pro-survival BCL2 family members and unleashing BAX/BAK [81]. Chemically engineered agents have been developed to block the cleft in which for instance BIM and BCL2 intimately interact to release BH3 domain-only proteins from anti-apoptotic proteins, making it available for interactions with BAX or BAK. Only three classes of inhibitors have entered clinical trials: GX15-070 (obatoclax mesylate), gossypol and its derivatives, and ABT-263 and ABT-199. GX15-070 (obatoclax) and gossypol (AT-101) were putative pan-BH3 mimetics but had many off-target effects, different modes of apoptosis induction and limited clinical efficacy [80]. The first BCL2 inhibitors with authentic BH3 mimicry (i.e. fully dependent from BAK/BAX) were ABT-

737 and its oral form navitoclax (ABT-263), which bind to BCL2, BCL-xL and Bcl-w but not to MCL1 or BFL1/A1, a binding profile reminiscent of the BH3-only protein BAD [81]. Navitoclax yielded an encouraging 35 % PR rate in R/R CLL and 22 % in NHL, with a median PFS of 16 months [82, 83]. Unfortunately, because platelet turnover is dictated by a fine tuning between BCL-xL and BAX, profound thrombocytopenia represents a major dose limiting toxicity of navitoclax [84]. With a better knowledge of platelet physiology, chemical modifications on the backbone of ABT-737 led to the development of ABT-199 (venetoclax) [85]. This compound does not bind to BCL-xL, and dose escalation cohorts became feasible without dose-limiting thrombocytopenia, leading to tremendous responses in CLL and encouraging response rates in other types of NHL. Its high potency, somewhat unexpected, led to fatal tumor lysis syndrome in two patients, prompting the adoption of a stepwise dose escalation schedule for most recent clinical trials. In a recent trial in CLL patients, ORR was 79 % (CR 22 %) with a median duration of response of 20.5 months for the 63 patients evaluable for efficacy [86]. In DLBCL and FL, response rates in phase I studies were lower, around 20–30 %, mostly occurring at higher doses [87]. Levels of BID or MCL1 are low in CLL cells, precluding buffering of BH3 mimetic agents and explaining exquisite sensitivity of CLL cells. By contrast, BIM/MCL1 ratio is low in many tumors including NHL [88], prompting the need to develop drugs that directly target MCL1 [89].

19.4 Targeting ALK in lymphoma

Anaplastic lymphoma kinase (ALK) is an orphan receptor tyrosine kinase expressed in the nervous system during embryonic development. ALK activation has been implicated in a number of cancers, both hematological (lymphoma) and nonhematological (including nonsmall-cell lung cancer, renal cancer, and neuroblastoma) [90]. In lymphoma, constitutive activation of ALK results from chromosomal rearrangements that generate oncogenic fusion kinases (a mechanism that is infrequent in lymphoma). ALK fusion proteins induce aberrant ALK signaling that ultimately drive oncogenesis and render transformed cells "addicted" to ALK.

In lymphoma, ALK expression is observed mostly in anaplastic large cell lymphoma (ALCL), especially in younger patients [91], where it usually results from the translocation t(2;5)(p23;q35) (≈ 75 % of the cases) that fuses ALK on chromosome 2 to NPM (nucleophosmin) on chromosome 5 [128]. The second most frequent rearrangement is t(1;2)(p25;q23) (≈ 20 % of the cases) which results in the fusion of TPM3 and ALK [90]. Other partner genes for ALK have been described including TFG, clathrin (CLTC), and ATIC. A rare variant of DLBCL (≈ 0.5 % of cases) can also express ALK [92], largely due to a rearrangement t(2;17)(p23;q23) that fuses CLTC with ALK (CLTC-ALK) [92].

Crizotinib, a competitive small molecule inhibitor of ALK and MET kinase activity, has demonstrated significant antitumor effect in patients harboring NSCLC with

ALK translocations [91, 94]. In ALK-positive ALCL, the efficacy of crizotinib was first reported in 2 patients with chemotherapy-refractory disease who displayed complete remission [95]. These encouraging results were later confirmed in additional studies in ALK-positive lymphomas, including pediatric patients [96–98].

19.5 Targeting immune cells in lymphoma

19.5.1 Lenalidomide

The interaction among cell types in the immune microenvironment is increasingly recognized as important to tumor cell recognition and destruction, as well as to protection of normal immune cells. Lenalidomide is an orally active immunomodulatory drug that has direct antineoplastic activity and indirect effects mediated through multiple types of immune cells found in the tumor microenvironment, including B, T, natural killer (NK), and dendritic cells. Recently, the E3 ubiquitin ligase cereblon has been identified as a molecular target in myeloma cells, which may underlie the effects of lenalidomide on lymphoma cells, as well as on cells in the tumor microenvironment. Cereblon expression was required for lenalidomide-induced toxicity in ABC DLBCL cells, which was associated with reduced levels of its target IRF4 and subsequently decreased NF-κB activity. In experimental models, lenalidomide seems to repair the immune synapse defects seen in NHL irrespective of the patient's level of disease [99]. These experimental findings have translated into the clinic, where its use as a single-agent showed durable responses in patients with R/R indolent or aggressive NHL in several phase II trials [100–102]. In particular, a recent study on 254 intent-to-treat MCL patients randomized 2:1 to lenalidomide or single-agent investigator's choice therapy showed that lenalidomide significantly improved PFS, as well as overall response rates (40 % versus 11 %) and complete response rates (5 % versus 0 %). Also, the preclinical evidence of the ability of lenalidomide to enhance the antitumor activity of rituximab led to clinical trials combining rituximab and lenalidomide (R^2) in first-line and relapsed settings. Enhanced activity has been observed with R^2 in MCL [103, 104], DLBCL [105–107], FL [108, 109], and indolent NHL [110, 111].

19.5.2 Immune checkpoints inhibitors

The immune system plays an important role in controlling and eradicating cancer, as tumor cells often escape immune surveillance through various mechanisms [112]. Immune interventions may thus restore the capacity of immune cells to eradicate cancer [113]. One important mechanism of immune escape in lymphoma results from the expression by the tumor cells or the microenvironment of inhibitory molecules (checkpoints) that prevent T-cell activation. Antibodies (Ab), or "checkpoint inhibitors"

(CPI), can block these inhibitory signals within the immune synapse [114]. To date, two types of CPIs have been approved for the treatment of cancer, all in melanoma: anti-CTLA4 (cytotoxic T lymphocyte-associated antigen-4) antibodies and anti-PD1 (programmed cell death-1, CD279) antibodies.

CTLA4 is a co-inhibitory receptor that regulates T cells at the stage of initial activation by competing with the CD28 co-stimulatory receptor for the binding of CD80 (B7-1) and CD86 (B7-2) expressed by antigen presenting cells [115]. Interestingly, T cells that infiltrate lymphoma tumors can express CTLA4, while tumor cells can express its ligands, CD80 and CD86, notably in Hodgkin and B-cell lymphomas [116]. A phase I trial evaluated ipilimumab in patients with B-cell lymphoma [117] and this agent was also tested in cancer patients presenting with relapsed or progressive disease following allogeneic stem cell transplantation [118].

PD1 receptor is another inhibitory T-cell receptor that is engaged by two ligands, PD-L1 (B7-H1, CD274) and PD-L2 (B7-DC, CD273). PD1 is primarily believed to inhibit effector T-cell activity in the effector phase at the tumor site [115]. Aberrant expression of PD-L1 by tumor cells has been reported in a number of human malignancies [119], and PD-L1 may also be expressed by normal immune cells within the tumor microenvironment. In lymphoma, PD-L1 (and PD-L2) expression on tumor cells is mostly observed in HL and primary mediastinal B-cell lymphoma (PMBL), as well as ABC DLBCL [124–127]. PD-L1 expression in lymphoma can result from amplification of the PD-L1 and PD-L2 genes at 9p24 [128], CIITA rearrangements [129, 130], constitutive AP-1 activity and/or EBV infection [131]. PD-L1 expression can also result from IFNγ secretion by the tumor microenvironment [132]. PD1/PD-L1 blockade resulted in tumor regression in several types of solid tumors [120–123], and three different anti-PD1 antibodies have been tested in patients with lymphoma to date: pidilizumab, nivolumab and pembrolizumab. Pidilizumab has been investigated in three clinical trials: a phase I trial comprising a variety of hematologic malignancies [133] and two phase II trials devoted to lymphoma [134, 135]. The first phase II trial evaluated pidilizumab in 66 patients with DLBCL following autologous stem cell transplantation (ASCT) [134]. At 16 months after the first treatment, PFS was 72% and compared favorably to historical controls. The second phase II study evaluated pidilizumab in rituximab-sensitive follicular lymphoma patients [135]. The proportion of patients who achieved objective and complete responses (66 % and 52%, respectively) compared favorably to historical controls. However, no definitive conclusions can be drawn from these two studies, as there was no direct comparison and because confounding factors (ASCT and rituximab) precluded the possibility to directly measure the efficacy of pidilizumab by itself. In fact, the most impressive results were seen with the two other anti-PD1 Abs, nivolumab and pembrolizumab, in patients with HL [136, 137]. In two small studies (n = 23 and 29 patients), nivolumab and pembrolizumab demonstrated remarkable anti-tumor activity. Although most patients had failed most effective, standard therapies (including brentuximab vedotin and ASCT), almost all of them experienced tumor regression. The objective response rate ranged between 66 % and 87 %, with roughly

20 % complete responses. Although these results warrant further evaluation, they suggest that CPIs will certainly be part of the armamentarium of lymphoma therapies in the future.

19.6 Targeting epigenetic deregulation

Epigenetic changes can contribute to lymphomagenesis and drug resistance by altering chromatin states and transcriptional regulation. The first therapeutic attempt to target epigenetic deregulation in lymphoma was the use of histone deacetylase (HDAC) inhibitors. Multiple clinical trials have tested various HDAC drug classes, with response rates of approximately 20–30 % across all lymphoma subtypes, but none included correlative biomarkers or genotype analyses. Currently, the main focus is peripheral T-cell lymphoma, where 3 HDAC inhibitors (vorinostat, romidepsin and belinostat) have been approved by the FDA for the treatment of R/R peripheral and/ or cutaneous T-cell lymphoma patients [138–140].

Recently, the importance of an epigenetic deregulation across all lymphoma subtypes was confirmed by NGS showing recurrent mutations in chromatin-modifying genes including MLL2, CREBBP, EP300, TET2, EZH2, and IDH2. For EZH2 and IDH2, these findings may now have therapeutic consequences since highly selective inhibitors of these enzymes have recently been developed by high throughput biochemical screenings [141–143].

19.6.1 EZH2 inhibitors

The histone lysine N-methyltransferase EZH2 is part of the polycomb repressive complex 2 (PRC2), which catalyzes trimethylation of lysine 27 on histone H3 (H3K27me3), a repressive chromatin mark. EZH2, in cooperation with BCL6, plays a critical role in the normal germinal center reaction by allowing the centroblast B cells to undergo the characteristic proliferative burst associated with somatic hypermutation [144]. EZH2 Mutations were detected in several types of germinal center-derived B-cell lymphomas including approximately 20 % GCB DLBCLs and 27 % of FLs. These aberrations represents gain of function events [146] and several small molecule inhibitors have been developed that are able to block both the wild-type and mutant forms of the enzyme [144, 145, 147, 148].

19.6.2 IDH2

The enzyme isocitrate dehydrogenase 2 (IDH2) is a component of the Krebs cycle that normally catalyzes the conversion of isocitrate to alpha-ketoglutarate (α-KG), an important metabolic intermediate for several enzymes involved in epigenetic regulation[149]. Gain-of-function mutations affecting IDH2 have been found in a variety of myeloid cancers and, more recently, in 20 % of angioimmunoblastic T-cell lymphomas (AITLs), but not in other peripheral T-cell lymphoma subtypes [152]. These events result in a neomorphic enzymatic activity that promotes the conversion of α-KG to the structurally related 2-hydroxyglutarate (2HG), an "oncometabolite" that may drive tumor progression by interfering with 2-oxoglutarate-dependent enzymes such as those involved in hypoxia signaling (prolyl hydroxylases), histone methylation, and DNA methylation (TET2) [150, 151]. Specific inhibitors of mutant IDH2 enzymes demonstrated promising preclinical activity in mouse models of *IDH2*-mutant AML and in primary human leukemia cells carrying *IDH2* mutations [153, 154]. The recent identification of *IDH2* mutations in a significant fraction of AITL warrants further testing in this disease.

19.7 Concluding remarks

The efforts of translating our understanding of the molecular pathogenesis of lymphoid malignancies into the clinic and into daily patient care are still evolving. Scientific programs within clinical trials are crucial to obtaining a better understanding of responses observed in patients treated with novel compounds. Although long duration responses have been observed with some of these new drugs as single agents, especially in CLL, response rates and equally important duration of response need to be improved. Lack of efficacy can potentially be explained by oncogenic cooperation, redundant signaling pathways, feedback mechanisms circumventing the drug target, or emergence of mutants abrogating drug-target interactions. These limitations can potentially be overcome by combining novel inhibitors. High throughput combination screenings *in silico* and *in vitro*, animal models, synthetic lethality drug/genetic screens should help to select the best possible combinations. However, validation in patients is mandatory to confirm efficacy and safety [155, 156]. This will be challenging given the multitude of potential combinations, the complexity of regulatory environments in clinical trials, and the high number of patients required to validate a given hypothesis. Thus, substantial preclinical work and a better understanding of the driving oncogenic events is required to identify the most suitable combinations. This might eventually pave the way to more effective and less toxic treatment protocols for patients with malignant lymphomas.

References

[1] Intlekofer AM, Younes A. Precision therapy for lymphoma–current state and future directions. Nat Rev Clin Oncol. 2014;11(10):585–596.

[2] Alizadeh AA, Eisen MB, Davis RE, Ma C, Lossos IS, Rosenwald A, et al. Distinct types of diffuse large B-cell lymphoma identified by gene expression profiling. Nature. 2000;403(6769):503–511.

[3] Rosenwald A, Wright G, Leroy K, Yu X, Gaulard P, Gascoyne RD, et al. Molecular diagnosis of primary mediastinal B cell lymphoma identifies a clinically favorable subgroup of diffuse large B cell lymphoma related to Hodgkin lymphoma. J Exp Med. 2003;198(6):851–862.

[4] Morin RD, Mendez-Lago M, Mungall AJ, Goya R, Mungall KL, Corbett RD, et al. Frequent mutation of histone-modifying genes in non-Hodgkin lymphoma. Nature. 2011;476(7360):298–303.

[5] Rossi D, Ciardullo C, Gaidano G. Genetic aberrations of signaling pathways in lymphomagenesis: revelations from next generation sequencing studies. Semin Cancer Biol. 2013;23(6):422–430.

[6] Compagno M, Lim WK, Grunn A, Nandula SV, Brahmachary M, Shen Q, et al. Mutations of multiple genes cause deregulation of NF-kappaB in diffuse large B-cell lymphoma. Nature. 2009;459(7247):717–721.

[7] Pasqualucci L, Trifonov V, Fabbri G, Ma J, Rossi D, Chiarenza A, et al. Analysis of the coding genome of diffuse large B-cell lymphoma. Nat Genet. 2011;43(9):830–837.

[8] Lohr JG, Stojanov P, Lawrence MS, Auclair D, Chapuy B, Sougnez C, et al. Discovery and prioritization of somatic mutations in diffuse large B-cell lymphoma (DLBCL) by whole-exome sequencing. Proc Natl Acad Sci USA. 2012;109(10):3879–3884.

[9] Love C, Sun Z, Jima D, Li G, Zhang J, Miles R, et al. The genetic landscape of mutations in Burkitt lymphoma. Nat Genet. 2012;44(12):1321–1325.

[10] Richter J, Schlesner M, Hoffmann S, Kreuz M, Leich E, Burkhardt B, et al. Recurrent mutation of the ID3 gene in Burkitt lymphoma identified by integrated genome, exome and transcriptome sequencing. Nat Genet. 2012;44(12):1316–1320.

[11] Okosun J, Bodor C, Wang J, Araf S, Yang CY, Pan C, et al. Integrated genomic analysis identifies recurrent mutations and evolution patterns driving the initiation and progression of follicular lymphoma. Nat Genet. 2014;46(2):176–181.

[12] Bea S, Valdes-Mas R, Navarro A, Salaverria I, Martin-Garcia D, Jares P, et al. Landscape of somatic mutations and clonal evolution in mantle cell lymphoma. Proc Natl Acad Sci USA. 2013;110(45):18250–18255.

[13] Rossi D, Trifonov V, Fangazio M, Bruscaggin A, Rasi S, Spina V, et al. The coding genome of splenic marginal zone lymphoma: activation of NOTCH2 and other pathways regulating marginal zone development. J Exp Med. 2012;209(9):1537–1551.

[14] Yoo HY, Sung MK, Lee SH, Kim S, Lee H, Park S, et al. A recurrent inactivating mutation in RHOA GTPase in angioimmunoblastic T cell lymphoma. Nat Genet. 2014;46(4):371–375.

[15] Palomero T, Couronne L, Khiabanian H, Kim MY, Ambesi-Impiombato A, Perez-Garcia A, et al. Recurrent mutations in epigenetic regulators, RHOA and FYN kinase in peripheral T cell lymphomas. Nat Genet. 2014;46(2):166–170.

[16] Niemann CU, Wiestner A. B-cell receptor signaling as a driver of lymphoma development and evolution. Semin Cancer Biol. 2013;23(6):410–421.

[17] Rickert RC. New insights into pre-BCR and BCR signalling with relevance to B cell malignancies. Nat Rev Immunol. 2013;13(8):578–591.

[18] Young RM, Staudt LM. Targeting pathological B cell receptor signalling in lymphoid malignancies. Nat Rev Drug Discov. 2013;12(3):229–243.

[19] Davis RE, Ngo VN, Lenz G, Tolar P, Young RM, Romesser PB, et al. Chronic active B-cell-receptor signalling in diffuse large B-cell lymphoma. Nature. 2010;463(7277):88–92.

[20] Pan Z, Scheerens H, Li SJ, Schultz BE, Sprengeler PA, Burrill LC, et al. Discovery of selective irreversible inhibitors for Bruton's tyrosine kinase. ChemMedChem. 2007;2(1):58–61.

[21] Honigberg LA, Smith AM, Sirisawad M, Verner E, Loury D, Chang B, et al. The Bruton tyrosine kinase inhibitor PCI-32765 blocks B-cell activation and is efficacious in models of autoimmune disease and B-cell malignancy. Proc Natl Acad Sci USA. 2010;107(29):13075–13080.

[22] Majerus PW, Ross TS, Cunningham TW, Caldwell KK, Jefferson AB, Bansal VS. Recent insights in phosphatidylinositol signaling. Cell. 1990;63(3):459–465.

[23] Vanhaesebroeck B, Guillermet-Guibert J, Graupera M, Bilanges B. The emerging mechanisms of isoform-specific PI3K signalling. Nat Rev Mol Cell Biol. 2010;11(5):329–341.

[24] Vanhaesebroeck B, Leevers SJ, Ahmadi K, Timms J, Katso R, Driscoll PC, et al. Synthesis and function of 3-phosphorylated inositol lipids. Annu Rev Biochem. 2001;70:535–602.

[25] Okkenhaug K, Bilancio A, Farjot G, Priddle H, Sancho S, Peskett E, et al. Impaired B and T cell antigen receptor signaling in p110delta PI 3-kinase mutant mice. Science. 2002;297(5583):1031–1034.

[26] Vanhaesebroeck B, Welham MJ, Kotani K, Stein R, Warne PH, Zvelebil MJ, et al. P110delta, a novel phosphoinositide 3-kinase in leukocytes. Proc Natl Acad Sci USA. 1997;94(9):4330–4335.

[27] Spaargaren M, Beuling EA, Rurup ML, Meijer HP, Klok MD, Middendorp S, et al. The B cell antigen receptor controls integrin activity through Btk and PLCgamma2. J Exp Med. 2003;198(10):1539–1550.

[28] de Gorter DJ, Beuling EA, Kersseboom R, Middendorp S, van Gils JM, Hendriks RW, et al. Bruton's tyrosine kinase and phospholipase Cgamma2 mediate chemokine-controlled B cell migration and homing. Immunity. 2007;26(1):93–104.

[29] de Rooij MF, Kuil A, Geest CR, Eldering E, Chang BY, Buggy JJ, et al. The clinically active BTK inhibitor PCI-32765 targets B-cell receptor- and chemokine-controlled adhesion and migration in chronic lymphocytic leukemia. Blood. 2012;119(11):2590–2594.

[30] Woyach JA, Smucker K, Smith LL, Lozanski A, Zhong Y, Ruppert AS, et al. Prolonged lymphocytosis during ibrutinib therapy is associated with distinct molecular characteristics and does not indicate a suboptimal response to therapy. Blood. 2014;123(12):1810–1817.

[31] Chang BY, Francesco M, De Rooij MF, Magadala P, Steggerda SM, Huang MM, et al. Egress of CD19(+)CD5(+) cells into peripheral blood following treatment with the Bruton tyrosine kinase inhibitor ibrutinib in mantle cell lymphoma patients. Blood. 2013;122(14):2412–2424.

[32] Wodarz D, Garg N, Komarova NL, Benjamini O, Keating MJ, Wierda WG, et al. Kinetics of CLL cells in tissues and blood during therapy with the BTK inhibitor ibrutinib. Blood. 2014;123(26):4132–4135.

[33] Cheson BD. Rethinking clinical response and outcome assessment in a biologic age. Curr Oncol Rep. 2015;17(6):27.

[34] Young RM, Shaffer AL, 3rd, Phelan JD, Staudt LM. B-cell receptor signaling in diffuse large B-cell lymphoma. Semin Hematol. 2015;52(2):77–85.

[35] Wilson W, Gerecitano J, Goy A, De Vos S, Kenkre V, Barr P, et al. The Burton's tyrosine kinase (BTK) inhibitor, ibrutinib (PCI-32765), has preferential activity in the ABC subtype of relapsed/refractory de novo diffuse large B-cell lymphoma (DLBCL): interim results of a multicenter, open-label, phase 2 study. Abs. 686. Blood. 2012.

[36] Ngo VN, Young RM, Schmitz R, Jhavar S, Xiao W, Lim KH, et al. Oncogenically active MYD88 mutations in human lymphoma. Nature. 2011;470(7332):115–119.

[37] Uddin S, Hussain AR, Siraj AK, Manogaran PS, Al-Jomah NA, Moorji A, et al. Role of phosphatidylinositol 3'-kinase/AKT pathway in diffuse large B-cell lymphoma survival. Blood. 2006;108(13):4178–4186.

[38] Pfeifer M, Grau M, Lenze D, Wenzel SS, Wolf A, Wollert-Wulf B, et al. PTEN loss defines a PI3K/AKT pathway-dependent germinal center subtype of diffuse large B-cell lymphoma. Proc Natl Acad Sci USA. 2013;110(30):12420–12425.

[39] Lenz G, Wright GW, Emre NC, Kohlhammer H, Dave SS, Davis RE, et al. Molecular subtypes of diffuse large B-cell lymphoma arise by distinct genetic pathways. Proc Natl Acad Sci USA. 2008;105(36):13520–13525.

[40] Schwartz S, Wongvipat J, Trigwell CB, Hancox U, Carver BS, Rodrik-Outmezguine V, et al. Feedback suppression of PI3Kalpha signaling in PTEN-mutated tumors is relieved by selective inhibition of PI3Kbeta. Cancer Cell. 2015;27(1):109–122.

[41] Chen L, Monti S, Juszczynski P, Ouyang J, Chapuy B, Neuberg D, et al. SYK inhibition modulates distinct PI3K/AKT- dependent survival pathways and cholesterol biosynthesis in diffuse large B cell lymphomas. Cancer Cell. 2013;23(6):826–838.

[42] Dreyling M, Morschhauser F, Bron D, Bouabdallah K, Vitolo U, Linton K, et al. Preliminary results of a phase Ii study of single agents BAY 80-6946, a novel P13K inhibitor, in patients with relapsed/refractory, indolent or aggressive Lymphoma. abs 87. Blood. 2013.

[43] Bartlett N, Laplant B, Qi J, Ansell S, Kuruvilla J, Reeder C, et al. Ibrutinib Monotherapy in Relapsed/Refractory Follicular Lymphoma (FL): Preliminary Results of a Phase 2 Consortium (P2C) Trial. Abs 800. Blood. 2014.

[44] Coelho V, Krysov S, Ghaemmaghami AM, Emara M, Potter KN, Johnson P, et al. Glycosylation of surface Ig creates a functional bridge between human follicular lymphoma and microenvironmental lectins. Proc Natl Acad Sci USA. 2010;107(43):18587–18592.

[45] Bouska A, McKeithan TW, Deffenbacher KE, Lachel C, Wright GW, Iqbal J, et al. Genome-wide copy-number analyses reveal genomic abnormalities involved in transformation of follicular lymphoma. Blood. 2014;123(11):1681–1690.

[46] Green MR, Gandhi MK, Courtney MJ, Marlton P, Griffiths L. Relative abundance of full-length and truncated FOXP1 isoforms is associated with differential NFkappaB activity in Follicular Lymphoma. Leuk Res. 2009;33(12):1699–1702.

[47] Yahiaoui OI, Nunes JA, Castanier C, Devillier R, Broussais F, Fabre AJ, et al. Constitutive AKT activation in follicular lymphoma. BMC Cancer. 2014;14:565.

[48] Rolf J, Bell SE, Kovesdi D, Janas ML, Soond DR, Webb LM, et al. Phosphoinositide 3-kinase activity in T cells regulates the magnitude of the germinal center reaction. J Immunol. 2010;185(7):4042–4052.

[49] Furman RR, Sharman JP, Coutre SE, Cheson BD, Pagel JM, Hillmen P, et al. Idelalisib and rituximab in relapsed chronic lymphocytic leukemia. N Engl J Med. 2014;370(11):997-1007.

[50] Flinn I, Oki Y, Patel A, Horwitz SM, Foss F, Sweeney J, et al. A Phase 1 Evaluation of Duvelisib (IPI-145), a PI3K-δ,γ Inhibitor, in Patients with Relapsed/Refractory iNHL. Abs 802. Blood. 2014.

[51] Treon SP, Xu L, Yang G, Zhou Y, Liu X, Cao Y, et al. MYD88 L265P somatic mutation in Waldenstrom's macroglobulinemia. N Engl J Med. 2012;367(9):826–833.

[52] Treon SP, Cao Y, Xu L, Yang G, Liu X, Hunter ZR. Somatic mutations in MYD88 and CXCR4 are determinants of clinical presentation and overall survival in Waldenstrom macroglobulinemia. Blood. 2014;123(18):2791–276.

[53] Cao Y, Hunter ZR, Liu X, Xu L, Yang G, Chen J, et al. The WHIM-like CXCR4(S338X) somatic mutation activates AKT and ERK, and promotes resistance to ibrutinib and other agents used in the treatment of Waldenstrom's Macroglobulinemia. Leukemia. 2015;29(1):169–1676.

[54] Treon SP, Tripsas CK, Meid K, Warren D, Varma G, Green R, et al. Ibrutinib in previously treated Waldenstrom's macroglobulinemia. N Engl J Med. 2015;372(15):1430–1440.

[55] Miletic AV, Anzelon-Mills AN, Mills DM, Omori SA, Pedersen IM, Shin DM, et al. Coordinate suppression of B cell lymphoma by PTEN and SHIP phosphatases. J Exp Med. 2010;207(11):2407–2420.

[56] Messmer BT, Albesiano E, Efremov DG, Ghiotto F, Allen SL, Kolitz J, et al. Multiple distinct sets of stereotyped antigen receptors indicate a role for antigen in promoting chronic lymphocytic leukemia. J Exp Med. 2004;200(4):519–525.
[57] Byrd JC, Furman RR, Coutre SE, Flinn IW, Burger JA, Blum KA, et al. Targeting BTK with ibrutinib in relapsed chronic lymphocytic leukemia. N Engl J Med. 2013;369(1):32–42.
[58] O'Brien S, Furman RR, Coutre SE, Sharman JP, Burger JA, Blum KA, et al. Ibrutinib as initial therapy for elderly patients with chronic lymphocytic leukaemia or small lymphocytic lymphoma: an open-label, multicentre, phase 1b/2 trial. Lancet Oncol. 2014;15(1):48–58.
[59] Farooqui MZ, Valdez J, Martyr S, Aue G, Saba N, Niemann CU, et al. Ibrutinib for previously untreated and relapsed or refractory chronic lymphocytic leukaemia with TP53 aberrations: a phase 2, single-arm trial. Lancet Oncol. 2015;16(2):169–176.
[60] Byrd JC, Brown JR, O'Brien S, Barrientos JC, Kay NE, Reddy NM, et al. Ibrutinib versus ofatumumab in previously treated chronic lymphoid leukemia. N Engl J Med. 2014;371(3):213–223.
[61] Brown JR, Barrientos JC, Barr PM, Flinn IW, Burger JA, Tran A, et al. The Bruton tyrosine kinase inhibitor ibrutinib with chemoimmunotherapy in patients with chronic lymphocytic leukemia. Blood. 2015;125(19):2915–2922.
[62] O'Brien S, Patel M, Kahl B, Horwitz SM, Foss F, Porcu P, et al. Duvelisib (IPI-145), a P13K-δ, γ inhibitor, is clinically active in patients with relapsed/refractory chronic lymphocytic leukemia. Abs 3334. Blood. 2014.
[63] Gockeritz E, Kerwien S, Baumann M, Wigger M, Vondey V, Neumann L, et al. Efficacy of phosphatidylinositol-3 kinase inhibitors with diverse isoform selectivity profiles for inhibiting the survival of chronic lymphocytic leukemia cells. Int J Cancer. 2015.
[64] Balakrishnan K, Peluso M, Fu M, Rosin NY, Burger JA, Wierda WG, et al. The phosphoinositide-3-kinase (PI3K)-delta and gamma inhibitor, IPI-145 (Duvelisib), overcomes signals from the PI3K/AKT/S6 pathway and promotes apoptosis in CLL. Leukemia. 2015.
[65] Hadzidimitriou A, Agathangelidis A, Darzentas N, Murray F, Delfau-Larue MH, Pedersen LB, et al. Is there a role for antigen selection in mantle cell lymphoma? Immunogenetic support from a series of 807 cases. Blood. 2011;118(11):3088–3095.
[66] Dasmahapatra G, Patel H, Dent P, Fisher RI, Friedberg J, Grant S. The Bruton tyrosine kinase (BTK) inhibitor PCI-32765 synergistically increases proteasome inhibitor activity in diffuse large-B cell lymphoma (DLBCL) and mantle cell lymphoma (MCL) cells sensitive or resistant to bortezomib. Br J Haematol. 2013;161(1):43–56.
[67] Wang ML, Rule S, Martin P, Goy A, Auer R, Kahl BS, et al. Targeting BTK with ibrutinib in relapsed or refractory mantle-cell lymphoma. N Engl J Med. 2013;369(6):507–516.
[68] Furtado M, Wang ML, Munneke B, McGreivy J, Beaupre DM, Rule S. Ibrutinib-associated lymphocytosis corresponds to bone marrow involvement in mantle cell lymphoma. Br J Haematol. 2015;170(1):131–134.
[69] Rahal R, Frick M, Romero R, Korn JM, Kridel R, Chan FC, et al. Pharmacological and genomic profiling identifies NF-kappaB-targeted treatment strategies for mantle cell lymphoma. Nat Med. 2014;20(1):87–92.
[70] Rudelius M, Pittaluga S, Nishizuka S, Pham TH, Fend F, Jaffe ES, et al. Constitutive activation of Akt contributes to the pathogenesis and survival of mantle cell lymphoma. Blood. 2006;108(5):1668–1676.
[71] Papadopoulos KP, Egile C, Ruiz-Soto R, Jiang J, Shi W, Bentzien F, et al. Efficacy, safety, pharmacokinetics and pharmacodynamics of SAR245409 (voxtalisib, XL765), an orally administered phosphoinositide 3-kinase/mammalian target of rapamycin inhibitor: a phase 1 expansion cohort in patients with relapsed or refractory lymphoma. Leuk Lymphoma. 2015;56(6):1763–1670.

[72] Kahl BS, Spurgeon SE, Furman RR, Flinn IW, Coutre SE, Brown JR, et al. A phase 1 study of the PI3Kdelta inhibitor idelalisib in patients with relapsed/refractory mantle cell lymphoma (MCL). Blood. 2014;123(22):3398–3405.
[73] Iyengar S, Clear A, Bodor C, Maharaj L, Lee A, Calaminici M, et al. P110alpha-mediated constitutive PI3K signaling limits the efficacy of p110delta-selective inhibition in mantle cell lymphoma, particularly with multiple relapse. Blood. 2013;121(12):2274–2284.
[74] Tabe Y, Jin L, Konopleva M, Shikami M, Kimura S, Andreeff M, et al. Class IA PI3K inhibition inhibits cell growth and proliferation in mantle cell lymphoma. Acta Haematol. 2014;131(1):59–69.
[75] Lopez J, Tait SW. Mitochondrial apoptosis: killing cancer using the enemy within. Br J Cancer. 2015;112(6):957–962.
[76] Tsujimoto Y, Cossman J, Jaffe E, Croce CM. Involvement of the bcl-2 gene in human follicular lymphoma. Science. 1985;228(4706):1440–1443.
[77] Hanada M, Delia D, Aiello A, Stadtmauer E, Reed JC. bcl-2 gene hypomethylation and high-level expression in B-cell chronic lymphocytic leukemia. Blood. 1993;82(6):1820–1828.
[78] Cimmino A, Calin GA, Fabbri M, Iorio MV, Ferracin M, Shimizu M, et al. miR-15 and miR-16 induce apoptosis by targeting BCL2. Proc Natl Acad Sci USA. 2005;102(39):13944–13949.
[79] Ghia P, Boussiotis VA, Schultze JL, Cardoso AA, Dorfman DM, Gribben JG, et al. Unbalanced expression of bcl-2 family proteins in follicular lymphoma: contribution of CD40 signaling in promoting survival. Blood. 1998;91(1):244–251.
[80] Zaman S, Wang R, Gandhi V. Targeting the apoptosis pathway in hematologic malignancies. Leuk Lymphoma. 2014;55(9):1980–1992.
[81] Oltersdorf T, Elmore SW, Shoemaker AR, Armstrong RC, Augeri DJ, Belli BA, et al. An inhibitor of Bcl-2 family proteins induces regression of solid tumours. Nature. 2005;435(7042):677–681.
[82] Wilson WH, O'Connor OA, Czuczman MS, LaCasce AS, Gerecitano JF, Leonard JP, et al. Navitoclax, a targeted high-affinity inhibitor of BCL-2, in lymphoid malignancies: a phase 1 dose-escalation study of safety, pharmacokinetics, pharmacodynamics, and antitumour activity. Lancet Oncol. 2010;11(12):1149–1159.
[83] Roberts AW, Seymour JF, Brown JR, Wierda WG, Kipps TJ, Khaw SL, et al. Substantial susceptibility of chronic lymphocytic leukemia to BCL2 inhibition: results of a phase I study of navitoclax in patients with relapsed or refractory disease. J Clin Oncol. 2012;30(5):488–496.
[84] Mason KD, Carpinelli MR, Fletcher JI, Collinge JE, Hilton AA, Ellis S, et al. Programmed anuclear cell death delimits platelet life span. Cell. 2007;128(6):1173–1186.
[85] Souers AJ, Leverson JD, Boghaert ER, Ackler SL, Catron ND, Chen J, et al. ABT-199, a potent and selective BCL-2 inhibitor, achieves antitumor activity while sparing platelets. Nat Med. 2013;19(2):202–208.
[86] Seymour J, Davids M, Pagel J, Kahk B, Werda W, Puvvada S, et al. ABT-199 (GDC-0199) in relapsed/refractory (R/R) chronic lymphocytic leukemia (CLL) ans a small lymphocytic lymphoma (SLL): high complete-response rate and durable disease control. Abs 7015. J Clin Oncol. 2014;32:5s.
[87] Davids M, Seymour J, Gerecitano J, Kahl B, Pagel J, Werda W, et al. Phase I study of ABT-199 (GDC-0199) in patients with relapsed/refractory non-Hodgkin lymphoma (NHL): responses observed in diffuse large B-cell (DLBCL) and follicular lymphoma (FL) at higher cohort doses. Abs 8522. J Clin Oncol. 2014;32(5s).
[88] van Delft MF, Wei AH, Mason KD, Vandenberg CJ, Chen L, Czabotar PE, et al. The BH3 mimetic ABT-737 targets selective Bcl-2 proteins and efficiently induces apoptosis via Bak/Bax if Mcl-1 is neutralized. Cancer Cell. 2006;10(5):389–399.

[89] Foight GW, Ryan JA, Gulla SV, Letai A, Keating AE. Designed BH3 peptides with high affinity and specificity for targeting Mcl-1 in cells. ACS Chem Biol. 2014;9(9):1962–1698.
[90] Webb TR, Slavish J, George RE, Look AT, Xue L, Jiang Q, et al. Anaplastic lymphoma kinase: role in cancer pathogenesis and small-molecule inhibitor development for therapy. Expert Rev Anticancer Ther. 2009;9(3):331–356.
[91] Sibon D, Fournier M, Briere J, Lamant L, Haioun C, Coiffier B, et al. Long-term outcome of adults with systemic anaplastic large-cell lymphoma treated within the Groupe d'Etude des Lymphomes de l'Adulte trials. J Clin Oncol. 2012;30(32):3939–3946.
[92] Laurent C, Do C, Gascoyne RD, Lamant L, Ysebaert L, Laurent G, et al. Anaplastic lymphoma kinase-positive diffuse large B-cell lymphoma: a rare clinicopathologic entity with poor prognosis. J Clin Oncol. 2009;27(25):4211–4216.
[93] Morris SW, Kirstein MN, Valentine MB, Dittmer KG, Shapiro DN, Saltman DL, et al. Fusion of a kinase gene, ALK, to a nucleolar protein gene, NPM, in non-Hodgkin's lymphoma. Science. 1994;263(5151):1281–1284.
[94] Camidge DR, Bang YJ, Kwak EL, Iafrate AJ, Varella-Garcia M, Fox SB, et al. Activity and safety of crizotinib in patients with ALK-positive non-small-cell lung cancer: updated results from a phase 1 study. Lancet Oncol. 2012;13(10):1011–1019.
[95] Gambacorti-Passerini C, Messa C, Pogliani EM. Crizotinib in anaplastic large-cell lymphoma. N Engl J Med. 2011;364(8):775–776.
[96] Gambacorti-Passerini C, al. E. Crizotinib in advanced, chemoresistant anaplastic lymphoma kinase-positive lymphoma patients. J Natl Cancer Inst. 2014;106(djt378).
[97] Wass M, Behlendorf T, Schadlich B, Mottok A, Rosenwald A, Schmoll HJ, et al. Crizotinib in refractory ALK-positive diffuse large B-cell lymphoma: a case report with a short-term response. Eur J Haematol. 2014;92(3):268–270.
[98] Mosse YP, Lim MS, Voss SD, Wilner K, Ruffner K, Laliberte J, et al. Safety and activity of crizotinib for paediatric patients with refractory solid tumours or anaplastic large-cell lymphoma: a Children's Oncology Group phase 1 consortium study. Lancet Oncol. 2013;14(6):472–480.
[99] Ramsay AG, Clear AJ, Fatah R, Gribben JG. Multiple inhibitory ligands induce impaired T-cell immunologic synapse function in chronic lymphocytic leukemia that can be blocked with lenalidomide: establishing a reversible immune evasion mechanism in human cancer. Blood. 2012;120(7):1412–1421.
[100] Wiernik PH, Lossos IS, Tuscano JM, Justice G, Vose JM, Cole CE, et al. Lenalidomide monotherapy in relapsed or refractory aggressive non-Hodgkin's lymphoma. J Clin Oncol. 2008;26(30):4952–4957.
[101] Goy A, Sinha R, Williams ME, Kalayoglu Besisik S, Drach J, Ramchandren R, et al. Single-agent lenalidomide in patients with mantle-cell lymphoma who relapsed or progressed after or were refractory to bortezomib: phase II MCL-001 (EMERGE) study. J Clin Oncol. 2013;31(29):3688–3695.
[102] Witzig TE, Vose JM, Zinzani PL, Reeder CB, Buckstein R, Polikoff JA, et al. An international phase II trial of single-agent lenalidomide for relapsed or refractory aggressive B-cell non-Hodgkin's lymphoma. Ann Oncol. 2011;22(7):1622–1627.
[103] Ruan J, Martin P, Shah B, al. e. Sustained remission with the combination biologic doublet of lenalidomide plus rituximab as initial treatment for mantle cell lymphoma: a multicenter phase II study report. Abs. 625. Blood. 2014;124.
[104] Wang M, Fayad L, Wagner-Bartak N, Zhang L, Hagemeister F, Neelapu SS, et al. Lenalidomide in combination with rituximab for patients with relapsed or refractory mantle-cell lymphoma: a phase 1/2 clinical trial. Lancet Oncol. 2012;13(7):716–723.

[105] Wang M, Fowler N, Wagner-Bartak N, Feng L, Romaguera J, Neelapu SS, et al. Oral lenalidomide with rituximab in relapsed or refractory diffuse large cell, follicular and transformed lymphoma: a phase II clinical trial. Leukemia. 2013;27(9):1902–1209.

[106] Zinzani PL, Pellegrini C, Derenzini E, Argnani L, Pileri S. Long-term efficacy of the combination of lenalidomide and rituximab in elderly relapsed/refractory diffuse large B-cell lymphoma patients. Hematol Oncol. 2013;31(4):223–224.

[107] Zinzani PL, Pellegrini C, Gandolfi L, Stefoni V, Quirini F, Derenzini E, et al. Combination of lenalidomide and rituximab in elderly patients with relapsed or refractory diffuse large B-cell lymphoma: a phase 2 trial. Clin Lymphoma Myeloma Leuk. 2011;11(6):462–466.

[108] Leonard J, Jung S, Johnson J, al. e. CALGB 50401: a randomized trial of lenalidomide alone versus lenalidomide plus rituximab in patients with recurrent follicular lymphoma. Abs 8000. J Clin Oncol. 2012;30.

[109] Martin P, Jung S, Johnson J, al. e. CALGB 50803 (ALLIANCE): a phase 2 trial of lenalidomide plus rituximub in patients with previuously untreated follicular lymphoma. Abs 063. Hematol Oncol (IMCL). 2013;31(suppl 1).

[110] Tuscano JM, Dutia M, Chee K, Brunson A, Reed-Pease C, Abedi M, et al. Lenalidomide plus rituximab can produce durable clinical responses in patients with relapsed or refractory, indolent non-Hodgkin lymphoma. Br J Haematol. 2014;165(3):375–381.

[111] Fowler NH, Davis RE, Rawal S, Nastoupil L, Hagemeister FB, McLaughlin P, et al. Safety and activity of lenalidomide and rituximab in untreated indolent lymphoma: an open-label, phase 2 trial. Lancet Oncol. 2014;15(12):1311–1318.

[112] Kerkar SP, Restifo NP. Cellular constituents of immune escape within the tumor microenvironment. Cancer Res. 2012;72(13):3125–3130.

[113] Chen DS, Mellman I. Oncology meets immunology: the cancer-immunity cycle. Immunity. 2013;39(1):1–10.

[114] Yao S, Zhu Y, Chen L. Advances in targeting cell surface signalling molecules for immune modulation. Nat Rev Drug Discov. 2013;12(2):130–146.

[115] Ribas A. Tumor immunotherapy directed at PD-1. N Engl J Med. 2012;366(26):2517–2519.

[116] Xerri L, Devilard E, Hassoun J, Olive D, Birg F. In vivo expression of the CTLA4 inhibitory receptor in malignant and reactive cells from human lymphomas. J Pathol. 1997;183(2):182–187.

[117] Ansell SM, Hurvitz SA, Koenig PA, LaPlant BR, Kabat BF, Fernando D, et al. Phase I study of ipilimumab, an anti-CTLA-4 monoclonal antibody, in patients with relapsed and refractory B-cell non-Hodgkin lymphoma. Clin Cancer Res. 2009;15(20):6446–6453.

[118] Bashey A, Medina B, Corringham S, Pasek M, Carrier E, Vrooman L, et al. CTLA4 blockade with ipilimumab to treat relapse of malignancy after allogeneic hematopoietic cell transplantation. Blood. 2009;113(7):1581–1588.

[119] Zou W, Chen L. Inhibitory B7-family molecules in the tumour microenvironment. Nat Rev Immunol. 2008;8(6):467–477.

[120] Powles T, Eder JP, Fine GD, Braiteh FS, Loriot Y, Cruz C, et al. MPDL3280A (anti-PD-L1) treatment leads to clinical activity in metastatic bladder cancer. Nature. 2014;515(7528):558-62.

[121] Topalian SL, Hodi FS, Brahmer JR, Gettinger SN, Smith DC, McDermott DF, et al. Safety, activity, and immune correlates of anti-PD-1 antibody in cancer. N Engl J Med. 2012;366(26):2443–2454.

[122] Robert C, Long GV, Brady B, Dutriaux C, Maio M, Mortier L, et al. Nivolumab in previously untreated melanoma without BRAF mutation. N Engl J Med. 2015;372(4):320–330.

[123] Hamid O, Robert C, Daud A, Hodi FS, Hwu WJ, Kefford R, et al. Safety and tumor responses with lambrolizumab (anti-PD-1) in melanoma. N Engl J Med. 2013;369(2):134–144.

[124] Andorsky DJ, Yamada RE, Said J, Pinkus GS, Betting DJ, Timmerman JM. Programmed death ligand 1 is expressed by non-hodgkin lymphomas and inhibits the activity of tumor-associated T cells. Clin Cancer Res. 2011;17(13):4232–4244.
[125] Chen BJ, Chapuy B, Ouyang J, Sun HH, Roemer MG, Xu ML, et al. PD-L1 expression is characteristic of a subset of aggressive B-cell lymphomas and virus-associated malignancies. Clin Cancer Res. 2013;19(13):3462–3473.
[126] Xerri L, Chetaille B, Serriari N, Attias C, Guillaume Y, Arnoulet C, et al. Programmed death 1 is a marker of angioimmunoblastic T-cell lymphoma and B-cell small lymphocytic lymphoma/chronic lymphocytic leukemia. Hum Pathol. 2008;39(7):1050–1058.
[127] Shi M, Roemer MG, Chapuy B, Liao X, Sun H, Pinkus GS, et al. Expression of programmed cell death 1 ligand 2 (PD-L2) is a distinguishing feature of primary mediastinal (thymic) large B-cell lymphoma and associated with PDCD1LG2 copy gain. Am J Surg Pathol. 2014;38(12):1715–1723.
[128] Green MR, Monti S, Rodig SJ, Juszczynski P, Currie T, O'Donnell E, et al. Integrative analysis reveals selective 9p24.1 amplification, increased PD-1 ligand expression, and further induction via JAK2 in nodular sclerosing Hodgkin lymphoma and primary mediastinal large B-cell lymphoma. Blood. 2010;116(17):3268–3277.
[129] Steidl C, Shah SP, Woolcock BW, Rui L, Kawahara M, Farinha P, et al. MHC class II transactivator CIITA is a recurrent gene fusion partner in lymphoid cancers. Nature. 2011;471(7338):377–381.
[130] Twa DD, Chan FC, Ben-Neriah S, Woolcock BW, Mottok A, Tan KL, et al. Genomic rearrangements involving programmed death ligands are recurrent in primary mediastinal large B-cell lymphoma. Blood. 2014;123(13):2062–2065.
[131] Green MR, Rodig S, Juszczynski P, Ouyang J, Sinha P, O'Donnell E, et al. Constitutive AP-1 activity and EBV infection induce PD-L1 in Hodgkin lymphomas and posttransplant lymphoproliferative disorders: implications for targeted therapy. Clin Cancer Res. 2012;18(6):1611–1618.
[132] Spranger S, Spaapen RM, Zha Y, Williams J, Meng Y, Ha TT, et al. Up-regulation of PD-L1, IDO, and T(regs) in the melanoma tumor microenvironment is driven by CD8(+) T cells. Sci Transl Med. 2013;5(200):200ra116.
[133] Berger R, Rotem-Yehudar R, Slama G, Landes S, Kneller A, Leiba M, et al. Phase I safety and pharmacokinetic study of CT-011, a humanized antibody interacting with PD-1, in patients with advanced hematologic malignancies. Clin Cancer Res. 2008;14(10):3044–3051.
[134] Armand P, Nagler A, Weller EA, Devine SM, Avigan DE, Chen YB, et al. Disabling immune tolerance by programmed death-1 blockade with pidilizumab after autologous hematopoietic stem-cell transplantation for diffuse large B-cell lymphoma: results of an international phase II trial. J Clin Oncol. 2013;31(33):4199–4206.
[135] Westin JR, Chu F, Zhang M, Fayad LE, Kwak LW, Fowler N, et al. Safety and activity of PD1 blockade by pidilizumab in combination with rituximab in patients with relapsed follicular lymphoma: a single group, open-label, phase 2 trial. Lancet Oncol. 2014;15(1):69–77.
[136] Moskowitz CH. PD-1 Blockade with the Monoclonal Antibody pembrolizumab (MK-3475) in patients with classical Hodgkin Lymphoma after Brentuximab Vedotin failure: preliminary results from a phase 1b study (KEYNOTE-013). Abs 290. Blood. 2014.
[137] Ansell SM, Lesokhin AM, Borrello I, Halwani A, Scott EC, Gutierrez M, et al. PD-1 blockade with nivolumab in relapsed or refractory Hodgkin's lymphoma. N Engl J Med. 2015;372(4):311–319.
[138] Piekarz RL, Frye R, Turner M, Wright JJ, Allen SL, Kirschbaum MH, et al. Phase II multi-institutional trial of the histone deacetylase inhibitor romidepsin as monotherapy for patients with cutaneous T-cell lymphoma. J Clin Oncol. 2009;27(32):5410–5417.

[139] Coiffier B, Pro B, Prince HM, Foss F, Sokol L, Greenwood M, et al. Results from a pivotal, open-label, phase II study of romidepsin in relapsed or refractory peripheral T-cell lymphoma after prior systemic therapy. J Clin Oncol. 2012;30(6):631–636.

[140] O'Connor OA, Horwitz S, Masszi T, Van Hoof A, Brown P, Doorduijn J, et al. Belinostat in Patients With Relapsed or Refractory Peripheral T-Cell Lymphoma: Results of the Pivotal Phase II BELIEF (CLN-19) Study. J Clin Oncol. 2015.

[141] Thomas RK, Baker AC, Debiasi RM, Winckler W, Laframboise T, Lin WM, et al. High-throughput oncogene mutation profiling in human cancer. Nat Genet. 2007;39(3):347–351.

[142] Quivoron C, Couronne L, Della Valle V, Lopez CK, Plo I, Wagner-Ballon O, et al. TET2 inactivation results in pleiotropic hematopoietic abnormalities in mouse and is a recurrent event during human lymphomagenesis. Cancer Cell. 2011;20(1):25–38.

[143] de Leval L, Gisselbrecht C, Gaulard P. Advances in the understanding and management of angioimmunoblastic T-cell lymphoma. Br J Haematol. 2010;148(5):673–689.

[144] Beguelin W, Popovic R, Teater M, Jiang Y, Bunting KL, Rosen M, et al. EZH2 is required for germinal center formation and somatic EZH2 mutations promote lymphoid transformation. Cancer Cell. 2013;23(5):677–692.

[145] Knutson SK, Wigle TJ, Warholic NM, Sneeringer CJ, Allain CJ, Klaus CR, et al. A selective inhibitor of EZH2 blocks H3K27 methylation and kills mutant lymphoma cells. Nat Chem Biol. 2012;8(11):890–896.

[146] Younes A. Novel treatment strategies for patients with relapsed classical Hodgkin lymphoma. Hematology Am Soc Hematol Educ Program. 2009:507–519.

[147] McCabe MT, Ott HM, Ganji G, Korenchuk S, Thompson C, Van Aller GS, et al. EZH2 inhibition as a therapeutic strategy for lymphoma with EZH2-activating mutations. Nature. 2012;492(7427):108-1-12.

[148] Qi W, Chan H, Teng L, Li L, Chuai S, Zhang R, et al. Selective inhibition of Ezh2 by a small molecule inhibitor blocks tumor cells proliferation. Proc Natl Acad Sci USA. 2012;109(52):21360–21365.

[149] Fowler N, al. e. A biologic combination of lenalidomide and rituximab for front-line therapy of indolent B-cell non Hodgkin's lymphoma. Abs 1714. Blood. 2009;114.

[150] Figueroa ME, Abdel-Wahab O, Lu C, Ward PS, Patel J, Shih A, et al. Leukemic IDH1 and IDH2 mutations result in a hypermethylation phenotype, disrupt TET2 function, and impair hematopoietic differentiation. Cancer Cell. 2010;18(6):553–567.

[151] Lu C, Ward PS, Kapoor GS, Rohle D, Turcan S, Abdel-Wahab O, et al. IDH mutation impairs histone demethylation and results in a block to cell differentiation. Nature. 2012;483(7390):474–478.

[152] Cairns RA, Iqbal J, Lemonnier F, Kucuk C, de Leval L, Jais JP, et al. IDH2 mutations are frequent in angioimmunoblastic T-cell lymphoma. Blood. 2012;119(8):1901–1903.

[153] Wang F, Travins J, DeLaBarre B, Penard-Lacronique V, Schalm S, Hansen E, et al. Targeted inhibition of mutant IDH2 in leukemia cells induces cellular differentiation. Science. 2013;340(6132):622–626.

[154] Chen C, Liu Y, Lu C, Cross JR, Morris JPt, Shroff AS, et al. Cancer-associated IDH2 mutants drive an acute myeloid leukemia that is susceptible to Brd4 inhibition. Genes Dev.

[155] Lopez-Lazaro M. How many times should we screen a chemical library to discover an anticancer drug? Drug Discov Today. 2015;20(2):167–169.

[156] Lopez-Lazaro M. Two preclinical tests to evaluate anticancer activity and to help validate drug candidates for clinical trials. Oncoscience. 2015;2(2):91–98.

Index

www.ingramcontent.com/pod-product-compliance
Lightning Source LLC
Chambersburg PA
CBHW080936240926
51849CB00004BA/474
* 9 7 8 3 1 1 0 3 7 1 1 9 2 *